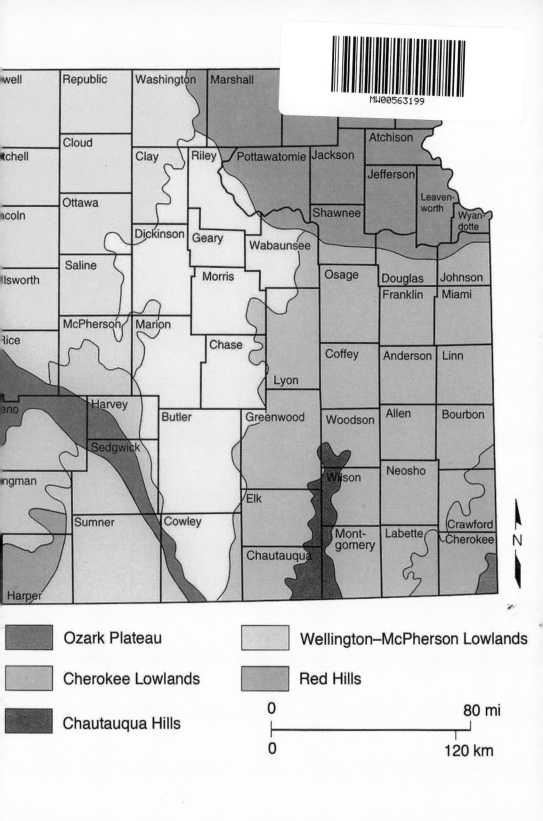

Ozark Plateau

Wellington–McPherson Lowlands

Cherokee Lowlands

Red Hills

Chautauqua Hills

0   80 mi

0   120 km

AMPHIBIANS, REPTILES, AND TURTLES IN KANSAS

Joseph T. Collins

Kansas Biological Survey; The University of Kansas; Lawrence, Kansas

Suzanne L. Collins

The Center for North American Herpetology; Lawrence, Kansas

Travis W. Taggart

Sternberg Museum of Natural History; Fort Hays State University; Hays, Kansas

WITH ARTWORK BY
Errol D. Hooper, Jr.
23523 State Highway AA; Greentop, Missouri

AND PHOTOGRAPHS BY
Suzanne L. Collins

Contribution from the
Sternberg Museum of Natural History
Fort Hays State University
Hays, Kansas 67601

This volume is the result of studies sponsored by the Sternberg Museum of Natural History, the Kansas Department of Wildlife and Parks, the U. S. Environmental Protection Ageny, and Westar Energy.

Citation for this book:
Collins, Joseph T., Suzanne L. Collins, and Travis W. Taggart. 2010. Amphibians, Reptiles, and Turtles in Kansas. Eagle Mountain Publishing, LC, Eagle Mountain, Utah. *xvi* + 312 pp.

Dedicated to

Kenneth Lee Brunson

in Recognition of his Commitment
to the Wildlife of Kansas

Co-Sponsored by:

Kansas Department of Wildlife & Parks
U. S. Environmental Protection Agency
Sternberg Museum of Natural History, Fort Hays State University
Touchstone Energy
Westar Energy

GREEN TEAM

The Center for North American Herpetology

This publication was funded in part by the Chickadee Checkoff of the Nongame Wildlife Improvement Program of the Kansas Department of Wildlife and Parks. All persons filing a Kansas income tax form have an opportunity to make a special contribution through the convenience of the tax form earmarked for the conservation of nongame wildlife. Do something wild! Make your mark on the tax form for nongame wildlife.

Although the information in this document has been funded wholly or in part by the U. S. EPA under assistance agreement X7 98759901-0 to *The Center for North American Herpetology,* it has not been subjected to the Agency's publications review process; thus, it may not neces-sarily reflect the views of the Agency and no official endorsement should be inferred.

TABLE OF CONTENTS

Preface ....*ix*

Introduction 1

• A condensed history of Kansas herpetology 3

• Taxonomy 6

• Explanation of species accounts...8

• Endangered and threatened species..9

• Alien species 10

• How to find and observe amphibians, reptiles, and turtles.............13

AMPHIBIANS (Class Amphibia) 15

Salamanders (Order Caudata)..15

A Phylogeny of Kansas Salamanders 15

Salamander species accounts:

Family Ambystomatidae..16

Barred Tiger Salamander (*Ambystoma mavortium*) 16

Smallmouth Salamander (*Ambystoma texanum*) 18

Eastern Tiger Salamander (*Ambystoma tigrinum*) 20

Family Salamandridae..21

Eastern Newt (*Notophthalmus viridescens*) 22

Family Plethodontidae 23

Longtail Salamander (*Eurycea longicauda*) 24

Cave Salamander (*Eurycea lucifuga*) 26

Grotto Salamander (*Eurycea spelaea*) 27

Family Proteidae 29

Red River Mudpuppy (*Necturus louisianensis*) 29

Mudpuppy (*Necturus maculosus*)..31

Frogs and toads (Order Anura) 32

A Phylogeny of Kansas Frogs and Toads...33

Frog and toad species accounts:

Family Scaphiopodidae..33

Plains Spadefoot (*Spea bombifrons*)....................................33

Family Bufonidae 35

American Toad (*Anaxyrus americanus*) 36

Great Plains Toad (*Anaxyrus cognatus*) 38

Green Toad (*Anaxyrus debilis*) 40

Fowler's Toad (*Anaxyrus fowleri*) 42

Red-spotted Toad (*Anaxyrus punctatus*).............................43

Woodhouse's Toad (*Anaxyrus woodhousii*) 45

Family Hylidae..47

Blanchard's Cricket Frog (*Acris blanchardi*) 47

Cope's Gray Treefrog (*Hyla chrysoscelis*) 50

Eastern Gray Treefrog (*Hyla versicolor*) 50
Spotted Chorus Frog (*Pseudacris clarkii*) 52
Spring Peeper (*Pseudacris crucifer*) 54
Boreal Chorus Frog (*Pseudacris maculata*) 56
Strecker's Chorus Frog (*Pseudacris streckeri*) 58
Family Ranidae...60
Crawfish Frog (*Lithobates areolatus*) 60
Plains Leopard Frog (*Lithobates blairi*) 62
Bullfrog (*Lithobates catesbeianus*)...64
Bronze Frog (*Lithobates clamitans*) 67
Pickerel Frog (*Lithobates palustris*) 68
Southern Leopard Frog (*Lithobates sphenocephalus*) 70
Family Microhylidae 72
Eastern Narrowmouth Toad (*Gastrophryne carolinensis*) 72
Great Plains Narrowmouth Toad (*Gastrophryne olivacea*)..................74

REPTILES (Class Reptilia) 76
Lizards (Order Squamata) 76
A Phylogeny of Kansas Lizards...77
Lizard species accounts:
Family Gekkonidae 77
Mediterranean Gecko (*Hemidactylus turcicus*) 77
Family Crotaphytidae 79
Eastern Collared Lizard (*Crotaphytus collaris*) 79
Family Phrynosomatidae 81
Lesser Earless Lizard (*Holbrookia maculata*) 81
Texas Horned Lizard (*Phrynosoma cornutum*) 83
Prairie Lizard (*Sceloporus consobrinus*) 86
Family Scincidae 88
Coal Skink (*Plestiodon anthracinus*)...88
Five-lined Skink (*Plestiodon fasciatus*) 89
Broadhead Skink (*Plestiodon laticeps*) 92
Great Plains Skink (*Plestiodon obsoletus*) 94
Southern Prairie Skink (*Plestiodon obtusirostris*) 96
Northern Prairie Skink (*Plestiodon septentrionalis*) 97
Ground Skink (*Scincella lateralis*) 99
Family Teiidae 101
Six-lined Racerunner (*Aspidoscelis sexlineata*) 101
Family Lacertidae 103
Western Green Lacerta (*Lacerta bilineata*)..103
Italian Wall Lizard (*Podarcis siculus*) 105
Family Anguidae..106
Western Slender Glass Lizard (*Ophisaurus attenuatus*) 107
Snakes (Order Squamata) 109
A Phylogeny of Kansas Snakes 109

Snake species accounts:

Family Leptotyphlopidae 110

   New Mexico Blind Snake (*Rena dissecta*) 110

Family Colubridae 112

   Eastern Glossy Snake (*Arizona elegans*)............................112

   Eastern Racer (*Coluber constrictor*) 114

   Prairie Kingsnake (*Lampropeltis calligaster*) 116

   Speckled Kingsnake (*Lampropeltis holbrooki*) 118

   Milk Snake (*Lampropeltis triangulum*) 120

   Coachwhip (*Masticophis flagellum*) 123

   Rough Green Snake (*Opheodrys aestivus*) 125

   Great Plains Rat Snake (*Pantherophis emoryi*) 127

   Gopher Snake (*Pituophis catenifer*)..................................129

   Longnose Snake (*Rhinocheilus lecontei*) 131

   Western Rat Snake (*Scotophis obsoletus*)..........................133

   Ground Snake (*Sonora semiannulata*)................................135

   Flathead Snake (*Tantilla gracilis*) 137

   Plains Blackhead Snake (*Tantilla nigriceps*) 139

Family Dipsadidae...140

   Western Worm Snake (*Carphophis vermis*)........................141

   Ringneck Snake (*Diadophis punctatus*)............................142

   Western Hognose Snake (*Heterodon nasicus*)...................145

   Eastern Hognose Snake (*Heterodon platirhinos*) 147

   Chihuahuan Night Snake (*Hypsiglena jani*) 150

Family Natricidae 151

   Plainbelly Water Snake (*Nerodia erythrogaster*)...............152

   Diamondback Water Snake (*Nerodia rhombifer*) 154

   Northern Water Snake (*Nerodia sipedon*).........................156

   Graham's Crayfish Snake (*Regina grahamii*).....................158

   Brown Snake (*Storeria dekayi*) 160

   Redbelly Snake (*Storeria occipitomaculata*) 162

   Checkered Garter Snake (*Thamnophis marcianus*)...........164

   Western Ribbon Snake (*Thamnophis proximus*) 166

   Plains Garter Snake (*Thamnophis radix*) 168

   Common Garter Snake (*Thamnophis sirtalis*) 170

   Lined Snake (*Tropidoclonion lineatum*) 172

   Rough Earth Snake (*Virginia striatula*) 174

   Smooth Earth Snake (*Virginia valeriae*) 176

Family Crotalidae 177

   Copperhead (*Agkistrodon contortrix*)...............................177

   Cottonmouth (*Agkistrodon piscivorus*)............................180

   Timber Rattlesnake (*Crotalus horridus*) 182

   Prairie Rattlesnake (*Crotalus viridis*) 185

   Massasauga (*Sistrurus catenatus*) 187

TURTLES (Class Chelonia) 190
A Phylogeny of Kansas Turtles 190
Straightneck Turtles (Order Cryptodira)
Turtle species accounts:
Family Chelydridae...191
Common Snapping Turtle (*Chelydra serpentina*) 191
Alligator Snapping Turtle (*Macrochelys temminckii*)........................194
Family Kinosternidae..196
Yellow Mud Turtle (*Kinosternon flavescens*) 196
Common Musk Turtle (*Sternotherus odoratus*)...................................199
Family Emydidae 201
Northern Painted Turtle (*Chrysemys picta*) 201
Common Map Turtle (*Graptemys geographica*) 203
Ouachita Map Turtle (*Graptemys ouachitensis*) 205
False Map Turtle (*Graptemys pseudogeographica*)..............................205
Eastern River Cooter (*Pseudemys concinna*) 208
Eastern Box Turtle (*Terrapene carolina*) 210
Ornate Box Turtle (*Terrapene ornata*) 212
Slider (*Trachemys scripta*) 215
Family Trionychidae..217
Smooth Softshell (*Apalone mutica*)..217
Spiny Softshell (*Apalone spinifera*) 220
Species of possible natural occurrence in Kansas 222
A technical key to adult Amphibia, Reptilia, and Chelonia in Kansas 225
Glossary 250
Metric conversion chart..252
Bibliography...253
Web Sites for Kansas Amphibians, Reptiles, and Turtles 253
Taxonomic and Phylogenetic References to Chordates 253
References to Fossil Amphibians, Reptiles, and Turtles............................254
Bibliography of Kansas Amphibians, Reptiles, and Turtles......................255
Index to standard common names and scientific names...................................306

PREFACE

Since the appearance of the first (1974) and second (1982) editions of *Amphibians and Reptiles in Kansas* by Joseph T. Collins, and third (1993) edition by Joseph T. Collins and Suzanne L. Collins (all now out-of-print), new information has been accumulated on the distribution and habits of the amphibians, reptiles, and turtles in Kansas. Through the efforts of many field herpetologists and our queries of herpetological museum collections worldwide, many new records of these animals have been documented in the state. Furthermore, extensive information on the natural history of these creatures has continued to appear in publications, as well as new state records for size maxima. All of this information appears in the text, maps, and bibliography of this book.

This guide is about the 99 species of amphibians, reptiles, and turtles now known to occur in Kansas. Of that total, only three lizards, the recently discovered Mediterranean Gecko (*Hemidactylus turcicus*), as well as the long-established Italian Wall Lizard (*Podarcis siculus*) and Western Green Lacerta (*Lacerta bilineata*), are not native to the state.

Since the publication of Collins and Collins (1993), Fowler's Toad (*Anaxyrus fowleri*), has been discovered to be a part of the Kansas fauna. In addition, the Kansas trilling chorus frogs are restricted to a single species, the Boreal Chorus Frog (*Pseudacris maculata*). The two species of Gray Treefrogs in our state remain visibly indistinguishable and are presented in a single species account pending more study. The False Map Turtle and Ouachita Map Turtle have proven so difficult to identify that we are uncertain where these species occur in the state, and therefore have chosen to display them in a single account pending future studies. Finally, the Oklahoma Salamander (*Eurycea tynerensis*), previously known in Kansas as the Many-ribbed Salamander (*Eurycea multiplicata*), is no longer recognized by us as a member of the state's herpetofauna; it has not been observed in Kansas for 40 years and until further evidence shows otherwise, we have declined to include it in this book. Thus, the herpetofaunal composition of Kansas has changed, and the number of taxa known to occur in the state has reached 99 species.

The distribution maps, bibliography, current scientific and standard common names, and endangered or threatened species designations provided in this book will be useful to biologists and non-biologists alike, but the bulk of this book is written for the people of Kansas in the hopes of answering their questions about amphibians, reptiles, and turtles.

During the last 40 years, we have (collectively or individually) visited all 105 counties in Kansas, viewing their varied environments and observing amphibians, reptiles, and turtles at every opportunity. We are indebted to many people for their assistance.

Our close friends and fellow herpetologists, Larry L. Miller (Kansas Heritage Photography, Wakarusa) and Curtis J. Schmidt (curator at the Sternberg Museum of

Natural History, Fort Hays State University), deserve our heartfelt thanks for their companionship and enthusiasm as, both individually and together, we traveled throughout Kansas looking for amphibians, reptiles, and turtles, and gathered information about them. Further, Jennifer Taggart showed tremendous patience toward our preoccupation with this project, and to her we are so grateful.

Over the last four decades, a number of our colleagues have gone to extraordinary lengths to assist us in the preparation of this book, and they deserve special recognition for their efforts: Mary Kate Baldwin, Ken Brunson, Keith Coleman, Jerry D. Collins, Frank B. Cross, Mark R. Ellis, Charles Ely, Henry S. Fitch, Eugene D. Fleharty, Daniel D. Fogell, Darrel R. Frost, Hank Guarisco, James E. Gubanyi, Eva A. Horne, Philip S. Humphrey, Kelly J. Irwin, Alan H. Kamb, Jay D. Kirk, James L. Knight, Walter E. Meshaka, Jr., Emily Moriarty Lemmon, Daniel G. Murrow, George R. Pisani, Dwight R. Platt, Robert Powell, R. Alexander Pyron, Derek Schmidt, Hobart M. Smith, and R. Bruce Taggart. Without the help and encouragement of these individuals, writing this book would have been much more difficult.

Many of our colleagues assisted us in a variety of ways, and we wish to thank Kraig Adler, Douglas W. Albaugh, Ray E. Ashton, David Barker, Robert R. Beatson, Deb Bennett, Thomas J. Berger, Timothy Broschat, Frank T. Burbrink, Donald R. Clark, David Cannatella, Jonathan Campbell, Sheila Fairleigh Collister, Kevin Comcowich, Brian I. Crother, Mary E. Dawson, Guntram Deichsel, David M. Dennis, Ray W. Drenner, Arthur C. Echternacht, Ruth Fauhl, Elmer Finck, Bert Fisher, Henry S. Fitch, Virginia Fitch, Neil Ford, John Frost, Gregory E. Glass, Steve Garber, Lisle Gibbs, Peter Gray, Owen Gorman, Wendy Gorman, Russell J. Hall, Stephen G. Haslouer, Robert W. Henderson, Aubry Heupel, David M. Hillis, Corson J. Hirschfeld, Wayne Hoffman, Tina Jackson, Dan Johnson, Randall Johnson, H. A. Kerns, David A. Kizirian, Rose Etta Kurtz, Amy Lathrop, Richard Lattis, Julian C. Lee, Alan Lemmon, Stuart C. Leon, Glenn Lessenden, Raymond Loraine, John D. Lynch, Luis Malaret, S. Ross McNearney, Joseph R. Mendelson III, Paul Mills, Paul Moler, Richard R. Montanucci, Randall L. Morrison, Randall E. Moss, Charles W. Myers, Marc Nadeau, Tim Noble, David Oldham, Robin Oldham, Robert Olley, Jeff Parmalee, Janice J. Perry, Nick Pipkin, Michael V. Plummer, Gregory Pregill, Rebecca Pyles, Gerald T. Regan, Randall S. Reiserer, Juan Renjifo, Stephen C. Richter, Daren Riedle, Jerome D. Robins, Steven M. Roble, Michael R. Rochford, Alan H. Savitzky, Terry D. Schwaner, Nancy Schwarting, Richard A. Seigel, Kathleen A. Shaw, Philip Simbotwe, A. K. Smith, Alan Smits, Robert Sprackland, William Stark, Eddie R. Stegall, John Stoklosa, Thomas Swearingen, Edward H. Taylor, R. Brent Thomas, Kevin R. Toal, James R. Triplett, Alberto Veloso, Jan Wagner, Richard J. Wassersug, Robert Waltner, Ginny N. Weatherman, Jeffrey F. Whipple, Chad Whitney, and Edward O. Wiley.

Many individuals in charge of museum and university herpetological collections generously supplied us with lists of their holdings of Kansas specimens of amphibians, turtles, and reptiles or lent us material for examination. We are indebted to Craig Guyer, Robert H. Mount, Mary A. Mendonca, and Ray Henry (Auburn University, Auburn, Alabama), Andrew Holycross (Arizona State University, Tempe, Arizona), Jimmy A. McGuire, Carol L. Spencer, David B. Wake, and Robert C. Stebbins (Museum of Vertebrate Zoology, University of California, Berkeley, California), Robert Bezy and John Wright (Los Angeles County Museum, Los Angeles, California), Bradford D. Hollingsworth (San Diego Natural History Museum, San Diego, California), Alan E. Leviton, Robert Drewes, Robin Lawson, Jens Vindum, and Jeff Wilkinson, and the late Joseph B. Slowinski (California Academy of Sciences, San

Francisco, California), Hobart M. Smith and Mariko Kageyama (University of Colorado Museum, Boulder, Colorado), Jacques Gauthier and Gregory Watkins-Colwell (Peabody Museum of Natural History, Yale University, New Haven, Connecticut), George R. Zug, W. Ronald Heyer, Kevin de Queiroz, and Addison Wynn (National Museum of Natural History, Smithsonian Institution, Washington, D. C.), F. Wayne King, Max A. Nickerson, and Kenny L. Krysko (Florida State Museum, Gainesville, Florida), M. Elizabeth McGhee (Georgia Museum of Natural History, University of Georgia, Athens, Georgia), Ronald A. Brandon (Southern Illinois University, Carbondale, Illinois), Tom Anton (Chicago Academy of Sciences, Chicago, Illinois), Harold Voris and Alan Resetar (Field Museum of Natural History, Chicago, Illinois), Chris Phillips (Illinois Natural History Survey, Urbana, Illinois), Jerry Choate, Eugene D. Fleharty and Curtis J. Schmidt (Sternberg Museum of Natural History, Fort Hays State University, Hays, Kansas), Rafe Brown (University of Kansas Natural History Museum, Lawrence, Kansas), Dwight R. Platt (Bethel College, North Newton, Kansas), Max C. Thompson (Southwestern College, Winfield, Kansas), Christopher Austin and Jeff Boundy (Museum of Natural Science, Louisiana State University, Baton Rouge, Louisiana), Harold A. Dundee (Museum of Natural History, Tulane University, New Orleans, Louisiana), James Hanken, Jonathan B. Losos, and Ernest E. Williams (Museum of Comparative Zoology, Harvard University, Cambridge, Massachusetts), Arnold Kluge, Ronald Nussbaum, and Greg Schneider (University of Michigan, Ann Arbor, Michigan), James Harding (The Museum, Michigan State University, East Lansing, Michigan), Kenneth Kozak (James Ford Bell Museum of Natural History, University of Minnesota, St. Paul, Minnesota), Harry W. Greene, Kelly Zamudio, and John P. Friel (Cornell University, Ithaca, New York), Darrel Frost, Christopher Raxworthy, and David Kizirian (American Museum of Natural History, New York, New York), Janalee P. Caldwell and Laurie J. Vitt (Sam Noble Museum of Natural History, University of Oklahoma, Norman, Oklahoma), Stanley F. Fox (Oklahoma State University, Stillwater, Oklahoma), Ross D. MacCulloch, Robert W. Murphy, and Amy Lathrop (Royal Ontario Museum, Toronto, Canada), Ned Gilmore and Thomas Uzzell, Jr. (Academy of Natural Sciences, Philadelphia, Pennsylvania), Steven P. Rogers (Carnegie Museum of Natural History, Pittsburgh, Pennsylvania), Jonathan Campbell and Paul Chippendale (University of Texas, Arlington, Texas), David C. Cannatella, David M. Hillis, and Travis LaDuc (Texas Memorial Museum, University of Texas, Austin, Texas), James R. Dixon, Lee A. Fitzgerald, Kathryn Vaughn, and Toby J. Hibbits (Texas Cooperative Wildlife Collection, Texas A & M University, College Station, Texas), Jerry D. Johnson and Robert G. Webb (Centennial Museum, University of Texas, El Paso, Texas), Jack W. Sites, Jr. (Monte L. Bean Life Science Museum, Brigham Young University, Provo, Utah), John M. Legler (Utah Museum of Natural History, University of Utah, Salt Lake City, Utah), and Robert W. Henderson and Robert W. Burgeois (Milwaukee Public Museum, Milwaukee, Wisconsin).

Many former students from Kansas and North American herpetology courses taught by Joseph T. Collins from 1998 to 2008 at Washburn University and the University of Kansas demonstrated immense enthusiasm and rendered much assistance during class field trips throughout Kansas. They are Caleb F. Acree, Jeannette Aranda, Keith R. Arkenberg, Mary Kate Baldwin, April A. Barker, Sarah Bellows-Blakely, Erika V. Bessey, Amanda D. Biester, Mary E. Bolfing, Jena M. Boucher, Ryan W. Bradbury, Jonna D. Bredemeier, Rachelle A. Brown, Donna L. Buchanan, Christopher Burdick, Aisha M. Butt, Rebecca J. Carttar, Jennifer A. Case, Diana L. Chamberlain,

Martha A. Cole, Matthew J. Conyac, Jeremiah Cripps, Cynthia A. Cummings, Christy Cunningham, Stephanie Cunningham, Jason M. Daniels, Karen L. Dedonder, Lynn Peter DeMarco, Erin G. Dugan, Anne L. Duperault, Emily S. Eastman, Craig A. Eddis, Lisa N. Ediger, Mark R. Ellis, Edward Eubanks, Franklyn F. Finks, Georgia M. Finks, Evan S. Fisher, William S. Frager, Meredith M. Fry, Teresa M. Frye, Jeremy R. Funk, Jason M. Gager, Elizabeth P. Ganser, Summer Gauger, Donald Gebers, Jaime Gilbert, Charles Glenn, Michael J. Goodwin, Dennis A. Gordinier, Erin Graham, Brendan D. Greening, Marla A. Gubanyi, Lacey Hamblet, Lindsey Hamblet, Brenda M. Hanna, Andrew Hare, Adam S. Harris, Victor C. Harris, April M. Hernandez, Jimmy G. Hawley, Zachary Hemmerling, Justin D. Hersh, Joyce Hesse, Joel M. Hicks, Margaret S. Hicks, Rebecca J. Hill-Larsen, Robert A. Hines, Wade T. Hoss, Ryan M. Huske, Jonathan M. Huss, Wendy L. Jenkins, Wayman J. Johnson, Matthew D. Jones, Carole L. Jontra, Amy K. Jungclaus, Jon D. Kee, Colin S. Kim, Tonya Kimbrell, Michael A. Kirby, Josie L. Kirk, Steven W. Klise, Carletha S. Kosky, Susan S. Labrador, Ricquelle Landis, Thomas O. Langer, Zackery N. Larsen, Katherine R. Larson, Brandy L. Lewis, Marc D. Linton, Lisa J. Locke, Janice R. Logan, Dominique Loreg, Candy M. Lotridge, Monica M. Marak, Thomas R. Marcellino, Brian D. Mathers, Sara M. Matthias, Kelly J. Maxwell, Andrea L. May, the late Edward J. May, Brian A. McCall, Luke E. McClurg, Richmond S. McDaniels, S. Ross McNearney, Dustin McPhail, T. Michael McRoberts, Lori Ann Meador, Matthew D. Miller, Kyle A. Montgomery, Morgann C. McMurry, Philip G. Newkirk, Sarah D. Noller, John T. O'Brien, Michael A. Odupitan, Kyle P. O'Neal, Daeyun Park, Jessica L. Peterson, John L. Petterson, Christine Porubsky, Chad J. Puff, Ben H. Rebein, Jay C. Reed, Ryan L. Rehmeier, Matt Renk, Shelley A. Robertson, Michael R. Rochford, Pamela Jean Rutter, Tanya Schlegel, Lisa J. Schmidt, Jennifer M. Seem, Jeffrey A. Shannon, Stephanie Sherraden, Travis C. Sieve, Andrew Sindorf, Ronald D. Smith, Linda K. Spaulding, LaBonna Speakman, Angela L. Spencer, Naomi Spitze, R. Gayan Stanley, Patricia Stein, Margaret E. Stewart, Dave T. Stineman, Gibran M. Suleiman, Seth M. Sundquist, Jeremiah Teller, Andrew A. Terhune, Sara E. Thielenhaus, Bradley A. Tobyne, Holly A. Tropf, Barbara Tucker, Richard L. Upshaw, Debra K. Vinning, Arrin A. Vrtiska, James R. Wadley, Stephen R. Wahle, Patrick Wakeman, Janeen A. Walters, Jordan P. Ward, Barbara Watkins, Ginny N. Weatherman, Amy K. Wethington, Carla M. Wheeler, Emilia S. Whiteaker, Jessica M. Willard, Donovan M. Wilson, Crystal L. Wipf, Adam J. Wolf, Cristine Yates, Michael Young, Tremaine C. Young, Cortney B. Zager, and Philip A. Zeikle.

Personnel (past and present) of the Kansas Department of Wildlife and Parks (Pratt), Kansas Biological Survey (Lawrence), U. S. Fish and Wildlife Service (Manhattan), U. S. Forest Service (Elkhart), U. S. Department of the Army, and other governmental agencies were most helpful during the gathering of information for this book, in particular Kyle Austin, Matt Bain, Kevin Becker, Tommie Berger, Robert Bergquist, Charles Bever, Mike Blair, C. Douglas Blex, Gene Brehm, Ralph Brooks, Troy Brown, William Busby, James Bussone, Janalee P. Caldwell, Jim Cherry, Jeff Chynoweth, Rick Cleveland, Jeff Clouser, Bob Culbertson, Craig Curtis, Lynn Davignon, Nate Davis, Beverly Downing, W. T. Edmonds, Robert Friggeri, Don George, Tom Glick, Mark Goldsberry, Carl Grover, Helen M. Hands, Ross Harrison, Robert Hartmann, Joe Hartman, Chris Havel, Mike Hayden, R. Lance Hedges, Kent Hensley, Dan Hesket, Bill Hlavachick, Jerry Horak, Don Huggins, Alan Hynek, Leonard Jirak, Eric Johnson, Joshua L. Jagels, Karl K. Karrow, Mark Kumberg, Murray Laubhan, William Layher, Paul M. Liechti, Chris Madson, Carl Magnuson, Chris Mammoliti, Rick Martin, Edward Martinko, Bob Mathews, Vic McLeran, Tim Menard, Ed Miller,

Tom Mosher, Dan Mosier II, Dan Mulhern, Greg Nichols, Don Nygren, Mike Nyhoff, Brad Odle, Matt Peek, Steve Price, Mike Rader, Johnny Ray, Kurt Reed, Randy Rogers, Richard Sanders, George Schlecty, Randall Schultz, Marvin Schwilling, Mark Sexson, Mark Shaw, Chris Shrack, John Silovsky, Bryan Simmons, Deb Simon, Brad Simpson, Steve Sorenson, Larry Stones, Gibran Suleiman, Tom Swan, Scott Thomasson, Larry Tiemann, Robert Todd, Manuel Torres, Rob Unruh, Mark Van Scoyoc, Ryan Waters, Sheila Wells, Eric Wickman, Roger Wolfe, Robert Wood, Michael Zajic, Amy Zavala, and Larry Zuckerman.

During our 40 years of accumulating information on the herpetofauna of Kansas, many persons extended a variety of courtesies and services. These included providing companionship, lodging, and collecting assistance, verifying data, finding needed references to amphibians, reptiles, and turtles, obtaining specimens, offering logistical support, and maintaining loyalty to our cause. We are thankful to Ted Abel, Dave Acker, Kathy Acuff, Laura Acuff, Rob Acuff, Dallas Adams, Dan Adams, John C. Adams, Michael Adams, Steve Adams, John Ahrens, Courtney Akers, Kevin Albright, Brad R. Anderson, C. L. Anderson, Harold Anderson, Lewis R. Anderson, Allen Andresen, Juli Armstrong, Matthew Avidan, Logan Babst, M. Roy Bachman, Charlie Baker, Cleda Baker, Richard J. Baldauf, Laura Baldwin, Lucia Baldwin, Will Baldwin, Robert L. Ball, Benjamin E. Bammes, John Banks, Roger W. Barbour, Brian C. Bartels, Erik Bartholomew, M. Neil Bass, Tori Bass, Sarah Bays, Kaela Beauclerc, Tom Beaver, Karl R. Bechard, Don Becke, Keith Becker, Kevin Becker, Roy J. Beckemeyer, R. Beer, Gregory Beilfuss, Gene Beilman, David Bender, Rick Bennett, Stasya Berber, Peter Berg, Byron Berger, Hillary Bernhardt, Jonathan Berry, Martha Bickford, Hank Bishop, Miles Bishop, Jeffrey H. Black, Junior Black, Warner Blackburn, Ron Blakely, Charlie Bliese, Jeff Blodig, Dave Blody, Melissa Boetig, James P. Bogart, Caleb Bond, Ronald Bonett, Lynett Bontrager, Ian Bork, John Bork, Brad Bott, Terry Bott, Jeff Boundy, Wes Bouska, Linda S. Boyd, Don Boyer, Brice Boyle, Ann Bradley, Sue Bradley, William Bradley, Ronald A. Brandon, Terence Brotherton, Mike Brown, Andrew Brungardt, Catherine Brungardt, Gerard Brungardt, Luke Brungardt, Petra Brungardt, Tom Brungardt, Andi Brunson, Jessica Brunson, Katelin Brunson, Lee Ann Brunson, M. Brunton, Jennifer Bryan, D. W. Brzoska, Angela G. Bulger, Mike Bumpus, Andrea Burcham, Jeffrey T. Burkhardt, Annie Burkhardt, Tim Burkhart, Andrew G. Burr, Calley Burr, Ann Bush, Jack Bush, Mark Butt, Ed Byrne, Larry Caldwell, Shane Callaway, Jill Capron, James Carlson, Dan Carpenter, Shelbi Carpenter, Letha Cartenson, Todd L. Carter, Steve Case, Harold Casey, Owen Casey, Shawn W. Casley, Phillip Cass, Ellen Censky, D. Chapman, Pamela Cheever, Sue Chesnut, Jason Chesser, Roger Christie, B. J. Clark, Gary K. Clarke, John W. Clarke, Linda Clarke, Eric Cleveland, Rick Cleveland, T. W. Cloutier, Michael Coker, Jay Cole, Matt Cole, Keith Coleman, Matt Combes, K. Comcowich, Roger Conant, Christopher Conrady, Sally Cool, Donna S. Cooper, Mark Cooper, Michael Cooper, Michael Corn, Maria Covell, Chad Cowan, Cara Cowger, Olivia Cowin, Jim Cox, Gary Cumro, Jeffrey W. Curts, Mark Dalton, David Dameron, J. Damm, James K. Daniel, Ken Davidson, Bruce Dawson, Kay Dawson, Mary E. Dawson, Brandon DeCavele, Austin DeGarmo, Patty Delmott, Joseph Delnero, Rosie Delnero, John Denison, William Denison, Mark Denney, Raul Diaz, Wayne Dickerson, S. Dietrich, Marla Dillard, Vikki Dillard, Tom F. Dillenbeck, James L. Dobie, Philip Doty, Cindy Downum, Pam Drenner, Joe Dreiling, Elysium Drumm, Andy Durbin, David A. Easterla, Mark Eberle, Alice F. Echelle, Shane Eckhardt, Corinne Edds, Kyle Edds, J. Edens, Richard A. Edgren, Jeff Ehlers, Victor B. Eichler, Maria Eifler, Dawn Eiland, Kathy Ellis, Connie Elpers, Mike T. Elson, Roy

Engeldorf, Ruth Engeldorf, Samuel Fairleigh, Greg H. Farley, Kami Farr, Rebecca Farr, Gerald Fay, Natalie Fayman, Lindsey Fender, Laura Fent, Gary W. Ferguson, Linda Ferguson, Dennis Ferraro, T. Feuerborn, H. I. Fisher, Virginia Fitch, M. Fitzgerald, Thomas L. Flowers, Cindy Ford, Mari-Jayne Fox, Seth Franklin, John Fraser, Z. Frazey, Kevin Freed, Arnold Froese, Cathi Fuller, James M. Fuller, T. L. Funk, Linda Fuselier, Katherine Garlinghouse, James Garvey, Goran Gasbarovic, Jack Gearhart, Galen Geiger, Nathan Geiger, Keith Geluso, Nicole Gerlanc, Brandon Geschwentner, Austin Gideon, Michelle Gilkerson, Ricky Ginn, Amy Glidden, Rene Gloshen, Tyler Gloshen, Ron Goellner, R. Goff, Nicholas J. Gomez, Lance Good, Holly Goodman, Karen Graham, Mike Graham, Alan Gravenstein, Tanner Gravenstein, Michael Graziano, Harry Gregory, Bob Gress, Nichole Grim, Kurt Grimm, David Grow, Dick Grusing, Julian Gubanyi, Marla A. Gubanyi, Laura Gunderson, R. B. Hager, Jenny Hall, B. Haller, S. Hamilton, Whitney Hamilton, Caleb Hardin, Dave Hardin, Holly Harding, Logan Harding, Brad Hare, Jerri Harland, Michael Harman, Trey Harrison, Michelle Hartman, Taberie Hartshorn, Richard Hatfield, Thea Haugen, Jeanie J. Hauser, Steven Hausler, Richard S. Hayes, Rod Heffley, Tom Hein, Kelvin L. Heitmann, Gordon Henley, Aaron Henton, Emily Heronemus, K. Hertel, Kyle Hesed, Dan E. Heskert, Troy Hibbitts, Kenneth Highfill, Scott Hillard, N. J. Hilscher, P. Hilt, Tracy Hirata-Edds, Dan Hodges, Karl Holdren, Cindy Hollis, John Holsonback, Carol Hoogheem, Irwin Hoogheem, Emily Hooser, Deb Horne, William House, B. Houser, John Howard, Rebecca Howard, Jeffrey S. Hubbard, S. Hubbard, Brian Hubbs, Rick Hudson, Scott Huettenmueller, Gus Huey, Arick E. Huggins, D. Hulsing, David Humenczuk, Tony Hurt, Chris Hutson, David Huyser, Duane Iles, G. Imel, Paul Ingram, Patrick Ireland, Francis Irish, Kris M. Irwin, Lisa Irwin, George Jackson, Garrett Jantz, David Jardine, Jay D. Jeffrey, Matthew P. Jeppson, Dean Jernigan, J. E. Jindra, Danica Johnson, Grace Anne Johnson, Jennifer Johnson, Troy Johnson, Daphne Jones, David P. Jones, Gary Jones, Jack Jones, Jason Jones, Jerry Jones, C. E. Judd, Terran Kallam, Al Kamb, Steve Kamb, M. Kane, Caleb Karch, Olin Karch, Richard Kazmaier, Keane Kearney, Steve Kearney, Bob Keller, Gary Keller, Jacob Keller, Richard Keller, Eric Kessler, Maura Kessler, Owen Kessler, Rebecca Kessler, Sherman Ketchum, Dean Kettle, R. Khowe, Jr., Levi Kinder, Ingrid M. Kircher-Nietfield, Steve Kite, John Kitterman, Harold Klaassen, John Klaus, Monica Klaus, Nick Kleiger, Amy Klein, Shelby Klima, Page Klug, William Knighton, Winifred Kucera Knoepker, Corey Koehn, Brent M. Konen, Allie Kossoy, Dana Kottman, Steve Kozak, Robert Krager, Kara Kramer, P. J. Kramer, Jill Krebs, Chris Kuhn, Mark Kumberg, Rose Etta Kurtz, Rob Ladner, Russell W. LaForce, G. Lairson, Gavin Lake, T. J. Lamb, Cindy Lang, Nathan Lang, Tara Lang, Richard L. Lardie, Karl Larson, Max Larson, Olaf Larson, Brogan Lasley, W. Leach, Ted Leonard, Bob Levin, A. Levings, Amy Li, Emmy Lieser, Cameron Liggett, Gregory Liggett, Jian Liu, Jill Lokke, John Lokke, Connie Long, Maria Lopez, D. Loring, Brandon Low, Denise Low, Judy Low, Todd Lowe, Ben Lowery, Jerry Lowry, T. Lucier, Carl Madorin, Lana Madson, Daniel Magill, Carol Mammoliti, Kirk Mammoliti, Bob Mangile, Ed Markel, James Markley, James H. Marlett, Patricia J. Marlett, Shauna Marquadt, Josh Marshall, Steve Marshall, Edward Maruska, the late Karl H. Maslowski, James Mason, Ralph Massoth, Jr., Dietrich Mast, Raymond S. Matlack, Jay E. Mattison, Richard Maxwell, Brad May, Zachary Mayers, AvNell Mayfield, C. McAfee, H. McArdle, Eric McCarrier, Dalton McCloud, Ian McCloud, Terry McCord, C. J. McCoy, Sharon McCue, Stephen McCue, Lori McElroy, F. McGinnis, Chris McMartin, Sarah McMartin, Louise Mead, Chris Meiers, Erin Melroy, Joseph R. Mendelson, Delfi Messinger, D. Meyer, Scott Meyer, Carl Michaels, Justin Michels, Evan Mielke, Alex Miller, Arlyn Miller, Carolyn

Miller, Clyde Miller, Daniel Miller, Laura Miller, Loretta Miller, Matt Miller, Seith Miller, Suzanne L. Miller, William Miller, Gregory Mills, Faye E. Minium, Peggy Minter, Wendy Misak, Lyndon A. Mitchell, Trevor Mitscher, Derek Moeller, Kellie Molone, Carolyn Moriarty, R. Morris, Carey Moss, Victor A. Moss, Robert H. Mount, Kirk Mullen, Bill Munholland, James B. Murphy, B. Musser, N. Musser, Steven Nagle, Tyler Nagle, Ian M. Nall, Kristie Nall, Brandy Nance, Susan Nazaran, A. Neal, I. Neal, Nathaniel A. Nelson, Al Neufeld, Tami Newkirk, Jocelyn Nichols, Max A. Nickerson, Beau Nissley, Rose Nissley, Lisa Nodolf, M. T. Nulton, Rhonda Oathout, Tammy Oathout, L. S. Oborny, Leah Ochs, Jackson Oldham, Tag Oldham, Eric Oligschlaeger, J. Oshell, Robert Otto, Ken Outt, Alissa Overall, Nicole Palenske, Laura Palmer, Jeffrey R. Parmalee, Kevin Parrish, Chris Parsons, Barbara Paschke, Tracy Patten, Michael Pearce, LeAnne Pelzel, Bonnie R. Perido, Leslie Perry, Todd W. Pesch, Emily Petersen, Erica Peterson, Jan Peterson, Kendra L. Phelps, Jacob Phillippi, Jeff Phillippi, Doris Phipps, James A. Pilch II, Ronald Pine, Mike Pingleton, Nick Pipkin, Galen Pittman, Richard Plumlee, James A. Pilch, Zac Plummer, Brian Poister, John Poister, Alexis Powell, John Powell, Robert Powell, Sarah Powell, Eric Priest, Nathan Priest, Paul Prewo, Tanner Procter, Rebecca Prosser, M. Puckett, Jean Purdy, Tisha Quick, Hugh Quinn, R. Racine, Jennifer Rader, Ann Randle, Eddy Rasmussen, Stan Rassmussen, George Ratzlaff, E. Ray, W. Reager, David Reberry, Sarah Reberry, Al Redmond, Sarah Reeb, Chris Reed, Rachel Reed, Todd W. Reeder, Ryan L. Rehmeier, Bill Reid, Randall Reiserer, Dayne Relihan, Carrie Remillard, D. Reser, Al Resetar, B. Rhodes, Sharon Richards, Kevin Ricke, Bill Ried, London Rief, Cherrie Riffey, R. Rinck, L. Roberson, M. Roberson, Eddie Roberts, Jennifer Rogers, Randy Rogers, Brian R. Roh, Rana Rollo, Shana Rollo, Nichole Rosencutter, Douglas A. Rossman, Benjie Rowe, Steve Royal, Michael Rush, Anne Russell, Dan Ryan, P. S. Salm, C. Sams, T. Scanlon, Rita Schartz, Eric Schenk, Pam Schiefelbein, Jane Schlapp, Michael Schlapp, Amanda Schmidt, Andy Schmidt, Avery Schmidt, Brett Schmidt, Jacob Schmidt, Kevin Schmidt, Loren Schmidt, Mary Schmidt, Mike L. Schmidt, Derek Van Schmus, Judy Schnell, Sally Schrank, Margaret Schulenberg-Ptacek, Caitlin Seals Schwanke, Brad Schwartz, Zachary J. Schwenke, Kevin Scott, Mark Scott, Wayne Seifert, Robert Seitz, Steven Seitz, Amanda Sevite, D. Sewing, Kathy Sexson, Mickey Shaffer, Zack Shaffer, Thomas G. Shane, Kylee Sharp, Scott Sharp, Mark Shaw, Jackson Sheets, Christopher Sheil, Cindy Shepherd, Matt Shepker, Stephanie Sherraden, D. Sherrer, Lenn Shipman, Paul A. Shipman, Becky Shoffner, Ryan Shofner, Jack L. Shumard, Greg Sievert, Lynnette Sievert, Shawn Silliman, Tom Sinclair, Matt Singer, Traci Sipple, Shelley Skie, Aaron Slife, Susan Slife, Joseph B. Slowinski, Donald D. Smith, Dorothy Smith, J. E. Smith, Matt A. Smith, Paige Smith, Philip W. Smith, Tim Smith, Troy D. Smith, Mitch Sommers, Bryan Sowards, David Spalsbury, Milen Spanowicz, Dale W. Sparks, Wally Spears, H. G. Spencer, Robert Sprackland, Andrea Stammler, Chris Stammler, Steve Starlin, Liz Stein, H. A. "Steve" Stephens, John Stephens, Peg Stephens, Kole Stewart, Margaret E. Stewart, Henry Stice, Ralph Stice, Charlie Stieben, Max Stieben, Jonathan Storm, Rick Strawn, Brad Strothkamp, Michael Stubbs, Delilah Subera, Steve Sullivan, Tom Sullivan, Charlie A. Swank, Adam Sweetman, Germaine Taggart, Jesse Taggart, Megan Taggart, Thomas Taggart, Shelly Apryl Tarbet, Anna Tarnowski, Mike Taylor, Butch Teppe, Scott Teppe, Daniel Thalman, Matt Thayer, Marc R. Thiry, Eric Thiss, Bryan Thomas, M. Thomas, Arcine Thompson, K. Thompson, Pat Tierce, A. Todd, B. L. Todd, Karen Toepfer, Russell Toepfer, George Toland, John Tollefson, D. Tolliver, Gus Tomlinson, J. C. Trager, Stanley Trauth, Austin Triboulet, Quoc Trinh, Gene Trott, Jody Trott, Quoc

V. Truong, Bern W. Tryon, Bennett Tuel, Hadley Tuel, Josh Tuel, Kelley Tuel, Peter Tuttle, Victor Tuttle, Fred Tweet, Leigh Tweet, Dean Ullrich, John Ulrey, Linda Ulrey, Mike Unruh, T. A. Unruh, Jonathan VanCampen, Thomas Van Devender, James Van Doren, Mark Van Doren, Mark Van Scoyoc, David Velasquez, Tino Velasquez, Chris Vitt, Joel Voelker, Marc Voiles, Allan Volkmann, Todd Volkmann, Joyce Volmut, Vincent Von Frese, Allison Viola, Warren Voorhees, Bill Vowinckel, Keith Waddington, Amy C. Waddle, A. Wagner, Steven Wahle, J. Waite, David B. Wake, Patrick A. Wakeman, Donald Walburn, Tom Walker, Doug Waller, Chris Waller, Debra Walta, Greg Walters, Ruth Walters, Cody Walton, Cory Ward, J. R. Ward, Karla Ward, Quinci Leighton Ward, Holly Warner, Michelle Warner, Jeremy Washburne, Michael Washburne, Erika Weber, Jamie Webster, Tyson Webster, Derek Welch, William G. Welch, D. Wells, Rob Wencel, Tyann Wencel, Eric Wenzl, Kristina Wenzl, Roy Wenzl, Sheryl Wenzl, John Werler, Lucy Werth, Rebecca Westblade, Luke Westerman, Jeffrey Whipple, E. L. Whitmer, Mike Wildgen, MacKenzie Wiley, Dustin Wilgers, Garrett Wilkinson, Victor Wilkinson, Claire Williams, Dani Williams, Megan Williams, Brad Williamson, Dan Williamson, LaRae Wilson, Denise Wittman, Brad Wolf, Chris Wolf, Henry Wolf, Karen Wolf, Curtis Wolfe, David Wolff, Bruce R. Wolhuter, Floyd Woods, Jana Woods, J. Word, John Wortman, Kenneth Wray, Timothy Wray, Jami Wyatt, Jeff Wyckoff, Larry Wyman, Travis York, John M. Young, Amy Zavala, Mike Zerwekh, Christy Zimmerman, David Zucconi, and Jon Zuercher.

The cooperative sponsorship of this book was brought about by Brad Loveless (Westar Energy, Topeka), Jeffrey Hohman (Touchstone Energy, Washington D. C.), Jason M. Daniels (U. S. Environmental Protection Agency, Kansas City), Jerry D. Choate and Reese Barrick (director emeritus and director, respectively, of the Sternberg Museum of Natural History, Fort Hays State University), and Mike Hayden, Keith Sexson, and Ken Brunson (Kansas Department of Wildlife and Parks, Pratt). To these colleagues we are so grateful.

The entire text was read and criticized by numerous reviewers. Our text profited from scrutiny by these individuals, but all errors of omission and commission that may have slipped into this book are our responsibility alone.

Joseph T. Collins
Suzanne L. Collins
Travis W. Taggart
30 June 2009
Lawrence, Kansas
Hays, Kansas

INTRODUCTION

Worldwide, approximately 13,458 species of amphibians, reptiles, turtles, and crocodilians are known to science (Frost et al., 2006; Zug et al., 2001). Currently, 626 native species of amphibians, reptiles, turtles, and crocodilians have been reported from the continental United States and Canada (Collins and Taggart, 2009, plus updates from *The Center for North American Herpetology* web site [www.cnah.org]). The native and non-native amphibians (salamanders, frogs, and toads), reptiles (lizards and snakes), and turtles found in Kansas consist of 99 species that are assembled in 25 families and 57 genera. Collectively, these animals constitute the herpetofauna of Kansas, their study is called herpetology, and the people who study them are herpetologists. Crocodilians are not native to Kansas and are not included in this book. Three species of alien lizards have been introduced into Kansas, and they are included in this book.

Amphibians, reptiles, and turtles are classified in the Phylum Chordata and are part of a subgroup called vertebrates, all of which have a backbone consisting of numerous bony vertebrae that provide their bodies with flexible support. They share this characteristic with some classes of fishes, as well as with mammals, crocodilians, and birds. Except for two kinds of completely aquatic salamanders and some turtles, the herpetofauna of Kansas is generally terrestrial and these animals usually live near or enter water only to catch food, escape predators, regulate their body temperatures and water balance, or in the case of amphibians, to lay eggs that later develop into an advanced stage. They differ from mammals and birds in lacking hair or feathers, and instead are covered with moist skin (amphibians), scales (reptiles) or shells and scales (turtles). Although not totally restricted to water like fishes, amphibians, reptiles, and turtles are similar to fishes in that they are poikilothermic—that is, their internal temperature varies and typically matches the ambient temperature of the surrounding environment. (This is not true of birds and mammals, which are homeothermic and can maintain a relatively stable body temperature despite the temperature of their surroundings.) Amphibians, reptiles, and turtles are able to regulate their internal temperature within the confines of their environment by basking in the sun to warm up or retiring to the shade to cool off. However, the temperature extremes experienced by these animals in Kansas requires certain other behavioral responses based on the season. For example, their inability to substantially regulate their own body temperature from that of the external environment causes them to be inactive during the cold winter months, a state called brumation. During this time, they retreat beneath the ground or into mud or leaves under water to avoid freezing temperatures, and those that do not retire deep enough will die. Similarly, during the heat of summer or dry periods, most of the Kansas herpetofauna will become dormant in a state called aestivation, which helps them avoid dehydration and high temperatures.

Amphibians, reptiles, and turtles are normally shy and retiring. Unlike other vertebrates, they are comparatively sedentary and incapable of migrating great distances (except sea turtles, sea snakes, and some crocodilians). Most are completely harmless. Only five of the 99 species found in Kansas can be considered dangerous to humans, and all are venomous snakes of the Family Crotalidae (Pitvipers).

Information on the distribution of all herpetofaunal species known to occur in Kansas was obtained by acquiring lists of the more than 51,000 preserved specimens from Kansas in the following 39 research collections:

- Academy of Natural Sciences, Philadelphia, Pennsylvania
- American Museum of Natural History, New York, New York
- Arizona State University, Tempe
- Auburn University Museum, Auburn, Alabama
- Bethel College, North Newton, Kansas
- California Academy of Sciences, San Francisco
- Carnegie Museum of Natural History, Pittsburgh, Pennsylvania
- Centennial Museum, University of Texas, El Paso
- Chicago Academy of Sciences, Illinois
- Cornell University, Ithaca, New York
- Field Museum of Natural History, Chicago, Illinois
- Florida State Museum, Gainesville
- Georgia Museum of Natural History, University of Georgia, Athens
- Illinois Natural History Survey, Urbana
- James Ford Bell Museum of Natural History, University of Minnesota, St. Paul
- Los Angeles County Museum of Natural History, California
- Milwaukee Public Museum, Wisconsin
- Monte L. Bean Life Science Museum, Brigham Young University, Provo, Utah
- Museum of Comparative Zoology, Harvard University, Cambridge, Massachusetts
- Museum of Natural History, Tulane University, New Orleans, Louisiana
- Museum of Natural Science, Louisiana State University, Baton Rouge
- Museum of Vertebrate Zoology, University of California, Berkeley
- Natural History Museum, University of Kansas, Lawrence
- Oklahoma State University, Stillwater
- Peabody Museum of Natural History, Yale University, New Haven, Connecticut
- Royal Ontario Museum, Toronto, Canada
- Sam Noble Museum of Natural History, University of Oklahoma, Norman
- San Diego Natural History Museum, California
- Southern Illinois University, Carbondale
- Southwestern College, Winfield, Kansas
- Sternberg Museum of Natural History, Fort Hays State University, Kansas
- Texas Cooperative Wildlife Collection, Texas A & M University, College Station
- Texas Memorial Museum, University of Texas, Austin
- The Museum, Michigan State University, East Lansing
- University of Colorado, Boulder
- University of Michigan Museum of Zoology, Ann Arbor
- University of Texas, Arlington
- Utah Museum of Natural History, University of Utah, Salt Lake City
- U. S. National Museum of Natural History, Washington, D. C.

To assist anyone using this book with the identification of amphibians, reptiles, or turtles from Kansas, we devised illustrated dichotomous keys (page 225) for each of these major groups. The keys consist of a series of numbered couplet statements. Each couplet has two statements (A and B) of characters useful in their identification. The animal in question will fit either statement A or B and lead the reader to the next couplet, until the scientific name of the animal is reached. At that point, the reader should consult the corresponding species account, as well as the range map, photographs, and description. The description contains a list of the characters that, in combination, will distinguish a particular species from all others. If all characters agree, the animal should be correctly identified. Note that the keys were prepared only for the identification of adult individuals. The keys will generally work for newborn or juvenile reptiles and turtles, but not for aquatic salamander larvae or frog and toad tadpoles. Except for the few biologists that specialize in studying amphibian larvae and tadpoles, their identification in the field is, at best, difficult.

Be sure to consult the brief glossary on page 250 to become familiar with the few technical terms used in this book. The glossary is meant to help the reader understand information about each kind of amphibian, reptile, or turtle found in Kansas.

A CONDENSED HISTORY OF KANSAS HERPETOLOGY

Historically, most of the initial biological surveys that took place in Kansas were associated with various U. S. Army explorations into the American West; they began after the Louisiana Purchase with the Lewis and Clark expedition in 1804 and lasted through 1863. Kansas was not declared a territory until 1854, and even then its borders enclosed a much larger area than they do today. Because of this, the historical record of some of the earliest students of Kansas herpetology is obscured by our inability to determine exactly where specimens were collected. Apparently, the first precise report on Kansas herpetofauna was published in 1823 and written by Thomas Say, who collected the type specimen of the Western Rat Snake (*Scotophis obsoletus*) near Leavenworth during a visit to the region in 1819–1820. Maximilian Prinz zu Wied obtained the type specimen of the Spring Peeper (*Pseudacris crucifer*) at Cantonment Leavenworth, with its description published in 1838 before Kansas was declared a territory. In 1857, Edward Hallowell, a 19th century naturalist, authored a report on the Kansas herpetofauna, but the precise localities for his specimens cannot be verified. Hallowell may never have visited Kansas and the specimens he wrote about were probably collected by William A. Hammond (an army medical director stationed at Fort Riley).

Kansas became a state in 1861, and nearly two decades passed before other works were published on the state's herpetofauna. In 1878, Annie E. Mozley, working only with snakes, listed 32 serpents from Kansas that were housed in a collection at Kansas State University. In addition, during the 1870s and 1880s herpetological specimens from Kansas made their way into museums back East. The most notable collectors at that time were students of Louis Agassiz (especially Joel A. Allen and Samuel W. Garman, both at the Harvard University Museum of Comparative Zoology) and the paid collectors of Edward D. Cope (which included several members of the Sternberg family, famous for their fossil discoveries in the western Kansas chalk). In 1881, Francis W. Cragin of Washburn University was the first biologist to compile a list of all amphibians, reptiles, and turtles known from

Kansas at that time. His *Preliminary catalogue of Kansas reptiles and batrachians* (= amphibians) recorded a robust 88 species and subspecies from the state; he continued to provide numerous notes on these animals until 1894. In 1904, Edwin B. Branson of the University of Kansas wrote his thesis on the snakes of Kansas, reporting 39 species and subspecies from the state.

After Branson's paper, little was reported on Kansas herpetology until 1916 when Edward H. Taylor, then a graduate student at the University of Kansas, completed his unpublished master's thesis, *The Lizards of Kansas*, in which he recorded 20 species and subspecies from the state. Taylor evidently finished his thesis in 1912 and left Kansas to take a position in the Philippines. During his absence, Victor H. Householder, also a graduate student at the University of Kansas, studied the lizards and turtles of the state and in 1917 completed his unpublished thesis, *The lizards and turtles of Kansas with notes on their distribution and habitat*. Householder reported 13 species of turtles and 19 species of lizards from Kansas. Taylor's thesis was published posthumously in 1993 by the Kansas Herpetological Society.

Once again, research on Kansas herpetology lay quiescent, but this time for only a decade. In the late 1920s, a tremendous burst of interest in Kansas herpetology began and has continued to this day. Hobart M. Smith (then a student of Edward H. Taylor at the University of Kansas), Charles E. Burt (a Topeka resident born in Neodesha), Howard K. Gloyd (a native Kansan born in DeSoto), Claude W. Hibbard (a native Kansan born in Toronto), and Joseph Tihen (a native Kansan) initiated studies that resulted in more than 30 notes and papers published between 1927 and 1950. Most notable of these were Burt's (1928) *The Lizards of Kansas* and Smith's (1934) *The Amphibians of Kansas*—the latter a sorely needed study of a group of animals almost completely ignored in the state before that time. Smith maintained his interest in Kansas herpetology and produced the *Handbook of Amphibians and Reptiles of Kansas*, which appeared in 1950 and was revised and re-issued as a second edition in 1956. In this handbook, Smith recorded 97 species from the state.

In the 1950s, the late Robert F. Clarke (Emporia State University) had numerous articles published on the ecology of amphibians, reptiles, and turtles in Osage, Chase, and Lyon counties, as well as on lizards in general. Richard A. Diener wrote three papers on Kansas snakes during 1956 and 1957. Hibbard revisited the state and contributed short reports in 1963 and 1964. By 1950, the late Henry S. Fitch had begun field work on the newly established Natural History Reservation of the University of Kansas in Douglas County. His excellent work and that of his many colleagues and students produced intensive ecological and natural history studies in northeastern Kansas. Fitch alone produced over 145 papers on the herpetofauna of the newly named Fitch Natural History Reservation. From the 1960s through the early 1980s, Eugene D. Fleharty was active in Kansas herpetology at Fort Hays State University, co-authoring with his students a number of papers on the distribution of these animals in the state.

Joseph T. Collins arrived in Kansas in 1968, and wrote *Amphibians and Reptiles in Kansas* (1974), confirming 91 species from the state. His second edition (1982) identified 92 species as occurring in Kansas. He and Suzanne L. Collins wrote a third edition in 1993, in which they recognized 97 kinds of amphibians, reptiles, and turtles as members of the state herpetofauna. Dwight R. Platt (Bethel College, Newton), along with members of the Conservation Committee of the Kansas Academy of

Science and other biologists, initiated (1974) a list of rare or endangered Kansas amphibians, reptiles, and turtles for the purpose of identifying animals in danger of extirpation. Their proposal and subsequent action by the Kansas Department of Wildlife and Parks resulted in an official endangered, threatened, or species in need of conservation (SINC) list, in 1978. Amphibians, reptiles, and turtles currently on this list are protected by law. In addition, starting in the 1960s, Platt wrote extensively about the herpetofauna of the sand prairies of southcentral Kansas, a topic he has continued to research.

The formation of the Kansas Herpetological Society was a significant development. Established in Lawrence in May of 1974 by Joseph T. Collins, with the help of some colleagues and friends, it rapidly developed into a strong statewide organization with active programs on the study and conservation of the native herpetofauna. Its members have contributed a great deal of distributional data and made significant observations on these animals throughout the state, many of which have been published in the *Kansas Herpetological Society Newsletter* (1974 to 2001) and its successor, the *Journal of Kansas Herpetology* (2002 to date), now the official periodical of the Society. Of special significance was the creation by Collins in 1989 of an annual KHS herpetofaunal count, initially held in Kansas during April and May (but now includes June) of each year (Rakestraw 1996). Patterned on the well-known Christmas bird counts conducted by ornithologists, this annual spring-summer herpetofaunal tally was the first of its kind in the nation and continues to date. These KHS counts have provided much data on comparative abundance, and stimulated a great deal of involvement by the interested public while teaching many of them about amphibians, reptiles, and turtles.

The Kansas Department of Wildlife and Parks, through voluntary contributions to its Chickadee Checkoff program, has supported many studies on some of the state's lesser known amphibians, reptiles, and turtles. The first of these was funded in 1981 (with the results reported in Collins, 1982b); such studies continue to be funded by KDWP.

Recently, strong interest in the herpetofauna of Kansas has re-emerged at the Sternberg Museum of Natural History, Fort Hays State University, where two of us (TWT and JTC) and Curtis J. Schmidt and Eugene D. Fleharty are curators of the herpetology collection, and are actively engaged in field surveys on these creatures throughout the state. The impact of some aspects of their research and field work is chronicled in a recent paper (Taggart et al., 2006). In addition, the following web sites on Kansas herpetology are excellent resources (see the bibliography for specific citations for these web sites):

Kansas Herpetofaunal Atlas: http://webcat.fhsu.edu/ksfauna/herps/
Kansas Herpetological Society: http://www.cnah.org/khs/
Kansas Anuran Monitoring Program: http://www.cnah.org/kamp/

Further, *The Center for North American Herpetology* (CNAH), established in 1994, maintains an academic herpetological web site that provides detailed information about the Kansas herpetofauna, as well that of the entire North American continent (north of Mexico). It can be accessed at:

http://www.cnah.org/

Users of this field guide are encouraged to consult these four web sites for additional information, including updated distributions maps, taxonomic changes, and important announcements about the profession of herpetology. All four web sites are coordinated with each other and updated daily. CNAH maintains the most stable and frequently accessed academic herpetological web site on the Internet.

TAXONOMY

The species accounts in this book are arranged by class, order, and family in a sequence generally similar to that used in Collins (1974, 1982) and Collins and Collins (1993). Within each family, generic and specific accounts appear in alphabetic order. Each species account supplies the basic information needed to identify an amphibian, reptile, or turtle, especially when used in conjunction with the key. It also contains color photographs of each species, a Kansas range map, the standard common and current scientific names, and information on identification, size, natural history, and pertinent literature references.

The authors of this field guide recognize species based on the *Evolutionary Species Concept* of Wiley (1978), and retain the traditional seven-part hierarchical taxonomy (Kingdom, Phylum, Class, Order, Family, Genus, and Species) in a manner that reflects evolutionary history based on current scientific evidence published in peer-reviewed outlets. A brief list of significant papers is included in a section called *Taxonomic and Phylogenetic References to Chordates*, which appears at the beginning of the *Bibliography* (see page 253). These references support a hierarchy showing the evolutionary history of vertebrates within the Phylum Chordata, as depicted in the hypothesis drawn below.

Over the past two decades, evidence has accumulated that traditional recognition of the Class Reptilia is not consistent with evolutionary history. Thus, certain nomenclatural changes were needed to reconcile the taxonomy with contemporary phylogenetic discoveries. We have adopted the phylogenetic hypothesis and taxonomy presented below.

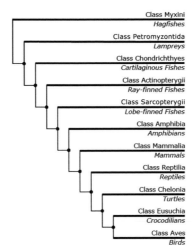

Class Myxini
Hagfishes

Class Petromyzontida
Lampreys

Class Chondrichthyes
Cartilaginous Fishes

Class Actinopterygii
Ray-finned Fishes

Class Sarcopterygii
Lobe-finned Fishes

Class Amphibia
Amphibians

Class Mammalia
Mammals

Class Reptilia
Reptiles

Class Chelonia
Turtles

Class Eusuchia
Crocodilians

Class Aves
Birds

A phylogeny of the Chordates (Phylum Chordata) that is consistent with evolutionary history based on current scientific evidence. See the section in the bibliography of this book entitled *Taxonomic and Phylogenetic References to Amphibians, Reptiles, and Turtles* for evidentiary support of this hypothesis.

The Class Reptilia had been traditionally composed of the Order Chelonia or Testudinata (the turtles), the Order Crocodylia (the crocodilians), the Order Squamata (the lizards, snakes, and amphisbaenids), and the Order Rhyncho- cephalia (the Tuataras). The discovery that birds (Class Aves) are the closest modern relative to crocodilians and turtles renders the Class Reptilia as an unnatural grouping. In order to bring the names of these animals in line with their hypothe- sis of evolutionary history (i.e. a name should include only a common ancestor and all of its descendants), two options were considered. The first was to move the Class Aves into the Class Reptilia, thereby removing the Class Aves from equal rank and making all birds reptiles (this has not received much support). The conse- quences of adopting this scenario are two-fold: (a) the Class Aves would no longer exist, and (b) the title of this book would need to be changed to "Amphibians and Reptiles (except Birds) in Kansas." The second option, illustrated in the phylogeny above, was to elevate crocodilians (under the more appropriate name Eusuchia) and turtles (under the name Chelonia) to Class status and restrict Class Reptilia to the Orders Squamata and Rhynchocephalia. We consider this arrangement to be more in harmony with traditional taxonomy and the most easily understood. Both options are supported by current scientific evidence, and we chose to adopt the second.

The scientific name is the formal name of each species and consists of two words. The first word always begins with a capital letter and identifies the genus (plural: genera), and the second never begins with a capital and is the species (plural: species). Both words are always italicized. A genus may contain a large number of species, but a species belongs to only one genus. Following the scien- tific name is the authority (the name of the individual[s] that first formally described the species as new to science), along with the year in which it was described (as printed on the original published description). If the name and date are in paren- theses, it means that the species was originally described in a different genus and subsequently transferred to the genus shown in this book; the absence of paren- theses indicates that the species was originally described in that genus. An exam- ple of the structure of the scientific name is as follows:

Genus—*Nerodia*
Species—*sipedon*
Describer—(Linnaeus, 1758)

Thus, the genus *Nerodia* refers to a large group of reptiles called water snakes and the species *sipedon* refers to a particular water snake within the genus called the Northern Water Snake. The describer, Linnaeus, originally placed this species in the snake genus *Coluber*, but it was eventually shown to belong to the genus *Nerodia*; thus, the parentheses around the name and date. The formal scientific name then appears as:

Nerodia sipedon (Linnaeus, 1758)

In addition, at the familial level, users of this field guide are advised to consult Frost et al. (2006) for amphibians, Frost and Etheridge (1989) and Frost et al. (2001) for lizards, Collins (2006) for snakes, and Iverson (1992) for turtles. Class and ordinal names are those that the authors feel are most logical; all are arranged in a manner supported by current scientific evidence. As with genera and species, all class, ordinal, and familial names have author attribution appended with the year- date published (as it appears on the publication).

EXPLANATION OF SPECIES ACCOUNTS

Common names: Each account begins with the standard common name (and scientific name, see above) for each kind of amphibian, reptile, or turtle. All common names are those standardized for U. S. species by Collins et al. (1978 et seq.) and perpetuated by Collins and Taggart (2009). To minimize confusion and promote easy recognition in reading and conversation, all schools, zoos, and museums should use only these standardized common names when referring to a particular amphibian, reptile, or turtle. In addition, use of these standard common names facilitates communication between the public and scientists. Common names are proper nouns and the first letter of each individual word is always capitalized.

Photographs: All photographs are of living amphibians, reptiles, and turtles, and were taken by Suzanne L. Collins unless otherwise noted in the caption. More than one photograph is used to show pattern variation by age and geography. All photographs of Kansas individuals are noted by including their county of origin in the caption. Photographs without such information are of animals from nearby states.

Maps: Each account includes a detailed Kansas distribution map showing a specific county or an outline of the state and each of its 105 counties. The symbols (solid red circles outlined in yellow) on each map indicate localities where that species has been collected and for which a preserved voucher specimen has been deposited in a university or museum herpetological collection. For information on the range of each species of amphibian, reptile, or turtle outside of Kansas consult the *Peterson Field Guide to Reptiles and Amphibians of Eastern and Central North America Third Edition Expanded* by Conant and Collins (1998) for the eastern and central United States and Canada. In each map caption, a brief statement of the range in Kansas is provided, with comments on peripheral or questionable records or where records from adjacent states suggest the species might be found.

Identification: Characters used to distinguish each species (when consulted after using the key), and a brief color description, are provided in this section. External characters, if any, for separating the sexes are mentioned here.

Size: The size of each species is given in three parts, the average total length, total length of the largest specimen from Kansas, and the maximum total length for the species throughout its range. Because it has not been generally accepted in the United States, the use of metric measurements has been kept to a minimum. The average total length and total length of the largest Kansas specimen are expressed in millimeters and inches. Snout-vent length or SVL (inches only) is given also for salamanders and lizards since both groups of animals are frequently missing part of the tail. Frog lengths are measured snout-vent (SVL) and this is given as the total length since they have no tail as adults. Turtle lengths are measured in a straight line from the front to the back of the upper shell, not across the curve. The total length of snakes is determined by measuring from the snout to the tip of the tail. All record size maxima include the county where collected, the date of collection, and the collector(s). The length of the largest Kansas example is based on preserved specimens deposited in herpetological collections at the Sternberg Museum of Natural History, Fort Hays State University (MHP) or the Natural History Museum, University of Kansas (KU). Where known, for some species the weight of the heaviest specimen from Kansas is given.

Natural History: The habitat, if known, is stated, and factors affecting daily and annual activity are presented. When known, such aspects of their lifestyle as territoriality, home range, winter retreat, and other pertinent information are given. Information on courtship, mating, egg deposition, nesting, hatching, and the birth of young are presented. We relied on Smith (1934) in conjunction with more recent information gleaned from the Kansas Anuran Monitoring Program and the Kansas Herpetofaunal Atlas for information on amphibian breeding, Fitch (1970) for reproductive data on lizards and snakes, and Ernst and Lovich (2009) for breeding information on turtles (again, the latter two in conjunction with the Kansas Herpetofaunal Atlas), as well as on data gathered by ourselves and our colleagues. The food preferences of each species of amphibian, reptile, or turtle, based on our personal observations and data from the literature, are detailed. In addition, information on predation, unusual or interesting behavior and observations, and, where appropriate, an endangered, threatened, or SINC designation (see below), are included.

Pertinent References: Listed here are the references we consulted in writing this book, and many of these will assist interested readers in acquiring more detailed information about a particular species of amphibian, reptile, or turtle. Emphasis has been placed on significant papers and notes, or those of historical interest.

ENDANGERED AND THREATENED SPECIES

Platt et al. (1974) originally prepared a list of amphibians, reptiles, and turtles from Kansas that may have been in danger of extirpation. Of the herpetofaunal species now known to occur in Kansas, none is considered endangered or threatened at the national level. In 1975, the Kansas Legislature passed the Kansas Nongame, Threatened and Endangered Species Act. This act vested the Kansas Department of Wildlife and Parks with the authority to establish an official list of endangered or threatened species in the state. Such a list was established in 1978 and currently (2010) includes 25 species (or around 25% of the herpetofauna of the state), as follows:

Endangered:
- Cave Salamander (*Eurycea lucifuga*)
- Grotto Salamander (*Eurycea spelaea*)

Threatened:
- Eastern Newt (*Notophthalmus viridescens*)
- Longtail Salamander (*Eurycea longicauda*)
- Green Toad (*Anaxyrus debilis*)
- Spring Peeper (*Pseudacris crucifer*)
- Strecker's Chorus Frog (*Pseudacris streckeri*)
- Bronze Frog (*Lithobates clamitans*)
- Eastern Narrowmouth Toad (*Gastrophryne carolinensis*)
- Broadhead Skink (*Plestiodon laticeps*)
- New Mexico Blind Snake (*Rena dissecta*)
- Longnose Snake (*Rhinocheilus lecontei*)
- Redbelly Snake (*Storeria occipitomaculata*)
- Checkered Garter Snake (*Thamnophis marcianus*)
- Smooth Earth Snake (*Virginia valeriae*)
- Common Map Turtle (*Graptemys geographica*)

Species in Need of Conservation (SINC):
- Red-spotted Toad (*Anaxyrus punctatus*)
- Crawfish Frog (*Lithobates areolatus*)
- Eastern Glossy Snake (*Arizona elegans*)
- Western Hognose Snake (*Heterodon nasicus*)
- Eastern Hognose Snake (*Heterodon platirhinos*)
- Chihuahuan Night Snake (*Hypsiglena jani*)
- Rough Earth Snake (*Virginia striatula*)
- Timber Rattlesnake (*Crotalus horridus*)
- Alligator Snapping Turtle (*Macroclemys temminckii*)

Endangered status indicates a higher priority than designation as threatened. The 25 species of amphibians, reptiles, and turtles listed above as endangered, threatened, or SINC are protected by state regulations, and a permit issued by the Kansas Department of Wildlife and Parks is required to collect them for scientific purposes. The list represents a major step by the people of Kansas toward conserving these creatures.

ALIEN SPECIES

One of the most disturbing herpetological phenomena, both in Kansas and nationally, is the introduction of alien (non-native) amphibians, reptiles, and turtles into areas where they do not occur naturally and where they generally cannot survive due to an adverse climate (Collins and Taggart, 2009; CNAH web site). Although Kansas has been relatively fortunate in having few of these introductions (compared to a state such as Florida, which has experienced dozens of such occurrences), at least three alien lizards, the Mediterranean Gecko (*Hemidactylus turcicus*), Italian Wall Lizard (*Podarcis siculus*), and Western Green Lacerta (*Lacerta bilineata*) have successfully established themselves in our state and are included in the species accounts in this guide. The following chronological list reports on eight other alien species known to have been introduced, unsuccessfully, into Kansas since 1940:

Green Anole (*Anolis carolinensis*). Brumwell (1942) reported the establishment of this lizard in Kansas on the basis of a specimen (KU 21450) he obtained in August of 1940 along the Missouri River bank at Fort Leavenworth, Leavenworth County. This specimen may have come from a population of two dozen individuals released in Leavenworth County in 1938 (Brumwell, 1951). Smith (1956) and Smith and Kohler (1977) included this species in Kansas as an introduced form. The Green Anole is normally found in the southeastern United States, and ranges as far north as southern Arkansas. Clarke (1965a, 1986b) reported this species as introduced in Johnson County (south of and adjacent to Leavenworth County). However, no evidence exists that any viable, breeding population has survived in Kansas. Due to harsh winter temperatures, this non-native lizard could not survive more than a few seasons in northeastern Kansas. This unsuccessful introduction was noted by Kraus (2009).

Wood Frog (*Lithobates sylvaticus*). Breukelman and Smith (1946) and Breukelman and Clarke (1951), reported a single specimen (KU 23149) of the Wood Frog from extreme southwestern Lyon County, which was collected by John Breukelman in 1942. Smith (1956) included this species as part of the Kansas fauna, probably based on the Lyon county specimen. Conant (1958) and Stebbins (1966) showed an isolated colony of this frog in the state. Collins (1974b) questioned whether the Wood Frog was a valid member of the state fauna, and Smith and Kohler (1977) and Kraus (2009) listed it as an introduced species in Kansas. No additional examples of this frog have been discovered in Kansas since 1942, and the specimen from Lyon County represents either an accidental (and unsuccessful) introduction or its locality has been mislabeled. No evidence exists that this species is native to the state; habitat for it in Kansas does not exist.

Northern Leopard Frog (*Lithobates pipiens*). A single specimen (KU 28642) of this frog was collected in 1950 in a pond at the University of Kansas campus, Douglas County (Kraus, 2009). This species has a natural range in states north and east of Kansas, and the Douglas County example undoubtedly represents an unsuccessful introduction, probably after its use in biology classes. No additional specimens have been discovered in Kansas in the intervening 55+ years.

Dusky Salamander (*Desmognathus* sp.). Salamanders of this genus are normally inhabitants of the eastern United States and range as far west as Arkansas and extreme eastern Oklahoma. A large series of salamanders of this genus was collected by the late Robert F. Clarke (Emporia State University) on 20 June 1960, from a cavelike excavation in Cherokee County. According to Clarke (pers. comm., 1973), these salamanders were introduced into the excavation by a commercial bait dealer. Evidently, a lack of moisture in the excavation created an inhospitable environment for the salamanders, and they eventually perished. No examples of this species have been found in Kansas since 1960. This unsuccessful introduction was noted by Kraus (2009).

Green Treefrog (*Hyla cinerea*). A Lawrence teacher and some students collected a single specimen (KU 176272) of this frog on 16 October 1974, around the pond of a commercial fish farm in Douglas County (Perry, 1974b). In addition, a field party to the same locality secured 37 specimens (KU 176273–309) of this amphibian on 14 October 1975. The Green Treefrog is normally an inhabitant of the southeastern United States and ranges as far northwest as the southeastern corner of Missouri. These frogs were presumably transported as tadpoles in fish shipments to Douglas County from somewhere in the southeastern United States. While they were able to metamorphose into adults during the warm Kansas summer, they did not survive winter temperatures. No examples of this frog have been found in Kansas since 1975.

Blanding's Turtle (*Emys blandingii*). A single specimen of the Blanding's Turtle was found during July of 1979, in a residential area of Lawrence in Douglas County. This turtle normally ranges from northcentral Nebraska east into the Great Lakes region. It was undoubtedly captured in that area and transported to Kansas, where it was released (or escaped). This unsuccessful introduction was noted by Kraus (2009).

Mohave Rattlesnake (*Crotalus scutulatus*). On 24 May 1980, three students at Bonner Springs High School collected an example of this extremely venomous and dangerous snake in a rock quarry in Leavenworth County near the Wyandotte County line. The specimen was preserved and brought to one of us (JTC) for examination. This rattlesnake normally is found in the southwestern United States and Mexico. The specimen from Leavenworth County was likely captured in the southwestern U. S., transported to Kansas, and released (or escaped). It probably would not have survived the winter temperatures of northeastern Kansas. This unsuccessful introduction was noted by Kraus (2009).

Western Diamondback Rattlesnake (*Crotalus atrox*). Hall and Smith (1947) reported this serpent from Crawford and Cherokee counties in southeastern Kansas, well outside of its known range and in habitat not suitable for this species; based on local newspaper accounts, the Cherokee County record was the result of snakes that escaped during an automobile accident near Weir, Kansas. The specimen from Crawford County cannot now be located. Because of the intense collecting activity in these two counties by biologists over the last half century, it seems implausible that this large, aggressive reptile exists there. Those records aside, starting in the 1980s, this venomous serpent was apparently brought into Kansas repeatedly and released in different areas of the state. Many of these introductions probably occurred as a result of individuals returning to Kansas with serpents that they captured during one of the infamous rattlesnake roundups in Oklahoma or Texas. Once they discovered that these creatures did not make good pets, they probably released them; others may have escaped from confinement. A particularly large number of these rattlesnakes were released in the Horsethief Canyon area at Kanopolis State Park in Ellsworth County, and a few of those animals might persist there today; however, there is no evidence of this reptile breeding in Kansas. The incidence of these snakes being introduced into Kansas has been documented or noted by Capron (1985b), Riedle (1996), Mosher (1997), Matlack and Rehmeier (2002), Taggart (2003d, 2006b), and Kraus (2009). In 2006, the state of Kansas passed legislation that resulted in a ban on the importation of nonnative venomous snakes into Kansas, except for research by accredited museums, universities, and zoos.

In addition to these eight examples of introduced amphibians, reptiles, and turtles, other introductions, such as waifs, have arrived from the tropics as stowaways in fruit, flower, or vegetable shipments. None of these creatures, however, could survive a Kansas winter.

These eight species were introduced into Kansas by a variety of means, most of which were accidental and might have been prevented. Introduction of alien amphibians, reptiles, and turtles may be detrimental to our native species and should be avoided. Documentation of the extent of these introductions into the United States is available in Conant and Collins (1998). Collins and Taggart (2009) listed those species that have established breeding populations in North America (north of Mexico) and Hawaii. *The Center for North American Herpetology* web site maintains a list of the herpetofauna introduced and established in the United States and Canada; it is updated daily and can be accessed at:

http://www.cnah.org/ex_nameslist.asp

HOW TO FIND AND OBSERVE AMPHIBIANS, REPTILES, AND TURTLES

Numerous techniques can be used for observing amphibians, reptiles, and turtles. With experience, each person usually becomes more competent at locating and observing them. The following paragraphs (modified from Smith, 1950, 1956, and Conant and Collins, 1998) describe the more common methods used to find and observe these interesting creatures.

By day in spring and fall (especially early October), and at night during hot summer months using the headlights of a car, much success can be obtained by slowly driving along rural roadways and carefully scrutinizing the area of the road. Even the smallest snakes can be seen by this means, and of course all manner of larger snakes, some lizards, and many amphibians may be found. Success varies with the speed of driving, alertness of the observers, intensity of the lights, nature of the road, darkness of the night (moonless nights are preferred), wind speed (windy nights are best), air temperature, humidity, and the type of country on either side of the road (i.e., rangeland is excellent; cropland is generally very poor). Careful observation of the relationship of these factors with abundance of animals discovered will increase an individual's enjoyment of this field technique. A word of caution: the tremendous overpopulation of our country by people has resulted in an enormous increase in automobile traffic—please use care when roadcruising and restrict your driving to little-used secondary and gravel roads. Your chances of safely seeing a living amphibian, turtle, or reptile will be much enhanced, and your safety will be enhanced as well.

Many amphibians are easily discovered at night when they are breeding or prowling for food, especially immediately following rains. Use a flashlight. In spring, the choruses of frogs and toads will lead you to them; at such times, it is profitable to drive about until voices of interest are heard. At other times, any pond, marsh, or other body of water may be expected to contain amphibians. Most species of salamanders breed early in the spring, even with a thin layer of ice present on the water. In western Kansas, Barred Tiger Salamanders have been found at the mouths of mammal burrows where they spend the day, but from which they may wander a short distance at night. All caves, especially those containing water, are likely places for amphibians and to a lesser extent for reptiles. In caves, take care to observe streams and pools closely for signs of small salamanders or salamander larvae. Flashlights are particularly valuable when observing nocturnal snakes on flat plains or prairies and along streams. In these open areas, watch far ahead for fleeting glimpses of wary snakes and near at hand for the motionless bodies of sluggish or temporarily blinded species. Snakes along streams at night often may remain motionless when approached; during the day they usually dive quickly and are seen only momentarily.

Another productive field technique is the seining of marshes, ponds, and streams. Turtles, snakes, frogs, and salamanders may be found by this means, and any closely packed debris or vegetation near the borders of the water should be hauled onto the shore and carefully inspected.

Cattle tanks, particularly in western Kansas, are important to herpetofauna. Often the spillover from these tanks creates small pools of water or semi-permanent ponds, and attracts many amphibians, reptiles, and turtles. Be sure to check in all cattle tanks, whether dry or full; if you find any of these creatures trapped in a cattle tank, be sure to remove it and let it go. Then do something to further prevent such incidents—prop a log or board at an angle down into the cattle tank, so that these animals have a way to escape a slow death by starvation or exposure. And don't forget to look under those cattle tanks when they are empty.

A most effective field technique is to keep a sharp eye on the entire surroundings and turn every conceivable type of cover. Stones, logs, cardboard, plywood, junk, sheet metal, and any other movable surface cover may conceal some seldom-seen reptile, turtle, or amphibian. An alert observer never leaves a stone unturned. In early spring and to a lesser extent in the fall, one may expect good results from this effort. In the summer, the ground under such cover is often too dry and will yield little. At such times, by far the best practice is to be about early in the morning, before the heat of the sun has penetrated through the cover to the ground, or at dusk as animals are becoming active to forage.

In certain areas, removing debris from the ground can be productive. Accumulations of leaves, twigs, and flood deposits often conceal reptiles or amphibians. Rotten logs and loose bark on logs or trees are also a favorite haunt; different species often prefer different types of logs. Always remember to lift the bark that remains on the ground, under which animals often seek protection. Bales of hay drying in fields may conceal lizards and snakes.

In the last decade or so, herpetologists across the United States have increased the productivity of their field work by placing large pieces of sheet metal and/or plywood on the ground in remote areas. Setting up this artificial cover (particularly during the winter when collecting is more difficult) creates hiding places for amphibians, lizards, snakes, and terrestrial turtles. Visiting these sites on a regular basis in spring, summer and fall, particularly after they have aged for a year or so, is usually very rewarding.

All frogs, toads, and salamanders occurring in Kansas may be captured and handled with safety. However, care should be taken to avoid any contact of their skin secretions with a person's eyes, nose, or mouth, because the skin secretions of certain salamanders, some frogs (for example, the Gray Treefrogs), and all toads (genus *Anaxyrus*) irritate mucous membranes and are poisonous if ingested.

No lizard in the state is venomous, but all can bite. Only the Eastern Collared Lizard, Great Plains Skink, and Broadhead Skink have jaws powerful enough to deliver a painful bite. All others have jaws so small that no precaution is necessary.

Snakes, however, include 5 dangerous, venomous species. Unless these can be recognized positively in advance, the novice should treat all snakes as venomous and observe them from a distance.

Harmless snakes can be temporarily captured for closer examination by picking them up quickly by any part of the body. However, any snake more than sixteen inches long may be capable of giving a painful bite, and some precaution may be desired. Wearing gloves when handling large harmless snakes will alleviate the possibility of feeling a bite and will save your hands from wear while turning rocks and other objects.

AMPHIBIANS
CLASS AMPHIBIA GRAY, 1825

Amphibians in Kansas consist of 9 families, 11 genera, and 31 species in 2 distinct Orders—salamanders, and frogs and toads. Fifteen families made up of 41 genera and 288 native species of amphibians live in the United States and Canada (Collins and Taggart, 2009, with updates from the CNAH web site).

Amphibians depend on water and moisture to exist, and they differ from turtles and reptiles in many ways. Amphibians have moist skin, whereas reptiles are covered with scales and turtles have shells (some stiff and hard; some flexible and leathery). Lacking a water-resistant skin, amphibians lose body moisture more easily and thus cannot move far from water or wet areas. Amphibians have no claws on their feet, whereas all turtles and reptiles (except snakes and limbless lizards) possess them. Amphibians also lay soft gelatinous eggs that lack a shell, depositing them in water or where they will remain wet. In Kansas, salamanders breed by internal fertilization and frogs and toads by external fertilization.

The eggs of most Kansas amphibians, unlike those of turtles and reptiles, hatch into aquatic larvae or tadpoles that spend their lives in water until they metamorphose into adults. Most adults are terrestrial, but in Kansas the Eastern Newt and the two species of mudpuppies are completely aquatic as adults. Adult salamanders differ from frogs and toads by the presence of a tail and by having front and hind limbs nearly equal in size. Frogs and toads lack tails, and their hind limbs are larger than their front limbs. In addition, adult male frogs and toads can chorus, calling throughout their annual breeding period (and some species beyond). Kansas salamanders make no sound. Amphibian larvae and tadpoles are extremely difficult to identify, and in this book no attempt is made to do so. Information about tadpole and larval identification can be accessed at links on the CNAH web site (http://www.cnah.org).

SALAMANDERS
ORDER CAUDATA SCOPOLI, 1777

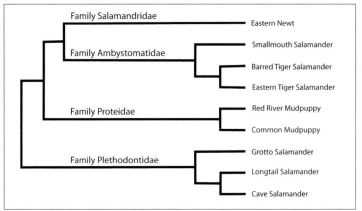

A phylogeny of Kansas Salamanders (Order Caudata) that is consistent with evolutionary history based on current scientific evidence.

Salamanders are classified in the Order Caudata, which in the United States and Canada consists of 8 families, 23 genera, and 188 species (Collins and Taggart, 2009, with updates from the CNAH web site). In Kansas, they comprise 9 species in 4 genera and 4 families.

Salamanders are generally restricted to the eastern third of Kansas, where higher rainfall and forests, forest ponds, and limestone streams provide a more moist and favorable habitat. Only the Barred Tiger Salamander has an extensive range in the state.

MOLE SALAMANDERS
FAMILY AMBYSTOMATIDAE GRAY, 1850

The family Ambystomatidae (Mole Salamanders) is represented in Kansas by 3 species in 1 genus, the Barred Tiger Salamander (*Ambystoma mavortium*), Smallmouth Salamander (*Ambystoma texanum*), and Eastern Tiger Salamander (*Ambystoma tigrinum*). This family contains 2 genera and 19 species in the United States and Canada; adults have lungs, grow up to 330 mm in total length, have short, blunt heads, robust bodies and limbs, and exhibit colors and patterns ranging from uniformly plain to brightly marked.

Barred Tiger Salamander
Ambystoma mavortium BAIRD, 1850

Identification: Body robust; 14 or fewer vertical grooves on each side of body between front and hind limbs; dorsum of body with 6 to 36 (mean 17) contrasting bright yellow or golden bars on black background; bars or blotches on sides of body may or may not extend onto belly, which is black or gray-black and mottled with yellow; chin usually yellow; during breeding season, females can be distinguished from males by their heavier bodies, whereas males have swollen cloacal lips; females have slightly longer bodies than males; males have proportionately longer tails than females.

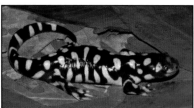

Adult: Cheyenne County, Kansas   Adult

Size: Adults normally 150–215 mm (6–8½ inches) in TL; largest terrestrial adult specimen from Kansas: female (KU 2593) from Morton County with SVL of 153 mm and TL of 290 mm (11½ inches) collected by Theodore H. White and Edward H. Taylor on 20 August 1926; largest neotenic specimen from Kansas: (KU 155356)

AMPHIBIANS

collected by T. Feuerborn and H. McArdle on 25 August 1974 in Wabaunsee County, measured 300 mm (11¾ inches) TL; maximum length throughout range: 12⅞ inches (Conant and Collins, 1998).

Adult   Larva

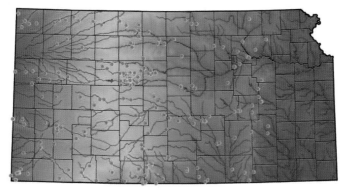

The **Barred Tiger Salamander** (*Ambystoma mavortium*) occurs primarily in the western three-quarters of Kansas and is the only salamander found in the western half of the state. Red dots indicate records backed by voucher specimens or images.

Natural History: Generally spends summer and winter beneath ground in caves or burrows of other animals, thereby avoiding temperature extremes and drought; frequently emerges from burrows at night or during rains, even when temperature approaches freezing; recorded active annually in Kansas from early February to mid-December, with a peak from mid-April to early July; little known of spring emergence or home range; breeds after sufficient rains, from December to March; seeks out breeding site, generally permanent shallow lakes, ponds, ditches, or backwater pools along rivers in open prairie; does not appear to mass migrate to breeding sites; length of breeding season in Kansas unknown, but probably lasts up to 3 months; courtship takes place in water and consists of males and females rubbing bodies with occasional "nips" at each other along with much lashing about of bodies and tails; this "foreplay" eventually stimulates male to swim in front of female, who follows with her snout near his cloaca; male deposits a spermatophore, which female swims over and mounts with her cloacal lips; eggs deposited singly or in small clumps of 2–3 and attach to sticks and weed masses along water's edge; females can lay up to 1,000 eggs, which hatch in a few weeks; gilled, pond-type larvae may metamorphose into adults during same summer, overwinter until second summer, or achieve sexual maturity as larvae and remain in that state their entire lives (latter condition, called neoteny, usually occurs when terrestrial conditions are harsh); eats grasshoppers, aquatic insects, terrestrial insects, earthworms, fishes, tadpoles, frogs, toads, other salamanders, and mice; predators include owls and snakes; maximum longevity: 12 years, 4 months, and 30 days (Snider and Bowler, 1992).

Pertinent References: Ballinger and Lynch (1983), Bartlett and Bartlett (2006a), Baxter and Stone (1980, 1985), Behler and King (1979), Bishop (1943), Black and Collins (1977), Black and Lardie (1976), Brennan (1934, 1937), Breukelman and Clarke (1951), Breukelman and Downs (1936), Burt (1927b, 1932, 1935), Busby and Parmelee (1996), Busby et al. (1994, 1996, 2005), Capron (1978b), Clarke (1970, 1980a), Clarke et al. (1958), Collins (1959, 1974a, 1974b, 1982a, 1985, 1990b), Collins and Collins (1991, 1993a, 1993b, 2006a), Collins et al. (1991, 1994), Conant (1958, 1975), Conant and Collins (1991, 1998), Cope (1867, 1889), Cragin (1881), Cross (1971), Dundee (1988), Dyrkacz (1981), Ellis and Henderson (1913), Fowler and Dunn (1917), Gehlbach (1956, 1967), Gier (1967), Gish (1962), Gress (2003), Guarisco (1981c), Hartman (1906), Hartmann (1986), Heinrich and Kaufman (1985), Holman (1976a), Ireland and Altig (1983), Irschick and Shaffer (1997), Irwin and Collins (1987), Karns et al. (1974), Kingsbury and Gibson (2002), Knight (1935), Knight and Collins (1977), Kretzer and Cully (2001), Kumberg (1996), Levell (1995, 1997), Maldonado-Koerdell (1950), Maldonado-Koerdell and Firschein (1947), Marr (1944), Miller (1976a, 1980a), Miller and Collins (1995), Mount (1975), Murrow (2008b, 2009a, 2009b), Perez (1985), Perry (1974a), Petranka (1998), Platt (1985b, 1998), Pound (1968), Preston (1982), Royal (1982), Rues (1986), Rundquist (1996a), Rush (1981), Rush et al. (1982), Schmidt (1953, 2001, 2003c, 2004l), Schmidt and Murrow (2007c), Simmons (1987c), Smith (1933b, 1934, 1950, 1956a, 1961), Stebbins (1951, 1954, 1966, 1985, 2003), Stejneger and Barbour (1943), Suleiman (2005), Taggart (1992b, 2001b, 2002c, 2004h, 2006b, 2006b, 2006c, 2007c), Taggart et al. (2007), Taylor (1929b), Terman (1988), Tiemeier (1962), Tihen (1937), Toepfer (1993a), Trott (1983), Uhler and Warren (1929, 2008), Van Doren and Schmidt (2000), Viets (1993), Wiley (1968), Wooster (1925), and Yarrow (1883).

Smallmouth Salamander
Ambystoma texanum (MATTHES, 1855)

Adult: Douglas County, Kansas

Adult: Linn County

Subadult: Miami County, Kansas.

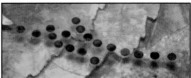

Eggs: Douglas County, Kansas.

Identification: Body robust; 14 or fewer vertical grooves on each side of body between front and hind limbs; dark belly; head, limbs, body, and tail uniform dark brown to black with irregular lichen-like gray or light brown mottling; belly gray-black, sometimes with tiny light flecks; females have slightly more robust bodies than males, but males have proportionately longer tails than females; during breeding, females have heavier bodies; males have swollen cloacal lips.

Size: Adults normally 100–140 mm (4–5½ inches) in TL; largest specimen from Kansas: female (KU 218583) from Douglas County with SVL of 92 mm and TL of 171 mm (6¾ inches) collected by Kevin R. Toal on 26 March 1991; maximum length throughout range: 7 inches (Conant and Collins, 1998).

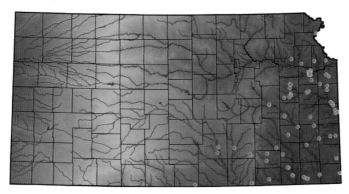

The **Smallmouth Salamander** (*Ambystoma texanum*) is abundant from the Missouri River bluffs (in northeastern Kansas) south and west to Harper County along the Oklahoma border. Red dots indicate records backed by voucher specimens or images.

Natural History: Generally spends summer and winter months beneath ground, where sufficient moisture and warmer temperatures prevent drying up or freezing; emerges from ground in early spring, breeds, and disperses to moist areas beneath rocks and logs or returns underground; recorded active annually in Kansas from early January to late November, with a peak from early March to late April; breeds after rains from mid-January to early April; migrates (usually at night) to and congregates at breeding sites such as shallow pasture streams that connect temporary grassy ponds, sloughs, rocky prairie streams, roadside ditches, and pools; from as few as ca. 90 up to ca. 900 eggs are laid singly, in small clumps, or in strings; some found as early as 25 January in Kansas; eggs hatch within a few weeks; gilled larvae remain in water after adults leave in late March or April; about two months after hatching, larvae metamorphose into young adults and emerge from water to begin terrestrial existence; primarily eats earthworms, but also centipedes, spiders, weevils, and beetles; exhibits defensive behavior when touched or molested by raising tail and waving it laterally, presumably to distract predator's attention from salamander's head (loss of tail to predator would allow more time to escape); maximum longevity: 13 years, 10 months, and 23 days (Snider and Bowler, 1992).

Pertinent References: Anderson (1967a), Ballinger and Lynch (1983), Bartlett and Bartlett (2006a), Behler and King (1979), Bishop (1943), Boyd (1988, 1991), Bragg (1949), Breukelman and Clarke (1951), Breukelman and Downs (1936), Busby and Pisani (2007), Capron et al. (1982), Clarke (1956c, 1958, 1970), Clarke et al. (1958), Collins (1959, 1974b, 1982a, 1985,

1990b), Collins and Caldwell (1978), Collins and Collins (1993a), Collins et al. (1991, 1994), Conant (1958, 1975), Conant and Collins (1991, 1998), Cook (1957), Cragin (1881, 1885a, 1885d), Cross (1971), Gier (1967), Gloyd (1928, 1932), Hall and Smith (1947), Hay (1892), Hooper and Whipple (1986), Hurter (1911), Ireland and Altig (1983), Jordan (1929), Karns et al. (1974), Kingsbury and Gibson (2002), Knight (1935), Kraus and Nussbaum (1989), Kraus and Petranka (1989), Kumberg (1996), Lardie (1990), Loomis (1956), Maldonado-Koerdell (1950), Miller (1976a), Minton (2001), Moriarty and Collins (1995a), Mount (1975), Murrow (2008a, 2009a, 2009b), Perry (1978), Petranka (1984, 1998), Plummer (1977a), Raffaeilli (2007), Riedle and Hynek (2002), Rundquist and Collins (1977), Schmidt (1953), Smith (1933b, 1934, 1950, 1956a), Stejneger and Barbour (1933), Taggart (1992a, 2001a, 2008), Titus (1991), Trauth et al. (2004), Trott (1983), Viets (1993), Wagner (1984), and Wilgers et al. (2006).

Eastern Tiger Salamander
Ambystoma tigrinum (GREEN, 1825)

Identification: Body robust; 14 or fewer vertical grooves on each side of body between front and hind limbs; dorsum of body with 15 to 58 (mean 30) sometimes indistinct dull yellow or olive spots, bars or blotches on dark background; light or light and dark mottled belly; bars or blotches on sides of body may or may not extend onto belly, which is black or gray-black and mottled with yellow; chin usually yellow; during breeding season, females can be distinguished from males by their heavier bodies, whereas males have swollen cloacal lips; females have slightly longer bodies than males; males have proportionately longer tails than females.

© Larry L. Miller

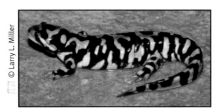
Adult: Shawnee County, Kansas

Adult: Douglas County, Kansas

Adult: Douglas County, Kansas

Aberrantly-colored adult:
Douglas County, Kansas.

Size: Adults normally 180–210 mm (7–8¼ inches) in TL; largest adult specimen from Kansas is unknown; maximum length throughout range: 13 inches (Conant and Collins, 1998).

Natural History: Generally spends much of summer and winter beneath ground in burrows of other animals, thus avoiding temperature extremes and drought;

frequently emerges from burrows at night or during rains; recorded active annually in Kansas from late February to late November, with a peak from early April to early June; little known of spring emergence or home range; breeds after sufficient rains from December to March; seeks out breeding site, generally permanent shallow lakes, ponds, ditches, or backwater pools along rivers in wooded regions; does not appear to migrate to breeding site; length of breeding season in Kansas unknown, but probably lasts up to 3 months; females probably lay up to 1,000 eggs, which hatch in a few weeks; gilled, pond-type larvae metamorphose into adults during same summer or overwinter until second summer; eats grasshoppers, aquatic insects, terrestrial insects, earthworms, fishes, tadpoles, frogs, toads, other sala-manders, and mice; maximum longevity: 20 years and 6 months (Snider and Bowler, 1992).

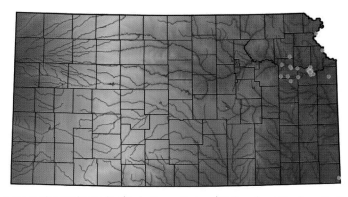

The **Eastern Tiger Salamander** (*Ambystoma tigrinum*) is found mostly along the Kansas and Missouri River drainages in the eastern fourth of the state, except for a single record for Cherokee County in southeastern Kansas. Red dots indicate records backed by voucher specimens or images.

Pertinent References: Ashton et al. (1974a), Ballinger and Lynch (1983), Bartlett and Bartlett (2006a), Behler and King (1979), Bishop (1943), Caldwell and Glass (1976), Clarke (1970), Clarke et al. (1958), Collins (1959, 1974b, 1982a), Collins and Collins (1993a), Collins et al. (1994), Conant (1958, 1975), Conant and Collins (1991, 1998), Cragin (1881), Cross (1971), Dundee (1988), Dunn (1918), Eddy and Babcock (1973), Fitch (1958b, 1991), Forney (1926), Gehlbach (1967), Karns et al. (1974), Knight (1935), Ireland and Altig (1983), Irschick and Shaffer (1997), Kingsbury and Gibson (2002), Kumberg (1996), Levell (1995, 1997), Loomis (1956), Moriarty and Collins (1995a), Mount (1975), Perez (1985), Petranka (1998), Rundquist and Collins (1977), Smith (1933b, 1934, 1950, 1956a), Spanbauer (1988), Taggart (2006b, 2008), Trauth et al. (2004), Wagner (1984), and Wassersug and Seibert (1975).

NEWTS
FAMILY SALAMANDRIDAE GOLDFUSS, 1820

The Eastern Newt (*Notophthalmus viridescens*) is the sole Kansas representative of the family Salamandridae (Newts). This salamander family consists of 2 genera and 6 species in the United States and Canada; adults are small (up to 110 mm), aquatic, and have well-developed limbs, lungs, dorsal body fins and tail fins.

Eastern Newt
Notophthalmus viridescens (RAFINESQUE, 1820)

Identification: Lacks vertical grooves on each side of body between front and hind limbs; distinct yellow belly with small black spots; body, head, limbs, and tail olive green to brown with small black spots; both efts and adults often exhibit red spots encircled by black on dorsum; during breeding season, females have more robust bodies, whereas males exhibit a high tail fin, a swollen cloacal opening, and horny black growths on inner surface of thighs and tips of hind toes.

Size: Adults normally 57–100 mm (2¼–4 inches) in TL; largest specimen from Kansas: female (KU 204158) from Cherokee County with SVL of 52 mm and TL of 110 mm (4⅜ inches) collected on 2 May 1984 by Stephen M. Reilly and David M. Hillis; maximum length throughout range: 5½ inches (Conant and Collins, 1998).

Eft-stage: Linn County, Kansas.　　Eft-stage: Cherokee County, Kansas.

 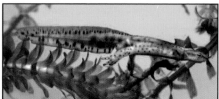

Adult: Bourbon County, Kansas.　　Adult male: Cherokee County, Kansas.

Natural History: Three distinct life stages of development (larvae, eft, and adult) as compared to other Kansas salamanders (larvae and adult); adults totally aquatic and inhabit ponds, small lakes, marshes, swamps, and ditches where they probably remain active the entire year (recorded active annually in Kansas from mid-February to late October, with a peak from late February to early April); courtship complex and takes place in water, as follows: male cautiously approaches female, who may initially dart rapidly away in state of excitement; male continues approach until female remains passive; male begins series of contortions and undulations (which last only a few seconds) and quickly mounts female, clasping her body tightly with his hind limbs; once in this position, female cannot dislodge him; both remain quiet for as long as several hours, except for fanning movement by the male's tail; fanning regular and rhythmic and probably stimulates both animals; female eventually responds by raising and fanning her tail; male then becomes more active for short time, his entire body quivering, while his cloacal opening swells as he drags female about water; at climax of courtship, male rapidly bends his body from side to side, leaves female, slightly raises his tail, and throws his whole body into series of rapid and strenuous undulations; if female follows and touches his tail or cloaca with her head, male deposits a spermatophore; female follows him, passes over spermatophore, and picks up some of the sperm with her cloacal lips; breeding and egg deposition in Kansas take place from mid-March to late April; eggs laid singly on aquatic vegetation and hatch in 3–5 weeks; larvae present from May to early July and metamorphose by August; after gilled, pond-type larvae metamorphose in fall, they enter eft stage of their lives (a terrestrial existence of 2–3 years); at end of the eft stage, they become sexually mature and return to water to begin completely aquatic adult existence; feed on insects, insect larvae, worms, tadpoles, small crustaceans, and mollusks; threatened species (protected by state law); maximum longevity: 9 years, 5 months, and 1 day (Snider and Bowler, 1992).

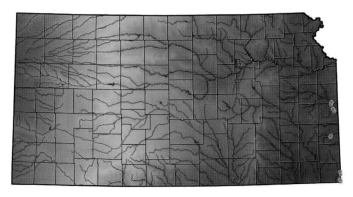

The distribution of the **Eastern Newt** (*Notophthalmus viridescens*) in Kansas is restricted to suitable aquatic situations along the Missouri-Kansas border, from Cherokee County north to southern Miami County. Red dots indicate records backed by voucher specimens or images.

Pertinent References: Ashton (1974b, 1977), Ballinger and Lynch (1983), Bartlett and Bartlett (2006a), Behler and King (1979), Bishop (1943), Breukelman and Smith (1946), Clarke (1970), Collins (1959, 1974b, 1982a, 1985, 1990b, 1994b), Collins and Collins (1993a), Collins et al. (1981, 1994, 1995), Conant (1958, 1975), Conant and Collins (1991, 1998), Cross (1971), Freeman and Kindscher (1992), Gloyd (1932), Guarisco et al. (1982), Harrison (1977), Henderson and Andelt (1982), Hlavachick (1978), Hunter et al. (1992), Ireland and Altig (1983), Irwin (1995a), Irwin and Collins (1994, 2005), Jagels (2003), Johnson (1974), Karns et al. (1974), Kingsbury and Gibson (2002), Klemens (1993), Knight (1935), Kumberg (1996), Lardie (1990), Layher et al. (1986), Levell (1997), Loomis (1956), Loraine (1983), Mecham (1967), Miller and Gress (2005), Moriarty and Collins (1995a), Mount (1975), Petranka (1998), Platt et al. (1974), Reilly (1990), Rundquist and Collins (1977), Rundquist et al. (1978), Schmidt and Murrow (2007c), Schwilling (1981), Schwilling and Schaefer (1983), Simmons (1989), Smith (1933b, 1934, 1950, 1956a, 1957), Spanbauer (1988), Stejneger and Barbour (1933), Taggart (2006b, 2007b), Trauth et al. (2004), Williams (1947), and Wood (1985).

LUNGLESS SALAMANDERS
FAMILY PLETHODONTIDAE GRAY, 1850

The family Plethodontidae (Lungless Salamanders) consists of 12 genera and 141 species in the United States and Canada. These salamanders lack lungs and

Kansas species are small (up to 166 mm in TL); they possess a pair of grooves that extend from each nostril down the upper lip. In Kansas, 1 species is a cave-dweller; the other 2 also frequent caves, but are equally at home in highly oxygenated streams in extreme southeastern Kansas (Cherokee County). Three species in 1 genus represent this family in Kansas: the Longtail Salamander (*Eurycea longicauda*), Cave Salamander (*Eurycea lucifuga*), and Grotto Salamander (*Eurycea spelaea*).

Longtail Salamander
Eurycea longicauda (GREEN, 1818)

Identification: Slender body; 14 or fewer vertical grooves on each side of body between front and hind limbs; dark areas on each side of body running from snout onto tail; back and head brownish yellow to bright yellow; black spots on back may be arranged in a double row or scattered irregularly; upper part of tail yellowish with few or no black spots; sides of body from snout onto tail dark brown or black, but this color fades toward underside of body; belly dull white; during breeding season, females have heavier bodies than males and males have swollen snouts.

Adult: Cherokee County, Kansas.   Subadult: Cherokee County, Kansas.

Adults: Cherokee County, Kansas.

Adult Longtail (**lower**) and Cave (**upper**) Salamanders: Cherokee County, Kansas.
Note distinct difference in color and pattern between the species.

Size: Adults normally 92–159 mm (3⅝–6¼ inches) in TL; largest specimen from Kansas: female (KU 51690) from Cherokee County with SVL of 55 mm and TL of 143 mm (5⅝ inches), collected by John M. Legler on 10 May 1958; maximum length throughout range: 7⅞ inches (Conant and Collins, 1998).

Natural History: Recorded active annually in Kansas from late February to late October, with a peak from early March to mid-April; lives along streams, underground, or in caves, depending on temperature and available moisture; observed under rocks at edge of spring-fed pools, beneath rocks along edge of small, intermittent, man-made streams that flow from caves, in cave streams, active along walls of moist cave tunnels, in cave twilight zones, in small springs, in shallow caves of human construction, in temporary springs, and inside Schermerhorn Cave (close to entrance); breeds from November to February; fertilization internal; averages 90 eggs per clutch; female produces 1 clutch per season; eggs attached in a single row to undersides of rocks; larvae have gills, live in streams, and metamorphose within 7 months of hatching; diet unknown in Kansas, but probably consists of small invertebrates; threatened species (protected by state law); maximum longevity: 5 years and 10 days (Snider and Bowler, 1992).

The distribution of the **Longtail Salamander** (*Eurycea longicauda*) in Kansas is restricted to southeastern Cherokee County in the extreme southeastern corner of the state. Red dots indicate records backed by voucher specimens or images.

Pertinent References: Ballinger and Lynch (1983), Bartlett and Bartlett (2006a), Beard (1986), Behler and King (1979), Bishop (1943), Clarke (1970), Collins (1959, 1974b, 1982a, 1982b, 1990b, 1994b), Collins and Collins (1993a), Collins et al. (1995), Conant (1958, 1960, 1975), Conant and Collins (1991, 1998), Cross (1971), Dunn (1926), Freeman and Kindscher (1992), Gray (1979), Guarisco (1980a), Guarisco et al. (1982), Harrison (1977), Ireland (1970, 1979), Ireland and Altig (1983), Irwin (1980), Johnson (1974, 2000), Karns et al. (1974), Kingsbury and Gibson (2002), Knight (1935), Kumberg (1996), Layher (2002), Layher et al. (1986), Levell (1997), Loomis (1956), Loraine (1983), Miller and Gress (2005), Moriarty and Collins (1995a), Mount (1975), Platt et al. (1974), Petranka (1998), Raffaeilli (2007), Rundquist (1996a), Rundquist and Collins (1977), Simmons (1989), Smith (1932, 1933b, 1934, 1950, 1956a), Spanbauer (1988), Stejneger and Barbour (1933, 1939, 1943), Taggart (1992a, 2000b, 2006b), Trauth et al. (2004), Young (1986), and Young and Beard (1993).

Cave Salamander
Eurycea lucifuga RAFINESQUE, 1822

Identification: Slender body, 14 or fewer vertical grooves on each side of body between front and hind limbs; body, head, limbs, and tail bright orange-yellow and covered with irregularly scattered black dots; belly white or yellowish; adult males have raised margins around cloacal opening at breeding time; average length of females slightly larger than males.

Size: Adults normally 100–152 mm (4–6 inches) in TL; largest specimen from Kansas: male (KU 23193) from Cherokee County with SVL of 69 mm and TL of 166 mm (6½ inches), collected by Claude W. Hibbard, E. W. Jameson, Jr., and Hobart M. Smith on 21 October 1945; maximum length throughout range: 7⅛ inches (Conant and Collins, 1998).

Adult: On wall of Schermerhorn Cave, Cherokee County, Kansas.

Adult: Cherokee County, Kansas.

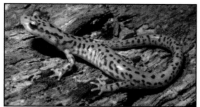

Adults: Cherokee County, Kansas.

Carver Hill Spring, Cherokee County, Kansas: Habitat of larval Cave Salamanders (*Eurycea lucifuga*).

Natural History: Recorded active annually in Kansas from late February to late October, with a peak from early March to mid-April; inhabits twilight zone of limestone caves, crevices in limestone rocks around springs, and beneath rocks and logs in moist forested areas near caves; has been discovered under stones or in leaf litter in twilight zone of caves, in splash zone beneath dripping, moss-covered, limestone rock overhangs at head of small streams, in small roadside springs, and in passages as much as 2,400 feet from cave entrance; optimal habitat probably consists of moist limestone caves and permanent cold springs; may breed twice a year; nothing known courtship in Kansas; female may lay 50–90 eggs attached to undersides of submerged rocks in cave streams; gilled, stream-type larvae metamorphose at length of 2¼–2½ inches; eats small insects; endangered species (protected by state law); maximum longevity: 9 years, 2 months, and 15 days (Snider and Bowler, 1992).

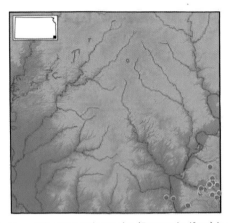

The Kansas distribution of the **Cave Salamander** (*Eurycea lucifuga*) is restricted to southeastern Cherokee County in the extreme southeastern corner of the state. Red dots indicate records backed by voucher specimens or images.

Pertinent References: Ballinger and Lynch (1983), Bartlett and Bartlett (2006a), Beard (1986), Behler and King (1979), Bishop (1943), Clarke (1970), Collins (1959, 1974b, 1982a, 1982b, 1985, 1990b, 1994b), Collins and Collins (1993a), Collins et al. (1981, 1991, 1995), Conant (1958, 1960, 1975), Conant and Collins (1991, 1998), Cross (1971), Freeman and Kindscher (1992), Gray (1979), Guarisco (1980a), Guarisco et al. (1982), Harrison (1977), Henderson and Andelt (1982), Henderson and Schwilling (1982), Hlavachick (1978), Hutchison (1956, 1958, 1966), Ireland (1970), Ireland and Altig (1983), Irwin (1980), Johnson (1974, 2000), Karns et al. (1974), Kingsbury and Gibson (2002), Kumberg (1996), Layher (2002), Layher et al. (1986), Levell (1997), Loomis (1956), Loraine (1983), Merkle and Guttman (1977), Miller and Gress (2005), Minckley (1959), Minton (1972, 2001), Moriarty and Collins (1995a), Mount (1975), Perry (1975), Petranka (1998), Platt et al. (1974), Raffaeilli (2007), Rundquist and Collins (1977), Schmidt (1953), Schwilling (1981), Schwilling and Schaefer (1983), Simmons (1989), Smith, (1946a, 1950, 1956a), Spanbauer (1988), Taggart (1992a, 2000b, 2006b), Trauth et al. (2004), Wiley and Mayden (1985), Wood (1985), Young (1986), and Young and Beard (1993).

Grotto Salamander
Eurycea spelaea (STEJNEGER, 1892)

Identification: Slender body; very small eyes in adults (eyelids partly or completely fused shut); body color brownish purple to pinkish white; 16–19 vertical grooves on each side of body between front and hind limbs; adult males exhibit swollen upper lip with two small, fleshy projections during breeding season; gravid females have heavier bodies than males.

Size: Adults normally 75–121 mm (3–4¾ inches) in TL; largest specimen from Kansas: larva (KU 153036) from Cherokee County with SVL of 42 mm and TL of 86 mm (3⅜ inches), collected by Ray E. Ashton on 3 March 1973; maximum length throughout range: 5⁵⁄₁₆ inches (Conant and Collins, 1998).

Larvae: Schermerhorn Cave, Cherokee County, Kansas.

Carver Hill Spring, Cherokee County, Kansas: Habitat of larval Grotto Salamanders (*Eurycea spelaea*).

Schermerhorn Cave, Cherokee County, Kansas, in the fall: Habitat of Grotto Salamanders (*Eurycea spelaea*).

The distribution of the **Grotto Salamander** (*Eurycea spelaea*) in Kansas is restricted to southeastern Cherokee County in the extreme southeastern corner of the state. Red dots indicate records backed by voucher specimens or images.

Natural History: Recorded active annually in Kansas from mid-February to early August, with a peak from early March to early April; adults blind and restricted to interiors of caves; larvae have functional eyes and frequently found in surface springs and small streams near caves; found in cave twilight zones, spring-fed pools, under rocks in small cave streams, small roadside spring at base of forested slope; larval examples have been observed in cave passage 2,300 feet from entrance; populations have been estimated at 5–10 larvae per square meter of stream; only 1 adult (KU 52241) has been discovered in Kansas; courtship and egg deposition have not been observed in Kansas; eggs probably deposited in or near water and attached to rocks; gilled, stream-type larvae transform into adults 2–3 years after hatching; feeds on flies, mosquito larvae, beetles, small snails, and centipedes; endangered species (protected by state law); maximum longevity: 9 years, 11 months, and 19 days (Snider and Bowler, 1992).

Pertinent References: Bartlett and Bartlett (2006a), Beard (1986), Behler and King (1979), Bishop (1943, 1944), Blair (1939), Brandon (1965, 1966, 1970), Clarke (1970), Cochran and Goin (1970), Collins (1959, 1974b, 1982a, 1982b, 1985, 1990b, 1994b), Collins and Collins (1993a), Collins et al. (1981, 1995), Conant (1958, 1975), Conant and Collins (1991, 1998), Cross (1971), Dowling (1956), Duellman and Berg (1962), Freeman and Kindscher (1992), Guarisco et al. (1982), Harrison (1977), Henderson and Andelt (1982), Hlavachick (1978), Ireland (1970), Ireland and Altig (1983), Irwin (1980), Johnson (1974, 2000), Karns et al. (1974), Kingsbury and Gibson (2002), Knight (1935), Kumberg (1996), Layher (2002), Layher et al. (1986), Levell (1997), Loomis (1956), Loraine (1983), Miller and Gress (2005), Mittleman (1950), Moriarty and Collins (1995a), Perry (1975), Petranka (1998), Platt et al. (1974), Raffaeilli (2007), Rudolph (1978), Rundquist and Collins (1977), Schmidt (1953), Schwilling (1981), Simmons (1989), Smith (1932, 1933b, 1934, 1950, 1956a), Spanbauer (1988), Stejneger and Barbour (1933, 1939, 1943), Taggart (1992a, 2000b, 2006b), Trauth et al. (2004), Wood (1985), Young (1986), and Young and Beard (1993).

MUDPUPPIES AND OLMS
FAMILY PROTEIDAE GRAY, 1825

The Red River Mudpuppy (*Necturus lousianensis*) and Common Mudpuppy (*Necturus maculosus*) are representatives of the family Proteidae in Kansas; this family consists of 1 genus and 6 species in the United States and Canada. These completely aquatic salamanders are distinguished by the presence of lungs, a distinct caudal fin, external bushy gills, and a large size (up to 385 mm).

Red River Mudpuppy
Necturus louisianensis VIOSCA, 1938

Identification: Large size; presence of bushy gills on each side of neck behind head; 4 toes on each foot; large, distinctive, reddish or maroon bushy gills; head, body, limbs, and tail yellowish brown to reddish; body and tail covered with many indistinct blue-black spots; center of belly grayish white with few or no spots and tinged with pink; adult males differ from females during breeding season by presence of swollen ridge around cloacal opening and crescent-shaped groove just behind it.

Spring River, Cherokee County, Kansas, a fall view: Habitat of the Red River Mudpuppy (*Necturus louisianensis*).

Red River Mudpuppies (*Necturus louisianensis*) congregate in highly-oxygenated areas below low-water dams, such as this one in southeastern Kansas.

Adult: Greenwood County, Kansas.

Adult

Size: Adults normally 200–250 mm (8–9¾ inches) in TL; largest Kansas specimen: male (MHP 7496) from Allen County with SVL of 211 mm and TL of 307 mm (12 inches) collected by Travis W. Taggart on 22 February 2003; exceeds maximum length throughout range, as reported in Conant and Collins (1998).

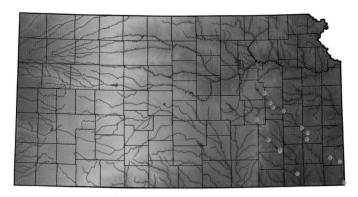

The **Red River Mudpuppy** (*Necturus louisianensis*) is a completely aquatic salamander found in the Verdigris, Neosho, and Spring river systems in the eastern third of the state. Red dots indicate records backed by voucher specimens or images.

Natural History: Nocturnal and totally aquatic; spends entire life in large streams and rivers; probably active year round (recorded active annually in Kansas from early February to late November, with a peak from late February through March); time of breeding unknown in Kansas (possibly in fall); male deposits spermatophore, which female grasps in her cloacal opening; between April and June female retires beneath log or rock on stream or river bottom, turns over on her back and lays eggs, attaching them singly to underside of shelter; eggs hatch within 5–7 weeks; larvae intermediate between pond and stream types, possessing large gills but with reduced fin confined to tail; larvae mature in 4–6 years; feed on fishes, fish eggs, crayfishes, insects, insect larvae, snails, leeches, worms, and other amphibians; snakes are known predators.

Pertinent References: Ballinger and Lynch (1983), Bartlett and Bartlett (2006a), Behler and King (1979), Bishop (1943), Breukelman and Clarke (1951), Breukelman and Downs (1936), Clarke (1970), Clarke et al. (1958), Collins (1959, 1974b, 1982a), Collins and Collins (1993a), Collins and Caldwell (1978), Conant (1958, 1975), Conant and Collins (1991, 1998), Cragin (1885a, 1885d), Cross (1971), Fassbender and Watermolen (2002), Guarisco (1979b), Hall and Smith (1947), Harkavy (1994), Hecht (1958), Hunter et al. (1992), Ireland and Altig (1983), Johnson (2000), Karns et al. (1974), Kingsbury and Gibson (2002), Kumberg (1996), Moriarty and Collins (1995a), Mount (1975), Murrow (2009a), Petranka (1998), Preston (1982), Raffaeilli (2007), Rundquist and Collins (1977), Smith (1933b, 1934, 1950, 1956a), Taggart (2003a, 2004d, 2006b, 2008), and Trauth et al. (2004).

Common Mudpuppy
Necturus maculosus (RAFINESQUE, 1818)

Identification: Large size; presence of bushy gills on each side of neck behind head; 4 toes on each foot; large, distinctive, bushy gills reddish or maroon; head, body, limbs, and tail reddish-brown, brown or gray; body and tail covered with a few indistinct blue-black spots; belly uniform gray or pale with spots; during breeding season, adult males differ from females in possessing swollen ridge around, and crescent-shaped groove behind, cloacal opening.

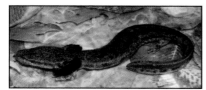

Adults: Franklin County, Kansas.

Adults: Franklin County, Kansas.

Size: Adults normally 200–330 mm (8–13 inches) in TL; largest specimen from Kansas: sex undetermined (KU 209746) from Osage County with SVL of 262 mm and TL of 385 mm (15⅛ inches) collected by Tom Mosher on 4 April 1988; maximum length throughout range: 19⅛ inches (Conant and Collins, 1998).

Natural History: Nocturnal and totally aquatic; spends entire life in suitable ponds, streams, and rivers; probably active year round (recorded active annually in Kansas from early February to late April, with a peak during March); time of breeding unknown in Kansas (possibly in fall); male deposits spermatophore, which female grasps in her cloacal opening; between April and June, female retires beneath log or rock on stream or river bottom, turns over on her back, and lays eggs, attaching them singly to underside of shelter; eggs hatch within 5–7 weeks; larvae intermediate between pond and stream types, possessing large gills but with reduced fin confined to tail; larvae mature in 4–6 years; feed on fishes, fish eggs, crayfishes, insects, insect larvae, snails, leeches, worms, and other amphibians; maximum longevity: 4 years (Snider and Bowler, 1992).

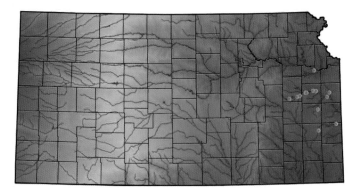

The completely aquatic **Common Mudpuppy** (*Necturus maculosus*) is found in the Marais des Cygnes, Little Osage, and Marmaton river systems in the eastern fourth of the state. Red dots indicate records backed by voucher specimens or images. The single record from the Kansas River in Douglas County was collected in 1916 and, although supported by more recent specimens found downstream in Missouri, needs corroboration.

Pertinent References: Ballinger and Lynch (1983), Bartlett and Bartlett (2006a), Behler and King (1979), Bishop (1943), Clarke (1956c, 1958, 1970), Collins (1959, 1974b, 1982a, 1985, 1990b), Collins and Caldwell (1978), Collins and Collins (1993a), Conant (1958, 1975), Conant and Collins (1991, 1998), Gier (1967), Gloyd (1928), Harkavy (1994), Hay (1892), Hecht (1958), Hunter et al. (1992), Ireland and Altig (1983), Johnson (2000), Karns et al. (1974), Kingsbury and Gibson (2002), Knight (1935), Kumberg (1996), Maldonado-Koerdell (1950), McKenna and Seabloom (1979), Minton (1972, 2001), Moriarty and Collins (1995a), Mount (1975), Petranka (1998), Preston (1982), Schmidt (1953), Shaffer and McCoy (1991), Simon (1988), Smith (1933b, 1934, 1950, 1956a), Taggart (2006b), Tihen (1937), Trauth et al. (2004), Wagner (1984), and Wheeler and Wheeler (1966).

FROGS AND TOADS
ORDER ANURA FISCHER VON WALDHEIM, 1813

Frogs and toads are classified in the Order Anura, which consists of 8 families in the United States and Canada, represented by 18 genera and 100 species (Collins and Taggart, 2009, with updates from the CNAH web site). The 22 species of frogs and toads known to occur in Kansas represent 5 families and 7 genera. Anurans have adapted more successfully than salamanders to the comparatively dry environment of Kansas. Nine of the 22 species of frogs and toads included in this book have a distribution statewide or nearly so. Annual calling activity for each species is based on personal observations by the authors, published reports, and data obtained from KAMP (the Kansas Anuran Monitoring Program). In addition, KAMP recordings of the calls of each Kansas species of frogs and toad can be heard at:

http://www.cnah.org/kamp/

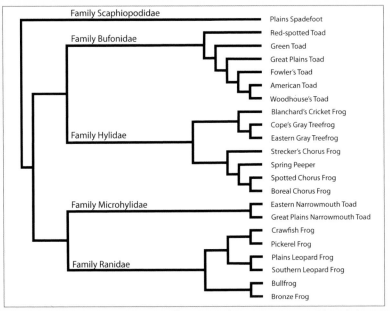

A phylogeny of Kansas Frogs and Toads (Order Anura) that is consistent with evolutionary history based on current scientific evidence.

NORTH AMERICAN SPADEFOOTS
FAMILY SCAPHIOPODIDAE COPE, 1865

The Plains Spadefoot (*Spea bombifrons*) is the single Kansas species of the family Scaphiopodidae. These anurans typically have short hind legs, smooth skin, and lead a burrowing existence.

Plains Spadefoot
Spea bombifrons (COPE, 1863)

Identification: Moist skin; round snout; pupils vertically slit when exposed to strong light; boss or raised bump between eyes; black spur at base of each hind foot; head, body, and limbs light to medium gray with irregular darker markings and two poorly-defined light stripes down back; belly white; males have dark throats; females have more robust bodies than males.

Adult   Adult: Ness County, Kansas.

Adult: Sheridan County, Kansas.   Adult: Leavenworth County, Kansas.

Adult: Finney County, Kansas.

Size: Adults normally 38–50 mm (1½–2 inches) in SVL; largest Kansas specimen: male (KU 20012) from Barber County with SVL of 64 mm (2½ inches) collected by Claude W. Hibbard on 29 August 1935; this is the maximum length throughout range (Conant and Collins, 1998).

Natural History: Inhabits prairies and open floodplains; prefers areas of loose earth or sand; burrows during day; emerges at night to forage, particularly after rain; scarce in wooded uplands; small mammal burrows are favored retreats; recorded active annually in Kansas from mid-January to mid-October, with a peak from early May through mid-July; opportunistic breeder, emerging to do so from underground retreats after first warm, heavy spring rain; breeds from late March to mid-August, with a peak from early May to late June (most robust choruses recorded during nocturnal hours from 10:00 pm to 3:00 am); males chorus loudly around temporary rainpools on floodplains or open prairie (reportedly a minimum of 3.5 inches of rainfall and 52°F are required to stimulate mating); females eventually join them; male mounts female, clasping her around groin with his front limbs; female deposits eggs and male arches his body and fertilizes them in water; female may deposit up to 2,000 eggs in masses of up to 250 each, attached to partly submerged vegetation or other objects; eggs hatch, metamorphosis occurs at varying intervals dependent on water temperature, oxygen content, and competition for available food among tadpoles; tadpoles cannibalistic under crowded conditions; adults eat beetles, crickets, grasshoppers, ants, and other small insects; domestic cats are known predators.

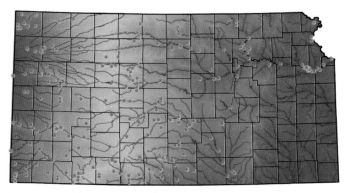

The **Plains Spadefoot** (*Spea bombifrons*) is an anuran found west of the Flint Hills, as well as east along the Kansas River floodplain to the Missouri border and north along the Missouri River floodplain. Red dots indicate records backed by voucher specimens or images.

Pertinent References: Ballinger and Lynch (1983), Bartlett and Bartlett (2006a), Behler and King (1979), Boyd (1988, 1991), Bragg (1944, 1965), Brennan (1934, 1937), Brown (1967), Burt and Burt (1929c), Busby and Parmelee (1996), Busby et al. (1994, 1996, 2005), Capron (1978b), Capron and Perry (1976), Chrapliwy (1956), Clarke (1980a, 1984), Collins (1959, 1974b, 1982a, 1985, 1990b, 2008a, 2009a), Collins and Collins (1991, 1993a, 1993b, 2006a), Collins et al. (1991), Conant (1958, 1975), Conant and Collins (1991, 1998), Duellman and Trueb (1985), Eddy and Hodson (1961), Ellis (2002), Fitch (1958b, 1991), Gier (1967), Gish (1962), Gress (2003), Heinrich and Kaufman (1985), Hoyt (1960), Hudson (1942), Irwin (1983), Irwin and Collins (1987), Karns et al. (1974), Kingsbury and Gibson (2002), Kluge (1966), Knight and Collins (1977), Kretzer and Cully (2001), Legler (1960b), Levell (1995, 1997), Livezey and Wright (1947), Loomis (1956), Marr (1944), Miller (1982a), Moriarty and Collins (1995a), Murrow (2009a), Perry (1974a), Royal (1982), Rundquist (1975), Rush (1981), Sattler (1985), Schmidt (2004l, 2008a), Schmidt and Murrow (2007b), Schwarting (1984), Simon (1988), Smith (1933b, 1934, 1950, 1956a), Smith and Leonard (1934), Stebbins (1951, 1954, 1966, 1985, 2003), Stejneger and Barbour, 1939, 1943), Taggart (1992b, 2001b, 2006b, 2006c, 2007c), Taggart et al. (2007), Tanner (1989), Taylor (1929b), Tihen (1937), Tihen and Sprague (1939), Toepfer (1993a), Trauth et al. (2004), Trott (1983), Trowbridge and Trowbridge (1937), Turner (1952), Van Doren and Schmidt (2000), Wagner (1984), Warner and Wencel (1978), Wassersug and Seibert (1975), Wiens (1989), Wiens and Titus (1991), and Wright and Wright (1949).

TRUE TOADS
FAMILY BUFONIDAE GRAY, 1825

The family Bufonidae (True Toads) is represented in Kansas by 6 species of the genus *Anaxyrus*. These anurans are characterized by thick, bumpy skin, short limbs, a terrestrial, burrowing existence, and a large gland on the head behind each eye.

American Toad
Anaxyrus americanus (HOLBROOK, 1836)

Identification: Relatively dry skin; round snout; enlarged kidney-shaped glands on neck behind each eye; bony crests present between and behind eyes; bony crests behind eyes either not touching enlarged, kidney-shaped gland or connected to it only by small spur; white belly generally covered with dark spotting (some individuals from the area between Chautauqua and Cherokee counties are smaller in size, often reddish in color, and exhibit reduced or absent breast spotting); dorsal color gray, light brown, brown, or reddish brown; body covered with light patches and/or dark brown or black spots; light stripe often present down back; 1–2 warts per spot on back; large warts on upper surface of hind limbs; during breeding season, males can be distinguished from females by their black throat and presence of an enlarged, horny pad on inner fingers of each hand; females larger than males.

Adult: Douglas County, Kansas.

Adult: Cherokee County, Kansas.

Adult: Douglas County, Kansas.

Adult: Cherokee County, Kansas.

Size: Adults normally grow 50–90 mm (2–3½ inches) in SVL. largest Kansas specimen: sex undetermined (KU 211394) from Neosho County with SVL of 102 mm (4 inches) collected by Travis W. Taggart and R. Bruce Taggart on 14 September 1988; maximum length throughout range: 4⅜ inches (Conant and Collins, 1998).

Natural History: Optimal habitat is rocky situations in open woods, woodland edge, hillside forests, and open cultivated fields; avoids fields with dense vegetation; lives under large, flat rocks having loose, damp soil beneath them; has minimal home range of 0.16 acres; during periods of high humidity, frequently wanders great distances; recorded active annually in Kansas from mid-March to early November, with a peak from mid-April to late June; during day, generally remains hidden; breeds from mid-March to early July, with a peak from early April to late

May; choruses recorded most robust during nocturnal hours from 8:00 pm to 1:00 am; buries beneath ground during temperature extremes of winter and summer; with onset of spring rains, adults congregate at breeding sites, usually shallow upland streams or ponds; males vigorously sing to attract mates; upon succeeding, male mounts female and clasps her body immediately behind front legs with his front limbs, pressing his thumbs into her armpits; they float in water and female lays eggs in a long, double string while male releases sperm on them; female may deposit 2,800–20,500 eggs; eggs hatch quickly and tiny black tadpoles grow until they metamorphose into small toadlets that disperse from pond within a week; eats beetles, crickets, leaf-hoppers, grasshoppers, spiders, and ants; like most anurans, beneficial to farmers due to large quantity of insects eaten; may hybridize with Great Plains and Woodhouse's Toads; maximum longevity: 4 years, 8 months, and 25 days (Snider and Bowler, 1992).

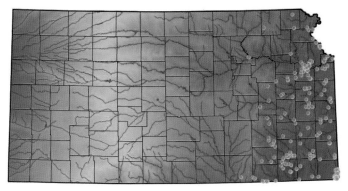

The distribution of the **American Toad** (*Anaxyrus americanus*) in Kansas is restricted to the eastern third of the state; a single specimen from Riley County (collected in 1937) needs corroboration. Red dots indicate records backed by voucher specimens or images.

Pertinent References: Anderson and Arruda (2006), Anderson et al. (1995), Ballinger and Lynch (1983), Bartlett and Bartlett (2006a), Behler and King (1979), Blair (1958, 1972), Boyd (1988, 1991), Bragg (1954), Breukelman and Clarke (1951), Breukelman and Downs (1936), Breukelman and Smith (1946), Burkett (1984), Busby and Pisani (2007), Busby et al. (1996, 2005), Caldwell and Glass (1976), Clarke (1958, 1984), Clarke et al. (1958), Cochran and Goin (1970), Coleman (1987), Collins (1959, 1974b, 1982a, 1985, 1989c, 1990b), Collins and Collins (1993a), Collins and Taggart (2006), Collins et al. (1994), Conant (1958, 1975), Conant and Collins (1991, 1998), Cope (1889), Cragin (1881), Cross (1971), Eddy and Babcock (1973), Ellis and Henderson (1913), Fitch (1951, 1956e, 1958b, 1963b, 1982, 1991), Fraser (1983a), Gier (1967), Gloyd (1928, 1932), Gray (1979), Grow (1976b), Hall and Smith (1947), Hallowell (1857a, 1857b), Hartman (1906), Hunter et al. (1992), Karns et al. (1974), Kingsbury and Gibson (2002), Klemens (1993), Knight (1935), Korb (2001), Levell (1995, 1997), Livezey and Wright (1947), Loomis (1956), Lynch (1964), Mansfield (1932), Masta et al. (2002), Miller (1976d), Minton (2001), Moriarty and Collins (1995a), Murrow (2008a, 2009a, 2009b), Parmelee and Fitch (1995), Pauly et al. (2004), Perry (1975, 1978), Riedle (1994a), Rundquist and Collins (1977), Rush and Fleharty (1981), Sanders (1987), Schmidt (1953), Schmidt and Murrow (2007c), Simon (1988), Simon and Dorlac (1990), Skie and Bickford (1978), Smith (1933b, 1934, 1950, 1956a), Smith et al. (1983), Spanbauer (1988), Taggart (1992a, 2000b, 2001a, 2003g, 2005, 2006b), Tihen (1937), Trauth et al. (2004), Viets (1993), Volpe (1955), Wilgers and Horne (2006), Wilgers et al. (2006), Wright and Wright (1949), and Yarrow (1883).

Great Plains Toad
Anaxyrus cognatus (SAY, 1823)

Identification: Relatively dry skin; round snout; enlarged kidney-shaped glands on neck behind each eye; bony crests between eyes, merging on snout to form flat, raised knob or boss; pairs of large black, dark green, or dark brown blotches outlined with cream or yellow on back and sides; numerous small warts in each blotch; sometimes narrow light stripe down back; hind limbs banded; front limbs banded or spotted; belly cream or yellow with few or no dark spots on chest; females larger than males; during breeding season, males have dark throats and horny pads on inner fingers of each hand.

Adult: Finney County, Kansas.

Adult: Ellis County, Kansas.

Adult: Clark County, Kansas.

Adult: Douglas County, Kansas.

Adult: Sheridan County, Kansas.

Adult: Shawnee County, Kansas.

Size: Adults normally 50–85 mm (2–3⅜ inches) in SVL; largest Kansas specimen: female (KU 186099) from Sumner County with SVL of 102 mm (4 inches) collected by Jeff Ehlers on 23 June 1980; maximum length throughout range: 4½ inches (Conant and Collins, 1998).

Distribution in Kansas: Throughout western two-thirds of state; east along Kansas River floodplain to Missouri border; north along Missouri River (apparently absent away from Missouri River floodplain); records for Marais des Cygnes River drainage in southern Douglas and northern Miami counties need corroboration; generally absent in eastern third of Kansas south of Kansas River floodplain.

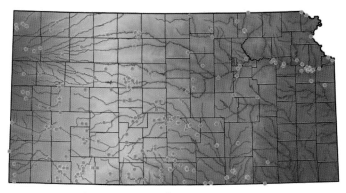

The **Great Plains Toad** (*Anaxyrus cognatus*) is found from the Flint Hills west, as well as east along the Kansas River floodplain to the Missouri border and north along the Missouri River (apparently absent away from the Missouri River floodplain). The records for the Marais des Cygnes River drainage in southern Douglas and northern Miami counties need corroboration; this toad is generally absent from the eastern quarter of Kansas south of the Kansas River floodplain. Red dots indicate records backed by voucher specimens or images.

Natural History: Resident of upland mixed-grass and short-grass prairies; rarely occurs in woodlands; also frequents open floodplains, at least in eastern portion of range; remains hidden beneath ground during day; emerges on summer nights to search for food; digs burrows in shape of a question-mark in bare, sandy areas with resting area at end of upper curve; sometimes found in caves; recorded active annually in Kansas from early March to early November, with a peak from early May to mid-July; spends cold winter months beneath ground; opportunistic breeder; congregates at suitable breeding sites such as temporary pools; recorded choruses from late March through August, with peak from early May to late June; most robust choruses recorded during nocturnal hours from 10:00 pm to 2:00 am; may produce two clutches per season; during breeding, males chorus and attract and mount females, clasping them beneath their front limbs; long strings of eggs laid and fertilized in water; female may produce up to 20,000 eggs; after hatching, tadpoles grow in pond until metamorphosis; eats insects; may hybridize with American and Woodhouse's Toads; maximum longevity: 10 years, 8 months, and 10 days (Snider and Bowler, 1992).

Pertinent References: Anderson et al. (1995), Ballinger and Lynch (1983), Bartlett and Bartlett (2006a), Baxter and Stone (1980, 1985), Behler and King (1979), Blair (1958), Bragg (1937, 1940a), Brennan (1934, 1937), Busby and Parmelee (1996), Busby et al. (1994, 1996, 2005), Capron and Perry (1976), Carpenter (1940), Clarke (1980a, 1984), Collins (1959, 1974b, 1982a, 1985, 1990b, 2009a), Collins and Collins (1991, 1993a, 1993b, 2006a), Collins and Taggart (2006), Collins et al. (1991, 1994), Conant (1958, 1975), Conant and Collins (1991, 1998), Cope (1889), Cragin (1881), Dickerson (1913), Ellis (2002), Ellis and Henderson (1913), Fleharty and Johnson (1975), Gier (1967), Gish (1962), Gloyd (1929), Gress (2003), Gress and Potts (1993), Guarisco (1980c), Hartman (1906), Hudson (1942), Irwin (1977, 1983), Irwin and Collins (1987), Karns et al. (1974), Kellogg (1932), Kingsbury and Gibson (2002), Knight

(1935), Knight and Collins (1977), Krupa (1990), Legler (1960b), Levell (1995, 1997), Livezey and Wright (1947), Loomis (1956), Marr (1944), Martin (1960), Miller (1980a, 1981c, 1982a), Moriarty and Collins (1995a), Murrow (2009a), Nulton and Rush (1988), Perry (1974a), Platt (1985b), Rogers (1973a, 1973b), Royal (1982), Rundquist (1975, 1996a), Rush (1981), Ruthven (1907), Schmid (1965), Schmidt (2004f, 2004l, 2008b), Schmidt and Murrow (2007b), Schwilling and Schaefer (1983), Smith (1933b, 1934, 1946b, 1950, 1956a, 1961), Stebbins (1951, 1954, 1966, 1985, 2003), Stejneger and Barbour (1917, 1923, 1933, 1939), Storer (1925), Taggart (1992b, 2006b, 2006c, 2007c), Taggart et al. (2007), Taylor (1929b), Taylor et al. (1975), Tihen (1937), Tihen and Sprague (1939), Toepfer (1993a), Trott (1977, 1983), Uhler and Warren (1929, 2008), Van Doren and Schmidt (2000), Wagner (1984), Warner and Wencel (1978), Wright and Wright (1949), and Yarrow (1883).

Green Toad
Anaxyrus debilis (GIRARD, 1854)

Identification: Dry skin; round snout; enlarged elongated glands behind each eye on neck; no bony crests between or behind eyes; distinctive color; head, body, and limbs green (male) or yellow (female) with black spots or streaks that may form network pattern; belly yellowish and may or may not be darkly spotted; males green and have dark throats; females yellow with yellowish throats, and larger than males.

Adult males: Logan County, Kansas.

Adult male: Logan County, Kansas.   Adult female: Wallace County, Kansas.

Adult female (**left**) and male (**right**): Wallace County, Kansas.

Size: Adults normally 32–50 mm (1¼–2 inches) in SVL; largest Kansas specimen: female (KU 5652) from Morton County with SVL of 44 mm (1¾ inches) collected by

Theodore E. White and Edward H. Taylor on 15 August 1928; maximum length throughout range: 2⅛ inches (Conant and Collins, 1998).

Natural History: Generally inhabits areas of rugged topography amidst open plains at elevations of 2,500 feet or higher with an average annual rainfall under 20 inches; prefers areas along intermittent streams and wet places below cattle tanks and dams of small ponds; very secretive and normally active only at night; recorded active annually in Kansas from mid-May to late August, with a peak from early June to early July; inhabits small temporary pools; often found beneath rocks; an opportunistic breeder; recorded chorusing from early June to mid-August, with a peak from early June to early July; breeds in shallow ditches, flooded fields, cattle tanks, and other temporary pools; eggs laid singly or in short strings (which form clumps) and hatch into tadpoles that later metamorphose; metamorphosis from egg to emergent toad may take less than a week; egg clutches in Kansas known to number up to 1,610 in a single female; young toads live in mud cracks on dry pond bottoms where there is more moisture and greater protection from predators; eats ants, small moths, beetles, and small grasshoppers; threatened species (protected by state law); in late 1980s, field work in Morton County failed to uncover this amphibian at sites where it was abundant before drought of 1930s, suggesting that it may have been extirpated; restocking program was initiated in Morton County in 1992 (success of this project unknown); predation on tadpoles and young by Plains Garter Snakes has been observed; maximum longevity: 5 years and 3 months (Snider and Bowler, 1992).

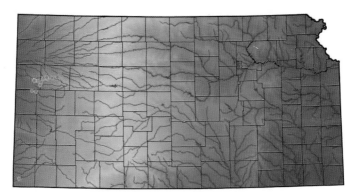

The **Green Toad** (*Anaxyrus debilis*) is a small anuran with a distribution in Kansas restricted to the western quarter of the state, from Wallace and Logan counties in the north to Morton County in the southwestern corner. This toad is particularly abundant in western reaches of Smoky Hill River; its status in Morton County is uncertain. Red dots indicate records backed by voucher specimens or images. The success of repatriation efforts for this species during 1992 in the Cimarron National Grasslands of Morton County is unknown.

Pertinent References: Ballinger and Lynch (1983), Bartlett and Bartlett (2006a), Behler and King (1979), Burkhart (1984), Clarke (1984), Cochran and Goin (1970), Collins (1959, 1974b, 1982a, 1990b), Collins and Collins (1991, 1993a), Collins and Taggart (2006), Collins et al. (1991, 1994, 1995), Conant (1958, 1975), Conant and Collins (1991, 1998), Conant et al. (1992), Cragin (1894), Ellis and Henderson (1913), Guarisco (1979b), Guarisco et al. (1982), Hammerson (1982, 1999), Hill (1931), Johnson (1974), Karns et al. (1974), Kellogg (1932),

Kingsbury and Gibson (2002), Knight (1935), Lardie and Black (1981), Layher et al. (1986), Levell (1997), Leviton (1971), Maslin (1959), Miller and Gress (2005), Moriarty and Collins (1995a), Platt et al. (1974), Roth and Collins (1979), Rundquist (1979), Sanders and Smith (1951), Savage (1954), Schmidt (1953, 2004l), Schwilling and Schaefer (1983), Simmons (1989), Smith (1933b, 1934, 1950, 1956a), Stebbins (1951, 1954, 1966, 1985, 2003), Stejneger and Barbour (1939, 1943), Taggart (1992a, 1992b, 1992c, 1994, 1997, 2006b), Taggart et al. (2007), Taylor (1929b), and Wright and Wright (1949).

Fowler's Toad
Anaxyrus fowleri (HINCKLEY, 1882)

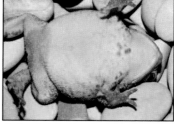

Adult: Cherokee County, Kansas.   Belly of adult: Cherokee County, Kansas.

Adults: Cherokee County, Kansas.

Adult and subadult: Cherokee County, Kansas.

Identification: Dry skin; round snout; enlarged kidney-shaped gland on neck behind each eye; bony crests between and behind eyes; bony crests touch enlarged, kidney-shaped gland behind each eye; light-colored stripe extends down back; dorsal color varies from gray or greenish gray to brown, with dark green-gray or dark brown spots; 2 or more warts per spot on back; large warts absent from upper surface of hind limbs; belly white or yellowish with few or no spots (sometimes single breast spot present); breeding males smaller than females, and have dark throats and an enlarged, horny pad on inner finger of each hand; rely on geography to distinguish from Woodhouse's Toad; rely on belly pattern to distinguish from American Toad.

Size: Adults normally 50–75 mm (2–3 inches) in SVL; largest Kansas specimen: female (MHP 11768) from Cherokee County with SVL of 62 mm (2⁷/₁₆ inches) collected by Travis W. Taggart and Richard Hayes on 19 July 2005; maximum length throughout range: 3¾ inches (Conant and Collins, 1998).

Natural History: Occurs anywhere that suitable habitat exists; appears to prefer areas of sand or loose soil; generally only toad found on floodplains of larger streams and rivers; remains hidden during day; emerges at night to hunt food; recorded active annually in Kansas between early March and mid-October, with a peak from early March to late April; breeds opportunistically on river floodplains; males congregate in small numbers at suitable breeding sites when rainfall and temperatures permit; chorusing begins, female is attracted, and male mounts her, clasping her behind front legs with his forelimbs; eggs hatch and free-swimming tadpoles metamorphose into adults; eats bees, beetles, insect larvae, spiders, and ants; maximum longevity: 4 years and 2 days (Snider and Bowler, 1992).

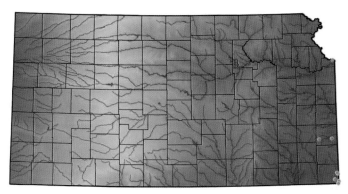

The distribution of **Fowler's Toad** (*Anaxyrus fowleri*) in Kansas is restricted to suitable habitat along the Missouri-Kansas border from Cherokee County north to southern Bourbon County. Red dots indicate records backed by voucher specimens or images.

Pertinent References: Bartlett and Bartlett (2006a), Collins (1974b, 1982a, 1989c), Collins and Collins (1993a), Collins and Taggart (2006), Conant (1958, 1975), Conant and Collins (1991, 1998), Cragin (1881), Gier (1967), Hall and Smith (1947), Masta et al. (2002), Pauly et al. (2004), Rundquist and Collins (1977), Schmidt (1953), Smith (1933b, 1934, 1950, 1956a), Taggart (2005, 2006b, 2008), and Trauth et al. (2004).

Red-spotted Toad
Anaxyrus punctatus (BAIRD AND GIRARD, 1852)

Identification: Dry skin; round snout; enlarged round gland on neck, behind each eye; no bony crests between or behind eyes; head, limbs, and body uniform brown or gray with small warts that may or may not be red-spotted; belly yellowish with small blackish spots; males have dark throat; females grow larger.

Adults.

Adult: Barber County, Kansas.   Adult: Barber County, Kansas.

Size: Adults normally 38–64 mm (1½–2½ inches) in SVL; largest Kansas specimen: female (KU 193290) from Barber County with SVL of 56 mm (2³⁄₁₆ inches) collected by Larry Miller on 14 June 1983; maximum length throughout range: 3 inches (Conant and Collins, 1998).

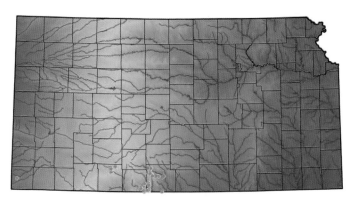

Two populations of the **Red-spotted Toad** (*Anaxyrus punctatus*) are known in Kansas. The first is restricted to the gypsum outcrops in Barber and Comanche counties and the second to the canyons along Bluff Creek in Clark County. A single record (taken in 1927) from Morton County needs corroboration. Red dots indicate records backed by voucher specimens or images.

Natural History: Inhabits rocky areas of dry prairies and canyons in southwestern Kansas; nocturnal; during day, hides beneath rocks where soil is fairly moist; recorded active in Kansas from early April to early September, with a peak from mid-May to mid-June; probably breeds opportunistically anytime after heavy rainfall during summer, when it congregates in small numbers around canyon pools and streams (recorded chorusing during June); males call at night to attract females; male mounts female and secures his front limbs around her groin; female deposits eggs singly as male fertilizes them; eggs adhere to plants, other objects, and sometimes to each other to form small, single-layered mass; after hatching, free swimming tadpoles metamorphose into adults; eats beetles, ants, and bees; SINC species (protected by state law); maximum longevity: 11 years, 4 months, and 11 days (Snider and Bowler, 1992).

Pertinent References: Ballinger and Lynch (1983), Bartlett and Bartlett (2006a), Behler and King (1979), Clarke (1984), Cochran (1961), Cochran and Goin (1970), Collins (1959, 1974b, 1982a, 1990b), Collins and Collins (1991, 1993a), Collins and Taggart (2006), Conant (1958, 1975), Conant and Collins (1991, 1998), Feder (1979), Guarisco (1981b), Guarisco et al. (1982), Hammerson (1999), Hibbard and Leonard (1936), Hill (1931), Johnson (1974), Karns et al. (1974), Kingsbury and Gibson (2002), Knight et al. (1973), Lardie (1990), Layher et al. (1986), Levell (1997), Leviton (1971), Livezey and Wright (1947), Miller (1980b, 1983c), Moriarty and Collins (1995a), Perry (1974a), Platt et al. (1974), Simmons (1989), Smith (1933b, 1934, 1950, 1956a), Stebbins (1951, 1954, 1966, 1985, 2003), Stejneger and Barbour (1939), Taggart (2006b), and Tihen (1937).

Woodhouse's Toad
Anaxyrus woodhousii (GIRARD, 1854)

Identification: Dry skin; round snout; enlarged kidney-shaped gland on neck behind each eye; bony crests between and behind eyes; bony crests touch enlarged, kidney-shaped gland behind each eye; white or yellowish belly with no spots or only one spot on breast; white or yellowish stripe extends down back; dorsal color varies from gray or greenish gray to brown, with dark green-gray or dark brown spots; 2 or more warts per spot on back; breeding males smaller than females and have dark throats and an enlarged, horny pad on inner finger of each hand.

Adult: Ellsworth County, Kansas.

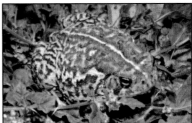
Adult: Hamilton County, Kansas.

Adult: Ellis County, Kansas.

Adult: Sheridan County, Kansas.

AMPHIBIANS

Bellies of an adult Woodhouse's Toad (**left**: *Anaxyrus woodhousii*) and an adult American Toad (**right**: *Anaxyrus americanus*), both from Jefferson County, Kansas.

Size: Adults normally 64–100 mm (2½–4 inches) in SVL; largest Kansas specimen: female (KU 158018) from Douglas County with SVL of 120 mm (4¾ inches) collected by Ken Davidson on 9 July 1975; maximum length throughout range: 5 inches (Conant and Collins, 1998).

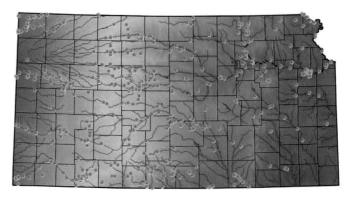

Woodhouse's Toad (*Anaxyrus woodhousii*) is found throughout most of Kansas, but is absent from the southeastern corner of the state where it is replaced by Fowler's Toad. Red dots indicate records backed by voucher specimens or images.

Natural History: Occurs anywhere that suitable habitat exists; appears to prefer lowlands and sandy areas; often found on floodplains of larger streams and rivers; remains hidden during day; emerges at night to hunt food; appears to use burrows of small mammals as retreats; uncommon in wooded uplands; active at higher temperatures than American Toad; recorded active annually in Kansas between mid-February and mid-October, with a peak from late May to early August; does not emerge as early as American Toad; breeds opportunistically in ditches, ponds, and on river floodplains; breeding choruses recorded from late March to late August, with a peak in May and June (most robust choruses recorded during nocturnal hours from 10:00 pm to 2:00 am); males congregate in small numbers at suitable breeding sites when rainfall and temperatures permit; when chorusing begins, female is attracted and male mounts her, clasping her behind front legs with his forelimbs; single female may lay up to 25,000 eggs; eggs hatch and free-swimming tadpoles metamorphose into adults; eats bees, beetles, insect larvae, spiders, and ants; economically important due to wide distribution in state and consumption of large numbers of insects; sometimes eats as much as two-thirds of its own weight in a single day; may hybridize with American and Great Plains Toads; maximum longevity: 13 years, 3 months, and 12 days (Snider and Bowler, 1992).

Pertinent References: Anderson et al. (1995), Atherton (1974), Ballinger and Lynch (1983), Bartlett and Bartlett (2006a), Behler and King (1979), Boyd (1988, 1991), Bragg (1940b), Bragg and Sanders (1951), Brennan (1934, 1937), Breukelman and Clarke (1951), Breukelman and Downs (1936), Breukelman et al. (1961), Burt (1932), Burt and Burt (1929a, 1929c), Busby and Parmelee (1996), Busby et al. (1994, 1996, 2005), Capron and Perry (1976), Capron et al. (1982), Carpenter (1940), Clarke (1980a, 1984), Clarke and Clarke (1984), Clarke et al. (1958), Coleman (1987), Collins (1959, 1974b, 1982a, 1985, 1989c, 1990b, 2006b, 2007b), Collins and Collins (1991, 1993a, 1993b, 2006a), Collins and Taggart (2006), Collins et al. (1991, 1994), Conant (1958, 1975, 1978b), Conant and Collins (1991, 1998), Cope (1889), Cross (1971), Dice (1923), Dickerson (1913), Ellis (2002), Fitch (1951, 1956e, 1958b, 1982, 1991), Gish (1962), Guarisco (1981c), Heinrich and Kaufman (1985), Hoyle (1936), Hudson (1942), Irwin (1977, 1983), Irwin and Collins (1987), Jewell (1927), Johnson (2000), Karns et al. (1974), Kellogg (1932), Kingsbury and Gibson (2002), Knight (1935), Knight and Collins (1977), Kretzer and Cully (2001), Levell (1995, 1997), Linsdale (1925, 1927), Livezey and Wright (1947), Loomis (1956), Maher (1967), Marr (1944), Masta et al. (2002), Meachen (1962), Miller (1976d, 1979, 1980a, 1981c, 1982a, 1985b), Moriarty and Collins (1995a), Mount (1975), Murrow (2008b, 2009a, 2009b), Pauly et al. (2004), Perry (1974a, 1975), Roth (1959), Royal (1982), Rues (1986), Rundquist (1975), Rundquist and Collins (1977), Rush (1981), Ruthven (1907), Sanders (1987), Schmidt (2001, 2004), Schmidt and Murrow (2006, 2007b, 2007c), Schwaner (1978), Schwarting (1984), Schwilling and Schaefer (1983), Sievert and Bailey (2000), Simmons (1987c), Simon (1988), Simon and Dorlac (1990), Smith (1933b, 1934, 1950, 1956a, 1961), Stebbins (1951, 1954, 1966, 1985, 2003), Stejneger and Barbour (1917, 1923, 1933, 1939, 1943), Storer (1925), Suleiman (2005), Sullivan et al. (1996), Taggart (1992a, 1992b, 2001b, 2002c, 2002d, 2004a, 2004h, 2006b, 2006c, 2007c, 2008), Taggart et al. (2005a, 2007), Taylor (1929b), Terman (1988), Tihen (1937), Tihen and Sprague (1939), Toepfer (1993a), Trott (1983), Uhler and Warren (1929, 2008), Van Doren and Schmidt (2000), Viets (1993), Wagner (1984), Warner and Wencel (1978), Washburne and Washburne (2003), Wassersug and Seibert (1975), Wilgers et al. (2006), Wright and Wright (1949), and Yarrow (1883).

TREEFROGS AND ALLIES
FAMILY HYLIDAE RAFINESQUE, 1815

The family Hylidae (Treefrogs) in Kansas is represented by 7 species assigned to 3 genera. These are a single species of the cricket frog genus *Acris*, 4 species of the chorus frog genus *Pseudacris*, and 2 species of the treefrog genus *Hyla*. In appearance, size, and habitat preference, frogs in this family are highly variable. Treefrogs are long-limbed and arboreal, and chorus frogs and the cricket frog are more terrestrial and have shorter limbs.

Blanchard's Cricket Frog
Acris blanchardi HARPER, 1947

Identification: Moist rough skin; round snout; dark triangular mark between eyes; stripe extends from behind triangular mark (where it is widest) down the back (where it is narrowest); irregular, black, lengthwise stripe on inside of each thigh; alternating light and dark bars on upper lips; body, head, and limbs gray or brown; triangular mark between eyes dark brown; stripe down back may be whitish gray, green, brown, or reddish; belly white; chin of males may be spotted and yellowish during breeding season.

Adult: Riley County, Kansas.   Adult: Wyandotte County, Kansas.

Adult: Butler County, Kansas.   Adult: Douglas County, Kansas.

Size: Adults normally 15–38 mm (⅝–1½ inches) in SVL; largest Kansas specimen: female (KU 215686) from Douglas County with SVL of 33 mm (1⁵⁄₁₆ inches) collected by Kevin R. Toal on 24 May 1990; maximum length throughout range: 1½ inches (Conant and Collins, 1998).

Natural History: Recorded active annually in Kansas from February to November, with a peak from late April to mid-August; during colder months, generally active only during day, but as months become warmer is active day and night; preferred habitat muddy, beachlike edges of small, shallow streams and ponds; avoids deep water; sometimes found around caves; evidently wanders great distances from water during both dry and wet weather; apparently, many die during these wanderings, keeping populations at optimal level; congregates to breed around lakes, ponds, marshes, roadside ditches, and streams from mid-March to late July, with a breeding peak from May to late June (most robust choruses recorded during hours from 1:00 pm to 3:00 am); warm temperatures stimulate chorusing; male choruses, attracts female and mounts her back, clasping her behind front limbs with his front legs; female lays eggs and male fertilizes them in water; females lay up to 400 eggs, deposited singly or in small clusters of 2–7 eggs each; eggs hatch after 3–4 days into tiny, solitary, secretive tadpoles, which metamorphose within 5–10 weeks; chorusing does not always indicate breeding activity; this species evidently choruses for other, unknown, reasons; females might lay eggs twice during spring-summer season; egg-laden females have been found from mid-April to mid-July; after metamorphosis, young undergo two distinct growth periods: juvenile period immediately after metamorphosis, from July to September (followed by period of little growth during winter because of low temperatures and scarcity of food), and a second period to adult size from March until coming breeding season when presence of food and temperatures reach an optimum; eats water beetle larvae, small spiders, midges, flies, water-boatmen, and small crayfishes; feeds underwater and on surface; predators include raccoons, skunks; tadpoles exhibit two types of tail coloration: those in ponds have primarily black tail tips, whereas those in lakes and creeks have mostly plain tails; tadpoles with black tail tips coincide with high density of predatory dragonfly larvae; in this situation, black tail directs attacks of dragonfly larvae toward tail and away from more vulnerable head and body of tadpole; predators of tadpoles in lakes and creeks are fishes, which swallow prey whole; thus, it is adaptive for tadpoles in this habitat to have plain tails, making them difficult for predators to observe; maximum longevity: 4 years, 11 months, and 25 days (Snider and Bowler, 1992).

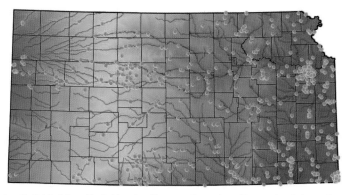

Blanchard's Cricket Frog (*Acris blanchardi*) occurs throughout Kansas, but is least abundant along the western border, where it is restricted to major river and stream drainages. Populations appear to be declining in the western quarter of the state. Red dots indicate records backed by voucher specimens or images.

Pertinent References: Anderson (1983), Anderson and Arruda (2006), Anderson et al. (1995), Ashton et al. (1974a), Ballinger and Lynch (1983), Bartlett and Bartlett (2006a), Behler and King (1979), Boyd (1988, 1991), Brennan (1934, 1937), Breukelman and Clarke (1951), Breukelman and Downs (1936), Breukelman et al. (1961), Burkett (1969, 1984b), Burt (1932), Burt and Burt (1929a), Busby and Parmelee (1996), Busby and Pisani (2007), Busby et al. (1994, 1996, 2005), Caldwell (1982), Caldwell and Glass (1976), Capron et al. (1982), Clarke (1958, 1980a, 1984), Clarke and Clarke (1984), Clarke et al. (1958), Cole (1966b), Coleman (1987), Collins (1959, 1974b, 1982a, 1983, 1985, 1990b, 2006b, 2007b), Collins and Collins (1991, 1993a, 1993b, 2006a), Collins et al. (1991, 1994), Conant (1958, 1975, 1978b), Conant and Collins (1991, 1998), Cope (1889), Cragin (1881), Cross (1971), Dessauer and Nevo (1969), Dickerson (1913), Dillenbeck (1988), Ditmars (1905a), Duellman (1970), Dunn (1938), Eddy and Babcock (1973), Ellis (2002), Ellis and Henderson (1913), Falls (1933), Fitch (1951, 1956e, 1958b, 1963b, 1965b, 1982, 1991), Fitch and Packard (1955), Forney (1926), Gamble et al. (2008), Gish (1962), Gloyd (1928, 1932), Gorman (1986), Gorman and Gaines (1987), Gray (1979), Grow (1976b), Guarisco (1980a), Hammerson (1999), Hammerson and Livo (1999), Hartman (1906), Heinrich and Kaufman (1985), Hoyle (1936), Irwin (1977, 1980, 1983), Irwin and Collins (1987), Jameson (1947), Jantzen (1960), Karns et al. (1974), Kingsbury and Gibson (2002), Knight (1935), Knight and Collins (1977), Korb (2001), Levell (1995, 1997),

Linsdale (1925, 1927), Livezey and Wright (1947), Loomis (1956), Marr (1944), Miller (1976a, 1979, 1980a, 1980c, 1981c, 1981d, 1982a, 1983a, 1985b, 1988, 1996a, 1996b, 2000c), Minton (2001), Moriarty and Collins (1995a), Mount (1975), Murrow (2008a, 2008b, 2009a, 2009b), Nevo (1973a, 1973b), Nevo and Capranica (1985), Nur and Nevo (1969), Perry (1975, 1978), Regan (1972), Riedle (1994a, 1998a), Riedle and Hynek (2002), Rundquist (1975, 1992), Rundquist and Collins (1977), Rush and Fleharty (1981), Salthe and Nevo (1969), Schmidt (2001, 2004l), Schmidt and Murrow (2006, 2007c), Shirer and Fitch (1970), Shoup (1996a), Simmons (1987c), Simon (1988), Simon and Dorlac (1990), Smith (1933b, 1934, 1950, 1956a, 1961), Stebbins (1951, 1954, 1966, 1985, 2003), Steiner and Lehtinen (2008), Suleiman (2005), Taggart (1992a, 2000b, 2001a, 2002d, 2003g, 2005), Taggart et al. (2007), Tihen (1937), Tihen and Sprague (1939), Toepfer (1993a), Trauth et al. (2004), Trott (1977, 1983), Uhler and Warren (1929, 2008), Van Doren and Schmidt (2000), Viets (1993), Wagner (1984), Warner and Wencel (1978), Wilgers and Horne (2006), Wilgers et al. (2006), Wright (1931), Wright and Wright (1949), and Yarrow (1883).

Cope's Gray Treefrog
Hyla chrysoscelis COPE, 1880

Eastern Gray Treefrog
Hyla versicolor LECONTE, 1825

Cope's Gray Treefrog (*Hyla chrysoscelis*) adults: Cherokee County, Kansas.

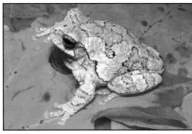

Gray Treefrog (*Hyla chrysoscelis* or *H. versicolor*) adults: Linn County, Kansas (**left**), Coffey County, Kansas (**right**).

Cope's and Eastern Gray Treefrogs are considered indistinguishable using external morphology. Hillis et al. (1987), based on chromosome number, defined both species more precisely in Kansas, as follows: Cope's Gray Treefrog apparently is more arboreal, has half as many chromosomes, and apparently requires less humid habitat than the Eastern Gray Treefrog; however, both can be identified with certainty only by an analysis of their calls or chromosomal material, or by determining the size of their red blood cells. The following information, except where noted, applies to both species.

Left: Adult Cope's Gray Treefrog (*Hyla chrysoscelis*) from Cherokee County, Kansas, showing the characteristic black and yellow pattern on the rear and underside of the thighs. **Right:** Gray Treefrog larvae (*Hyla chrysoscelis* or *H. versicolor*): Douglas County, Kansas.

Identification: Moist skin; round snout; conspicuously enlarged pads on toes of both front and hind feet; distinct colors; only amphibians in Kansas that can change color, presumably to blend better with habitat; head, body, and limbs vary from bright green to tan or light gray; green phase normally uniform, with no pattern, but light gray phase may have darker blotches and spots on back, head, and limbs; distinct white spot on upper lip directly below eye; inside and underside of rear limbs bright yellow-orange; belly white; females have greater SVL length and, during breeding season, heavier bodies than males.

Size: Adults normally 32–58 mm (1¼–2¼ inches) in SVL; largest Kansas specimen: female *Hyla chrysoscelis* (MHP 8430) from Miami County with SVL of 54 mm (2⅛ inches) collected by Keith Coleman on 20 March 2004; maximum length throughout range: 2⅜ inches (Conant and Collins, 1998).

Natural History: Inhabits trees and low shrubs in woodland and woodland edge areas; more arboreal than other Kansas frogs; enlarged adhesive toe pads allow it to cling to surfaces; appears tolerant of high temperatures; recorded active annually in Kansas from early March to early October, with a peak from late April to late June; frequently calls from treetops on warm, humid summer nights; active in summer during high humidity; sometimes retreats into small holes and niches in sun-heated limestone rocks, but can maintain body temperatures lower than rocks; during summer can be found on low-hanging branches in trees, perched in direct sunlight with no apparent discomfort (body of frog can reach temperatures as high as 90° F); males territorial and maintain their territories by using encounter calls and fighting when another frog approaches too closely; normally breeds from mid-March to mid-July (with a peak from late April to mid-June) in permanent, semipermanent, or temporary pools with mud bottoms or weedy vegetation in or near woodlands (most robust choruses recorded during nocturnal hours from 10:00 pm to 3:00 am); male mounts female in water, clasping her behind her forelegs with his front limbs; females may lay from 700–3,800 eggs in small floating masses; eggs hatch in 4–5 days; free-swimming tadpoles metamorphose within 2 months; young frogs evidently remain near breeding site during first season following metamorphosis; few differences are known in breeding site selection in Kansas between the two species; both were taken at same breeding site on same date in Douglas, Elk, and Franklin counties; the Eastern Gray Treefrog is apparently restricted to breeding sites within heavily forested areas, whereas Cope's Gray Treefrog used a wider variety of sites, forested or not; breeding areas varied from temporary flooded ditches to rain pools in cultivated fields to permanent ponds, and ranged in depth from 8 feet to a few inches; only a few sites contained fish predators, and none was more than 500 feet from forest; eats both terrestrial and flying insects; maximum longevity: 7 years, 9 months, and 20 days (Snider and Bowler, 1992).

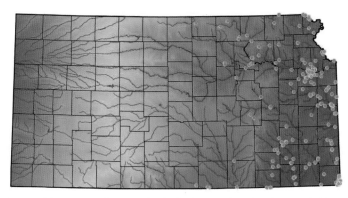

Distribution of the **Gray Treefrog** complex (*Hyla chrysoscelis* and *Hyla versicolor*) in Kansas. These frogs occur primarily in the eastern quarter of the state, and have been expanding their range west along riparian corridors. The **Eastern Gray Treefrog** (*Hyla versicolor*) may have a more restricted range in the state than **Cope's Gray Treefrog** (*Hyla chrysoscelis*). Red dots indicate records backed by voucher specimens or images.

Pertinent References: Anderson and Arruda (2006), Anderson et al. (1995), Ballinger and Lynch (1983), Bartlett and Bartlett (2006a), Behler and King (1979), Blair (1958), Bogart (1980), Boyd (1991), Bragg (1948a), Breukelman and Clarke (1951), Burt (1928e), Busby and Parmelee (1996), Busby and Pisani (2007), Busby et al. (1994, 1996, 2005), Caldwell and Glass (1976), Clarke (1958, 1984), Clarke et al. (1958), Coleman (2004), Collins (1959, 1974b, 1982a, 1985, 1990b, 1993a), Collins and Collins (1993a), Collins and Hillis (1985), Collins et al. (1991, 1994), Conant (1958, 1975), Conant and Collins (1991, 1998), Cragin (1881), Cross (1971), Dice (1923), Dickerson (1913), Eddy and Babcock (1973), Fitch (1951, 1956e, 1958b, 1982, 1991), Fitch and Packard (1955), Fleharty and Johnson (1975), Gier (1967), Gloyd (1928, 1932), Godwin and Roble (1983), Gress and Potts (1993), Hall and Smith (1947), Hartman (1906), Heinrich and Kaufman (1985), Hillis et al. (1987), Hudson (1942), Hunter et al. (1992), Johnson (1961, 2000), Jordan (1929), Karns et al. (1974), Kingsbury and Gibson (2002), Klemens (1993), Knight (1935), Korb (2001), Legler (1960b), Linsdale (1925, 1927), Livezey and Wright (1947), Loomis (1956), McKenna and Seabloom (1979), Miller (1983a, 2002a, 2009), Minton (2001), Mittleman (1945a, 1945b), Moriarty and Cannatella (2004), Moriarty and Collins (1995a), Mount (1975), Murrow (2009a, 2009b), Perry (1978), Ralin (1968, 1977), Reichman (1987), Riedle (1994a), Riedle and Hynek (2002), Ritke et al. (1991), Roble (1979, 1985), Rundquist and Collins (1977), Schmidt (2004e), Schmidt and Murrow (2007c), Shoup (1996a), Simmons (1987b), Simon (1988), Simon and Dorlac (1990), Smith (1933b, 1934, 1950, 1956a), Suleiman (2005), Taggart (1992a, 2000b, 2003g, 2005), Tanner (1992), Trauth et al. (2004), Viets (1993), Wagner (1984), and Wright and Wright (1949).

Spotted Chorus Frog
Pseudacris clarkii (BAIRD, 1854)

Identification: Moist skin; round snout; light line along upper lips; pattern of dark-edged pale green spots (sometimes merging into stripes) scattered irregularly on back and limbs; head, body, and limbs gray; green spots on back and limbs are edged with dark gray or black; some individuals have a green triangular mark between eyes; greenish stripe passes through each eye; belly white; males differ from females by their dark throats.

Adult: Barber County, Kansas.

Adult: Chautauqua County, Kansas.

Adult: Harper County, Kansas.

Adult: Sumner County, Kansas.

Size: Adults normally 19–28 mm (¾–1⅛ inches) in SVL; largest Kansas specimen: male (KU 4515) from Rush County with a total length of 31 mm (1¼ inches) collected by Theodore E. White on 12 August 1927; maximum length throughout range: 1¼ inches (Conant and Collins, 1998).

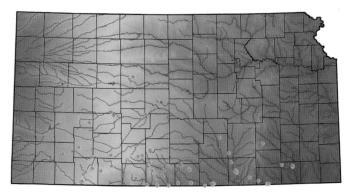

The **Spotted Chorus Frog** (*Pseudacris clarkii*) is abundant in south-central Kansas in the Arkansas and Cimarron river drainages; it is also known from the Vergidris River drainage in the southeastern part of state. Specimens from Ellis County at the northern edge of the range are the only records for the Kansas River drainage and need corroboration; records for Barton and Rush counties (collected in 1942 and 1927, respectively) also need corroboration. Red dots indicate records backed by voucher specimens or images.

Natural History: Inhabits open prairie grasslands and edges of woodlands; most active during spring and summer rains; becomes very secretive during dry weather, probably retreating to underground burrows; recorded active annually in Kansas from late March to late September, with a peak from mid-April to early June; breeds opportunistically from late March to late August (depending on rainfall and suitable temperatures), with a peak from late April to early June; reproduces in open, shallow, temporary rain pools, although mating sometimes occurs in permanent ponds; males chorus and attract females; amplexus may occur for entire day; females deposit up to 1,000 eggs, which are laid in small masses of 6–30 each on grasses and stems just below water surface; eggs hatch in 2–10 days; tadpoles remain in pond until metamorphosis in late summer; eats small insects; predators include larger frogs, large birds, and snakes.

Pertinent References: Ashton et al. (1974b), Bartlett and Bartlett (2006a), Behler and King (1979), Bragg (1943b), Breukelman and Clarke (1951), Caldwell and Collins (1977), Capron (1978b), Clarke (1980a, 1984), Clarke et al. (1958), Cochran and Goin (1970), Collins (1959, 1974b, 1982a, 1985, 1990b), Collins and Collins (1993a, 1993b, 2006a), Collins et al. (1991, 1994), Conant (1958, 1975), Conant and Collins (1991, 1998), Duellman (1977), Eaton and Imagawa (1948), Fleharty and Ittner (1967), Fleharty and Johnson (1975), Fraser (1983a), Gier (1967), Gray and Stegall (1979), Grow (1977), Guarisco (1980a), Irwin (1983), Irwin and Collins (1987), Karns et al. (1974), Kingsbury and Gibson (2002), Knight (1935), Lardie (1990), Lemmon et al. (2007), Levell (1995, 1997), Leviton (1971), Livezey and Wright (1947), Loomis (1956), Miller (1979, 1980a, 1982a, 1988), Moriarty and Cannatella (2004), Moriarty and Collins (1995a), Pierce and Whitehurst (1990), Schmidt (1953), Schmidt and Murrow (2007b), Shoup (1996a), Simmons (1987c), Smith (1933b, 1934, 1950, 1956a), Stejneger and Barbour (1939, 1943), Taggart (2008), Tihen (1937), Toepfer (1993a), Trott (1983), Van Doren and Schmidt (2000), Warner and Wencel (1978), Wiley (1968), and Wright and Wright (1949).

Spring Peeper
Pseudacris crucifer (Wied-Neuwied, 1838)

Adult: Miami County, Kansas.

Adult male chorusing: Shoal Creek oxbow, Cherokee County, Kansas.

Adults: Cherokee County, Kansas.

Identification: Moist skin; round snout; slightly enlarged toe pads on front and hind feet; head, body, and limbs light brown or gray, with a darker brown or grayish X-shaped mark on back; limbs narrowly banded with brown or gray; belly yellowish with no pattern; males differ from females by having dark throat; females reach slightly larger size than males.

Size: Adults normally 19–32 mm (¾–1¼ inches) in SVL; largest Kansas specimen: female (KU 186100) from Cherokee County with SVL of 30 mm (1¼ inches) collected by Chris Stammler and Dan Hodges on 30 July 1980; maximum length throughout range: 1½ inches (Conant and Collins, 1998).

Vernal pool in Cherokee County, Kansas, typical habitat for the Spring Peeper.

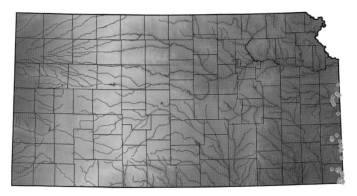

The **Spring Peeper** (*Pseudacris crucifer*) is a small anuran with a Kansas distribution restricted to the eastern edge of the state in counties that border Missouri, from Cherokee in the south to Miami in the north; within and between those counties, it inhabits the Marais des Cygnes, Little Osage, Marmaton, and Spring river drainages. The type locality (place where originally discovered) for this species is Fort Leavenworth, Leavenworth County, Kansas, but no type specimen exists and the species has never been collected in that county since the original description in 1838; the type locality is not plotted on the map. Red dots indicate records backed by voucher specimens or images.

Natural History: Inhabits woodland areas near small intermittent ponds or marshes; recorded active annually in Kansas from late February through July, with a peak from late February through March; males territorial and defend their territory by giving encounter calls and fighting when another frog approaches too closely; breeds in small, temporary, shallow ponds and marshes adjoining woodlands (most robust choruses recorded during nocturnal hours from 10:00 pm to midnight); prefers breeding site containing abundant, low-standing, aquatic vegetation in still water; males call to attract females and mount them in the water; eggs (up to 700–750) laid singly and hatch in 4–5 days, and free-swimming tadpoles metamorphose in 90–100 days; breeding choruses and congregations, ranging in size from a single calling male to over 100 adults, have been documented in southeastern Kansas from late February to late May, with a peak during March; feeds on small insects; threatened species (protected by state law); maximum longevity: 2 years, 2 months, and 9 days (Snider and Bowler, 1992).

Pertinent References: Anderson and Arruda (2006), Ballinger and Lynch (1983), Bartlett and Bartlett (2006a), Behler and King (1979), Clarke (1984), Coleman (2000), Collins (1959, 1974b, 1982a, 1982b, 1990b, 1994b), Collins and Collins (1993a, 1999, 2002), Collins et al. (1991, 1994, 1995), Conant (1958, 1975), Conant and Collins (1991, 1998), Cross (1971), Duellman (1977), Freeman and Kindscher (1992), Gloyd (1932), Guarisco et al. (1982), Harper (1939), Hunter et al. (1992), Johnson (1974), Karns et al. (1974), Kingsbury and Gibson (2002), Klemens (1993), Knight (1935), Layher et al. (1986), Levell (1997), Loomis (1956), Loraine (1983, 1984), Mansueti (1941), Miller and Gress (2005), Minton (1972, 2001), Mitchell (1979), Moriarty and Cannatella (2004), Moriarty and Collins (1995a), Mount (1975), Platt et al. (1974), Rundquist (1978), Rundquist and Collins (1977), Schulenberg-Ptacek (1984a, 1984b), Schwilling and Schaefer (1983), Shoup (1996a), Simmons (1989), Simon (1988), Smith (1933b, 1934, 1946a, 1950, 1956a), Spanbauer (1988), Stejneger and Barbour (1917, 1923, 1933, 1939, 1943), Taggart (2006b, 2007b, 2008), Trauth et al. (2004), and Wright and Wright (1949).

Boreal Chorus Frog
Pseudacris maculata (AGASSIZ, 1850)

Adult: Leavenworth County, Kansas.

Adult: Anderson County, Kansas.

Adult: Douglas County, Kansas.

Eggs: Shawnee County, Kansas.

Identification: Moist skin; round snout; light line along upper lips; 5 dark stripes or rows of spots on back and sides; head, body, and limbs gray or gray-brown; stripes and/or spots on back and sides, as well as bands on limbs, dark gray or black; dark gray or black stripe passes through eyes on each side of head; sometimes similarly-colored dark triangle between eyes; belly white, sometimes with few dark flecks or spots; females grow larger than males; males have dark gray or brown throat during breeding season.

Size: Adults normally 19–39 mm (¾–1½ inches) in SVL; largest Kansas specimen: female (KU 184955) from Douglas County with SVL of 39 mm (1½ inches) collected by Steven M. Roble on 2 April 1980; maximum length throughout range.

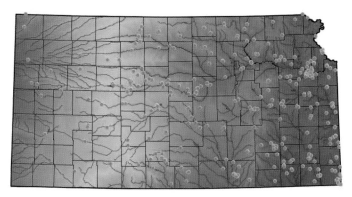

The **Boreal Chorus Frog** (*Pseudacris maculata*) occurs throughout most of the state, but is scarce in the arid western third of Kansas where aquatic situations are limited. Red dots indicate records backed by voucher specimens or images.

Natural History: Lives in damp meadows and pastures, streams and water-filled ditches, around edges of temporary or permanent ponds and lakes, on floodplains, and in moist woods; recorded active annually in Kansas from early February to mid-November, with a peak from early April to late June; retires to underground bur-rows of other animals during weather extremes; emerges earlier in spring than other anurans in Kansas; choruses from late February to through August with a peak from early April to early June; after first early rains of year, males congregate in roadside ditches, small ponds, lakes, marshes or swamps, or along slow-moving streams, and begin chorusing to attract females; choruses may occur at temperatures as low as 35°F and are most robust during nocturnal hours from 7:00 pm to 3:00 am; when mating, male mounts female, clasping her behind front limbs, and eggs deposited and fertilized in water; females may deposit up to 1,500 eggs in small clutches; clutches attached to plant stems in water and hatch in about two weeks; free-swim-ming tadpoles metamorphose within two months; wide fluctuations in abundance of these frogs occurs, apparently correlated with previous season breeding suc-cess; eats ants, grubs, beetles, and spiders; predators include large birds, small mammals, snakes, and other frogs.

Pertinent References: Anderson and Arruda (2006), Anderson et al. (1995), Ballinger and Lynch (1983), Bartlett and Bartlett (2006a), Behler and King (1979), Boyd (1988, 1991), Bragg (1948b), Brennan (1934), Breukelman and Clarke (1951), Breukelman and Downs (1936),

AMPHIBIANS

57

Breukelman et al. (1961), Burkett (1984), Burt and Burt (1929c), Busby (1997), Busby and Parmelee (1996), Busby and Pisani (2007), Busby et al. (1994, 1996, 2005), Caldwell and Collins (1977), Caldwell and Glass (1976), Capron (1978b), Capron and Perry (1976), Clarke (1958, 1984), Clarke et al. (1958), Cochran and Goin (1970), Coleman (1987, 2002), Collins (1959, 1974b, 1982a, 1985, 1990b, 2006b, 2007b), Collins and Collins (1993a, 1993b, 2006a), Collins et al. (1991, 1994), Conant (1958, 1975), Conant and Collins (1991, 1998), Cope (1889), Coues and Yarrow (1878), Cragin (1881), Cross (1971), Eddy and Babcock (1973), Ellis (2002), Ellis and Henderson (1913), Fitch (1951, 1956e, 1958b, 1965b, 1991), Fitch and Hall (1978), Fitch and Packard (1955), Gerlanc (1999), Gerlanc and Kaufman (1997), Gier (1967), Gish (1962), Gloyd (1928, 1932), Hammerson (1999), Hardy (1959), Hartman (1906), Heinrich (1985), Heinrich and Kaufman (1985), Hurter (1911), Hurter and Strecker (1909), Irwin (1983), Irwin and Collins (1987), Johnson (2000), Karns et al. (1974), Kingsbury and Gibson (2002), Knight (1935), Knight and Collins (1977), Korb (2001), Lemmon et al. (2007, 2008), Linsdale (1925, 1927), Livezey and Wright (1947), Loomis (1956), Miller (1985b, 1988, 1996a, 2000c), Minton (1972, 2001), Moriarty and Cannatella (2004), Moriarty and Collins (1995a), Mount (1975), Murrow (2008b, 2009a, 2009b), Nulton and Rush (1988), Perry (1975), Parmelee and Fitch (1995), Platt (1998), Platz (1989), Platz and Lathrop (1993), Pope (1944b), Powell (1980b), Reichman (1987), Riedle and Hynek (2002), Roble (1985), Rundquist (1975, 1996a), Rundquist and Collins (1977), Schmidt (2004l), Shoup (1996a), Simon (1988), Simon and Dorlac (1990), Skie and Bickford (1978), Smith (1933b, 1934, 1950, 1956a, 1956b, 1957, 1961), Smith and Smith (1952a), Spanbauer (1988), Stebbins (1951, 1954, 1966, 1985, 2003), Suleiman (2005), Taggart (1992a, 1992b, 2003g, 2005), Taggart et al. (2007), Tihen (1937), Toepfer (1993a), Trauth et al. (2004), Van Doren and Schmidt (2000), Viets (1993), Wagner (1984), Wassersug and Seibert (1975), Wilgers and Horne (2006), Wilgers et al. (2006), Wright and Wright (1949), and Yarrow (1883).

Strecker's Chorus Frog

Pseudacris streckeri WRIGHT AND WRIGHT, 1933

Adult: Harper County, Kansas.

Adult

Adult males: Harper County, Kansas.

Identification: Moist skin; round snout; light line along upper lips; dark spots irregularly scattered on back and sides (sometimes forming two elongate dark bars on back), dark spot below and just in front of each eye; short head; squat body; thick limbs that are brown, gray, green, or hazel; spots and bars on back dark brown or black; dark stripe passes through each eye and curves down onto shoulder; dark triangular or V-shaped marking may or may not be present between eyes; belly light-colored; females normally grow larger than males; males have dark throat during breeding season.

Size: Adults normally 25–41 mm (1–1⅝ inches) in SVL; largest Kansas specimen: male (KU 195621) from Harper County with SVL of 38 mm (1½ inches) collected by Suzanne L. Collins, Joseph T. Collins, and Larry Miller on 7 April 1984; maximum length throughout range: 1⅞ inches (Conant and Collins, 1998).

Natural History: Primarily resides in open areas such as sand prairies and culti-vated fields, and in ditches, streams, marshes, and pools located in these habitats; recorded active annually in Kansas from mid-March to late May, with a peak from mid-March to mid-April; breeds opportunistically during rains from late February to late May, with a peak from late March to late April; may call at night or during day, depending on temperature; prefers breeding sites with unpolluted water, absence of fish predators, and presence of some vascular water plants; males chorus to attract females to aquatic situations for breeding; eggs laid in clumps of 2–100 and attached to aquatic plants and twigs; female may lay up to 700 eggs in as many as 75 clumps; eggs probably hatch within a week; tadpoles metamorphose during summer; feeds on small insects; threatened species (protected by state law).

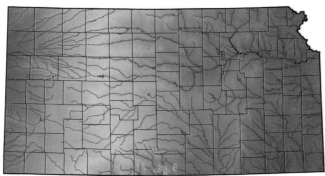

Strecker's Chorus Frog (*Pseudacris streckeri*) is a chubby little anuran with a distribution in Kansas restricted to the sandy soils of four counties in the extreme south-central part of the state, along the Oklahoma border between the Chikaskia and Medicine Lodge rivers. Red dots indicate records backed by voucher specimens or images.

Pertinent References: Bartlett and Bartlett (2006a), Clarke (1984), Collins (1974b, 1979, 1982a, 1983, 1990b), Collins and Collins (1993a), Collins et al. (1994, 1995), Conant and Collins (1991, 1998), Davis and Taggart (2004), Davis et al. (2004), Gray (1978, 1982), Gray and Stegall (1986), Guarisco (1979b), Guarisco et al. (1982), Kingsbury and Gibson (2002), Layher et al. (1986), Lardie (1990), Levell (1997), Miller (1987a, 1988, 2004b), Miller and Gress (2005), Moriarty and Cannatella (2004), Moriarty and Collins (1995a), Rundquist et al. (1978), Schwilling and Schaefer (1983), Shoup (1996a), Simmons (1989), Stegall (1977), Taggart (2006b), and Trauth et al. (2004, 2007).

TRUE FROGS
FAMILY RANIDAE RAFINESQUE, 1814

The family Ranidae (True Frogs) in Kansas is represented by 6 species assigned to the genus *Lithobates*. This genus contains some of the largest known anurans (up to 300 mm) and its species are distinctive because of their smooth skin, webbed feet, and long, muscular legs.

Crawfish Frog
Lithobates areolatus (BAIRD AND GIRARD, 1852)

Adult: Bourbon County, Kansas.

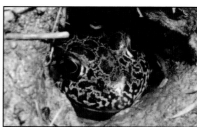
Adult female in burrow: Linn County, Kansas.

Subadult: Lyon County, Kansas.

Egg mass: Wilson County, Kansas.

Adult: Cherokee County, Kansas.

Adult: Anderson County, Kansas.

Identification: Moist skin; round snout; raised fold of skin on each side of back running from behind eye down to thigh; reticulate pattern of light and dark flecks and lines between spots on back and sides; head, body, and front limbs gray or light brown and covered with numerous blue-brown or dark brown light-edged spots; reticulate pattern appears between and around these spots; rear limbs have dark brown crossbands; belly whitish; males have enlarged inner finger on each hand, and females grow larger than males.

Size: Adults normally 57–75 mm (2¼–3 inches) in SVL; largest Kansas specimen: female (MHP 10447) from Bourbon County with SVL of 122 mm (4¾ inches) collected by Derek Welch and Curtis J. Schmidt on 31 March 2005; exceeds maximum length throughout range as reported in Conant and Collins (1998).

Natural History: Inhabits floodplains, moist native hay meadows, pastures, cultivated fields, as well as prairie bottomland; prefers crayfish burrows (those used by this frog usually have a flattened mud platform near entrance and are 3–5 feet deep); undoubtedly also uses burrows of small mammals; has been taken beneath logs, in holes along roadside ditches, and in sewers; probably uses these burrows as retreats during winter; considered by some most secretive frog in Kansas; young probably retreat to underground burrows immediately after metamorphosis; recorded active annually in Kansas from early March to late September, with a peak from late March through April; choruses recorded from early March to early July, with a peak from late March to late April (choruses most robust during nocturnal hours from 10:00 pm to 2:00 am); breeding activities appear to be correlated with rainfall and air temperatures 50° F or above; migrates to temporary rain-filled pools in lowlands or floodplains, where males commence chorusing (often with Southern Leopard Frogs); males mount females, clasping them behind their forelimbs, and eggs laid and fertilized in water; females may deposit up to 7,000 eggs in shallow water, in masses 5–6 inches in diameter around plants and stems; eggs hatch and tadpoles metamorphose during summer; adults active from early March to late September; feeds on beetles, spiders, crickets, ants, and possibly crayfish; species in need of conservation (protected by state law); may be disappearing from previously suitable habitat in Kansas; very susceptible to fluctuations of water tables caused by draining of wetlands.

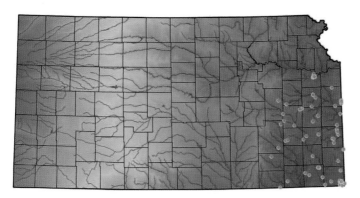

The distribution of the **Crawfish Frog** (*Lithobates areolatus*) in Kansas is restricted to the eastern fourth of the state south of the Kansas River. Red dots indicate records backed by voucher specimens or images.

Pertinent References: Altig and Lohoefener (1983), Anderson and Arruda (2006), Bartlett and Bartlett (2006a), Bailey (1943), Behler and King (1979), Boyd (1988, 1991), Bragg (1953), Burr and Burr (2003), Busby (1990), Busby and Brecheisen (1997), Caldwell (1977), Caldwell and Collins (1977), Caldwell and Glass (1976), Clarke (1984), Clarke et al. (1958), Collins (1959, 1974b, 1982a, 1982b, 1985, 1990b, 1994b, 2009a), Collins and Collins (1993a), Collins and Dawson (1975), Collins and Taggart (2006), Collins et al. (1981, 1994), Conant (1958, 1975), Conant and Collins (1991, 1998), Cross (1971), Flowers (2005), Frederickson (1987), Freeman and Kindscher (1992), Gier (1967), Gloyd (1928), Goin and Netting (1940), Guarisco et al. (1982), Hall and Smith (1947), Harrison (1977), Hartman (1906), Henderson and Andelt (1982), Herin (1987), Hillis (1985), Hillis and Wilcox (2005), Hillis et al. (1983), Hlavachick (1978), Johnson (1974, 2000), Karns et al. (1974), Kingsbury and Gibson (2002), Knight (1935), Layher et al. (1986), Levell (1997), Livezey and Wright (1947), Loomis (1956), Mansfield (1932), Moriarty and Collins (1995a), Mount (1975), Murrow (2009a), Platt et al. (1974), Riedle and Hynek (2002), Rundquist and Collins (1977), Schmidt (1953), Schmidt et al. (2005), Schroeder and Baskett (1965), Schwilling (1981), Schwilling and Nilon (1988), Schwilling and Schaefer (1983), Simmons (1989), Simon (1988), Smith (1933b, 1934, 1950, 1956a, 1973, 1977), Smith et al. (1948), Stejneger and Barbour (1943), Stuart et al. (2008), Taggart (1992a, 2006b, 2008), Taggart and Schmidt (2005b, 2005c), Trauth et al. (2004), Tweet and Tweet (2006), Volpe (1957), Von Achen (1987), Wagner (1984), Welch and Schmidt (2006), Wood (1985), Wright and Wright (1949), and Young and Crother (2001).

Plains Leopard Frog
Lithobates blairi (MECHAM, LITTLEJOHN, OLDHAM, BROWN, AND BROWN, 1973)

Adults: Douglas County, Kansas.

Adult: Franklin County, Kansas.　　　　Adult: Doniphan County, Kansas.

Identification: Moist skin, round snout, irregular pattern of distinct scattered spots on back and sides with few or no flecks or network pattern between them, a raised fold or ridge of skin on each side of back running from behind eyes down to thighs where it is broken, posterior portion being set inward toward middle of back; head, body, and limbs gray or tan with dark gray, brown, or black spots; spots sometimes narrowly edged with black; hind limbs darkly banded; raised fold or ridge on each side of back light yellow, gray, or whitish; belly white, although area around groin may be yellowish; females grow larger than males.

Adult Plains Leopard Frog (*Lithobates blairi*), Miami County, Kansas, (**left**) and adult Southern Leopard Frog (*Lithobates sphenocephalus*), Linn County, Kansas (**right**).

Size: Adults normally 51–95 mm (2–3¾ inches) in SVL; largest Kansas specimen: sex undetermined (MHP 3131) from Montgomery County with SVL of 107 mm (4¼ inches) collected by L. S. Oborny on 15 April 1966; maximum length throughout range: 4⅜ inches (Conant and Collins, 1998).

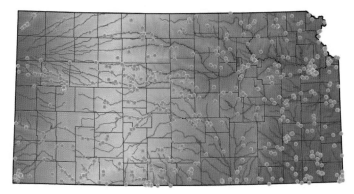

The **Plains Leopard Frog** (*Lithobates blairi*) is found throughout Kansas. Red dots indicate records backed by voucher specimens or images.

Natural History: Found in every permanent or temporary aquatic situation; wanders great distances from water; evidently has no home range or territory; recorded active annually in Kansas during every month of year within a wide range of temperatures; peak of activity occurs from late April to early August; digs into mud and leaves of lake and stream bottoms during winter months and remains inactive until temperatures are favorable; may freeze to death during rapid weather changes; choruses recorded from early March to late August with a peak from mid-April to mid-June (choruses are most robust during hours from noon to 2:00 am); evidence of fall breeding has been observed; mass migrations of newly metamorphosed young and adults occur, but many perish from failure to find suitable aquatic habitat during their wanderings; during breeding season, males chorus, attract females, and mount them by clasping them behind their front limbs; females may lay up to 6,500 eggs in small masses attached to stems and plants just below water's surface; eggs hatch within three weeks and tadpoles metamorphose during summer or overwinter and transform the following spring; feeds primarily on nonaquatic insects including beetles, grasshoppers, and crickets as well as worms and water snails; known to hybridize with Southern Leopard Frogs (*Lithobates sphenocephala*) in southeastern Kansas (particularly Woodson County); such hybridization probably occurs at or near impoundments of human construction where environmental barriers are eliminated that would normally separate both species; predators include raccoons, opossums, and skunks.

Pertinent References: Anderson et al. (1995), Axtell (1976), Ballinger and Lynch (1983), Bartlett and Bartlett (2006a), Behler and King (1979), Boulenger (1920), Boyd (1991), Brennan (1934), Breukelman and Clarke (1951), Breukelman and Downs (1936), Breukelman et al. (1961), Brown (1958, 1992), Burkett (1984), Burt and Burt (1929a), Busby and Parmelee (1996), Busby and Pisani (2007), Busby et al. (1994, 1996, 2005), Caldwell and Glass (1976), Capron (1978b), Capron and Perry (1976), Capron et al. (1982), Carpenter (1940), Clarke (1958, 1980a, 1984), Clarke et al. (1958), Cloutman (1982), Coleman (1987), Collins (1959, 1974b, 1982a, 1982b, 1985, 1990b, 2006b, 2007b), Collins and Collins (1991, 1993a, 1993b, 2006a), Collins and Taggart (2006), Collins et al. (1994), Conant (1958, 1975), Conant and Collins (1991, 1998), Di Candia and Routman (2007), Dice (1923), Dunlap and Kruse (1976), Eddy and Babcock (1973), Ellis (2002), Ellis and Henderson (1913), Falls (1933), Fitch (1951, 1956e, 1958b, 1982, 1987, 1991), Fitch and Packard (1955), Fleharty (1995), Forney (1926), Funk (1975), Gish (1962), Gloyd (1928), Goldberg et al. (2000), Gress (2003), Guarisco (1979b, 1980a, 1981c), Gunthorp (1923), Hartman (1906), Heinrich and Kaufman (1985), Hillis (1981, 1985), Hillis and Wilcox (2005), Hillis et al. (1983), Hoyle (1936), Irwin (1977), Irwin and Collins (1987), Jantzen (1960), Johnson (2000), Karns et al. (1974), Kingsbury and Gibson (2002), Knight (1935), Knight and Collins (1977), Levell (1995, 1997), Linsdale (1925, 1927), Loomis (1956), Lynch (1978), Mansfield (1932), Marr (1944), Mecham et al. (1973), Merrell (1965), Miller (1976a, 1976b, 1976c, 1977a, 1979, 1980a, 1981c, 1981d, 1982a, 1983a, 1985b, 1988, 1996a, 1996b, 2000c), Minton (2001), Moore (1944), Moriarty and Collins (1995a, 1995b), Murrow (2008b, 2009a, 2009b), Pace (1974), Perry (1975, 1978), Platt (1985b, 1998), Riedle and Hynek (2002), Rues (1986), Rundquist (1975, 1992, 1996a), Rundquist and Collins (1977), Rush and Fleharty (1981), Sage and Selander (1979), Schmidt (2000, 2001, 2004d, 2004l), Schmidt and Murrow (2006, 2007b, 2007c), Schmidt and Taggart (2004b), Shirer and Fitch (1970), Simmons (1987c), Simon (1988), Simon and Dorlac (1990), Skie and Bickford (1978), Smith (1933b, 1934, 1950, 1956a, 1961), Stebbins (1951, 1954, 1966, 1985, 2003), Suleiman (2005), Taggart (1992b, 2000b, 2001b, 2002c, 2002d, 2003g, 2004h, 2005), Taggart et al. (2007), Taylor (1929b), Tihen (1937), Tihen and Sprague (1939), Toepfer (1993a), Trauth et al. (2004), Trott (1977, 1983), Tucker (1976), Uhler and Warren (1929, 2008), Van Doren and Schmidt (2000), Viets (1993), Wagner (1984), Warner and Wencel (1978), Wassersug (1976), Wassersug and Seibert (1975), Wiley (1968), Wilgers and Horne (2006), Wilgers et al. (2006), Wright and Wright (1949), and Zimmerman (1990).

Bullfrog
Lithobates catesbeianus (SHAW, 1802)

Identification: Moist skin, round snout, generally uniform color, prominent raised fold of skin runs from behind eye and down around back of distinct circular tympanic membrane; head, body, and front limbs green, olive, or brown, sometimes with large or small indistinct darker spots or blotches; hind limbs darkly banded; throat and belly white or yellow with gray mottling; males have much larger tympanic membranes and mature earlier than females.

Adult: Douglas County, Kansas.   Adult: Jefferson County, Kansas.

Tadpole: Douglas County, Kansas.   Adult: Anderson County, Kansas.

Adult: Montgomery County, Kansas.   Aberrantly-colored adult: Graham County, Kansas.

Size: Largest frog in state; adults normally 90–152 mm (3½–6 inches) in SVL; largest Kansas specimen: female (KU 181593) from Chase County with SVL of 185 mm (7¼ inches) collected by B. Haller on 26 May 1979; maximum length throughout range: 8 inches (Conant and Collins, 1998); heaviest Kansas specimen weighed slightly over 580 grams (about 1 pound, 4 ounces).

Natural History: Recorded active annually in Kansas from mid-February to early December, with a peak from late April to late July; restricted to permanent lakes, rivers, streams, and swamps where deep water is available; may also live near permanently-filled stock tanks; apparently spends winter months burrowed in mud beneath water of lakes and rivers; generally less tolerant of cold temperatures than other frogs and toads; drought can cause heavy fluctuation in population numbers; can regulate body temperature by posturing and basking; small individuals active at lower temperatures than adults; adults respond to distress calls of immatures and other species to prey on them; breeds later than other frogs and toads (from early March to mid-August with a peak from mid-May to late June), chorusing around permanent deep lakes, ponds, swamps, and backwater sloughs (most robust choruses recorded from noon to 3:00 am); males territorial during breeding season and defend their territory by kicking, bumping, and biting other males; male mounts female, clasping her with his front limbs behind her front limbs; females lay up to 48,000 eggs in masses from 1–2½ square feet; females prefer to lay eggs in one place, rarely moving during this activity; eggs hatch in 4–5 days and free-swimming tadpoles spend 3–14 months in water before metamorphosing; eats earthworms, leeches, insects, centipedes and millipedes, spiders, crayfishes, snails, salamanders, tadpoles, frogs, fishes, small turtles, lizards, snakes, shrews, moles, bats, and birds; in northeastern Kansas, small frogs composed 80% (by volume) of diet; opportunistic predator, eating anything it can swallow, including road-killed animals; classified as game animal in Kansas and can be hunted, with a valid Kansas fishing license, only from 1 July to 30 September; hind limbs are well-known "frog-legs" served in restaurants; predators include opossums, raccoons, and skunks; maximum longevity: 7 years, 3 months, and 24 days (Snider and Bowler, 1992).

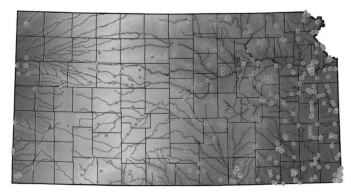

The **Bullfrog** (*Lithobates catesbeianus*) is a large amphibian found throughout Kansas, but is least abundant in the western quarter of the state. Red dots indicate records backed by voucher specimens or images.

Pertinent References: Anderson and Arruda (2006), Anderson et al. (1995), Austin and Zamudio (2008), Ballinger and Lynch (1983), Bartlett and Bartlett (2006a), Behler and King (1979), Boyd (1991), Brennan (1934, 1937), Breukelman and Clarke (1951), Breukelman and Downs (1936), Breukelman et al. (1961), Burkett (1984), Burt and Burt (1929a), Bury and Whelan (1984), Busby and Parmelee (1996), Busby and Pisani (2007), Busby et al. (1994, 1996, 2005), Caldwell and Glass (1976), Capron (1986d), Capron and Perry (1976), Capron et al. (1982), Clarke (1958, 1980a, 1984), Clarke et al. (1958), Cloutman (1982), Coleman (1987), Collins (1959, 1974b, 1982a, 1982b, 1985, 1990b, 2006b, 2007b, 2009a), Collins and Collins (1991, 1993a, 1993b, 2006a), Collins and Taggart (2006), Collins et al. (1994), Conant (1958, 1975), Conant and Collins (1991, 1998), Cox (1990), Cragin (1881), Cross (1971), Dyche (1914), Eddy and Babcock (1973), Ellis (2002), Falls (1933), Ferrier (1997), Fitch (1956e, 1958b, 1963b, 1965b, 1982, 1991), Fitch and Packard (1955), Fleharty (1995), Flowers (2005), Forney (1926), Gablehouse et al. (1982), Gish (1962), Gloyd (1928, 1932), Grow (1976b, 1976c), Guarisco (1980a, 1985), Gunthorp (1923), Hall and Smith (1947), Hammerson (1999), Hartman (1906), Heinrich and Kaufman (1985), Hillis (1985), Hillis and Wilcox (2005), Hoyle (1936), Hunter et al. (1992), Irwin (1977, 1983), Irwin and Collins (1987), Jinks and Johnson (1970), Jordan (1929), Karns et al. (1974), Kingsbury and Gibson (2002), Klemens (1993), Knight (1935), Knight and Collins (1977), Korb (2001), Kraus (2009), Levell (1995, 1997), Linsdale (1925, 1927), Loomis (1956), Marr (1944), Maslowski (1964), McKay (2006), McLeran (1971), Miller (1976a, 1977a, 1979, 1980a, 1980c, 1981c, 1981d, 1982a, 1983a, 1988, 1996a,

1996b), Miller and Collins (1995), Minton (2001), Moriarty and Collins (1995a), Mount (1975), Murrell (1993), Murrow (2008a, 2008b, 2009a, 2009b), O'Roke (1922), Parmelee and Fitch (1995), Perry (1974a, 1978), Platt (1998), Reichman (1987), Riedle (1994a), Riedle and Hynek (2002), Rues (1986), Rundquist (1975), Rundquist and Collins (1977), Schmidt (1953, 2004l, 2007), Schmidt and Murrow (2006, 2007c), Schumann (1923), Schwenke (2008), Shirer and Fitch (1970), Shoup (1996a), Simmons (1987c), Simon (1988), Simon and Dorlac (1990), Smith (1933b, 1934, 1950, 1956a, 1973, 1977), Sparks et al. (1999), Stebbins (1951, 1954, 1966, 1985, 2003), Suleiman (2005), Taggart (1992b, 2000b, 2001a, 2002c, 2002d, 2003g, 2005, 2006b, 2006c, 2007c), Taggart et al. (2005b, 2007), Terman (1988), Tiemeier (1962), Tihen (1937), Tihen and Sprague (1939), Toepfer (1993a), Trauth et al. (2004), Trott (1983), Van Doren and Schmidt (2000), Viets (1993), Wagner (1984), Warner and Wencel (1978), Whipple and Collins (1990b), Wilgers et al. (2006), Willis et al. (1956), Wright and Wright (1949), Zimmerman (1990), and Zuckerman (1988).

Bronze Frog
Lithobates clamitans (LATREILLE, 1801)

Identification: Moist skin; round snout; generally uniform brown or gray-brown color sometimes marked with small, irregular, indistinct black spots; raised fold or ridge of skin on each side of back running from behind eye down back; head, body, and limbs olive to olive-brown, sometimes with small black spots on back; limbs have dark, indistinct narrow bands; belly and throat white (latter sometimes flecked with gray); inner finger of hands of males swollen at base; females heavier during breeding season.

Adults: Cherokee County, Kansas.

Adults: Cherokee County, Kansas.

Size: Adults normally 54–90 mm (2⅛–3½ inches) in SVL; largest Kansas specimen: female (KU 17474) from Cherokee County with SVL of 88 mm (3½ inches) collected by Edward H. Taylor and Hobart M. Smith on 25 March 1933; maximum length throughout range: 4¼ inches (Conant and Collins, 1998).

AMPHIBIANS

67

The distribution of the **Bronze Frog** (*Lithobates clamitans*) in Kansas is restricted to southeastern Cherokee County in the extreme southeastern corner of the state. An old (1911) record for Miami County cannot be verified (may have been the result of an accidental introduction); the specimen (KU 9281) is lost and, until its presence in Miami County is verified by the collection of specimens, we have not mapped it. Red dots indicate records backed by voucher specimens or images.

Natural History: Prefers still water with lush aquatic vegetation and frequents edges of swamps, marshes, ponds, ditches, stripmine ponds, and backwater sloughs of streams and rivers; solitary except during breeding season; recorded active annually in Kansas from mid-January to late October, with a peak from mid-January to late March; males territorial during breeding season and defend their territory from other frogs by kicking, bumping, and biting; eggs hatch into tadpoles that metamorphose after approximately 3 months; feeds primarily on terrestrial insects; threatened species (protected by state law).

Pertinent References: Ballinger and Lynch (1983), Bartlett and Bartlett (2006a), Behler and King (1979), Clarke (1984), Collins (1959, 1974b, 1980b, 1982a, 1982b, 1990b, 1994b), Collins and Collins (1993a), Collins and Taggart (2006), Collins et al. (1994, 1995), Conant (1958, 1975), Conant and Collins (1991, 1998), Cross (1971), Freeman and Kindscher (1992), Gier (1967), Guarisco et al. (1982), Hillis (1985), Hunter et al. (1992), Johnson (1974), Karns et al. (1974), Kingsbury and Gibson (2002), Knight (1935), Korb (2001), Layher et al. (1986), Levell (1997), Livezey and Wright (1947), Loomis (1956), Mansueti (1941), Mecham (1954), Meshaka et al. (2009), Miller (1985a), Miller and Gress (2005), Moriarty and Collins (1995a), Mount (1975), Platt et al. (1974), Rundquist and Collins (1977), Schmidt (1953), Schwilling and Schaefer (1983), Simmons (1989), Smith (1932, 1933b, 1934, 1950, 1956a, 1957), Spanbauer (1988), Stebbins (1951), Stejneger and Barbour (1939, 1943), Stewart (1983), Taggart (2006b), Trauth et al. (2004), and Wright and Wright (1949).

Pickerel Frog
Lithobates palustris (LeConte, 1825)

Identification: Moist skin; round snout; distinct spots on back arranged approximately in 2 rows between raised folds or ridges of skin on each side of back and running from behind eyes down to thighs; bright yellow-orange color on undersides of thighs; head, body, and limbs tan or light brown, with dark brown or red-brown lightly outlined spots on head and back; hind limbs darkly banded; raised ridges of skin on each side of back may be light yellow, gray, or white; belly white; males distinguished from females by enlarged finger on each hand during breeding season; females heavier during breeding season.

Adults.

Adult.   Subadult.

Size: Adults normally 44–75 mm (1¾–3 inches) in SVL; largest Kansas specimen: female (KU 17471) from Cherokee County with a total length of 71 mm (2⅞ inches) collected by Edward H. Taylor and Hobart M. Smith on 25 March 1932; maximum length throughout range: 3⁷⁄₁₆ inches (Conant and Collins, 1998).

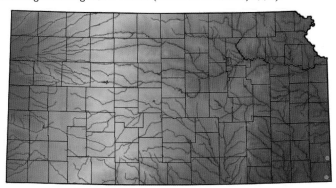

The **Pickerel Frog** (*Lithobates palustris*) is a rare frog with a distribution in Kansas that once was apparently restricted to southeastern Cherokee County in the extreme southeastern corner of state; a record from Crawford County (collected in 1911) needs verification it (may be result of accidental introduction during fish-stocking); reintroduced in Cherokee County in 1999 and 2000, but results of this repatriation effort are unknown. Red dots indicate records backed by voucher specimens or images.

Natural History: Prefers cool, clear streams and pools associated with limestone crevices and caves; within caves may remain active during the entire year; 2 of 3

Kansas examples found prior to introduction efforts (in 1999–2000) were in steep-sided deep pools of cool streams in Cherokee County during March and April; no breeding observations in Kansas have been made; probably emerges from winter inactivity in early spring and breeds during March and April; eggs laid and hatch in water; presumably tadpoles metamorphose during following summer; feeds primarily on terrestrial arthropods; has ability to secrete an irritating, somewhat toxic substance from skin, an adaptation that undoubtedly provides protection from potential predators; secretion is toxic to other amphibians and some reptiles; persons handling this anuran should be careful to keep their hands away from their mouth and eyes; maximum longevity: 2 years, 5 months, and 1 day (Snider and Bowler, 1992).

Pertinent References: Ballinger and Lynch (1983), Bartlett and Bartlett (2006a), Behler and King (1979), Clarke (1984), Collins (1974b, 1982a, 1990b), Collins and Collins (1993a), Collins and Taggart (2006), Conant (1958, 1975), Conant and Collins (1991, 1998), Cross (1971), Guarisco et al. (1982), Hillis (1985), Hunter et al. (1992), Johnson (1974), Karns et al. (1974), Kirk (2001a), Klemens (1993), Knight (1935), Korb (2001), Layher et al. (1986), Loraine (1983), Mansueti (1941), Minton (2001), Moriarty and Collins (1995a), Mount (1975), Platt et al. (1974), Pope (1944b), Rundquist and Collins (1977), Schaaf and Smith (1970, 1971), Smith (1932, 1933b, 1934, 1950, 1956a, 1957), Spanbauer (1988), Stejneger and Barbour (1939, 1943), Taggart (2006b), Thompson and Grigsby (1971), Trauth et al. (2004), and Wright and Wright (1949).

Southern Leopard Frog
Lithobates sphenocephalus (COPE, 1886)

Adult: Elk County.

Adult: Neosho County.

Adult in defensive posture: Linn County.

Eggs: Cherokee County.

Identification: Moist skin; round snout; irregular pattern of distinct, scattered, sometimes elongate spots on back and sides with few or no flecks or reticulate patterns between them; unbroken, raised fold or ridge of skin on each side of back running from behind eyes down to thighs; head, body, and limbs green, greenish brown, or brown, but some green coloration normally present on dorsum; spots dark gray, brown, or black; hind limbs darkly banded; raised fold or ridge on each side of back yellow or tan; belly white; adult females grow larger than males.

Size: Adults normally 51–90 mm (2–3½ inches) in SVL; largest Kansas specimen: female (KU 9462) from Montgomery County with a total length of 87 mm (3¼ inches) collected by Theodore E. White and Edward H. Taylor in August 1926; maximum length throughout range: 5 inches (Conant and Collins, 1998).

Natural History: Found in marshes, swamps, lakes, ponds, sloughs, rivers, and creeks; during summer, may wander great distances from water; evidently has no home range or territory; recorded active annually in Kansas from mid-February to late October, with a peak from late March through May; may be active during winter in proximity of caves; retreats into mud and leaves of lake and stream bottoms during cold winter months and remains inactive until advent of favorable spring temperatures; in Kansas, chorusing recorded from late February to mid-July (with a peak from late April through May); most robust choruses recorded from 8:00 pm to 2:00 am; during breeding season, males chorus to attract females, mount them, and clasp them behind their front limbs; females may lay up to 5,000 eggs in numerous round clumps; rate of development until hatching is dependent on water temperature and may take up to 3 weeks; tadpoles metamorphose during late summer; feeds on wide variety of insects and other small invertebrates; occasionally hybridizes with Plains Leopard Frog.

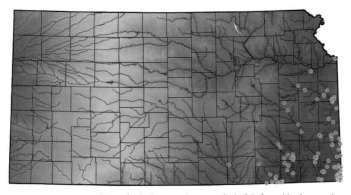

The sleek **Southern Leopard Frog** (*Lithobates sphenocephalus*) is found in the southeastern part of the state, from Wyandotte County in the northeast thence southwestward to Elk County in the south; the record for Cowley County in the Arkansas River drainage needs corroboration. Red dots indicate records backed by voucher specimens or images.

Pertinent References: Anderson and Arruda (2006), Anderson et al. (1995), Axtell (1976), Ballinger and Lynch (1983), Bartlett and Bartlett (2006a), Behler and King (1979), Boyd (1991), Busby and Pisani (2007), Caldwell (1986), Clarke (1984), Cochran and Goin (1970), Collins (1959, 1974b, 1982a, 1985, 1990b), Collins and Collins (1993a), Collins and Taggart (2006), Collins et al. (1994), Conant (1958, 1975), Conant and Collins (1991, 1998), Cross (1971), Dunlap and Kruse (1976), Gier (1967), Gloyd (1928, 1932), Gray (1979), Grow (1976b), Guarisco (1979b), Hall and Smith (1947), Hillis (1981, 1985), Hillis and Wilcox (2005), Johnson

(2000), Kingsbury and Gibson (2002), Knight (1935), Loomis (1956), Minton (2001), Moriarty and Collins (1995a), Mount (1975), Murrow (2008a), Pace (1974), Perry (1975), Riedle (1994a), Rundquist and Collins (1977), Sage and Selander (1979), Schmidt and Murrow (2007c), Smith (1933b, 1934, 1950, 1956a), Stebbins (1954), Stejneger and Barbour (1943), Taggart (2000b, 2005), Trauth et al. (2004), Wassersug (1976), and Wright and Wright (1949).

MICROHYLID FROGS AND TOADS
FAMILY MICROHYLIDAE GUNTHER 1843

The family Microhylidae (Microhylid Frogs and Toads) in Kansas is represented by 2 species assigned to 1 genus: the Eastern Narrowmouth Toad (*Gastrophryne carolinensis*) and the Great Plains Narrowmouth Toad (*Gastrophryne olivacea*). In Kansas, this genus is characterized by its small size, narrow, pointed snout, and presence of a fold of skin across the back of its head.

Eastern Narrowmouth Toad
Gastrophryne carolinensis (HOLBROOK, 1836)

 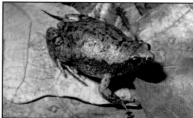

Adults: Cherokee County, Kansas.

Adult: Cherokee County, Kansas.   Eggs: Cherokee County, Kansas.

Identification: Moist skin; fold of skin across back of head behind eyes; very pointed snout; small head compared to rest of body; spotted or mottled belly; head, limbs, and body range from uniform brown to reddish brown; narrow light stripe down back may or may not be present; belly whitish with dark gray spots or mottling; females grow larger than males; males differ from females by having dark throats.

Size: Adults normally 22–32 mm (⅞–1¼ inches) in SVL; largest Kansas specimen: male (KU 218746) from Cherokee County with a total length of 37 mm (1⁷⁄₁₆ inches) collected by Travis W. Taggart, Shane Eckhardt, Kelly J. Irwin, and Joseph T. Collins on 28 September 1991; maximum length throughout range: 1½ inches (Conant and Collins, 1998).

Color and pattern comparison of adult Eastern Narrowmouth Toad from Cherokee County, Kansas (**right**) and adult Great Plains Narrowmouth Toad from western Kansas (**left**).

Natural History: Spends much of year in loose, moist soil beneath large rocks and logs shaded by trees in vicinity of temporary woodland pools, ponds, and ditches; recorded active annually in Kansas from late April to late September, with a peak from late April to mid-May; in Kansas, apparently restricted to suitable habitat along Spring River and its larger tributaries in Cherokee County; late spring and early summer breeder in clear temporary pools or ditches that are free of predators and have lush vegetation; chorusing recorded from late May to late July, with a peak during early June; males chorus to attract females to water, mount them, and clasp them securely behind front limbs; female lays eggs in water, where they are fertilized by male; eggs (up to 850 per female) form surface mass in water and quickly hatch into small tadpoles which metamorphose within 2 months; feeds almost exclusively on ants; threatened species (protected by state law); Water Snakes (*Nerodia*) and Western Ribbon Snakes are possible predators; maximum longevity: 6 years, 9 months, and 16 days (Snider and Bowler, 1992).

The **Eastern Narrowmouth Toad** (*Gastrophryne carolinensis*) is a small amphibian with a distribution in Kansas restricted to eastern Cherokee County in the extreme southeastern corner of the state. Red dots indicate records backed by voucher specimens or images.

Pertinent References: Anderson and Arruda (2006), Ballinger and Lynch (1983), Bartlett and Bartlett (2006a), Behler and King (1979), Clarke (1984), Collins (1959, 1974b, 1980b, 1982a, 1982b, 1990b, 1994b), Collins and Collins (1993a), Collins et al. (1994, 1995), Conant (1958, 1975), Conant and Collins (1991, 1998), Cross (1971), Freeman and Kindscher (1992),

Guarisco et al. (1982), Karns et al. (1974), Kingsbury and Gibson (2002), Layher et al. (1986), Levell (1997), Loomis (1956), Miller (1985a, 1991), Miller and Gress (2005), Moriarty and Collins (1995a), Mount (1975), Nelson (1972a, 1972c, 1973b), Platt et al. (1974), Porter (1972), Rundquist and Collins (1977), Rush and Fleharty (1980), Schmidt (1953), Schwilling and Schaefer (1983), Simmons (1989), Smith (1933a, 1933b, 1934, 1947b, 1950, 1956a, 1957), Spanbauer (1988), Taggart (1992a, 2006b), and Trauth et al. (2004).

Great Plains Narrowmouth Toad
Gastrophryne olivacea (HALLOWELL, 1856)

Identification: Moist skin, fold of skin across back of head behind eyes; very pointed snout; small head compared to rest of body; unspotted belly; head, body, and limbs uniform gray or light tan; belly whitish; females grow larger than males; males have dark throats.

Adult: Douglas County, Kansas.   Adult: Sumner County, Kansas.

Adult   Adult with aberrant color and pattern:
                          Wilson County, Kansas.

Size: Adults normally 22–38 mm (⅞–1½ inches) in SVL; largest Kansas specimen: female (MHP 9099) from Lincoln County with a total length of 43 mm (1¹¹⁄₁₆ inches) collected by Curtis J. Schmidt and Richard Hayes on 13 July 2004; exceeds maximum length throughout range, as reported in Conant and Collins (1998).

Natural History: Commonly found beneath flat limestone rocks (which have good drainage, loose soil, and few twigs or leaves) on open grassy slopes during April; also inhabits dry, rocky, upland area in open woods or woodland edge but tolerant of wide variety of habitats, including river floodplains and cultivated fields; where surface rocks are not available to hide under, may use rodent burrows; very secretive and spends most of life beneath ground; much less tolerant of cold temperatures than other amphibians in Kansas; recorded active annually in Kansas from mid-March to early November, with a peak from early May to early July; nocturnal, preferring to emerge from beneath ground on humid or rainy nights; opportunistic breeder; sufficient rainfall and optimal temperatures from early April and late July (with a peak in June) will initiate breeding; males may breed twice in one season; after rain, migrates to temporary pools where males begin to chorus, normally at night (most robust chorusing recorded from 9:00 pm to midnight); male mounts female and clasps her behind front legs with his forelimbs; egg-laying occurs 1–2 days after frogs have reached breeding pool; single female may produce up to 600 eggs, which hatch in 2 days; free-swimming tadpoles metamorphose into adults in 20–30 days and disperse from breeding site at first rainfall; sexual maturity reached within 1–2 years; feeds almost exclusively on ants; predators include shrews, larger frogs, and snakes.

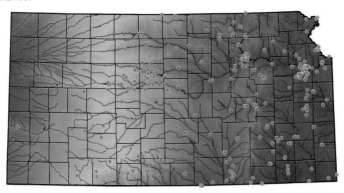

The **Great Plains Narrowmouth Toad** (*Gastrophryne olivacea*) is a widespread anuran that occurs throughout the eastern three-quarters of the state; it is absent from the western fourth of Kansas. The single record for Hamilton County (collected in 1955) near the Colorado border needs corroboration. Red dots indicate records backed by voucher specimens or images.

Pertinent References: Anderson (1942), Ballinger and Lynch (1983), Bartlett and Bartlett (2006a), Behler and King (1979), Boyd (1991), Brennan (1934, 1937), Breukelman and Clarke (1951), Breukelman and Downs (1936), Busby and Parmelee (1996), Busby and Pisani (2007), Busby et al. (1994, 1996, 2005), Caldwell and Glass (1976), Clarke (1980a, 1984), Clarke et al. (1958), Collins (1959, 1974b, 1982a, 1985, 1990b, 2007b, 2009a), Collins and Caldwell (1978), Collins and Collins (1993a), Collins et al. (1994), Conant (1958, 1975), Conant and Collins (1991, 1998), Cross (1971), Dice (1923), Ellis (2002), Fitch (1951, 1956a, 1956d, 1956e, 1958b, 1982, 1991), Freiburg (1951), Fraser (1983a), Gier (1967), Gish (1962), Gomez (2008a), Gress (2003), Guarisco (1981c), Hall (1968), Hammerson (1999), Hecht and Matalas (1946), Heinrich and Kaufman (1985), Karns et al. (1974), Kingsbury and Gibson (2002), Knight (1935), Linsdale (1925, 1927), Livezey and Wright (1947), Loomis (1945, 1956), Low (2008), Miller (1980a, 1981d, 1982a, 1985b, 1996a), Moriarty and Collins (1995a), Murrow (2008a, 2008b, 2009a, 2009b), Nelson (1972b, 1972c, 1973a, 1973b), Porter (1972), Reichman (1987), Riedle (1994a), Riedle and Hynek (2002), Roth (1959), Rundquist and Collins (1977), Schmidt (1953), Schmidt and Hayes (2004), Schmidt and Murrow (2006, 2007c), Schmidt and Taggart (2002a), Schwarting (1984), Schwilling and Schaefer (1983), Simon (1988), Smith (1933a, 1933b, 1934, 1950, 1956a), Stebbins (1951, 1954, 1966, 1985, 2003), Stejneger and Barbour (1933, 1939, 1943), Stuart and Painter (1996), Suleiman (2005), Taggart (1992a, 2001a, 2003g, 2004h, 2008), Taggart and Schmidt (2005c), Tanner (1950), Tihen (1937), Tihen and Sprague (1939), Toepfer (1993a), Trauth et al. (2004), Turner (1960), Van Doren and Schmidt (2000), Viets (1993), Wagner (1984), Warner and Wencel (1978), Wilgers and Horne (2006), Wilgers et al. (2006), Wright (1931), and Wright and Wright (1949).

AMPHIBIANS

REPTILES
CLASS REPTILIA LAURENTI, 1768

Lizards and snakes (and a tropical group called amphisbaenians, which are not native to Kansas) are members of the Class Reptilia, which consists in Kansas of 13 families (2 are not native to the state), 35 genera (3 are not native), and 54 species (3 are not native). Twenty native families in the reptilian Order Squamata are represented in the United States and Canada by 80 genera and 274 species (Collins and Taggart, 2009, with updates from the CNAH web site). The other reptilian Order, Rhynchocephalia (the lizard-like Tuataras), is restricted to islands off the coast of New Zealand.

Kansas reptiles, unlike amphibians, have skin consisting of dry scales and most lizards have claws on their toes and fingers. Unlike turtles, reptiles do not have shells. Reptiles are primarily terrestrial, and can stray far from water because their skin is better able to contain body moisture; thus they are much less likely to dry up and die than amphibians. Unlike amphibians in Kansas, male reptiles fertilize females internally with a paired copulatory organ during mating (unlike turtles, which have a single copulatory organ). Furthermore, reptiles do not return to water or moist areas to bear their young or lay eggs. Approximately 86% of the species of turtles in Kansas are semi-aquatic, whereas less than 8% of the species of reptiles (Crayfish Snakes and Water Snakes) spend much time in water.

Reptiles have 2 methods of reproduction. All Kansas lizards plus some snakes lay eggs on land. The eggs, unlike those of amphibians, have shells that prevent the loss of moisture. In addition, some snakes forego egg-laying and give direct birth to young, a mode of reproduction that eliminates the need for egg concealment and the hazards associated with predation on them. Young reptiles hatch or are born as small duplicates of adults. Some may have a different pattern and color, but all generally look like their parents. Reptiles do not have a larval or tadpole stage.

The 2 main groups of reptiles found in Kansas can be easily distinguished from each other. Lizards have eyelids that shut, and all but 1 species (the Western Slender Glass Lizard) have front and hind limbs. Like all snakes, those in Kansas have no limbs and their eyelids are immovable.

LIZARDS
ORDER SQUAMATA OPPEL, 1811

Lizards (along with snakes) are members of the Order Squamata. Lizards are a highly variable group that in the United States and Canada consists of 11 families, 26 genera and 119 native species (Collins and Taggart, 2009, with updates from the CNAH web site). The 13 native species of lizards known to occur in Kansas represent 5 families and 8 genera; 3 non-native, alien species in 2 families and 3 genera have been introduced in Kansas. Thus, the lizard fauna of Kansas now consists of 16 species.

Although Kansas has only 13 native species of lizards, these are very abundant in the state. Eight of these species have a wide distribution in Kansas, and the remaining 4 are found in the eastern third of the state. The non-native Italian Wall Lizard, Western Green Lacerta, and Mediterranean Gecko have only been documented in urban areas.

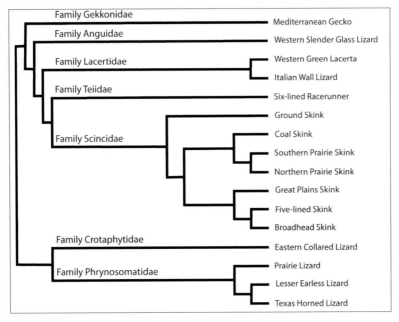

A phylogeny of Kansas Lizards (Order Squamata) that is consistent with evolutionary history based on current scientific evidence.

GECKOS
FAMILY GEKKONIDAE GRAY, 1825

Within the United States, the family Gekkonidae (Geckos) is composed of 14 genera and 24 species, all non-native except for 6 species. A single alien lizard of this family, the Mediterranean Gecko (*Hemidactylus turcicus*), was introduced and established in Kansas, probably from individuals transported north on trucks from a southern state. These lizards are characterized by being nocturnal, and having toe pads that enable them to crawl up the walls of buildings and across ceilings.

Mediterranean Gecko
Hemidactylus turcicus (LINNAEUS, 1758)

Identification: Not native to Kansas; four limbs; ear opening present on each side of head; pupils of eyes become vertical slits when exposed to light; head, neck, and upper body pink to pale yellow with small irregular brown spots; body covered with tubercles and appears very warty; females and males similar in size.

Belly of adult female with eggs: Johnson County, Kansas.

Adult: Johnson County, Kansas.

Adult

Adult: Johnson County, Kansas. Note stump tail; lizard will grow a new one, but not as long as the original.

Size: Adults normally 75–110 mm (4–4⅜ inches) in TL; largest specimen from Kansas: female (MHP 12335) from Johnson County with SVL of 56 mm and TL of 115 mm (4⁹⁄₁₆ inches) collected by Travis W. Taggart, Dan Murrow, and Chad Whitney on 16 June 2006; maximum length throughout range: 5 inches (Conant and Collins, 1998).

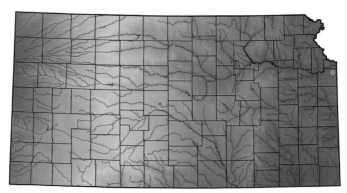

The **Mediterranean Gecko** (*Hemidactylus turcicus*) is a non-native lizard that was introduced in the state within the city limits of Lenexa, Johnson County, where it has apparently established a breeding colony; its natural range is Eurasia and northern Africa. Red dots indicate records backed by voucher specimens or images.

Natural History: In urban Lenexa, Kansas, found on buildings, particularly those lighted at night when lizard is active and foraging on walls from dusk to midnight; will quickly retreat into nearest crack when chased; generally active on outside of buildings from April to October, temperatures permitting; apparently spends rest of year inside buildings; males territorial; mating occurs opportunistically throughout warm season; females have multiple clutches of 2 eggs each throughout warm season; feeds on wide variety of insects; predators probably include domestic cats and birds; introduced colony in Lenexa discovered in 2005, probably from individuals transported north on trucks from Florida, Texas, or Mexico.

Pertinent References: Hare (2006), Jadin and Coleman (2007), and Kraus (2009).

COLLARED AND LEOPARD LIZARDS
FAMILY CROTAPHYTIDAE SMITH AND BRODIE, 1982

The family Crotaphytidae (Collared and Leopard Lizards) is restricted to North America and consists of 2 genera and 8 species (Collins and Taggart, 2009). One genus, with a single species, the Eastern Collared Lizard (*Crotaphytus collaris*), represents this family in Kansas. It is characterized by a large head, robust body, an ability to run on 2 hind legs, and its large size (up to 350 mm).

Eastern Collared Lizard
Crotaphytus collaris (SAY, 1823)

Adult male: Wabaunsee County, Kansas.

Adult male: Barber County, Kansas.

Adult male: Riley County, Kansas.

Adult male (**lower**) and female: Russell County, Kansas.

Adult male: Kiowa County, Kansas.

Juvenile: Barber County, Kansas.

REPTILES

Identification: Four limbs; ear opening present on each side of head; smooth granular scales on back; 1 or 2 dark bands on neck; body, large head, limbs, and tail are bluish green or green in males and light brown in females; both sexes have small dark and light spots; belly white; throat yellow or orange in males; males have larger heads, grow to larger size, and more brightly colored than females; during pregnancy, females have orange-red spots or bars on sides of body.

Size: Adults normally 203–306 mm (8–12 inches) in TL; largest specimen from Kansas: male (KU 84623) from Chase County with an SVL of 110 mm and TL of 302 mm (12 inches) collected by Charles J. Cole on 27 August 1963; maximum length throughout range: 14 inches (Conant and Collins, 1998).

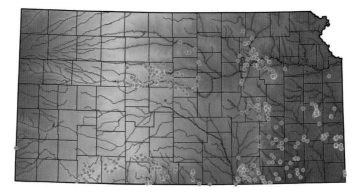

The **Eastern Collared Lizard** (*Crotaphytus collaris*) is found throughout much of the state in three widely scattered populations: (1) the eastern third of Kansas except the northeastern corner (north of the Kansas River), (2) north-central Kansas in the Saline, Soloman, and Smoky Hill river drainages, and (3) south-central Kansas along the Oklahoma border (west of the Chikaskia River). This lizard is absent from many northern border counties and the western quarter of the state. The single record for Johnson County in the northeast needs corroboration. Red dots indicate records backed by voucher specimens or images.

Natural History: Restricted to rocky areas on plains and near woodlands; adults recorded active annually in Kansas from late March to early November, with a peak from early May to late July; active only during day, perching on rocks to bask in sun and watch for enemies; when disturbed, may run with front feet held off ground, using strong hind limbs to race to burrows or crevices beneath large rocks; territorial, with home range of up to several thousand square feet; males will defend territory from other males; burrows are small chambers 8–12 inches below ground beneath large rocks with small, tunnels leading to edge of each rock; breeds during May and June after emerging from winter inactivity; courtship involves male displaying bright throat and body colors to female while he struts about her; after displaying, male moves around and over female in relaxed, slithering walk, rapidly bobbing head up and down; if female is receptive, male grasps skin of her neck with his mouth, bends rear part of his body under her tail, and copulation occurs; after becoming pregnant, female develops orange-red spots or bars on sides; female seeks nest of moist earth beneath large rock or in burrow; 2–11 eggs laid within 3 weeks after mating; hatching occurs in 2–3 months; females frequently have second clutch each season; hunts food by sight, running swiftly from perches to grab observed prey; grasshoppers, moths, beetles, spiders, wasps, and cicadas eaten, as well as smaller lizards; 95% of food in Kansas specimens consisted of grasshoppers and moths; chief predators hawks and snakes.

Pertinent References: Anderson (1965), Anderson et al. (1995), Ballinger and Lynch (1983), Bartlett and Bartlett (2006b), Behler and King (1979), Blair and Blair (1941), Brennan (1934, 1937), Breukelman and Clarke (1951), Breukleman and Downs (1936), Breukelman et al. (1961), Burt (1927b, 1928a, 1928b, 1928c, 1929b, 1933, 1935), Burt and Burt (1929b, 1929c), Burt and Hoyle (1934), Busby and Parmelee (1996), Busby and Pisani (2007), Busby et al. (1994, 1996, 2005), Campbell (2008), Capron (1975a, 1978b, 1986b), Capron et al. (1982), Clarke (1965a, 1980a), Clarke et al. (1958), Cole (1966a), Collins (1959, 1974b, 1982a, 1983, 1985, 1990b, 2006b, 2007b), Collins and Collins (1993a), Collins et al. (1991), Conant (1958, 1975), Conant and Collins (1991, 1998), Cooper and Ferguson (1972a, 1972b, 1973), Cope (1900), Cragin (1881), Cross (1971), Dice (1923), Diener (1957a), Ferguson (1976), Fitch (1956c, 1956e, 1958b, 1963b, 1967, 1970, 1985b, 1991, 2006a), Fitch and Tanner (1951), Fraser (1983a, 1983b), Garman (1884), Gish (1962), Gress (2003), Guarisco (1979a, 1980a, 1981c), Hall and Smith (1947), Hallowell (1857a, 1857b, 1859), Hammerson (1982, 1999), Hartman (1906), Heinrich and Kaufman (1985), Householder (1917), Hoyle (1936, 1937), Hurter (1897, 1911), Hutchison and Larimer (1960), Ingram and Tanner (1971), Irwin (1983), Johnson (2000), Karns et al. (1974), Kingsbury and Gibson (2002), Knight (1935), Knight et al. (1973), Lee (1967), Legler and Fitch (1957), Loomis (1956), Malaret (1979, 1985), McAllister (1983, 1985), McGuire (1996), Mielke (2002), Miller (1976a, 1981a, 1983a, 1985b, 1988, 1996a), Montanucci (1974), Montanucci et al. (1975), Murrow (2008a, 2009a, 2009b), Perry (1974a, 1975), Platt (2003), Pleshkewych (1964), Reichman (1987), Riedle (1994a), Robison and Tanner (1962), Roth (1959), Rundquist (1992), Ruthven (1907), Schmidt and Murrow (2006, 2007b), Simmons (1987c), Skie and Bickford (1978), Smith (1946c, 1950, 1956a), Smith and Tanner (1974), Sperry (1960), Sprackland (1985, 1993), Stebbins (1954, 1966, 1985, 2003), Stejneger and Barbour (1933, 1939, 1943), Suleiman (2005), Taggart (2000a, 2002d, 2005, 2008), Taylor (1916, 1993), Tihen (1937), Tihen and Sprague (1939), Toepfer (1993a), Trauth (1978), Trauth et al. (2004), Viets (1993), Wells (1997), Werth (1969, 1972, 1974), Wilgers et al. (2006), Wolfenbarger (1951, 1952), Yarrow (1883), Yedlin and Ferguson (1973), and Young and Thompson (1995).

SAND AND SPINY LIZARDS
FAMILY PHRYNOSOMATIDAE FITZINGER, 1843

The family Phrynosomatidae (Sand and Spiny Lizards) consists of 9 genera and 46 species in the United States and Canada (Collins and Taggart, 2009). Three genera, each with a single species, represent the family in Kansas: the Lesser Earless Lizard (*Holbrookia maculata*), Prairie Lizard (*Sceloporus consobrinus*), and Texas Horned Lizard (*Phrynosoma cornutum*). These lizards are cryptically colored and patterned. In Kansas, 1 genus (*Sceloporus*) is semi-arboreal (in the eastern quarter of the state) and terrestrial (in the western three-quarters), 2 genera possess spiny scales, and all 3 are active, diurnal, and small (up to 165 mm TL).

Lesser Earless Lizard
Holbrookia maculata GIRARD, 1851

Identification: Four limbs; lacks an ear opening on each side of head; head, body, limbs, and tail light gray or gray-brown; upper surface of neck, back, and base of tail

REPTILES

81

covered with 9–14 dark brown spots; belly grayish and unmarked except for 2 or 3 blue-bordered, short, black bars, which barely extend onto sides; adult males have grayer throats than females; pregnant females develop orange coloration on sides.

Size: Adults normally 100–130 mm (4–5⅛ inches) in TL; largest specimen from Kansas: female (MHP 6675) from Trego County with SVL of 60 mm and TL of 122 mm (4⅝ inches) collected by Dan E. Hesket on 22 March 1986; maximum length throughout range: 5⅛ inches (Conant and Collins, 1998).

Adult: Sumner County, Kansas.

Adult: Chase County, Kansas.

Adult: Sumner County, Kansas.

Adult female: Cheyenne County, Kansas.

Adult: Sumner County, Kansas.

Adult male: Sumner County, Kansas.

Natural History: Restricted to flat, sandy, cultivated, clay, or gravel areas of loose soil with little or no vegetation; distribution in eastern Kansas spotty, because continuous habitat for lizard apparently does not occur; often abundant around prairie dog towns, but also found in sandy areas around impoundments and along rivers; recorded active annually in Kansas from late March to late October, with a peak from mid-May to mid-July; spends colder months beneath ground to avoid adverse temperatures; active only during day, basking and foraging for food; like many lizards, has home range and is territorial; home range may contain numerous individuals of both sexes, but 1 male is normally dominant; males display dominance by executing "push-ups" or "bobbing" in distinct cadence; courting male rapidly nods head upon approaching female and may nudge her on side or beneath tail with his nose; male then grasps receptive female by loose skin between her shoulders and curls rear of his body beneath her tail until their cloacae meet; copulation lasts at least 20 seconds; upon dismounting from female, male elevates rear of body and holds his tail up in an arching curve; females lay 1–10 eggs during May or June; eggs hatch in 1–2 months, depending on air temperatures during incubation; 75% of diet consists of grasshoppers and true bugs; species consumes large numbers of harmful insects and is beneficial to farmers; predators include birds, small mammals, snakes, and larger lizards.

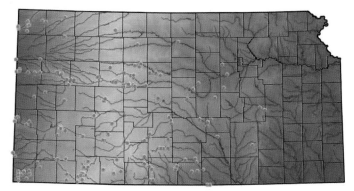

The **Lesser Earless Lizard** (*Holbrookia maculata*) is a small lizard found in the western two-thirds of the state. Red dots indicate records backed by voucher specimens or images.

Pertinent References: Axtell (1954, 1956, 1958), Ballinger and Lynch (1983), Behler and King (1979), Brennan (1934, 1937), Bartlett and Bartlett (2006b), Baxter and Stone (1980, 1985), Breukelman and Clarke (1951), Breukleman and Downs (1936), Burt (1928a, 1928b, 1933, 1935, 1936a), Burt and Hoyle (1934), Carpenter (1940), Clarke (1965a, 1965b, 1980a), Clarke et al. (1958), Cochran and Goin (1970), Collins (1959, 1974b, 1982a, 1985, 1990b), Collins and Caldwell (1978), Collins and Collins (1991, 1993a), Conant (1958, 1975), Conant and Collins (1991, 1998), Cope (1900), Cragin (1881), Crampton (1983), Cross (1971), Ditmars (1915), Ellis and Henderson (1913), Fitch (1970, 1985b), Gier (1967), Gish (1962), Guarisco (1980a), Hammerson (1982, 1999), Hartman (1906), Householder (1917), Hoyle (1936), Johnson (1987), Jones and Ballinger (1987), Karns et al. (1974), Kingsbury and Gibson (2002), Knight (1935), Knight and Collins (1977), Kretzer and Cully (2001), Lardie and Black (1981), Marr (1944), Miller (1980a, 1981d, 1982a), Murrow (2008b, 2009a, 2009b), Nickerson and Krager (1973), Platt (1975, 1984a, 2003), Royal (1982), Rush (1981), Schmidt (1922, 1953, 2001, 2004l), Schoewe (1937), Schwaner (1978), Schilling and Schaefer (1983), Smith (1946c, 1950, 1956a, 1961), Stebbins (1954, 1966, 1985, 2003), Taggart (1992b, 2000a, 2006a, 2006b, 2006c, 2007c, 2009a), Taggart et al. (2007), Taylor (1916, 1929b, 1993), Tihen (1937), Tihen and Sprague (1939), Tinkle (1972), Toepfer (1993a), Trott (1983), Werth (1969, 1972), and Wooster (1925).

Texas Horned Lizard
Phrynosoma cornutum (HARLAN, 1825)

Identification: Four limbs; ear opening present on each side of head; extremely rough, raised scales on body, large spines projecting out from back of head; general color ranges from yellowish brown to reddish brown; dark brown blotch present on each side of neck; series of dark spots present on each side of back and separated by yellow or white line; belly white or yellow with a few small gray spots; external sexual differences few, but females grow slightly larger than males.

REPTILES

Adult: Sumner County, Kansas.

Adult: Comanche County, Kansas.

Adult: Comanche County, Kansas.

Adult: Ellsworth County, Kansas.

Adult: Stanton County, Kansas.

Juvenile: Comanche County, Kansas.

Size: Adults normally 64–100 mm (2½–4 inches) in TL; largest specimen from Kansas: female (MHP 7469) from Stevens County with SVL of 90 mm and a TL of 123 mm (4⅞ inches) collected by Travis W. Taggart and Curtis J. Schmidt on 31 May 2002; maximum length throughout range: 7⅛ inches (Conant and Collins, 1998).

Natural History: Generally inhabits dry, flat areas with sandy, loamy, or rocky surfaces and little vegetation; recorded active annually in Kansas from early April to early October, with a peak from early May to late July; strictly diurnal, spending day basking in sun, foraging for food, or hiding just below soil surface; body coloration makes it difficult to observe; little known of daily cycle, but temperature preference may be higher than that of many other lizard species; not territorial; although well known to people in Kansas, few observations have been made on its breeding habits; males apparently emerge from winter inactivity earlier than females; mating probably occurs no earlier than May or June; courtship unknown; females probably lay single clutch of eggs (average 23) per season, in nest dug in loose soil or under rocks; incubation requires 1–2 months; ants make up major part of diet, but other small insects and spiders also eaten; difficult to maintain in captivity and most individuals die from improper care; when disturbed, sometimes squirts small stream or several drops of blood from eyes [this ability presumed to be a defensive mechanism because blood is considered offensive to a potential predator, but even most recent evidence to support this idea is suspect; alternative explanation is that behavior occurs because of handling; these lizards shunt blood to achieve optimal overall temperature when head (exposed to sun) reaches higher temperature than that of body (buried beneath sand); when handled or touched under these conditions, any slight pressure may induce an overheated lizard to squirt blood; researchers have never observed this lizard squirting blood unless it was being handled or manipulated in some way]; maximum longevity: 1 year, 1 month, and 24 days (Snider and Bowler, 1992).

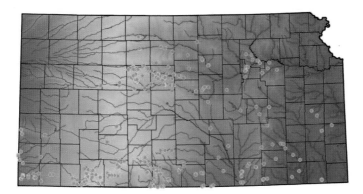

The **Texas Horned Lizard** (*Phrynosoma cornutum*) is a spiny reptile that is abundant in the two tiers of counties along the Oklahoma border (west of Chautauqua County) and in the Saline, Soloman, and Smoky Hill drainages west of the Flint Hills. Elsewhere it is confined to isolated local populations. Red dots indicate records backed by voucher specimens or images.

Pertinent References: Anderson (1965), Ballinger (1974), Ballinger and Lynch (1983), Bartlett and Bartlett (2006b), Behler and King (1979), Bender and Stark (2007), Brennan (1934, 1937), Brest van Kempen (2006), Breukelman and Clarke (1951), Breukleman and Downs (1936), Burt (1928a, 1928b, 1933, 1935, 1936a), Burt and Hoyle (1934), Busby and Parmelee (1996), Busby et al. (1994, 1996, 2005), Caldwell and Collins (1977), Capron (1975a, 1978b), Capron et al. (1982), Carpenter (1940), Carpenter and Vaughn (1992), Clarke (1965a, 1980a), Clarke et al. (1958), Cochran and Goin (1970), Collins (1959, 1974b, 1981, 1982a, 1985, 1990b, 1994a, 2006b, 2007b, 2009b), Collins and Collins (1991, 1993a), Collins et al. (1991, 2002), Conant (1958, 1975), Conant and Collins (1991, 1998), Cook (1942), Cope (1900), Cragin (1881), Cross (1971), Dice (1923), Ellis and Henderson (1913), Fitch (1985b), Fleharty (1995), Gish (1962), Gress (2003), Gress and Potts (1993), Guarisco (1980a, 1981c), Hall and Smith (1947), Hallowell (1857a, 1857b, 1859), Hammerson (1999), Hartman (1906), Heinrich and Kaufman (1985), Householder (1917), Hoyle (1936), Hurter (1911), Hvass (1964), Irwin (1983), Johnson (1974, 1986, 2000), Jordan (1929), Karns et al. (1974), Kingsbury and Gibson (2002), Knight (1935), Knight and Collins (1977), Kraus (2009), Kretzer and Cully (2001), Legler (1960b), Martin (1960), Martof et al. (1980), Miller (1976a, 1980a, 1981c, 1981d, 1982a, 1985b, 1988), Murrow (2008b, 2009a, 2009b), Owens et al. (2002), Platt (1975, 2003), Platt et al. (1974), Pope (1956), Price (1990), Reeve (1952), Reichman (1987), Roth (1959), Royal (1982), Rundquist (1992), Sattler and Ries (1995), Schmidt (1953, 2004d), Schmidt and Murrow (2007b), Sherbrooke (1981), Simmons (1987c), Sites et al. (1992), Smith (1946c, 1950, 1956a), Stebbins (1954, 1966, 1985, 2003), Stejneger and Barbour (1917, 1923, 1933, 1939, 1943), Suleiman (2005), Taggart (1992a, 2000a, 2006b, 2002c, 2008), Taggart and Schmidt (2005a, 2005e), Taylor (1916, 1929b, 1993), Tihen (1937), Tihen and Sprague (1939), Toepfer (1993a), Trauth et al. (2004), Viets (1993), Wilgers and Horne (2006), Wilgers et al. (2006), Wooster (1925), and Yarrow (1883).

Prairie Lizard
Sceloporus consobrinus BAIRD & GIRARD, 1853

Juvenile: Russell County, Kansas.

Adult: Clark County, Kansas.

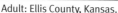
Adult: Ellis County, Kansas.

Adult: Cherokee County, Kansas.

Identification: Four limbs; ear opening present on each side of head; rough, raised scales on body; lack of horns sticking out of back of head; head, body, limbs, and tail vary from gray to brown, with pattern of narrow, dark irregular crossbands, dark spots, or dark and light stripes on back; belly and throat gray-white, sometimes with bright bluish green or blue patches on each side (patches more brilliant in males and indistinct or absent in females); females grow larger than males; 2 distinct variants of this lizard in Kansas (with no evidence of gene exchange between them), 1 in southeastern forests along Oklahoma border from Chautauqua to Cherokee counties (and possibly single record to north in Johnson County) and another throughout rest of state; they are easily distinguished, as follows:

Southeastern Kansas: irregular, transverse dark crossbands on back; males with bright blue, black-bordered patches on belly and throat.

Remainder of Kansas: 2 distinct light stripes running down each side of back; dark spots on back between light stripes; belly and throat of males with pale blue-green indistinct patches.

Size: Adults normally 90–180 mm (3½–7 inches) in TL; largest specimen from Kansas: sex undetermined (MHP 4254) from Cherokee County with SVL of 75 mm and TL of 165 mm (6⅝ inches) collected by L. H. Panks on 29 April 1967; maximum length throughout range: 7½ inches (Conant and Collins, 1998).

Natural History: Lives in wide variety of habitats. Prefers dry, open forests in eastern Kansas and readily climbs trees to escape enemies; west of Flint Hills, inhabits low, open, sandy regions and frequently found along sandstone and limestone outcrops; recorded active annually in Kansas from early March to early November, with a peak from late April to early July; during winter months, burrows beneath ground to avoid cold temperatures; active during day; in western Kansas, has small home range of ⅒ acre; males generally have harem of 2–3 females; when disturbed near trees, climbs quickly, always keeping trunk between it and disturbance; spends much time basking on logs and rocks; breeding occurs during warm months, from May to August; females probably produce 2–3 clutches of eggs per season; courtship involves male quickly approaching female, mounting her, and curling hindquarters beneath her tail until their cloacae meet and copulation occurs; average of 7 eggs laid in nests beneath ground, in loose soil, and hatch in about 2 months; primarily feeds on wide variety of insects but also eats spiders and other invertebrates; predators include larger lizards, snakes, birds, and small mammals; maximum longevity: 1 year, 1 month, and 4 days (Snider and Bowler, 1992).

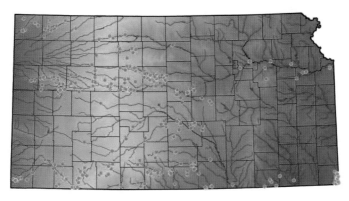

The **Prairie Lizard** (*Sceloporus consobrinus*) is widespread throughout much of the western two-thirds of Kansas, but is inexplicably absent from the Arkansas River valley in the southwestern part of the state. It is absent from the eastern third of Kansas except for populations along the riparian corridor of the Kansas River in the north and the lower Verdigris and Spring river corridors along the Oklahoma border. Red dots indicate records backed by voucher specimens or images.

Pertinent References: Bell (1996), Ballinger et al. (1981), Ballinger and Lynch (1983), Bartlett and Bartlett (2006b), Baxter and Stone (1980, 1985), Behler and King (1979), Brennan (1934, 1937), Breukelman and Clarke (1951), Burt (1928a, 1928b, 1933, 1935, 1936a), Burt and Hoyle (1934), Busby et al. (1996, 2005), Carpenter (1940), Clarke (1965a, 1980a), Cochran and Goin (1970), Collins (1959, 1974b, 1982a, 1985, 1990b, 2006b), Collins and Collins (1991, 1993a), Conant (1958, 1975), Conant and Collins (1991, 1998), Cope (1900), Cragin (1881, 1885a, 1885d), Cross (1971), DeMarco et al. (1985), Derickson (1976a, 1976b), Ferguson and Bohlen (1978), Ferguson and Brockman (1980), Ferguson and Snell (1986), Ferguson et al. (1980, 1982, 1983), Fitch (1967, 1970, 1978, 1985b), Fleharty and Ittner (1967), Fleharty and Johnson (1975), Fuller (1966), Gier (1967), Gish (1962), Gray (1979), Guarisco (1981c), Hall and Smith (1947), Hammerson (1999), Hartman (1906), Heger and Sherrin (1991), Householder (1917), Hoyle (1936), Hudson (1942), Hurter (1911), Irwin (1983), Iverson (1976), Johnson (2000), Jones and Ballinger (1987), Jordan (1929), Karns et al. (1974), Kingsbury and Gibson (2002), Knight (1935), Knight and Collins (1977), Leaché and Reeder (2002), Legler (1960b), Lemos-Espinal et al. (1996), Loewen (1941), Loomis (1956), Marr (1944), Martin (1960), McCauley (1945), McCoy (1961), Miles et al. (2002), Miller (1976a, 1979, 1980a, 1981a, 1981c, 1985b), Miller and Williams (2002), Minton (1972, 2001), Mitchell (1994), Mount (1975), Nulton and Rush (1988), Perry (1974a), Platt (1975, 1984a, 1985b, 1998,

2003), Royal (1982), Rundquist (1975, 1992), Schmidt (1953, 2001, 2004l), Schmidt and Murrow (2007b), Schmidt and Taggart (2005a), Shaffer and McCoy (1991), Simmons (1987c), Smith (1938, 1946c, 1950, 1956a), Smith et al. (1991, 1992a, 1995, 2002), Sparks et al. (1999), Stebbins (1954, 1966, 1985, 2003), Stejneger and Barbour (1939, 1943), Suleiman (2005), Taggart (1992b, 2000a, 2000b, 2002c, 2004h, 2008), Taggart et al. (2005f, 2007), Taylor (1916, 1929b, 1965, 1993), Tihen (1937), Tihen and Sprague (1939), Tinkle and Dunham (1986), Tinkle and Hadley (1975), Toepfer (1993a), Trauth et al. (2004), Trott (1977, 1983), Tubbs and Ferguson (1976), Warner and Wencel (1978), Werth (1969, 1972), Wolfenbarger (1951, 1952) and Yarrow (1883).

SKINKS
FAMILY SCINCIDAE GRAY, 1825

The lizard family Scincidae (Skinks) in North America is composed of 2 genera and 15 species (Collins and Taggart, 2009). Seven species assigned to both genera represent this family in Kansas: the Ground Skink (*Scincella lateralis*) and 6 members of the genus *Plestiodon*. The family is characterized by being covered with smooth, shiny scales and having elongate, depressed bodies, small, flattened heads, and being generally small to moderate in size (up to 13⅜ inches in total length).

Coal Skink
Plestiodon anthracinus BAIRD, 1849

Identification: Four limbs; ear opening present on each side of head; smooth, flat, shiny scales on body; broad, dark brown, or black stripe on each side of body from eye onto tail bordered by a thin light stripe; no light stripes on head; back, head, limbs, and tail medium olive-brown to brown; belly gray; in males, chin and jaws orange-red during breeding season.

Adult male: Douglas County, Kansas.

Adult: Cherokee County, Kansas.

Adult: Cherokee County, Kansas.

Adult male: Douglas County, Kansas.

Size: Adults normally 125–178 mm (5–7 inches) in TL; largest specimen from Kansas: female (KU 88527) from Miami County with SVL of 59 mm and TL of 176 mm (7 inches) collected by Jack E. Joy in May 1949; maximum length throughout range: 7 inches (Conant and Collins, 1998).

Natural History: Inhabits leaf-littered rocky slopes of forests as well as floodplains; recorded active annually in Kansas from early March to early October, with a peak during April, but little known of its habits; retreats beneath ground during winter, presumably in crevices or burrows of other animals; breeds in spring; egg clutches laid in May or June and hatch in June or July; females protect eggs by curling on top of them; feeds on small insects; predators include small mammals, snakes, and larger lizards.

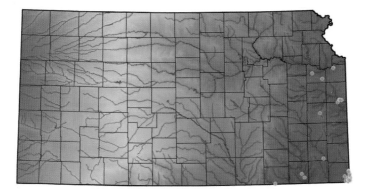

Isolated populations of the **Coal Skink** (*Plestiodon anthracinus*) are generally found in the forested areas of the eastern part of the state, south of the Kansas River and east of the Arkansas River drainage. Red dots indicate records backed by voucher specimens or images.

Pertinent References: Anderson (1965), Ballinger and Lynch (1983), Bartlett and Bartlett (2006b), Behler and King (1979), Burt (1928a, 1933, 1936a), Burrage (1962), Clarke (1965a), Cochran and Goin (1970), Collins (1959, 1974b, 1975, 1982a, 1982b, 1990b), Collins and Collins (1993a), Conant (1958, 1975), Conant and Collins (1991, 1998), Cross (1971), Fitch (1991), Fleharty and Ittner (1967), Fleharty and Johnson (1975), Gloyd (1928, 1932), Guarisco et al. (1982), Householder (1917), Irwin (1982b), Johnson (2000), Karns et al. (1974), Kingsbury and Gibson (2002), Knight (1935), Leviton (1971), Loomis (1956), McCauley (1945), Mitchell (1994), Mount (1975), Rush and Fleharty (1981), Schmidt (1953), Smith (1946a, 1946c, 1950, 1956a), Smith and Smith (1952b), Stejneger and Barbour (1933, 1939, 1943), Taggart (2006b, 2007b), Taylor (1916, 1936, 1993), Trauth (1994), Trauth et al. (2004).

Five-lined Skink
Plestiodon fasciatus (Linnaeus, 1758)

Identification: Four limbs; ear opening present on each side of head; flat, smooth, shiny scales covering body; a fifth scale (counting back from nose) on upper lip, which extends up to edge of eye; adults exhibit different colors at various life stages: newborns, juveniles, and young adults black with yellow stripes on back, sides, and head, and have bright blue tails; older females brownish with yellowish stripes that may fade to brown or gray and have gray tail; old males uniform olive or tan and lack stripes; during breeding season, heads of older males orange-red; adult males grow slightly longer than females.

Subadult male: Osage County, Kansas.

Juvenile: Cherokee County, Kansas.

Old adult male: Jefferson County, Kansas.

Adult female: Cherokee County, Kansas. Note stump tail, re-grown much shorter than the original.

Adult female guarding her eggs: Johnson County, Kansas.

Egg clutch: Cherokee County, Kansas.

Size: Adults normally 125–178 mm (5–7 inches) in TL; largest specimen from Kansas: male (KU 288632) from Wyandotte County with unknown SVL and TL of 222 mm (8½ inches) collected by Daniel G. Murrow, James Markley, and Matt Singer on 30 April 1998; exceeds maximum length throughout range, as reported in Conant and Collins (1998).

Natural History: Lives in open, rocky, well-drained, cut-over forests; patchy forest leafcover preferred; basks in spots where sun reaches ground; also abundant around sawmills or artificial piles of rocks and logs; prefers humid environment and obtains water by lapping dew from plant leaves; recorded active annually in Kansas from mid-January to late October, with a peak from early April to early June; active diurnally in home range that varies 30–90 feet in diameter; not territorial; at night, retires beneath rocks and logs; during winter, retreats beneath ground to avoid cold temperatures; in spring, about 3 weeks after emerging from winter inactivity, becomes sexually active; at this time, males develop bright orange-red heads and are extremely aggressive, continually fighting with other males; orange-red head of males permits quick sex recognition; males search for and find females by sight and smell, but exhibit no courtship behavior; males pursue females, bite and hold them by their loose shoulder skin, and bend beneath their tail until cloacae meet; copulation lasts about 5 minutes; a few days after copulation, females become hostile to males and in May or June and become extremely secretive; at this time, they hide beneath rocks or in rotten logs and stumps and dig nests; females lay only 1 clutch of 4–14 eggs per year, averaging 8–9 in number; females remain on nest with eggs to protect them; eggs hatch within 1–2 months, depending on available moisture and prevailing temperatures; eats spiders, roaches, crickets, locusts, small grasshoppers, moths, beetles, snails, smaller lizards, and newborn mice; predators include hawks, opossums, armadillos, skunks, moles, shrews, snakes, and other larger lizards; juveniles and young adults have bright blue tails until second year of life; aggressive adult males are inhibited from attacking lizards with blue tails; blue tail possibly evolved for this purpose; blue tail has also been considered decoy to distract predators from head and body of lizard; maximum longevity: 2 years, 7 months, and 28 days (Snider and Bowler, 1992).

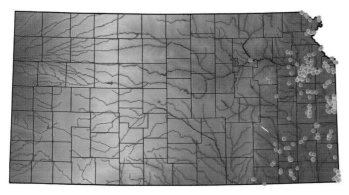

The distribution of the **Five-lined Skink** (*Plestiodon fasciatus*) in Kansas is generally restricted to the eastern quarter of the state. It occurs westward along forested river valleys; the record for Riley County (collected in 1933) needs corroboration. Red dots indicate records backed by voucher specimens or images.

Pertinent References: Anderson et al. (1995), Ashton et al. (1974a), Ballinger and Lynch (1983), Bartlett and Bartlett (2006b), Behler and King (1979), Boyd (1991), Breukelman and Clarke (1951), Breukleman and Downs (1936), Burt (1928a, 1928b, 1933, 1935), Burt and Hoyle (1934), Burton and Vensel (1966), Busby and Pisani (2007), Busby et al. (1996, 2005), Caldwell and Glass (1976), Clark and Hall (1970), Clarke (1958, 1965a), Clarke et al. (1958), Coleman (1987), Collins (1959, 1974b, 1982a, 1983, 1985, 1990b), Collins and Collins (1993a), Conant (1958, 1975), Conant and Collins (1991, 1998), Cope (1900), Cragin (1881), Cross (1971), Duvall et al. (1980), Edgren (1959), Fitch (1951, 1954, 1956b, 1956e, 1958b, 1963b, 1965b, 1967, 1970, 1982, 1985b, 1991, 2006a), Fitch and Fitch (1967), Fitch and von Achen (1977), Fraser (1983a), Gier (1967), Gloyd (1928, 1932), Grant (1927), Gray (1979), Hall (1968), Hall and Smith (1947), Householder (1917), Howes et al. (2006), Hoyle (1936), Hudson (1942), Johnson (2000), Jones et al. (1982), Karns et al. (1974), Kingman (1932), Kingsbury and Gibson (2002), Klemens (1993), Knight (1935), Linsdale (1925, 1927), Loomis

(1956), Miller (1976a, 1983a, 1996b), Minton (2001), Mitchell (1994), Mount (1975), Murrow (2008a), Noble and Mason (1933), Pearce (1991), Perry (1975), Riedle (1994a), Rues (1986), Schmidt (1953), Schmidt and Murrow (2007c), Sever (1966), Smith (1946c, 1950, 1956a, 1957), Suleiman (2005), Taggart (2000b, 2001a, 2003g, 2005), Taylor (1916, 1936, 1993), Tihen (1937), Trauth et al. (2004), Viets (1993), Vitt and Cooper (1986), Von Achen and Rakestraw (1984), von Frese (1973), and Wagner (1984).

Broadhead Skink
Plestiodon laticeps (SCHNEIDER, 1801)

Identification: Four limbs; ear opening present on each side of head; flat, smooth, shiny scales covering body; a sixth and/or seventh scale (counting back from nose) on upper lip, which extends up to edge of eye; exhibits same color changes with age as Five-lined Skink (adults of these 2 species are extremely difficult to tell apart); males grow larger than females and develop orange-red head during breeding.

Adult male showing bright head color during the mating season.

Adult female.

Adult male: Crawford County, Kansas.

Adult female: Crawford County, Kansas.

Adult female guarding eggs: Cherokee County, Kansas.

Size: Adults normally 165–230 mm (6½–9 inches) in TL; largest specimen from Kansas: male (KU 222265) from Linn County with SVL of 116 mm and TL of 287 mm (11¼ inches) collected by Kelly J. Irwin, Emily C. Moriarty, Suzanne L. Collins, and Joseph T. Collins on 7 May 1994; maximum length throughout range: 12⅝ inches (Conant and Collins, 1998).

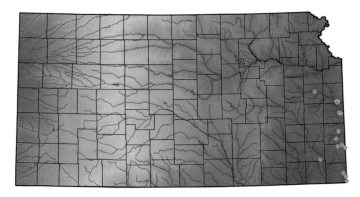

The **Broadhead Skink** (*Plestiodon laticeps*) is found along the forested rivers and larger streams of the extreme eastern portion of Kansas, from Cherokee and Neosho counties in the south to Franklin and Miami counties in the north. Red dots indicate records backed by voucher specimens or images.

Natural History: Inhabits forested regions in eastern Kansas, particularly near lowland aquatic situations; spends much time on ground around logs and brush piles near standing trees, but will ascend trees to escape danger and may use abandoned woodpecker holes as retreats; recorded active annually in Kansas from early March to mid-July; daily activity cycles similar to those of Five-lined Skink; little known of breeding habits in Kansas; courtship, mating, and nesting presumably similar to that of Five-lined Skink; number of eggs laid by females cannot be estimated with accuracy because of history of confusion with Five-lined Skink; apparently lays more eggs than Five-lined Skink; single clutch of 8 eggs recorded from verified Kansas specimen; eats insects, smaller lizards and their eggs, small snakes, and newborn mice; threatened species (protected by state law); maximum longevity: 7 years, 8 months, and 22 days (Snider and Bowler, 1992).

Pertinent References: Anderson (1950, 1965), Ballinger and Lynch (1983), Bartlett and Bartlett (2006b), Behler and King (1979), Clarke (1965a), Cochran and Goin (1970), Collins (1959, 1974b, 1982a, 1982b, 1985, 1990b, 1994b), Collins and Collins (1993a), Collins et al. (1991, 1994, 1995), Conant (1958, 1975), Conant and Collins (1991, 1998), Cross (1971), Freeman and Kindscher (1992), Guarisco et al. (1982), Irwin (1995b), Irwin and Collins (1994, 2005), Johnson (1974, 2000), Karns et al. (1974), Kingsbury and Gibson (2002), Lardie (1990), Layher et al. (1986), Legler (1960b), Levell (1997), Loomis (1956), Miller and Collins (1993), Miller and Gress (2005), Minton (1972, 2001), Mitchell (1994), Moehn (1981a, 1981b), Mount (1975), Platt et al. (1974), Schmidt (1953), Schilling and Schaefer (1983), Shaffer and McCoy (1991), Simmons (1989), Smith (1956a), Taggart (2001a, 2006b, 2008), Taggart and Collins (2000), Taylor (1993), Trauth et al. (2004), Welch (2005), and Whipple and Collins (1990a).

Great Plains Skink

Plestiodon obsoletus BAIRD AND GIRARD, 1852

Identification: Four limbs; ear opening present on each side of head; flat, smooth, shiny scales covering body; pattern of dark spots on body create striped appearance; scale rows on each side of body between front and hind limbs slant upward from front to back; head, body, limbs, and tail gray with dark borders on each scale that create light and dark striped pattern; unpatterned belly whitish gray; young jet black with blue tails and small bluish white to orange spots on sides of head; males and females extremely difficult to distinguish; during breeding season, heads of males slightly swollen.

Young: Finney County, Kansas.

Newborn: Finney County, Kansas.

Adult: Finney County, Kansas.

Old adult: Atchison County, Kansas.

Adult: Cloud County, Kansas.

Adult: Trego County, Kansas.

Size: Largest member of genus *Plestiodon* in North America; adults normally 165–230 mm (6½–9 inches) in TL; largest specimen from Kansas: female (KU 189186) from Cheyenne County with SVL of 133 mm and TL of 350 mm (13⅝ inches) collected by Brad Anderson and John Fraser on 2 May 1981; greatest weight from Kansas: 49.2 grams (1⅝ ounces) from Jefferson County collected by Robert R. Fleet and Russell J. Hall on 28 April 1968; maximum length throughout range: 13⅝ inches (Conant and Collins, 1998).

Natural History: Inhabits open rocky hillsides with low vegetation; recorded active annually in Kansas during every month, with a peak from late April to late July; during colder days of winter, burrows beneath soil or into crevices deep enough to avoid freezing temperatures; adult males emerge earlier in spring than females; diurnal but rarely basks in open sunlight, evidently obtaining sufficient heat beneath sun-warmed rocks; average home range about 50 feet in diameter, but may wander longer distances to establish new home range when conditions become unfavorable in original area; breeds in May after brief courtship; male approaches female and touches her with flicks of his tongue; after pursuit, male grasps female with mouth and bites loose skin on her shoulder, and loops hindquarters beneath hers; copulation occurs for several minutes; evidence indicates some females do not breed each year; pregnant females dig deep burrows beneath large boulders and lay 5–32 eggs, with an average of 12, and remain with eggs during 1–2 month incubation period; young hatch and may require several years to reach sexual maturity; eats beetles, roaches, grasshoppers, spiders, and snails; in captivity eats small rodents and other lizards; predators include snakes, birds, and small mammals; mortality from predation evidently slight (probably because this lizard remains hidden beneath rocks during much of its yearly activity period); maximum longevity: 6 years, 2 months, and 19 days (Snider and Bowler, 1992).

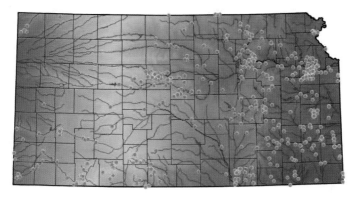

The **Great Plains Skink** (*Plestiodon obsoletus*) is found throughout most of the state, but appears to be least abundant in the northwestern part of Kansas except for the Republican River drainage. Red dots indicate records backed by voucher specimens or images.

Pertinent References: Anderson (1942, 1965), Anderson et al. (1995), Bartlett and Bartlett (2006b), Behler and King (1979), Belfit and Belfit (1985), Boyd (1991), Brennan (1934, 1937), Breukelman and Clarke (1951), Breukleman and Downs (1936), Breukelman et al. (1961), Burt (1927b, 1928a, 1928b, 1929a, 1933, 1935, 1936a), Burt and Burt (1929b, 1929c), Burt and Hoyle (1934), Busby and Parmelee (1996), Busby and Pisani (2007), Busby et al. (1994, 1996, 2005), Caldwell and Glass (1976), Capron (1978b), Capron et al. (1982), Cavitt (2000a), Clarke (1958, 1965a, 1980a), Clarke et al. (1958), Collins (1959, 1974b, 1979, 1982a, 1985, 1990b, 2001c, 2006b, 2007b, 2009a), Collins and Collins (1991, 1993a, 1993b, 2006a), Conant (1958, 1975), Conant and Collins (1991, 1998), Cope (1900), Cragin (1881, 1885a, 1885d), Cross (1971), Dice (1923), Ditmars (1907, 1915), Eddy and Babcock (1973), Ellis (2002), Ellis and Henderson (1913), Evans (1959), Fitch (1951, 1955, 1956e, 1958b, 1963b, 1967, 1970, 1982, 1985b, 1991, 2006a), Fitch and Fitch (1967), Fitch and Hall (1978), Fleet and Hall (1969), Fraser (1983a, 1983b), Garman (1884), Gish (1962), Gloyd (1928), Gress (2003), Guarisco (1981c, 1982a), Hall (1968, 1971, 1972, 1976), Hall and Fitch (1972), Hall and Smith (1947), Hallowell (1857a), Hammerson (1982, 1999), Hartman (1906), Heinrich and Kaufman (1985), Hoover (1931), Householder (1917), Hoyle (1936), Hutchison and

Larimer (1960), Irwin (1977, 1983), Irwin and Collins (1987), Johnson (1987, 2000), Jordan (1929), Karns et al. (1974), Kingman (1932), Kingsbury and Gibson (2002), Knight (1935), Knight and Collins (1977), Kretzer and Cully (2001), Loomis (1956), Marr (1944), Martin (1960), Mayers (2008), Miller (1976a, 1983a, 1985b, 1996a, 1996b), Mitchell (1965), Murrow (2008a, 2008b, 2009a, 2009b), Nulton and Rush (1988), Perry (1975), Platt (1984a, 1985b, 1998, 2003), Reichman (1987), Riedle and Hynek (2002), Roth (1959), Royal (1982), Rundquist (1992), Rush (1981), Schmidt (1953, 2000), Schmidt and Murrow (2006, 2007b), Schoewe (1937), Schwaner (1978), Simbotwe (1978, 1981), Simon and Dorlac (1990), Skie and Bickford (1978), Smith (1946c, 1950, 1956a), Stebbins (1954, 1966, 1985, 2003), Stejneger and Barbour (1917, 1923, 1933, 1939, 1943), Suleiman (2005), Taggart (1992b, 1992e, 2000a, 2001a, 2001b, 2002c, 2003g, 2006b, 2006c, 2007c, 2008), Taylor (1916, 1929b, 1936, 1993), Terman (1988), Toepfer (1993a), Trauth et al. (2004), Viets (1993), von Frese (1973), Watkins and Hinesley (1970), Wilgers and Horne (2006), Wilgers et al. (2006), Wolfenbarger (1951, 1952) and Yarrow (1883).

Southern Prairie Skink
Plestiodon obtusirostris (BOCOURT, 1879)

Identification: Four limbs; ear opening present on each side of head; flat, smooth, shiny scales covering body; usually uniform brown dorsum or sometimes with single faint middorsal stripe; stripes usually restricted to sides of body; widest dark stripe on each side of body always bordered above and below by distinct light stripes; head, body, limbs, and tail olive and tan; widest dark stripe on each side of body extends onto tail; belly uniform gray; young have blue tails; males develop reddish orange chins and lips during breeding season.

Adult male showing bright head color during the mating season: Clark County, Kansas.

Adult female: Kiowa County, Kansas.

Adult female: Barber County, Kansas.

Adult female: Sumner County, Kansas.

Size: Adults normally 124–178 mm (5–7 inches) in TL; largest specimen from Kansas: male (KU 221779) from Sumner County with an SLV of 74 mm and TL of 201 mm (7¹⁵⁄₁₆ inches) collected by Larry L. Miller on 16 April 1994; exceeds maximum length throughout range, as reported in Conant and Collins (1998).

Natural History: Frequents open, grass-covered, rocky hillsides near streams, but occasionally found in forests or at forest edge; will enter water to escape enemies; recorded active annually in Kansas from late March to late September, with a peak from mid-April to mid-May; like most skinks, highly secretive, spending much of day beneath flat rocks; during winter, apparently burrows deeply into earth to avoid temperature extremes; probably mates during May and June; nothing known of courtship; presumably, female digs shallow nest in loose, moist soil beneath objects such as logs, boards, or rocks; female deposits clutch of eggs during late June; eggs hatch in 1–2 months and young reach sexual maturity in 2 years; feeds on insects, snails, spiders, and smaller lizards; predators include larger lizards, snakes, large birds, and small mammals.

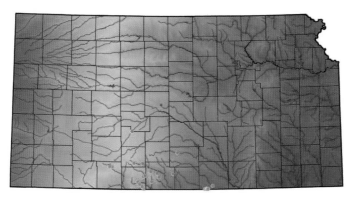

The distribution of the **Southern Prairie Skink** (*Plestiodon obtusirostris*) in Kansas is restricted to the south-central part of the state, south and west of the Arkansas River. Red dots indicate records backed by voucher specimens or images.

Pertinent References: Anderson (1965), Ballinger and Lynch (1983), Bartlett and Bartlett (2006b), Behler and King (1979), Burt (1928a), Cochran and Goin (1970), Collins (1959, 1974b, 1979, 1982a, 2006b), Collins and Collins (1993a), Conant (1958, 1975), Conant and Collins (1991, 1998), Crampton (1983), Cross (1971), Fuerst and Austin (2004), Gier (1967), Grow (1977), Guarisco et al. (1982), Johnson (2000), Karns et al. (1974), Knight (1935), McKenna and Seabloom (1979), Rundquist (1975), Smith (1946c, 1950, 1956a), Smith and Slater (1949), Somma and Cochran (1989), Stains and Ozment (1962), Stejneger and Barbour (1917, 1923, 1933, 1939, 1943), Taylor (1936), Tihen (1937), Trauth et al. (2004), and Trott (1983).

Northern Prairie Skink
Plestiodon septentrionalis BAIRD, 1858

Identification: Four limbs; ear opening present on each side of head; flat, smooth, shiny scales covering body; 7 light stripes alternating with 6 or 8 dark stripes on back and sides; widest dark stripe on each side of body always bordered above and below by distinct light stripes; head, body, limbs, and tail olive and tan; widest dark stripe on each side of body extends onto tail; belly uniform gray; young have blue tails; males develop reddish orange chins and lips during the breeding season.

Adult male: Shawnee County, Kansas.   Adult female: Franklin County, Kansas.

Adult male: Jefferson County, Kansas.   Adult female: Geary County, Kansas.

Size: Adults normally 124–178 mm (5–7 inches) in TL; largest specimen from Kansas: female (KU 206280) from Jackson County with SVL of 87 mm and TL of 224 mm (8⅝ inches) collected by Al Kamb and Steve Kamb on 26 May 1986; maximum length throughout range: 8⅝ inches (Conant and Collins, 1998).

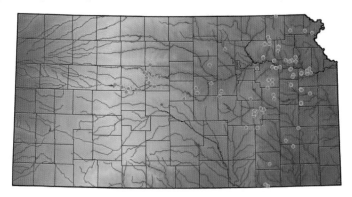

The **Northern Prairie Skink** (*Plestiodon septentrionalis*) is found throughout the eastern two-thirds of Kansas, north and east of the Arkansas River valley. Red dots indicate records backed by voucher specimens or images.

Natural History: Frequents open, grass-covered, rocky hillsides near streams, but occasionally found along streams in forests or at forest edge; will enter water to escape enemies; recorded active annually in Kansas from early April to early October, with a peak from mid-April to mid-May; highly secretive, spending much of day beneath flat rocks; during winter, apparently burrows deeply into earth to avoid temperature extremes; mates during May and June, several weeks later than Five-lined Skink; nothing known of courtship; presumably, female digs a shallow nest in loose, moist soil beneath objects such as logs, boards, or rocks, and deposits clutch of 5–18 eggs during late June; eggs hatch in 1–2 months and young attain sexual maturity in 2 years; feeds on insects, snails, spiders, and smaller lizards; predators include larger lizards, snakes, large birds, and small mammals.

Pertinent References: Anderson (1965), Ballinger and Lynch (1983), Bartlett and Bartlett (2006b), Behler and King (1979), Breukelman and Clarke (1951), Breukleman and Downs (1936), Breukelman et al. (1961), Burt (1928b, 1933), Burt and Hoyle (1934), Busby and Pisani (2007), Busby et al. (1996, 2005), Cavitt (2000a), Clarke (1955, 1956c, 1958, 1965a), Clarke et al. (1958), Collins (1959, 1974b, 1982a, 1990b, 2007b), Collins and Collins (1993a), Collins et al. (1991), Conant (1958, 1975), Conant and Collins (1991, 1998), Cope (1900), Coues and Yarrow (1878), Cragin (1881), Ditmars (1907, 1915), Fitch (1956e, 1958b, 1967, 1970, 1985b, 1991, 2006a), Fleharty and Ittner (1967), Fuerst and Austin (2004), Gier (1967), Gravenstein and Gravenstein (2003), Gress (2003), Guarisco et al. (1982), Hooper (2001), Householder (1917), Hoyle (1936), Hudson (1942), Iverson (1976), Johnson (2000), Jordan (1929), Karns et al. (1974), Kingman (1932), Kingsbury and Gibson (2002), Knight (1935), Loomis (1956), McKenna and Seabloom (1979), Murrow (2009a, 2009b), Peterson (1992), Preston (1982), Schmidt (1953, 2003d), Smith (1946c, 1950, 1956a), Smith and Slater (1949), Somma and Cochran (1989), Stejneger and Barbour (1917, 1923, 1933, 1939, 1943), Suleiman (2005), Taggart (2000a, 2008), Taylor (1916, 1936, 1993), Tihen (1937), Toepfer (1993a), Viets (1993), Washburne (2003b), Wheeler and Wheeler (1966), and Yarrow (1883).

Ground Skink
Scincella lateralis (SAY, 1823)

Identification: Four tiny limbs; ear opening present on each side of head; flat, smooth, shiny scales covering body; back brown with occasional dark flecks on back and lower sides; wide, dark stripe on each side of upper body running from eye onto tail; belly gray; adult females slightly larger than males.

Adult: Cherokee County, Kansas.   Adult: Crawford County, Kansas.

Adult.   Adult.

Size: Adults normally 75–125 mm (3–5 inches) in TL; largest specimen from Kansas: female (KU 28795) from Miami County with SVL of 57 mm and TL of 145 mm (5⅝ inches) collected by I. M. Claiborne on 30 April 1950; maximum length throughout range: 5⅝ inches (Conant and Collins, 1998).

Natural History: Lives in wooded areas; spends most of its time among leaf litter of forest floor; recorded active annually in Kansas from early March to late November, with a peak from early April to mid-May; appears to tolerate lower temperatures than many other species of lizards; primarily active during day, but evidently moves about at night on occasion; rarely basks in sun; instead, prefers to forage beneath rocks and leaf litter, usually in shaded areas; does not hesitate to enter water to escape enemies; breeding occurs in March or April, shortly after spring emergence; females lay 2 clutches of eggs per season, in late April and again in June or July; clutches 1–7 with an average of 3–4 eggs; eggs contain well-developed embryos when laid and incubation may be 22–33 days; eggs laid in rotten logs, stumps, or beneath leaf litter; nothing is known of courtship; insects are main diet, although spiders and earthworms are sometimes eaten; predators include other lizards, snakes, birds, and small mammals; maximum longevity: 2 years, 6 months, and 16 days (Snider and Bowler, 1992).

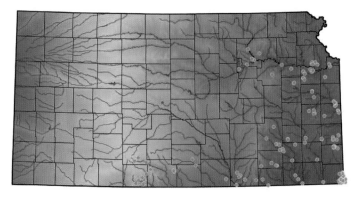

The **Ground Skink** (*Scincella lateralis*), the smallest of the skinks, is found along river systems in the eastern and southern part of the state, southeast of a line from Riley County in the northeast to Comanche and Kiowa counties in the southwest. It is also apparently absent from the northeastern part of the state. Red dots indicate records backed by voucher specimens or images.

Pertinent References: Anderson (1965), Ballinger and Lynch (1983), Bartlett and Bartlett (2006b), Behler and King (1979), Breukelman and Clarke (1951), Breukleman and Downs (1936), Brooks (1975), Burt (1928a, 1928b, 1933, 1936a), Burt and Burt (1929b), Burt and Hoyle (1934), Busby and Parmelee (1996), Busby et al. (1994, 1996, 2005), Caldwell and Glass (1976), Clarke (1965a), Clarke et al. (1958), Cochran and Goin (1970), Collins (1959, 1974b, 1982a, 1990b, 2006b), Collins and Collins (1993a), Conant (1958, 1975), Conant and Collins (1991, 1998), Cook (1942), Cope (1900), Cragin (1881), Cross (1971), Fitch (1956e, 1958b, 1965, 1967, 1970, 1985b, 1991, 2006a), Fitch and Greene (1965), Fitch and von Achen (1977), Gier (1967), Gloyd (1928, 1932), Gray (1979), Greer (1974), Guarisco et al. (1982), Hall and Smith (1947), Hay (1892), Householder (1917), Hurter (1911), Johnson (2000), Karns et al. (1974), Kingsbury and Gibson (2002), Knight (1935), Lardie (1990), Leviton (1971), Loomis (1956), McCauley (1945), Minton (1972, 2001), Mitchell (1994), Mount (1975), Murrow (2008a), Parmelee and Fitch (1995b), Perry (1978), Riedle (1994a), Rundquist (1992), Schmidt (1953), Schmidt and Murrow (2006, 2007c), Smith (1946c, 1950, 1956a), Stains and Ozment (1962), Stejneger and Barbour (1933, 1939, 1943), Suleiman (2005), Taggart (2000b, 2004e, 2005, 2008), Taylor (1916, 1993), Tihen (1937), Trauth et al. (2004), Wilgers and Horne (2006), Wilgers et al. (2006), and Yarrow (1883).

RACERUNNERS AND WHIPTAILS
FAMILY TEIIDAE GRAY, 1827

In North America, the family Teiidae (Racerunners and Whiptails) is composed of 1 genus and 21 species (Collins and Taggart, 2009). The Six-lined Racerunner (*Aspidoscelis sexlineata*) is the single Kansas representative of this family, which is generally characterized by large head scales, tiny granular dorsal body scales, and large ventral body scales.

Six-Lined Racerunner
Aspidoscelis sexlineata (LINNAEUS, 1766)

Identification: Four limbs; ear opening present on each side of head; smooth tiny granular scales on back, much smaller than those on belly; scales on belly in 8 rows; 7 light stripes running down back, 1 in the middle and 3 on each side; stripes range in color from greenish blue to yellow and may be indistinct in older adult males; areas between stripes vary in color from brown to dark green; throat, chest, and forward sides of body suffused with light green or bluish green, particularly in males; tail brownish or gray; adult males have broader head than females.

Adult female: Cherokee County, Kansas.

Adult female: Finney County, Kansas.

Adult male: Franklin County, Kansas.

Adult male: Sheridan County, Kansas.

Adult male: Ellsworth County, Kansas.

Belly of adult: Neosho County, Kansas.

Size: Adults normally 152–203 mm (6–8 inches) in TL; largest specimen from Kansas: male (KU 220236) from Cherokee County with SVL of 82 mm and TL of 233 mm (9⅛ inches) collected by Randall Reiserer on 18 May 1992; maximum length throughout range: 10½ inches (Conant and Collins, 1998).

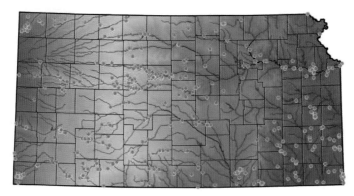

The swift **Six-lined Racerunner** (*Aspidoscelis sexlineata*) occurs throughout Kansas, but is least abundant in the northeastern part of the state and the central Flint Hills. Red dots indicate records backed by voucher specimens or images.

Natural History: Prefers dry, open, sandy areas with little leafy vegetation; also inhabits open, rocky, grazed, and cultivated regions (populations decrease when amount of available open area is reduced by leafy plant growth); requires warmer weather than most other Kansas lizards; recorded active annually in Kansas from early March to mid-October, with a peak from early May to late July; seasonally active only during day; spends cold winter days inactive beneath ground; in western Kansas, exhibits peak of activity from noon to 2:00 pm; in northeastern part of state, daily activity period extends from 8:00 am to 3:00 pm and depends on prevailing temperature; during extreme heat, rests in shade or in small burrows; in eastern Kansas, has home range up to ¼ acre; not territorial; mates in May and June; courtship initially involves male displaying brightly colored throat and chest; male bites female on neck or flank, pinions her, and twists his rear body around and under her tail until their cloacal openings meet; copulation is brief, lasting only few minutes; after male releases female, he may follow her with his tail arched and cloaca pressed to ground; females nest in June or July and may produce more than 1 clutch of eggs per season; eggs buried a few inches beneath sandy soil and hatch in 2 months; females lay 1–8 white eggs with average of 3 per clutch; spiders, snails, and small insects are eaten (grasshoppers, moths, and spiders make up 75% of diet); predators include snakes, birds, and small mammals.

Pertinent References: Anderson (1942, 1965), Anderson et al. (1995), Ballinger and Lynch (1983), Barden (1942), Bartlett and Bartlett (2006b), Baxter and Stone (1980, 1985), Behler and King (1979), Bender and Stark (2007), Brennan (1934, 1937), Breukelman and Clarke (1951), Breukleman and Downs (1936), Burt (1928a, 1928b, 1931b, 1933, 1935), Burt and Hoyle (1934), Busby and Parmelee (1996), Busby and Pisani (2007), Busby et al. (1994, 1996, 2005), Carpenter (1940), Caldwell and Glass (1976), Clarke (1958, 1965a, 1980a), Clarke et al. (1958), Collins (1959, 1974b, 1982a, 1985, 1990b, 2001c, 2006b), Collins and Collins (1991, 1993a, 1993b, 2006a), Collins et al. (1991), Conant (1958, 1975), Conant and Collins

(1991, 1998), Cope (1900), Cragin (1881, 1885a, 1885d), Crampton (1983), Cross (1971), Duellman and Zweifel (1962), Ellis (2002), Ellis and Henderson (1913), Falls (1933), Fitch (1951,
1956e, 1958a, 1958b, 1967, 1970, 1985b, 1991, 2006a), Fleharty and Hulett (1988), Fraser
(1983a, 1983b), Gish (1962), Gloyd (1928), Gress (2003), Grow (1976b), Guarisco (1981c),
Hall and Smith (1947), Hallowell (1857a, 1857b), Hammerson (1982, 1999), Hardy (1962),
Hartman (1906), Hay (1892), Householder (1917), Hoyle (1936, 1937), Irwin (1983), Irwin
and Collins (1987), Karns et al. (1974), Kingsbury and Gibson (2002), Knight (1935), Knight
and Collins (1977), Kretzer and Cully (2001), Loomis (1956), Lowe (1966), Marr (1944),
McLeran (1974b), Miller (1976a, 1979, 1980a, 1981a, 1981c, 1981d, 1982a, 1983a, 1985b,
1988, 1996a), Minton (2001), Minton and Brown (1997), Mitchell (1994), Mount (1975), Murrow (2008a, 2008b, 2009a), Peterson (1992), Platt (1975, 1984a, 1985b, 1998, 2003), Pope
(1944b), Riedle (1994a), Riedle and Hynek (2002), Royal (1982), Rundquist (1975, 1992),
Rush (1981), Schmidt (2001, 2004l), Schmidt and Murrow (2007b), Schoener (1977), Schoewe
(1937), Schwaner (1978), Schwarting (1984), Schilling and Schaefer (1983), Simmons
(1987c), Smith (1946c, 1950, 1956a, 1961), Stebbins (1954, 1966, 1985, 2003), Strimple
(1988), Suleiman (2005), Taggart (1992b, 2000a, 2001a, 2001b, 2002c, 2004h, 2005, 2006b,
2006c, 2007c, 2008), Taggart et al. (2007), Taylor (1916, 1929b, 1938b, 1940, 1993), Tihen
(1937), Tihen and Sprague (1939), Toepfer (1993a), Trauth (1992), Trauth and McAllister
(1996), Trauth et al. (2004), Trott (1983), Turner (1977), Vance (1983), Van Doren and Schmidt
(2000), Viets (1993), Warner and Wencel (1978), Werth (1969, 1972), Wilgers et al. (2006),
Wolfenbarger (1951, 1952), Wright and Vitt (1993), and Zimmerman (1990).

WALL LIZARDS AND LACERTAS
FAMILY LACERTIDAE COPE, 1864

The family Lacertidae (Wall Lizards and Lacertas) is native to the eastern hemisphere and is composed of 25 genera and 200 species. Two alien species, the Italian Wall Lizard (*Podarcis siculus*) and the Western Green Lacerta (*Lacerta bilineata*)
were introduced and established in Kansas when individuals were initially released
by or escaped from an animal dealer in Topeka. These reptiles are very similar in
appearance to native lizards of the Family Teiidae, characterized by large head
scales, granular dorsal body scales, and large ventral body scales.

Western Green Lacerta
Lacerta bilineata (DAUDIN, 1802)

Adult: Shawnee County, Kansas.   Juvenile: Shawnee County, Kansas.

Identification: Not native to Kansas; four limbs; ear opening present on each side of
head; smooth tiny granular scales on back, much smaller than those on belly; scales
on belly in 4 rows; neck and upper body with uniform green speckling; dorsum of

head darker with fewer speckles; throat white; chest yellowish; belly uniform yellowish or cream, sometimes with dark spots; tail green or olive; adult males display blue color on upper and lower lips, extending back to angle of jaw and below ear; females may be brownish or more subdued green color than males, and generally smaller; juveniles brown on dorsum with pale greenish or yellowish on lower sides.

Adult: Shawnee County, Kansas. Adult: Shawnee County, Kansas.

Size: Adults normally 150–280 mm (6–11 inches) in TL; largest specimen from Kansas: male (MHP 7927) from Shawnee County with SVL of 105 mm and TL of 320 mm (12⅝ inches) collected by James Gubanyi and Carl Michaels on 22 April 2002; maximum length throughout native range: 17⅝ inches (Corti and Cascio, 2002).

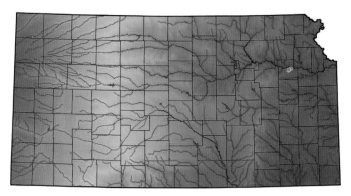

The **Western Green Lacerta** (*Lacerta bilineata*) is a non-native lizard that was introduced within the city limits of Topeka, Shawnee County, where a colony is well-established; its natural range is Europe. Red dots indicate records backed by voucher specimens or images.

Natural History: In Europe, prefers wide variety of habitats, including ruins, parks, gardens, and open grassy areas; in urban Topeka, chooses piles of rubble, wood, and other detritus, often in close proximity to buildings and automobile traffic; active by day, foraging around rock rubble and along walls of buildings; will quickly retreat into nearest crack, beneath shelter (e.g., ground air conditioning units), or up trees when chased; recorded active annually in Kansas from late April to mid-October, with a peak from mid-May to early September; apparently spends rest of year inactive beneath rubble, cement slabs, or in foundation cracks of buildings; males territorial; mating occurs in May and early June; females probably nest in June or July and may have second clutch in fall; eggs presumably buried in cavity beneath soil and hatch in 1–2 months; females lay 5–20 white eggs; feeds on wide variety of insects; predators probably include domestic cats and other small mammals, and birds; colony in Topeka introduced and established in 1950s, from individuals that escaped from a commercial animal dealer.

Pertinent References: Bartlett and Bartlett (2006b), Behler and King (1979), Burke and Deichsel (2008), Clarke (1986b), Collins (1974b, 1982a), Collins and Collins (1993a), Corti and Lo Cascio (2002), Deichsel and Miller (2000), Gubanyi (1996, 2000a, 2000b, 2001, 2002c, 2003e), Kraus (2009), Miller (1997b), Smith and Kohler (1977), and Taggart (2006b).

Italian Wall Lizard
Podarcis siculus (RAFINESQUE, 1810)

Identification: Not native to Kansas; four limbs; ear opening present on each side of head; smooth tiny granular scales on back that are much smaller than those on belly; scales on belly in 6 rows; head, neck, and upper body most often green with brownish middorsal stripe or row of spots, though some individuals may exhibit more subdued pattern of dark reticulations on green, olive, yellow or light brown ground color; blue spots sometimes present on shoulder; throat and belly are uniform white or gray; tail brown or gray; adult males longer and have larger heads than females.

Adult: Shawnee County, Kansas.

Juvenile: Shawnee County, Kansas.

Adult: Douglas County, Kansas.

Adult: Douglas County, Kansas.

Size: Adults normally 140–203 mm (5½–8 inches) in TL; largest specimen from Kansas: male (KU 223462) from Shawnee County with SVL of 76 mm and TL of 212 mm (8⁵⁄₁₆ inches) collected by James Gubanyi in July 1996; maximum length throughout native range: 9 ½ inches (Conant and Collins, 1998).

Natural History: In Europe, prefers a wide variety of habitats, including ruins and open grassy areas. In urban areas, chooses piles of rubble and other detritus, often in close proximity of buildings and automobile traffic; diurnal, foraging around rock rubble and along walls of buildings, and will quickly retreat into nearest crack or beneath shelter when chased; recorded active annually in Kansas from early March

REPTILES

105

to mid-October, with a peak from early April to early September; spends rest of year inactive beneath rubble, cement slabs, or in foundation cracks of buildings; are fast, nervous animals; movements greatly resemble those of native Six-lined Racerunner; nothing known of courtship and mating behavior in Kansas; probably mates in May and early June; males reported to intimidate each other with postures that include tilting head down, pushing throat out, and flattening trunk of body; females probably nest in June or July, and eggs presumably buried a few inches beneath soil and hatch in 1–2 months; females lay 1–4 white eggs and may produce more than 1 clutch per year; eats small insects; predators probably include domestic cats and other small mammals, and birds; colony in Topeka introduced and established in 1950s from individuals that escaped from commercial animal dealer; since then, range of this reptile within city has expanded considerably.

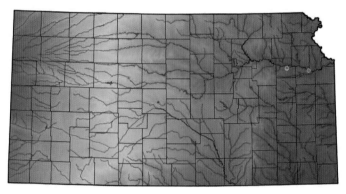

The non-native **Italian Wall Lizard** (*Podarcis siculus*) was introduced within the city limits of Topeka, (Shawnee County), and has spread to Lawrence (Douglas County) and Hays (Ellis County); it has well-established colonies in these three urban areas. This species is native to Europe. Red dots indicate records backed by voucher specimens or images.

Pertinent References: Bartlett and Bartlett (2006b), Burke and Deichsel (2008), Burke et al. (2007), Clarke (1986b), Collins (1974b, 1982a, 2005a), Collins and Collins (1993a, 2006c), Conant and Collins (1991, 1998), Corti and Lo Cascio (2002), Dugan (2006), Gisser (2000), Gubanyi (2002a, 2003d), Kingsbury and Gibson (2002), Kraus (2009), Miller (2004a, 2005), Taggart (2004c, 2006b, 2007d), and Tucker (1998).

ANGUID LIZARDS
FAMILY ANGUIDAE GRAY, 1827

The legless Western Slender Glass Lizard (*Ophisaurus attenuatus*) is the single Kansas representative of the family Anguidae (Anguid Lizards), which consists of 3 genera and 9 species in the United States and Canada (Collins and Taggart, 2009). These lizards are characterized by a groove that extends along the side of the body, a variable size (up to 1,080 mm in TL), and a carnivorous diet; limbs are absent in many species.

Western Slender Glass Lizard
Ophisaurus attenuatus COPE, 1880

Identification: Largest lizard in Kansas; lacks limbs; ear openings present on each side of head; often mistakenly identified as snake because limbs absent, but ear openings and eyelids distinguish it from snakes; body brown with dark stripes on back and sides; belly white or yellow; tail makes up two-thirds of entire length, but many lizards have much shorter, broken tails; older adult males have white flecks and black speckling on tan dorsolateral part of body (particularly anterior half) and grow larger than females; females generally lack dorsolateral markings; in young males they are subdued.

Size: Adults normally 560–900 mm (22–36 inches) in TL; largest specimen from Kansas: male (KU 207280) from Douglas County with SVL of 240 mm and TL of 762 mm (30 inches) collected by Lance Good and John Kitterman on 13 May 1987; maximum length throughout range: 46½ inches (Conant and Collins, 1998).

Adult: Douglas County, Kansas.

Adult: Franklin County, Kansas.

Adult: Douglas County, Kansas.

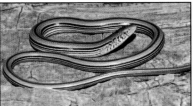
Adult: Ellis County, Kansas.

Natural History: Inhabits tallgrass prairie, sand prairie, open woodlands, and woodland edge, frequently near streams and ponds; particularly thrives in areas with abundant abandoned rodent burrows that it uses for shelter; recorded active annually in Kansas from early March to early December, with a peak from early May to late July; mostly diurnal and, during warm summer months, active in mornings and late afternoons, basking in sun, foraging for food, or moving about in tall grass; occasional individuals found active shortly after dusk; afternoons (1:00 pm to 5:00 pm) generally spent resting under shelter except during inclement weather; coloration and pattern blend extremely well in grassy areas (easily remains hidden provided it makes no movement); when disturbed, often escapes into underground burrows of small mammals; maximum home range of about 1 acre; evidently not territorial; population density in northeastern Kansas estimated at 26–41 per acre;

preferred air temperature range 61–75°F; with onset of winter, remains in summer areas and retreats underground into abandoned rodent burrows, deep enough to reach safety below frost line; breeds during May; courtship probably involves male pinning female by biting her on neck or back of head before copulation; females produce 1 clutch of eggs per year, normally from late June to mid-July; clutch size 5–17 eggs; eggs laid in nest and female remains with them until they hatch, an incubation period of about 7 weeks; females generally reach sexual maturity in 3–4 years; eats insects, spiders, snails, frogs, snakes, and newborn of small mammals; habitually loses tail when grabbed; sometimes tail breaks into several squirming parts and predator's attention focuses on wiggling tail while lizard escapes; will grow new tail, but never as long as original; predators include snakes, large birds, and mammals; minimum natural longevity of 9 years; maximum captive longevity: 4 years and 18 days (Snider and Bowler, 1992).

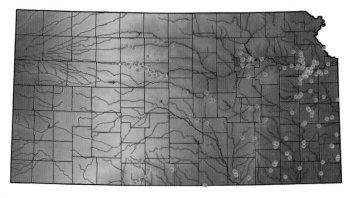

The legless **Western Slender Glass Lizard** (*Ophisaurus attenuatus*) is found in the eastern two-thirds of the state, but apparently is absent from the northern border counties. Red dots indicate records backed by voucher specimens or images.

Pertinent References: Anderson (1965), Ballinger and Lynch (1983), Bartlett and Bartlett (2006b), Behler and King (1979), Brennan (1934), Breukelman and Clarke (1951), Breukleman and Downs (1936), Burt (1928a, 1928b, 1933), Burt and Hoyle (1934), Busby and Parmelee (1996), Busby et al. (1994, 1996, 2005), Capron (1975a), Cavitt (2000a), Clarke (1958, 1965a), Clarke et al. (1958), Cochran and Goin (1970), Collins (1959, 1974b, 1982a, 1985, 1990b, 1995a), Collins and Collins (1993a), Conant (1958, 1975), Conant and Collins (1991, 1998), Cope (1900), Cragin (1881), Cross (1971), Eddy and Babcock (1973), Fitch (1951, 1956e, 1958b, 1967, 1970, 1982, 1985b, 1989, 1991, 2006a), Fitch and Hall (1978), Gish (1962), Gloyd (1928, 1932), Gress (2003), Hall and Smith (1947), Hartman (1906), Heinrich and Kaufman (1985), Holman (1971), Householder (1917), Hurter (1911), Johnson (2000), Johnson and LaDuc (1994), Karns et al. (1974), Kingsbury and Gibson (2002), Knight (1935), Loomis (1956), McConkey (1954), Miller (1976a, 1983a), Minton (2001), Mitchell (1994), Mount (1975), Murrow (2008a, 2009a), Ortenburger and Freeman (1930), Parmelee and Fitch (1995), Peterson (1992), Pisani (1974), Platt (1975, 1984a, 1985b, 1998, 2003), Riedle and Hynek (2002), Schmidt (2004h), Schmidt and Murrow (2007c), Simmons (1987c), Skie and Bickford (1978), Smith (1946c, 1950, 1956a), Stains (1954), Suleiman (2005), Taggart (1991a, 1992a, 2000a, 2000b, 2003g, 2004f, 2004h, 2005, 2008), Tanner (1978), Taylor (1916, 1993), Toepfer (1993a), Trauth et al. (2004), Von Achen and Rakestraw (1984), Wagner (1984), Wilgers and Horne (2006), Wilgers et al. (2006), Wooster (1925), and Yarrow (1883).

SNAKES
ORDER SQUAMATA OPPEL, 1811

Snakes (along with lizards) are members of the Order Squamata. Snakes are a highly diverse group and in the United States and Canada consist of 8 families, 53 genera, and 158 species (Collins and Taggart, 2009; Collins, 2006, with updates from the CNAH web site). The 38 species found in Kansas represent 5 families and 25 genera.

Snakes are the most diverse group of reptiles in Kansas. Fourteen of the 38 species have a nearly statewide distribution. Nine species have a primarily western distribution in the state, and 8 species are restricted to the eastern third of Kansas.

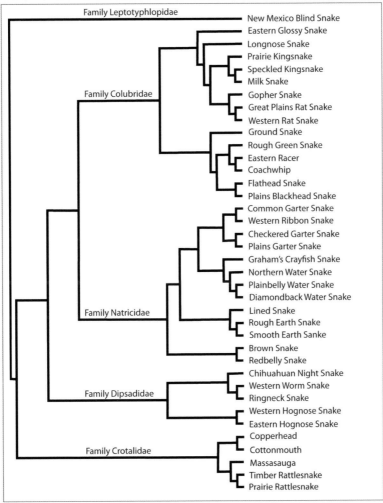

A phylogeny of Kansas Snakes (Order Squamata) that is consistent with evolutionary history based on current scientific evidence.

REPTILES

SLENDER BLIND SNAKES
FAMILY LEPTOTYPHLOPIDAE STEJNEGER, 1891

The family Leptotyphlopidae (Slender Blind Snakes) consists of 1 genus and 3 species in the United States (Collins and Taggart, 2009). In Kansas, the New Mexico Blind Snake (*Rena dissecta*) is the single representative of this family, which is generally characterized by 14 uniformly shaped rows of scales around the body, an enlarged anal scale, small size (normally 300 mm or less in TL), tiny eyes, burrowing habits, and an egg-laying reproductive mode.

New Mexico Blind Snake
Rena dissecta (COPE, 1896)

Identification: Very small; smooth body scales; belly scales same size as scales on rest of body; scale rows number 14 around body; eyes tiny black dots; uniform pinkish tan color; males with slightly longer tails than females.

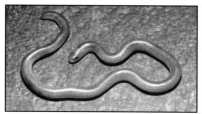
Juvenile: Clark County, Kansas.

Two adults showing the same size of dorsal and ventral scales: Barber County, Kansas.

Adult: Clark County, Kansas.

Adult: Sumner County, Kansas.

Adult showing small eye located beneath head scales: Clark County, Kansas.

Adult in defensive posture: Clark County, Kansas. Note elevated tail, which greatly resembles shape of the head.

© George R. Pisani

REPTILES

Size: Adults normally 125–203 mm (5–8 inches) in TL; largest specimen from Kansas: female (KU 206223) from Clark County with TL of 270 mm (10⅝ inches) collected by Larry Miller on 7 June 1986; maximum length throughout range: 11½ inches (Boundy, 1995).

Natural History: Very secretive; fossorial; frequents ant burrows; prefers moist areas; sometimes found in loose soil or sand beneath rocks; often observed on ground surface after rainfall; recorded active annually in Kansas from mid-April to early September, with a peak from early May to early June; nocturnal; because of subterranean habits, nothing known of courtship and mating; females brood eggs in small colonies within cavities beneath ground at depths of 18–30 inches; clutches of 5–6 eggs recorded in Kansas; feeds on ant eggs and termites; threatened species (protected by state law); predators include other snakes, birds, and small mammals; uses interesting defensive technique to withstand attacks by ants; if bitten or attacked, assumes a ball-like coil and writhes, smearing cloacal fluid over body; fluid repels further ant attacks.

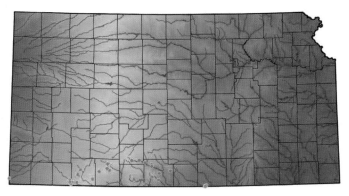

The Kansas distribution of the tiny **New Mexico Blind Snake** (*Rena dissecta*) is generally restricted to the southern border counties of the state, from Sumner County in the east to Morton County in the west; it has been recorded as far north as southern Kiowa County. Red dots indicate records backed by voucher specimens or images.

Pertinent References: Ball (1992a), Ballinger and Lynch (1983), Bartlett and Bartlett (2006c), Bartlett and Tennant (2000), Behler and King (1979), Burt (1935), Clarke (1980b), Cochran and Goin (1970), Collins (1959, 1974b, 1982a, 1983, 1985, 1990b, 2006b), Collins and Collins (1991, 1993a, 2006b, 2009a), Collins et al. (1995), Conant (1958, 1975), Conant and Collins (1991, 1998), Ditmars (1939), Ernst and Ernst (2003), Fitch (1970), Grow (1977), Guarisco et al. (1982), Hahn (1979, 1980), Hammerson (1999), Hibbard (1937, 1964), Irwin (1982a, 1983), Johnson (1974), Karns et al. (1974), Kingsbury and Gibson (2002), Klauber (1940), Knight et al. (1973), Lardie (1990), Lardie and Black (1981), Layher et al. (1986), Levell (1997), Leviton (1971), List (1966), Long (1992), Miller (1981a, 1982a, 1982b, 1987a), Miller and Gress (2005), Moeller et al. (2000), Mulvaney (1983), Perkins (1940), Platt et al. (1974), Rossi and Rossi (1995), Royal (1982), Rundquist (1998a), Rundquist et al. (1978), Schmidt (1953), Schmidt and Davis (1941), Schmidt and Murrow (2007b), Schilling and Nilon (1988), Scott (1996), Shaw and Campbell (1974), Simmons (1989), Smith (1950, 1956a), Smith and Sanders (1952), Stebbins (1954, 1966, 1985, 2003), Stejneger and Barbour (1939, 1943), Taggart (2006b), Taggart and Schmidt (2002b), Taylor (1939), Tennant and Bartlett (1999), Tihen (1937), Tihen and Sprague (1939), and Wright and Wright (1952, 1957).

REPTILES

111

HARMLESS EGG-LAYING SNAKES
FAMILY COLUBRIDAE OPPELL, 1811

The family Colubridae (Harmless Egg-laying Snakes) consists of 26 genera and 69 species in the United States (Collins and Taggart, 2009, with updates from the CNAH web site). In Kansas, 14 species assigned to 11 genera represent this family, which is characterized by an egg-laying reproductive mode, a smaller number of dorsal scales through reduction of the rows on the sides, and smooth dorsal scales (except *Opheodrys* and *Pituophis*). This family in Kansas consists of the genera *Arizona*, *Coluber*, *Lampropeltis*, *Masticophis*, *Opheodrys*, *Pantherophis*, *Pituophis*, *Rhinocheilus*, *Scotophis*, *Sonora*, and *Tantilla*.

Eastern Glossy Snake
Arizona elegans KENNICOTT, 1859

Juvenile: Finney County, Kansas.

Head of adult showing enlarged nasal scale used for burrowing through sand: Morton County, Kansas.

Adult: Finney County, Kansas.

Adult burrowing in sand: Finney County, Kansas.

Adult: Finney County, Kansas.

Adult: Morton County, Kansas.

Identification: Smooth body scales; single anal scale; 2 rows of scales on underside of tail; uniform white belly; pattern of 39–69 distinct dark gray or brown, black-edged blotches on body; smooth, glossy appearance; in addition to blotches on back, 2 alternating rows of dark spots present on each side of body; dark line extends from angle of each jaw forward through eyes; males have slightly longer tails than females.

Size: Adults normally 688–900 mm (27–36 inches) in TL; largest specimen from Kansas: female (MHP 7244) from Morton County with TL of 1,186 mm (46¼ inches) collected by Phillip Cass on 1 June 2002; maximum length throughout range: 55⅝ inches (Conant and Collins, 1998).

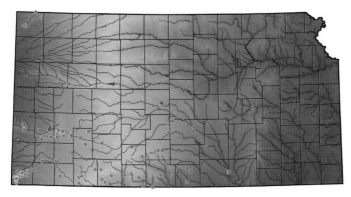

The **Eastern Glossy Snake** (*Arizona elegans*) is found throughout the southwestern part of the state in the Arkansas and Cimarron river drainages, from Hamilton County south to Morton County and thence eastward to Harvey and Sumner counties. Isolated colonies of this species occur in Cheyenne and Gove counties, and it was recently discovered in Chase County, well outside its range and in habitat not normally frequented by this serpent. Red dots indicate records backed by voucher specimens or images.

Natural History: Found in dry, open, sandy areas; little known of habits; recorded active annually in Kansas from late April to early October, with a peak from late May to early July; primarily nocturnal, prowling for food; during day, retires beneath rocks or into burrows to avoid heat and predators; abundant in sand prairies; mates during May, June, or July, following emergence from winter inactivity; courtship not observed; female evidently lays single clutch of eggs during summer; number of eggs per clutch 3–23, with an average of 8; eggs hatch in 2–3 months; kills prey by constriction; feeds primarily on lizards and small rodents; species in need of conservation (protected by state law); predators probably include snakes, mammals, and owls.

Pertinent References: Ball (1990, 1992a), Ballinger and Lynch (1983), Bartlett and Bartlett (2006c), Bartlett and Tennant (2000), Behler and King (1979), Burt (1933, 1935), Caldwell and Collins (1977), Clark (1966b), Clarke (1980a, 1980b), Cochran and Goin (1970), Collins (1959, 1974b, 1982a, 1985, 1990b), Collins and Collins (1991, 1993a, 2002, 2006b, 2009a), Collins et al. (1991), Conant (1958, 1975), Conant and Collins (1991, 1998), Davenport (1943), Dixon (1959), Dixon and Fleet (1976), Dyrkacz (1981), Ernst and Ernst (2003), Fitch (1993b), Fleet and Dixon (1971), Guarisco et al. (1982), Hammerson (1999), Irwin (1983), Karns et al. (1974), Kingsbury and Gibson (2002), Klauber (1946), Knight and Collins (1977), Layher et al. (1986), Levell (1997), Lynch (1985), Marr (1944), Miller (1980a), Murrow (2009a), Nulton and Rush (1988), Perkins (1940), Platt (1975, 1984a, 1985b, 1998, 2003), Platt and Rousell (1963), Pope (1937), Rodríguez-Robles et al. (1999), Rossi and Rossi (1995), Royal (1982), Rush (1981), Rush and Fleharty (1981), Schmidt (1953, 2004l), Schmidt and Davis (1941), Shaw and Campbell (1974), Sievert et al. (2006), Simmons (1987c, 1989), Smith (1950, 1956a), Sparks et al. (1999), Stebbins (1954, 1966, 1985, 2003), Stejneger and Barbour (1939, 1943), Taggart (2002a, 2002c, 2006b, 2006c, 2007c), Taggart et al. (2007), Taylor (1929a, 1929b), Tennant and Bartlett (1999), Tihen (1937), Tihen and Sprague (1939), and Wright and Wright (1952, 1957).

REPTILES

Eastern Racer
Coluber constrictor LINNAEUS, 1758

Identification: Smooth body scales; divided anal scale; 2 rows of scales on under-side of tail; uniformly cream or yellow belly; uniformly blue-gray, greenish blue, brown, or sometimes dark gray or black body; young have pattern of large, light-edged blotches on back alternating with smaller spots on sides (pattern distinct on front half of body and fades toward rear); young also have scattered dark speck-les on belly (as they grow older, young Eastern Racers lose all pattern and attain uni-form appearance of adults); adult males have slightly longer tails than females; females grow slightly larger than males.

Adult: Linn County, Kansas.

Two adults: Logan County, Kansas.

Adult: Ellsworth County, Kansas.

Adult: Wallace County, Kansas.

Adult: Ellis County, Kansas.

Juvenile: Morris County, Kansas.

Size: Adults normally 580–1,270 mm (23–50 inches) in TL; largest specimen from Kansas: female (KU 192239) from Jefferson County with TL of 1,413 mm (55½ inches) collected by Rick Strawn on 31 May 1982; maximum length throughout range: 75 inches (Boundy, 1995); maximum weight for Kansas specimen: 538 grams (1 pound, 3 ounces).

Natural History: Prefers open grassland, pasture, and prairie areas during sum-mer; generally found on rocky wooded hillsides in spring and fall; recorded active annually in Kansas from early January to early November, with a peak from late April to early August; diurnal, spending day basking in sun or gliding swiftly over ground in search of food; has an average home range of 25 acres but does not appear territorial; estimates indicate density of up to 6 individuals per acre in northeastern Kansas; during winter, crawls deep into rock crevices on wooded hillsides, remaining inactive until spring; known to co-habit in winter retreat with Prairie Kingsnakes, Gopher Snakes, Speckled Kingsnakes, Great Plains Rat Snakes, and Plains Garter Snakes; sometimes killed by controlled prairie fires when emerging from winter retreats; sometimes killed by swathing (mowing) machines during agricultural harvests; breeding occurs in May; during courtship, male positions himself alongside female and ripples his body spasmodically as he positions his cloaca beneath her tail; copulation lasts several minutes; while copulating, female may crawl forward, dragging attached male with her; females lay eggs from mid-June to early August, usually in tunnels or burrows of small mammals, such as moles; more than 1 female may use the same nest site at the same time; number of eggs per clutch 5–31, with an average of 11–12; incubation takes 2–3 months; relies primarily on sight to capture food; pursues and eats any small animal that moves; eats insects, frogs, lizards, other snakes, birds, bird eggs, and small mammals; primary predators hawks and small mammals; one of the fastest snakes in Kansas, but relies on distraction, as well as speed, to escape predators; a surprised Eastern Racer will thrash vigorously and attract a predator's vision to that area, and then quickly and quietly glide away into rocks, grass, and brush.

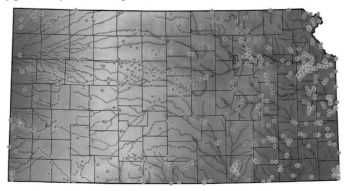

The swift **Eastern Racer** (*Coluber constrictor*) is found throughout Kansas. Red dots indicate records backed by voucher specimens or images.

Pertinent References: Anderson (1996), Ashton et al. (1974b), Auffenberg (1955), Ball (1992a), Ballinger and Lynch (1983), Bartlett and Bartlett (2006c), Bartlett and Tennant (2000), Baxter and Stone (1980), Behler and King (1979), Berglund (1967), Boyd (1991), Branson (1904), Brennan (1934, 1937), Breukelman and Clarke (1951), Breukelman and Downs (1936), Breukelman et al. (1961), Bugbee (1945), Burbrink et al. (2008), Burt (1931a, 1933, 1935), Burt and Burt (1929b), Burt and Hoyle (1934), Busby and Parmelee (1996), Busby and Pisani (2007), Busby et al. (1994, 1996, 2005), Caldwell and Glass (1976), Capron (1975a, 1976, 1977, 1985a, 1985c), Capron et al. (1982), Carpenter (1940), Cavitt (2000a, 2000b), Cink (1995b), Clark (1966b), Clark and Callison (1967), Clarke (1958, 1980a, 1980b), Clarke et al. (1958), Collins (1974b, 1982a, 1985, 1990b, 2001c, 2006b, 2007b, 2008e), Collins and Collins (1991, 1993a, 1993b, 1996, 2001, 2006a, 2006b, 2009a), Collins and Taggart (2008b), Collins et al. (1991, 1994), Conant (1958, 1975), Conant and Collins (1991, 1998), Cooper (1849), Cope (1900), Cragin (1881), Cross (1971), Dice (1923), Dixon and Werler (2005), Dunn

and Wood (1939), Dyrkacz (1981), Eddy and Babcock (1973), Ellis (2002), Ellis and Henderson (1913), Ernst and Ernst (2003),Falls (1933), Fitch (1951, 1956e, 1958b, 1960a, 1963a,
1970, 1980a, 1982, 1985b, 1987, 1991, 1992, 1993b, 1999, 2000b, 2004a, 2004b, 2005,
2006a), Fitch and Echelle (2006), Fitch and Fitch (1967), Fitch and Hall (1978), Fitch and Shirer
(1971), Fitch et al. (1981, 2003), Fleharty and Hulett (1988), Fraser (1983a), Garrigues (1962),
Gish (1962), Gloyd (1928, 1932), Grant (1937), Greene (1984), Guarisco (1981c, 1982b, 2001),
Guillette and Sullivan (1983), Hall (1968), Hall and Smith (1947), Hallowell (1857a, 1857b),
Halpin (1983), Hammerson (1999), Heinrich and Kaufman (1985), Henderson (1974), Hudson and Carl (1983), Hunter et al. (1992), Hurter (1911), Irwin and Collins (1987), Johnson
(1987, 2000), Karns et al. (1974), Kingsbury and Gibson (2002), Knight (1935), Knight and
Collins (1977), Kretzer and Cully (2001), Lillywhite (1985c), Linsdale (1925, 1927), Loomis
(1956), Loraine (1980), Marr (1944), Martin (1956, 1960), Miller (1976a, 1979, 1980a, 1981d,
1982a, 1983a, 1985b, 1988, 1996a, 1996b), Minton (2001), Mitchell (1994), Mount (1975),
Mozley (1878), Murrow (2008a, 2008b, 2009a, 2009b), Ortenburger (1926a), Parker and
Brown (1973), Parmelee and Fitch (1995b), Perry (1978), Platt (1975, 1984a, 1984b, 1985b,
1989, 1998, 2003), Prophet (1957), Pyron (2007), Rainey (1956), Reichman (1987), Riedle
(1994a), Riedle and Hynek (2002), Rosen (1991), Royal (1982), Rundquist (1975, 1992, 1998b,
1998d), Rush (1981), Rush and Fleharty (1981), Rush et al. (1982), Schmidt (1938b, 2001,
2004l), Schmidt and Murrow (2006, 2007c), Schoener (1977), Schwaner (1978), Simmons
(1987c, 1998), Smith (1950, 1956a, 1961), Smith et al. (1983, 1992b), Stebbins (1954, 1966,
1985, 2003), Stewart (1960), Swain and Smith (1978), Suleiman (2005), Taggart (1992a,
1992b, 2000a, 2001a, 2001b, 2002c, 2002d, 2003b, 2003g, 2004h, 2005, 2006b, 2006c,
2006d, 2007c, 2008), Taggart and Schmidt (2002a), Taggart et al. (2007), Taylor (1929a,
1929b), Tennant and Bartlett (1999), Tihen (1937), Tihen and Sprague (1939), Tinkle and Gibbons (1977), Toepfer (1993a), Toland and Dehne (1960), Trauth et al. (2004), Trott (1977),
Turner (1977), Uhler and Warren (1929, 2008), Van Doren and Schmidt (2000), Viets (1993),
Von Achen and Rakestraw (1984), Wagner (1984), Warner and Wencel (1978), Werth (1969),
Westerman (2006), Wilgers and Horne (2006, 2007), Wilgers et al. (2006), Wolfenbarger
(1951, 1952), Wooster (1925), Wright and Wright (1952, 1957), Yarrow (1883), and Yarrow
and Henshaw (1878).

Prairie Kingsnake
Lampropeltis calligaster (HARLAN, 1827)

Adult: Jefferson County, Kansas.   Adult: Seward County, Kansas.

Identification: Smooth body scales; single anal scale; 2 rows of scales on underside
of tail; light and dark pattern on belly; 40–78 dark gray or brown blotches on back;
head, body, and tail brown or gray; dark gray or brown blotches on back edged with
black; 2–3 series of small brown or gray spots on sides alternate with blotches on
back; belly white with dark irregular markings; adult males have slightly longer tails
and grow larger than females.

Adult: Barber County, Kansas.

Adult: Douglas County, Kansas.

Subadult: Shawnee County, Kansas.

Egg clutch: Osage County, Kansas.

Size: Adults normally grow 760–1,067 mm (30–42 inches) in TL; largest specimen from Kansas: male (KU 192452) from Miami County with TL of 1,324 mm (52 inches) collected by R. B. Hager on 23 September 1982; maximum length throughout range: 56¼ inches (Bird et al., 2005); maximum weight for Kansas specimen: 405 grams (slightly over 14 ounces).

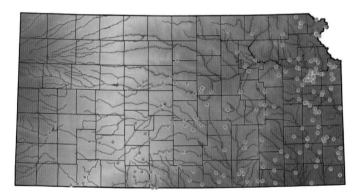

The **Prairie Kingsnake** (*Lampropeltis calligaster*) is found throughout the southeastern half of the state, from Washington County in the northeast and thence southwestward to Kearny and Seward counties in the southwest; it is absent from northwestern Kansas. Red dots indicate records backed by voucher specimens or images.

Natural History: Inhabits variety of areas, including rocky hillsides with open woods, prairie grassland, and sand prairies; very secretive; when not actively prowling for food, retreats beneath rocks or down burrows of other animals; recorded active annually in Kansas from late January to late November, with a peak from late April to mid-July; becomes nocturnal during hot summer months; during winter, retreats beneath ground, probably in small mammal burrows; breeds in early spring following emergence from winter inactivity, but some females may not breed every season; eggs laid in June or July and hatch in late August or September; number of eggs per clutch 5–17 and averages 9–11; eggs hatch in 1–3 months; young attain

REPTILES

117

sexual maturity in second or third season; courtship behavior unknown; constricts its prey; feeds primarily on small mammals, small snakes, and lizards; predators include owls, hawks, mammals, and other large snakes; when disturbed, rapidly vibrates tail, producing buzzing sound when tail vibrated amidst leaves or resonant objects; males apparently engage in combat dances that involve biting; maximum captive longevity: 23 years, 8 months, and 23 days (Snider and Bowler, 1992).

Pertinent References: Anderson (1965), Ballinger and Lynch (1983), Bartels (2007), Bartlett and Bartlett (2006c), Behler and King (1979), Blanchard (1921), Blaney (1973, 1979), Boyd (1991), Branson (1904), Brennan (1934), Breukelman and Clarke (1951), Breukleman and Downs (1936), Burt (1931a, 1933, 1935), Burt and Burt (1929b), Burt and Hoyle (1934), Busby (1997), Busby and Parmelee (1996), Busby and Pisani (2007), Busby et al. (1994, 1996, 2005), Capron (1978b, 1986c), Cavitt (2000a, 2000b), Cink (1995b), Clark (1966b), Clark et al. (1983), Clarke (1954, 1958, 1980a, 1980b), Clarke et al. (1958), Collins (1959, 1974b, 1982a, 1985, 1990b, 2001c, 2006b, 2007b), Collins and Collins (1993a, 1993b, 1996, 2001, 2006a, 2006b, 2009a), Collins et al. (2002), Conant (1958, 1975), Conant and Collins (1991, 1998), Cook (1954), Cope (1900), Cragin (1881), Cross (1971), Ditmars (1907, 1915, 1939), Ellis (2002), Ernst and Ernst (2003), Fitch (1951, 1956e, 1958b, 1979, 1982, 1985b, 1991, 1992, 1993b, 1999, 2000b, 2006a), Fitch and Echelle (2006), Fitch and Hall (1978), Fitch and Shirer (1971), Fitch et al. (2003), Fraser (1983a, 1987), Garrigues (1962), Gish (1962), Gloyd (1928), Guarisco (1981c), Hall and Smith (1947), Hallowell (1857a, 1857b), Halpin (1983), Henderson (1974), Hoyle (1936), Hudson (1942), Hurter (1911), Irwin (1983), Irwin and Collins (1987), Jan and Sordelli (1867), Johnson (2000), Karns et al. (1974), Kennicott (1859), Kingsbury and Gibson (2002), Knight (1935), Loomis (1956), Lueth (1941), Markel (1972, 1979, 1990), Markel and Bartlett (1995), McLeran (1973), Miller (1976a, 1981c, 1982a, 1983a), Minton (2001), Mitchell (1994), Mount (1975), Mozley (1878), Murrow (2008a, 2009a, 2009b), Nulton and Rush (1988), Perkins (1940), Pilch (2004), Pisani (2003), Platt (1975, 1984a, 1985b, 1998, 2003), Riedle and Hynek (2002), Rossi (1992), Rossi and Rossi (1995), Schmidt (1953), Schmidt and Davis (1941), Schmidt and Murrow (2007b), Simmons (1987c), Smith (1950, 1956a), Stebbins (1954), Stejneger and Barbour (1933, 1939, 1943), Taggart (2000a, 2001a, 2002d), Taylor (1929a), Tennant and Bartlett (1999), Tihen (1937), Tihen and Sprague (1939), Toepfer (1993a), Trauth et al. (2004), Trott (1977), Van Doren and Schmidt (2000), Viets (1993), Von Achen and Rakestraw (1984), Wagner (1984), Werth (1969), Wilgers et al. (2006), Wolfenbarger (1951, 1952), Wright and Wright (1952, 1957), and Yarrow (1883).

Speckled Kingsnake
Lampropeltis holbrooki STEJNEGER, 1902

Adult female giving birth to egg from her clutch: Barton County, Kansas.

Subadult: Greenwood County, Kansas.

Identification: Smooth body scales; single anal scale; 2 rows of scales on underside of tail; black head, body, and tail profusely speckled with small yellow (sometimes

REPTILES

cream or white) spots; belly yellow and irregularly patterned with black; occasionally yellow speckling on back fuses to form narrow bars in adults; these bars always present in young individuals; adult males have slightly longer tails and grow larger than females.

Adult: Ellis County, Kansas.

Subadult: Riley County, Kansas.

Adult: Ellsworth County, Kansas.

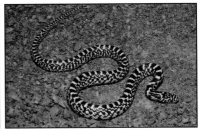
Newborn: Barber County, Kansas.

Size: Adults normally grow 900–1,220 mm (36–48 inches) in TL; largest specimen from Kansas: male (KU 182290) from Ford County with TL of 1,264 mm (49⅝ inches) collected by Jeffrey T. Burkhart, T. Lucier, and D. Tolliver on 28 October 1978; maximum length throughout range: 72 inches (Conant and Collins, 1998).

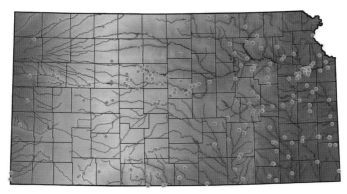

The **Speckled Kingsnake** (*Lampropeltis holbrooki*) is found throughout Kansas. Red dots indicate records backed by voucher specimens or images.

Natural History: Generally inhabits moist areas of open woodland, woodland edge, or lowlands, but has been found in open prairie; in forested areas, frequents rocky ledges on hillsides whereas on open prairie retreats to small mammal burrows; fairly common on open prairies near streams; often discovered by rolling over hay bales; recorded active annually in Kansas from early April to late November, with a peak from late April to early July; during spring and fall, active primarily during the day; high air temperatures of summer cause it to become nocturnal; during winter, crawls into burrows deep beneath ground to avoid cold temperatures; mates in spring and lays eggs in summer; egg clutches 2–17 and generally hatch in fall; little known of courtship, but apparently male bites female to hold her during mating; uses constriction to overpower prey; varied diet includes rodents, small birds, eggs (bird and reptile), lizards, and snakes; predators consist chiefly of large birds and mammals; like many snakes, rapidly vibrates tail when threatened; males engage in elaborate combat dance to establish dominance over other males; maximum captive longevity: 14 years, 5 months, and 22 days (Snider and Bowler, 1992).

Pertinent References: Anderson (1965), Ball (1992a, 1992b), Ballinger and Lynch (1983), Bartlett and Bartlett (2006c), Behler and King (1979), Berglund (1967), Blanchard (1921), Blaney (1973, 1977), Boyd (1991), Branson (1904), Brennan (1934, 1937), Breukelman and Clarke (1951), Breukleman and Downs (1936), Bugbee (1945), Burt (1933, 1935), Burt and Hoyle (1934), Busby and Parmelee (1996), Busby and Pisani (2007), Busby et al. (1994, 1996, 2005), Capron (1978b, 1985c), Capron and Perry (1976), Carpenter (1940), Cavitt (2000a, 2000b), Clark (1966b), Clarke (1956c, 1958, 1980a, 1980b), Clarke et al. (1958), Collins (1959, 1974b, 1982a, 1985, 1990b, 2001c, 2006b, 2007b), Collins and Collins (1993a, 1993b, 1996, 2001, 2006a, 2006b, 2009a), Conant (1958, 1975), Conant and Collins (1991, 1998), Cope (1900), Cragin (1881), Cranston (1994), Cross (1971), Dessauer and Pough (1975), Ditmars (1939), Ernst and Ernst (2003), Falls (1933), Fitch (1956e, 1985b, 1991, 1992, 1993b, 1999), Fraser (1983a), Garrigues (1962), Gish (1962), Gloyd (1928), Guarisco (1981c), Hall and Smith (1947), Hammerson (1999), Heinrich and Kaufman (1985), Hoyle (1936), Hubbs (2009), Hurter (1911), Irwin (1983), Irwin and Collins (1987), Johnson (2000), Karns et al. (1974), Kingsbury and Gibson (2002), Knight (1935), Krysko and Judd (2006), Lokke (1984), Loomis (1956), Lynch (1985), Markel (1972, 1979, 1990), Markel and Bartlett (1995), McLeran (1973), Miller (1976a, 1979, 1981c, 1982a, 1983a, 1996a), Minton (2001), Mitchell (1994), Mount (1975), Mozley (1878), Murrow (2008a, 2008b, 2009a, 2009b), Platt (1975, 2003), Pyron (2009), Pyron and Burbrink (2009a, 2009b), Riedle (1994a), Riedle and Hynek (2002), Rossi (1992), Rossi and Rossi (1995), Schmidt (1953, 2004l), Schmidt and Davis (1941), Schmidt and Murrow (2006), Schwaner (1978), Schwarting (1984), Secor (1987), Simmons (1987c), Smith (1950, 1956a, 1961), Stebbins (1954, 1966, 1985, 2003), Taggart (1991b, 1992a, 2000a, 2002b, 2002d, 2005, 2008, 2009b), Taggart et al. (2006, 2007), Taylor (1929a), Tennant and Bartlett (1999), Tihen (1937), Tihen and Sprague (1939), Toepfer (1993a), Trauth et al. (2004), Van Doren and Schmidt (2000), Viets (1993), von Frese (1973), Wagner (1984), Westerman (2006), Whitney (2009), Wilgers and Horne (2006), Wilgers et al. (2006), Wolfenbarger (1951, 1952), Wright and Wright (1952, 1957), Yarrow (1883), and Zimmerman (1990).

Milk Snake
Lampropeltis triangulum (LACÉPÈDE, 1788)

Adult: Linn County, Kansas.

Adult: Ellis County, Kansas.

Identification: Smooth body scales; single anal scale; two rows of scales on underside of tail; bold white and black pattern on belly; brilliantly contrasting colors on head, body, and tail; individuals from eastern Kansas have pattern of black-bordered bright red bands separated by narrower yellow, white, or cream bands; top of head red; Milk Snakes from western two-thirds of Kansas have same basic pattern, but tops of heads are black or orange, and large body bands are sometimes orange instead of red; adult males grow larger than females.

Adult: Franklin County, Kansas.

Adult: Comanche County, Kansas.

Adult: Finney County, Kansas.

Juvenile: Cheyenne County, Kansas.

Size: Adults normally grow 410–710 mm (16–28 inches) in TL; largest specimen from Kansas: male (KU 193235) from Anderson County with TL of 858 mm (33⅝ inches), collected by Harold A. Dundee on 9 May 1982; maximum length throughout range: 52 inches (Conant and Collins, 1998); maximum weight for Kansas specimen: 105 grams (just under 4 ounces).

Natural History: In eastern Kansas, frequents rocky hillsides of open woods and woodland edge; in western Kansas, inhabits rocky ledges of prairie canyons and edges of streams; recorded active annually in Kansas from late March to late November, with a peak from late April to early July; normally prowls actively by day, but during hot summer may become nocturnal, particularly in western Kansas; rarely basks in sun, preferring to remain hidden beneath sun-warmed rocks to maintain optimal body temperature; during winter, retreats into dens on rocky wooded hillsides or into mammal burrows to avoid cold temperatures; mating occurs during spring after emergence from winter inactivity; little known about courtship except that male bites female on neck a few inches behind head to hold her during copulation; eggs laid during June or July; 3–24 eggs per clutch with an average of 5–7; eggs hatch in August or September; incubation requires about 40–65 days, depending on prevailing air temperature; prey killed by constriction; feeds primarily on small lizards, snakes, and newborn mice; predators include birds, mammals, and larger snakes; maximum captive longevity: 21 years, 4 months, and 14 days (Snider and Bowler, 1992).

REPTILES

121

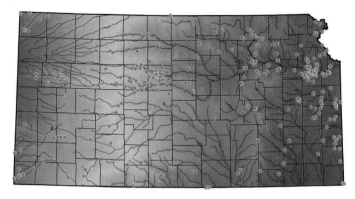

The **Milk Snake** (*Lampropeltis triangulum*) is generally found throughout the state but is rare in south-central Kansas. Red dots indicate records backed by voucher specimens or images.

Pertinent References: Anderson (1965), Ball (1992a), Ballinger and Lynch (1983), Bartlett and Bartlett (2006c), Bartlett and Tennant (2000), Baxter and Stone (1980, 1985), Behler and King (1979), Blanchard (1921), Blaney (1973), Boyd (1991), Branson (1904), Brennan (1934, 1937), Breukelman and Clarke (1951), Breukleman and Downs (1936), Bryson et al. (2007), Bugbee (1945), Burt (1933, 1935), Burt and Burt (1929b), Burt and Hoyle (1934), Busby and Parmelee (1996), Busby et al. (1994, 1996, 2005), Caldwell and Glass (1976), Capron et al. (1982), Carpenter (1940), Cavitt (2000a, 2000b), Clarke (1958, 1980a, 1980b), Clarke et al. (1958), Cochran and Goin (1970), Coleman (1987), Collins (1959, 1974b, 1979, 1982a, 1985, 1990b, 2006b, 2007b), Collins and Collins (1991, 1993a, 2006b, 2009a), Collins et al. (1991), Conant (1958, 1975), Conant and Collins (1991, 1998), Cook (1954), Cope (1900), Cox (1982), Cragin (1881), Cross (1971), Ditmars (1907, 1939), Eddy and Babcock (1973), Ellis and Henderson (1913), Ernst and Ernst (2003), Fitch (1951, 1956e, 1958b, 1980a, 1982, 1985b, 1991, 1992, 1993b, 1999, 2000b, 2006a), Fitch and Echelle (2006), Fitch and Fleet (1970), Fitch et al. (2003), Fogell (2009), Fraser (1983b), Garrigues (1962), Gier (1967), Gish (1962), Gloyd (1928, 1932), Guarisco (1980b, 1981c, 1982a), Hall (1968), Hall and Smith (1947), Hallowell (1857a, 1857b), Hammerson (1999), Hawthorn (1971), Heinrich and Kaufman (1985), Henderson (1974), Hunter et al. (1992), Hurter (1911), Irwin (1983), Iverson (1976, 1978), Johnson (1974, 2000), Jordan (1929), Kamb (1978a, 1978b), Karns et al. (1974), Kingsbury and Gibson (2002), Knight (1935), Knight and Collins (1977), Kretzer and Cully (2001), Lardie and Black (1981), Li (1978a, 1978b), Linsdale (1925, 1927), Loomis (1956), Lueth (1941), Lund (1974), Markel (1972, 1979, 1990), Markel and Bartlett (1995), Marr (1944), McLeran (1973), Miller (1980a, 1982a, 1982b, 1985b, 1996a, 2004d), Minton (2001), Mitchell (1994), Mount (1975), Mozley (1878), Murrell (1993), Murrow (2008a, 2008b, 2009a, 2009b), Perkins (1940), Perry (1974a, 1978), Platt (1984b), Platt et al. (1974), Rossi (1992), Rossi and Rossi (1995), Royal (1982), Rundquist (1994b), Rush (1981), Rush et al. (1982), Schmidt (1953, 2004l, 2006), Schmidt and Davis (1941), Schmidt and Murrow (2006, 2007a, 2007b, 2007c), Schwarting (1984), Shirer and Fitch (1970), Smith (1950, 1956a, 1957, 1961), Smith and Stephens (2003), Stebbins (1954, 1966, 1985, 2003), Stejneger (1891), Stejneger and Barbour (1933, 1939, 1943), Suleiman (2005), Taggart (1992a, 1992b, 1992i, 2000a, 2001a, 2003e, 2003g, 2004h, 2005, 2006d, 2008), Taggart et al. (2007), Tanner and Loomis (1957), Taylor (1929a, 1929b), Tennant and Bartlett (1999), Tihen (1937), Toepfer (1993a), Trauth et al. (1994, 2004), Tryon and Murphy (1982), Viets (1993), Wagner (1984), Whipple and Collins (1988), Wilgers and Horne (2006), Wilgers et al. (2006), Williams (1970, 1978, 1988, 1994), Wooster (1925), Wright and Wright (1952, 1957), and Yarrow (1875, 1883).

Coachwhip
Masticophis flagellum (SHAW, 1802)

Identification: Smooth body scales; divided anal scale; 2 scales bordering front edge of each eye; 10 or more scales on each lower lip; very slender body; body scales have braided appearance; dorsal color varies west to east: in western two-thirds of state, adults normally uniform light yellowish brown along entire length of body (belly may be whitish with indistinct small spots); in eastern third of state (from Cowley County eastward), adults jet black or black-brown on front half, gradually becoming yellowish brown on rear, or completely black (belly may be black on front half and light on rear, or all black); juveniles yellowish brown with dark brown crossbands on front of body that fade and disappear on rear (belly white with 2 rows of brown spots on front half); adult males have longer tails than females.

Adult: Hamilton County, Kansas.

Adult: Sumner County, Kansas.

Adult: Kiowa County, Kansas.

Adult: Wilson County, Kansas.

Juvenile: Finney County, Kansas.

Juvenile: Cowley County, Kansas.

Size: Adults normally grow 1,067–1,520 mm (42–60 inches) in TL; largest specimen from Kansas: male (KU 224649) from Comanche County with a total length of 1,822 mm (71⅝ inches) collected by James Gubanyi and Keith Coleman on 24 May 1997; maximum length throughout range: 102 inches (Conant and Collins, 1998).

Natural History: Found in widely varying habitat, from open grassland prairies in western Kansas to rocky hillsides in open woodlands in southeastern part of the

REPTILES

123

state; recorded active annually in Kansas from mid-March to early December, with a peak from late April to early July; completely diurnal, foraging for food even during hottest hours of day; in vegetated areas, climbs into bushes; observed with approximately 8 inches of head and body extended vertically from a hole in a rock (when still, snake resembled dead yucca stalks that surrounded rock); slithers away from an intruder with considerable speed and, when cornered, will rapidly vibrate tail and strike repeatedly; during winter, enters deep crevices on rocky hillsides or small mammal burrows on open prairies to avoid cold weather; little observed of breeding habits in Kansas; courtship unknown; mating probably occurs in April or May, followed by egg-laying in summer and hatching in fall; eggs laid beneath loose soil, up to a foot below surface; feeds on bats, mice, birds, lizards, and smaller snakes; although probably fastest snake in Kansas, cannot move as fast as people; maximum captive longevity: 18 years, 1 month, and 29 days (Snider and Bowler, 1992).

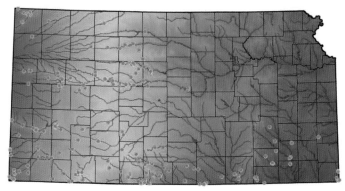

The fast-moving **Coachwhip** (*Masticophis flagellum*) ranges throughout the western half of the state, and southeastward through Ellsworth County to Cherokee County in southeastern Kansas. It is apparently absent from all of northeastern Kansas. Red dots indicate records backed by voucher specimens or images.

Pertinent References: Anderson (1965), Ball (1992a), Ballinger and Lynch (1983), Bartlett and Bartlett (2006c), Bartlett and Tennant (2000), Behler and King (1979), Branson (1904), Brennan (1934, 1937), Breukelman and Clarke (1951), Bugbee (1945), Burt (1933, 1935), Burt and Hoyle (1934), Capron (1975a, 1978b, 1983), Capron et al. (1982), Carpenter (1940), Clarke (1980a, 1980b), Cochran and Goin (1970), Collins (1959, 1974b, 1982a, 1985, 1990b, 2006a, 2006b), Collins and Collins (1991, 1993a, 1996, 2001, 2006b, 2009a), Collins and Taggart (2008b), Conant (1958, 1975), Conant and Collins (1991, 1998), Crampton (1983), Cross (1971), Davenport (1943), Ditmars (1907, 1939), Ellis and Henderson (1913), Ernst and Ernst (2003), Fitch (1993b), Fleharty and Hulett (1988), Fraser (1983a), Garrigues (1962), Gier (1967), Gish (1962), Guarisco (1981c), Hoyle (1936), Hudson (1942), Irwin (1983), Jordan (1929), Karns et al. (1974), Kingsbury and Gibson (2002), Knight (1935), Knight and Collins (1977), Marr (1944), Maslin (1953), McLeran (1974b), Miller (1982b), Mount (1975), Mozley (1878), Murrow (2009a), Ortenburger (1926a), Perkins (1940), Pope (1937), Pyron (2007), Rossi (1992), Rossi and Rossi (1995), Royal (1982), Rush (1981), Rush and Fleharty (1981), Rush et al. (1982), Schmidt (1953, 2000, 2001, 2004l), Schmidt and Davis (1941), Schmidt and Murrow (2007b), Scott (1996), Simmons (1987c), Smith (1950, 1956a), Snelson ands Palmer (1966a), Stebbins (1954, 1966, 1985, 2003), Stejneger and Barbour (1933, 1939, 1943), Taggart (1992b, 2000a, 2000b, 2001b, 2004h, 2006b, 2006c, 2007c, 2008), Taggart et al. (2005g, 2007), Taylor (1929a, 1929b), Tennant and Bartlett (1999), Tihen (1937), Tihen and Sprague (1939), Toepfer (1993a), Trauth et al. (2004), Trott (1983), Van Doren and Schmidt (2000),

Werth (1969), Wilgers et al. (2006), Wilson (1970, 1973a, 1973b), Wolfenbarger (1951, 1952) and Wright and Wright (1952, 1957).

Rough Green Snake
Opheodrys aestivus (LINNAEUS, 1766)

Identification: Keeled body scales; divided anal scale; bright green head, body, and tail; belly uniform white or cream; chin may be light yellow; adult males have slightly longer tail than females.

Size: Adults normally grow 560–810 mm (22–32 inches) in TL; largest specimen from Kansas: female (KU 45334) from Elk County with TL of 877 mm (34½ inches) collected by Henry S. Fitch on 8 July 1957; maximum length throughout range: 45⅝ inches (Conant and Collins, 1998).

Adult: Linn County, Kansas.

Adult: Cherokee County, Kansas.

Adult: Linn County, Kansas.

Adult: Bourbon County, Kansas.

Adult: Linn County, Kansas.

Adult: Cherokee County, Kansas.

Natural History: Inhabits trees, brush, shrubbery, and bushes near aquatic situations, particularly small lakes; when not gliding through vegetation in search of food, retires beneath rocks or leaves, or in hollow logs; blends perfectly with green

REPTILES

125

plants and extremely difficult to observe; exhibits another interesting adaptation for concealment in thick vegetation: when breeze causes stems of plants to bend and wave, an individual sitting in their midst will bend and wave in unison with vegetation; recorded active annually in Kansas from late March to late October, with a peak from mid-April to late May; strictly diurnal; small home range; at night, sleeps in shrubs and bushes; during winter retreats deep into leaf piles, burrows beneath ground, or hollow logs to avoid cold temperatures; mates during spring, but courtship unknown; during June or July, females lay clutches of 1–10 eggs, with an average of 5–6; eggs hatch in 2 months; primarily eats insects, readily consuming grasshoppers, caterpillars, and crickets, but will also eat spiders; an active day-time predator that locates prey by sight, slowly approaches it, and grabs it with short rapid strike; due to diet of insects and spiders, highly susceptible to harm from agricultural use of insecticides; predators include other snakes; maximum captive longevity: 6 years, 8 months, and 21 days (Snider and Bowler, 1992).

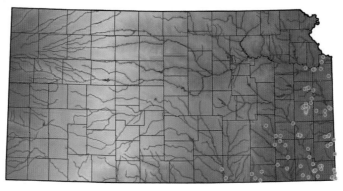

The graceful **Rough Green Snake** (*Opheodrys aestivus*) is found in well-vegetated forests along the rivers and larger streams of the southeastern part of the state, east of a line from Wyandotte County in the northeast to eastern Sumner County along the Oklahoma border. Red dots indicate records backed by voucher specimens or images.

Pertinent References: Anderson (1965), Ballinger and Lynch (1983), Bartlett and Bartlett (2006c), Behler and King (1979), Branson (1904), Burt (1933), Burt and Hoyle (1934), Caldwell and Glass (1976), Clark (1966b), Clarke (1956c, 1980b), Cochran and Goin (1970), Collins (1959, 1974b, 1982a, 1985, 1990b), Collins and Collins (1993a, 2006b, 2009a), Collins et al. (1991, 1994), Conant (1958, 1975), Conant and Collins (1991, 1998), Cook (1954), Cope (1900), Cragin (1885a, 1885d, 1885e), Cross (1971), Davenport (1943), Ditmars (1939), Ernst and Ernst (2003), Fitch (1985b, 1993b), Gier (1967), Gloyd (1928), Gray (1979), Grobman (1984), Guarisco et al. (1982), Hall and Smith (1947), Haltom (1931), Hay (1892), Hurter (1911), Johnson (2000), Jordan (1929), Karns et al. (1974), Kingsbury and Gibson (2002), Knight (1935), Lardie and Black (1981), Leviton (1971), Linzey and Clifford (1981), Loomis (1956), Lueth (1941), Mansueti (1941), McCauley (1945), Minton (1972, 2001), Mitchell (1994), Mount (1975), Murrow (2008a), Perkins (1940), Platt (2003), Plummer (1987), Pope (1937), Riedle (1994a), Riedle and Hynek (2002), Rossi (1992), Rossi and Rossi (1995), Rush and Fleharty (1981), Schmidt (1953), Schmidt and Davis (1941), Schmidt and Murrow (2007c), Shaffer and McCoy (1991), Smith (1950, 1956a), Snelson ands Palmer (1966a), Stejneger and Barbour (1917, 1923, 1933, 1939, 1943), Taggart (1992a, 2000b, 2001a, 2005), Taylor (1892, 1929a), Tennant and Bartlett (1999), Tihen (1937), Trauth et al. (2004), Wagner (1984), Wright and Wright (1952, 1957), and Yarrow (1883).

Great Plains Rat Snake
Pantherophis emoryi (BAIRD & GIRARD, 1853)

Identification: Weakly keeled body scales; divided anal scale; 2 rows of scales on underside of tail; pattern of 25–45 squarish brown blotches on grayish back; checkerboard pattern of white and black or dark gray markings on belly; young resemble adults and almost indistinguishable from young Western Rat Snakes; adult males have longer tails than females.

Size: Adults normally 610–1,220 mm (24–48 inches) in TL; largest specimen from Kansas: male (KU 192376) from Lyon County with TL of 1,343 mm (52⅝ inches) collected by H. A. Stephens on 15 May 1982; maximum length throughout range: 60¼ inches (Conant and Collins, 1998).

 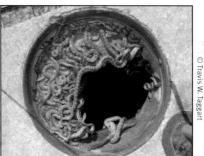

© Travis W. Taggart

| Adult: Ellis County, Kansas. | Den site: Ellis County, Kansas. |

Adults: Ellis County, Kansas.

| Adult: Linn County, Kansas. | Juvenile: Barber County, Kansas. |

Natural History: Roams rocky hillsides and canyons, and frequently inhabits caves in western Kansas; in eastern Kansas, chooses similar habitat in open woods or along woodland edge, avoiding heavily forested regions; recorded active annually in Kansas from late February to early November, with a peak from late April to mid-July; primarily nocturnal, prowling for food; during day, remains hidden beneath rocks or in caves and crevices; nothing known of home range or population density; sometimes killed by controlled prairie fires; three-phase courtship: male chases female, then, jerking and writhing, aligns his body with hers, and finally they copulate for 15 to 30 minutes; egg clutches probably 15–30, with an average of 13; mating probably occurs during spring, after emergence from winter inactivity; constricts prey and feeds primarily on small rodents and birds; in areas where it frequents caves, eats bats; predators include hawks, owls, mammals, and larger snakes; maximum captive longevity: 21 years, 1 month and 25 days (Snider and Bowler, 1992).

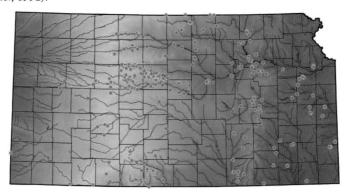

The **Great Plains Rat Snake** (*Pantherophis emoryi*) is found throughout much of Kansas, but is scarce in western fifth of the state. It has not been discovered in the northeastern part of Kansas nor in the extreme southeastern corner; there is an isolated population in Stanton County on the Colorado border. Red dots indicate records backed by voucher specimens or images.

Pertinent References: Ball (1992a), Ballinger and Lynch (1983), Bartlett and Bartlett (2006c), Bartlett and Tennant (2000), Behler and King (1979), Branson (1904), Brennan (1934, 1937), Breukelman and Clarke (1951), Breukleman and Downs (1936), Breukelman et al. (1961), Burbrink (2002), Burt (1931a, 1933, 1935), Burt and Hoyle (1934), Busby and Parmelee (1996), Busby et al. (1994, 1996, 2005), Caldwell and Glass (1976), Capron (1978b), Capron et al. (1982), Cavitt (2000a, 2000b), Clark (1953), Clarke (1958, 1980a, 1980b), Clarke et al. (1958), Cochran and Goin (1970), Collins (1959, 1974b, 1982a, 1985, 1990b, 2006b, 2007b), Collins and Collins (1993a, 1996, 2001, 2006a, 2006b, 2009a), Conant (1958, 1975), Conant and Collins (1991, 1998), Cope (1900), Cragin (1881), Cross (1971), Davenport (1943), Ditmars (1907, 1915, 1939), Dixon and Werler (2005), Dowling (1951, 1952), Ernst and Ernst (2003), Fitch (1970, 1985b, 1991, 1992, 1993b, 1999), Fleharty and Hulett (1988), Fraser (1983a), Garrigues (1962), Gillingham (1979), Gish (1962), Gloyd (1928, 1932, 1935), Guarisco (1981c), Hammerson (1999), Heinrich and Kaufman (1985), Hibbard (1934), Hoyle (1936), Hudson (1942), Hurter (1911), Hurter and Strecker (1909), Iverson (1976), Johnson (2000), Jordan (1929), Karns et al. (1974), Kingsbury and Gibson (2002), Knight (1935), Knight et al. (1973), Linsdale (1925, 1927), Lokke (1983b), Loomis (1956), Loraine (1980), McCoy (1975b), McLeran (1973), Miller (1985b), Mitchell (1979, 1994), Mount (1975), Murrow (2008b, 2009a, 2009b), Rossi (1992), Rossi and Rossi (1995), Roth (1959), Rush and Fleharty (1981), Schmidt (2004l), Schmidt and Murrow (2006, 2007b), Schultz (1996), Schwarting (1984), Schilling and Schaefer (1983), Simmons (1987c), Smith (1950, 1956a, 1961), Smith et al. (1994), Staszko and Walls (1994), Stebbins (1954, 1966, 1985, 2003), Stejneger and Barbour (1917, 1923, 1933, 1939, 1943), Suleiman (2005), Taggart (1992b, 1993, 2000a, 2001a, 2002d, 2003b, 2004g, 2004h, 2005, 2008), Taggart et al. (2007), Taylor (1929a), Tennant and Bartlett

(1999), Toepfer (1993a, 1993b), Trauth et al. (2004), Twente (1955), Viets (1993), Van Doren and Schmidt (2000), Wilgers and Horne (2006), Wilgers et al. (2006), Woodbury and Woodbury (1942), Wright and Wright (1952, 1957), and Yarrow (1883).

Gopher Snake (*aka* Bullsnake)
Pituophis catenifer (BLAINVILLE, 1835)

Identification: Keeled body scales; single anal scale; pattern of 33–73 large brown or black blotches on brownish-yellow body; tail has alternating yellow and black bands; belly yellowish with variable black mottling; adult males have longer tails than females.

Adult: Morton County, Kansas.

Adult: Hamilton County, Kansas.

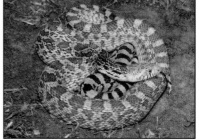

Adult: Ellis County, Kansas.

Adult: Trego County, Kansas.

Adult: Finney County, Kansas.

Subadult: Ellis County, Kansas.

Size: Largest snake in Kansas; adults normally grow 948–1,830 mm (37–72 inches) in TL; largest specimen from Kansas: female (KU 189258) from Harper County with TL of 2,260 mm (88⅝ inches) and weight of about 3,630 grams (8 pounds) collected by M. Kane on 21 May 1981; maximum length throughout range: 105 inches (Devitt et al., 2007).

REPTILES

129

Natural History: Lives in open grasslands, open woodland, and woodland edge; common in cultivated fields with an abundance of rodents, its preferred food; recorded active annually in Kansas from late March to mid-November, with a peak from mid-May to mid-August; generally diurnal, although may be active at night; during day, basks in sun or forages for food; uses snout like a spade to shovel loose soil into curve of head and neck, and then dumps soil away from excavation; this activity may be useful in digging retreats or searching for food; with approach of winter, seeks out deep crevices on rocky hillsides (eastern Kansas) or burrows of small mammals (western Kansas) where normally it remains inactive until spring; may opportunistically emerge from winter retreat whenever temperatures are favorable; mates during April and May after emergence from winter inactivity; courtship involves male crawling along and over female until he eventually rests almost entirely on top of her; during this process, male exhibits jerking body movements; female remains passive, except for elevating and waving tail; just prior to copulation, male may seize female with mouth, biting her on head or neck; then male curls tail beneath hers until their cloacal openings meet; copulation commences and may last over an hour; eggs laid from May to July in soft earth beneath large rocks or logs; egg clutches 3–22, with an average of 13; eggs hatch in August and September; probably most economically beneficial snake in Kansas, consuming large quantities of rodents and saving farmers from much grain loss; eats pocket gophers, rats, mice, rabbits, and ground squirrels; also occasionally eats birds and bird eggs; like Eastern and Western Hognose Snakes, emits loud "hiss" when disturbed or frightened; predators include large carnivorous birds and mammals; young may also be eaten by larger snakes; maximum captive longevity: 33 years and 10 months (Snider and Bowler, 1992).

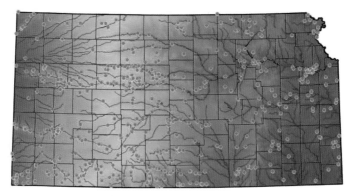

Distribution of the **Gopher Snake** (*Pituophis catenifer*) in Kansas. Also known as the Bullsnake, this large serpent is found throughout the state. Red dots indicate records backed by voucher specimens or images.

Pertinent References: Anderson (1942, 1965), Ball (1992a), Ballinger and Lynch (1983), Bartlett and Bartlett (2006c), Bartlett and Tennant (2000), Baxter and Stone (1980, 1985), Behler and King (1979), Berglund (1967), Branson (1904), Brennan (1934, 1937), Breukelman and Clarke (1951), Breukleman and Downs (1936), Breukelman et al. (1961), Bugbee (1945), Burt (1931a, 1933, 1935), Burt and Burt (1929b, 1929c), Burt and Hoyle (1934), Busby and Parmelee (1996), Busby and Pisani (2007), Busby et al. (1994, 1996, 2005), Caldwell and

Glass (1976), Capron (1985a, 1986c), Capron and Perry (1976), Capron et al. (1982), Carpenter (1940), Cavitt (2000a, 2000b), Clark (1966b), Clarke (1958, 1980a, 1980b), Clarke et al. (1958), Collins (1959, 1974b, 1978b, 1982a, 1985, 1990b, 2001c, 2004, 2006b, 2007b), Collins and Collins (1991, 1993a, 1993b, 1996, 2001, 2006a, 2006b, 2009a), Collins et al. (1991), Conant (1958, 1975), Conant and Collins (1991, 1998), Cragin (1881, 1885a, 1885d), Cross (1971), Dice (1923), Ellis (2002), Ellis and Henderson (1913), Ernst and Ernst (2003), Falls (1933), Fitch (1951, 1956e, 1958b, 1982, 1985b, 1991, 1992, 1993b, 1999, 2000b, 2006a, 2006b), Fitch and Echelle (2006), Fitch and Shirer (1971), Fitch et al. (2003), Fleharty (1995), Fleharty and Hulett (1988), Garrigues (1962), Gish (1962), Gloyd (1928, 1932), Grant (1937), Gress (2003), Hall and Smith (1947), Halpin (1983), Hammerson (1999), Heinrich and Kaufman (1985), Hisaw and Gloyd (1926), Hoyle (1936, 1937), Irwin (1983), Irwin and Collins (1987), Jantzen (1960), Karns et al. (1974), Kauffeld (1969), Kingsbury and Gibson (2002), Klauber (1947), Knight (1935), Knight and Collins (1977), Kretzer and Cully (2001), Linsdale (1925, 1927), Loomis (1956), Marr (1944), Martin (1960), McDowell (1951), McLeran (1973, 1974b, 1975a), Mehrtens (1952), Miller (1976a, 1980a, 1981a, 1981b, 1982a, 1983a, 1985b, 1987b), Minton (2001), Mitchell (1994), Mount (1975), Mozley (1878), Murrell (1993), Murrow (2008a, 2008b, 2009a, 2009b), Parks (1969), Peabody (1958), Perry (1974a), Peterson (1992), Platt (1975, 1984a, 1984b, 1985b, 1989, 1998, 2003), Pope (1927, 1937), Portel (1977), Rainey (1956), Reichling (1995), Reichman (1987), Riedle (1994a, 1998b), Riedle and Hynek (2002), Rossi and Rossi (1995), Royal (1982), Rundquist (1975, 1996a), Rush (1981), Schmidt (2001, 2004i, 2004l), Schmidt and Davis (1941), Schmidt and Murrow (2006, 2007b), Schwaner (1978), Schwardt (1938), Schilling and Schaefer (1983), Shantz et al. (1959), Shirer and Fitch (1970), Shoup (1996b), Simmons (1987c), Smith (1950, 1956a, 1957, 1961), Stebbins (1954, 1966, 1985, 2003), Stull (1940), Suleiman (2005), Sweet and Parker (1990), Taggart (1992a, 1992b, 2000a, 2001a, 2001b, 2002c, 2002d, 2004h, 2005, 2006b, 2006c, 2006d, 2007c, 2008), Taggart et al. (2007), Taylor (1929a, 1929b), Tennant and Bartlett (1999), Terman (1988), Tihen (1937), Tihen and Sprague (1939), Toepfer (1993a), Uhler and Warren (1929, 2008), Van Doren and Schmidt (2000), Viets (1993), Wagner (1984), Warner and Wencel (1978), Werth (1969), Wilgers and Horne (2006), Wilgers et al. (2006), Wolfenbarger (1951, 1952), Wooster (1925), Wright (1996), Wright and Wright (1952, 1957), Yarrow (1883), and Zimmerman (1990).

Longnose Snake
Rhinocheilus lecontei BAIRD AND GIRARD, 1853

Adult: Finney County, Kansas. Adult: Hamilton County, Kansas.

Identification: Smooth body scales; single anal scale; generally white, unpatterned belly but some individuals exhibit dark squarish marks; at least half scales on underside of tail not divided into 2 rows (a character not found in any other harmless snake in Kansas); ground color yellowish or cream with 18–35 black blotches on body separated by pink or reddish interspaces; tail with 6–17 blotches; belly cream or white and usually unpatterned, but occasionally with black or brown markings; adult males have slightly longer tails and grow larger than females.

Adult: Morton County, Kansas.   Adult eating a Six-lined Racerunner (*Aspidos-celis sexlineata*): Barber County, Kansas.

Size: Adults normally grow 560–760 mm (22–32 inches) in TL; largest specimen from Kansas: male (KU 288638) from Barber County with a TL of 877 mm (34½ inches) collected by Steve Kamb on 15 May 1993; maximum length throughout range: 41 inches (Conant and Collins, 1998).

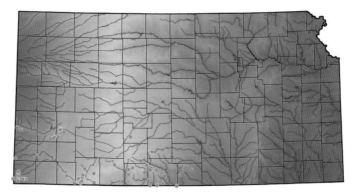

The distribution of the **Longnose Snake** (*Rhinocheilus lecontei*) in Kansas is restricted to the southwestern part of the state, south and west of the Arkansas River; there is an isolated record in Logan County to the north. Red dots indicate records backed by voucher specimens or images.

Natural History: Recorded active annually in Kansas from early March to early September (with a peak from mid-May to early July) on open prairies, sandy regions, and beneath rocks along slopes of canyons; nocturnal, retreating to underground burrows during day; during winter, avoids cold temperatures by burrowing deep beneath ground; mating occurs during spring after emergence from winter inactivity; females lay 4–9 eggs, with an average of 6, in underground nests; eggs hatch in 2–3 months; feeds on small rodents, lizards and lizard eggs, small snakes, and occasional insects; threatened species (protected by state law); predators include large birds, mammals, and other snakes; maximum captive longevity: 19 years, 10 months, and 22 days (Snider and Bowler, 1992).

Pertinent References: Ball (1992a), Ballinger and Lynch (1983), Bartlett and Bartlett (2006c), Bartlett and Tennant (2000), Behler and King (1979), Branson (1904), Burt (1935), Clark (1966b), Clarke (1980a, 1980b), Cochran and Goin (1970), Collins (1959, 1974b, 1982a, 1985, 1990b), Collins and Collins (1991, 1993a, 2006b, 2009a), Collins et al. (1995), Conant (1958, 1975), Conant and Collins (1991, 1998), Cragin (1885c, 1886c), Davenport (1943), Ditmars (1907, 1915, 1939), Ernst and Ernst (2003), Fitch (1993b), Guarisco et al. (1982), Gubanyi

(1992), Hammerson (1999), Irwin (1983), Johnson (1974), Karns et al. (1974), Kingsbury and Gibson (2002), Klauber (1941), Knight (1935), Lardie (1975a), Layher et al. (1986), Levell (1997), Leviton (1971), Medica (1975, 1980), Miller (1981a, 1983c, 1987a), Miller and Gress (2005), Ortenburger and Freeman (1930), Perkins (1940), Platt et al. (1974), Rodríguez-Robles et al. (1999), Rossi and Rossi (1995), Royal (1982), Rush (1981), Rush et al. (1982), Ruthven (1907), Scheffer (1911), Schmidt (1953, 2004l), Schmidt and Davis (1941), Schilling and Nilon (1988), Scott (1996), Simmons (1989), Smith (1942, 1950, 1956a), Smith et al. (1993), Stebbins (1954, 1966, 1985, 2003), Stejneger and Barbour (1943), Taggart (2002c, 2006b, 2006c, 2007c), Taggart et al. (2002, 2007), Taylor (1929a), Tennant and Bartlett (1999), Tihen and Sprague (1939), and Wright and Wright (1952, 1957).

Western Rat Snake
Scotophis obsoletus (SAY, 1823)

Adults: Cherokee County, Kansas.

Adult: Crawford County, Kansas.　　　　Adult: Sumner County, Kansas.

Identification: Large size; keeled body scales; divided anal scale; 2 rows of scales on underside of tail; generally uniform dark brown or black color on head, body, and tail (some adults exhibit indistinct pattern of dark blotches and adults in south-central Kansas may be yellowish with distinct dark brown blotches edged with red); belly cream or yellow-white with large indistinct darker areas; young individuals patterned and colored like Great Plains Rat Snakes, but lose this pattern as they grow older; adult males have slightly longer tails and grow larger than females.

Size: Adults normally grow 1,067–1,830 mm (42–72 inches) in TL; largest specimen from Kansas: male (KU 216168) from Jefferson County with TL of 1,912 mm

(75⅛ inches) and a weight of 1,729 grams (3 pounds, 13 ounces), collected by Roger Christie on 9 July 1990; maximum length throughout range: 101 inches (Boundy, 1995).

Adult: Crawford County, Kansas. Note blue color of eye, characteristic of a snake preparing to shed its skin.

Juvenile: Douglas County, Kansas.

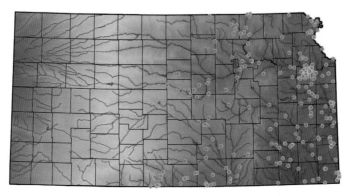

The large and semi-arboreal **Western Rat Snake** (*Scotophis obsoletus*) is found in the eastern half of Kansas, east of a line from Republic County on the Nebraska border southwest through Russell County and thence south to Barber County along the Oklahoma border. Red dots indicate records backed by voucher specimens or images.

Natural History: Inhabits forested areas, particularly rocky hillsides of open woodlands; along western edge of range, frequents wooded areas of streams and rivers; recorded active annually in Kansas from early March to early December, with a peak from late April to mid-July; primarily active by day during spring and fall, but becomes more nocturnal during hot summer; home range 25–30 acres; frequently climbs trees in search of food; population density approximately 1 snake per 3 acres; during fall, retreats to den sites on rocky, wooded hillsides, which it may share with many other kinds of snakes; spends winter months in burrows on hillsides avoiding cold temperatures; usually mates in spring, but occasional unions may occur throughout annual activity period; exhibits 3-phase courtship behavior similar to that of the Great Plains Rat Snake except that males may bite females prior to copulation; egg clutches number 6–44, laid beneath or in logs or in moist soil under rocks during June or July; eggs hatch in 1–2 months; constricts prey; feeds on frogs, lizards, snakes, bird eggs, birds, and small mammals; sometimes robs bird nests and provokes heckling by concentrations of birds; Blue Jays known to attack Western Rat Snakes during nest robbing; hawks main predator in Kansas (mammals and other snakes prey on it as well); hawks, in particular, easily locate this snake when it is being heckled by other birds during nest raiding; sometimes killed by swathing (mowing) machines during agricultural harvests from May to mid-July in southern Kansas; maximum captive longevity: 22 years, 11 months, and 30 days (Snider and Bowler, 1992).

Pertinent References: Anderson (1965), Anderson et al. (1995), Ballinger and Lynch (1983), Bartlett and Bartlett (2006c), Behler and King (1979), Boyd (1991), Branson (1904), Brennan (1934), Breukelman and Clarke (1951), Breukleman and Downs (1936), Breukelman et al. (1961), Burbrink (2001), Burbrink et al. (2000), Burt (1933, 1935), Burt and Hoyle (1934), Busby and Parmelee (1996), Busby and Pisani (2007), Busby et al. (1994, 1996, 2005), Caldwell and Glass (1976), Capron (1978b, 1985a, 1986c), Capron et al. (1982), Cavitt (2000a), Cink (1991), Clark (1966b), Clarke (1958, 1980b), Clarke et al. (1958), Cochran and Goin (1970), Coleman (1987), Collins (1959, 1974b, 1982a, 1984a, 1985, 1990b, 2001b), Collins and Collins (1993a, 1996, 2001, 2005, 2006b, 2009a), Conant (1958, 1975), Conant and Collins (1991, 1998), Cope (1900), Cragin (1881), Cross (1971), Danforth (1956), Ditmars (1915, 1939), Dowling (1951, 1952), Eddy and Babcock (1973), Ellis (2002), Ernst and Ernst (2003), Fitch (1951, 1956e, 1958b, 1963c, 1965b, 1970, 1982, 1985b, 1987, 1991, 1992, 1993b, 1999, 2000b, 2006a), Fitch and Bare (1978), Fitch and Echelle (2006), Fitch and Fitch (1967), Fitch and Hall (1978), Fitch and Shirer (1971), Fitch et al. (2003), Fraser (1983a), Gans (1975), Garrigues (1962), Gillingham (1979), Gloyd (1928, 1932), Grow (1976b), Guarisco (1980b, 1981a), Hall and Smith (1947), Hawthorn (1972), Heinrich and Kaufman (1985), Henderson (1974), Hensley (1959), Hoyle (1936), Hurter (1911), Irwin et al. (1992), Jackson (1976), Johnson (1987, 2000), Karns et al. (1974), Kingsbury and Gibson (2002), Klemens (1993), Knight (1935), Lardie (1999), Lillywhite (1985a, 1985b), Lillywhite and Smith (1981), Linsdale (1925, 1927), Linzey and Clifford (1981), Lokke (1982, 1983b), Loomis (1956), Mann (2007), Martin (1956), McCauley (1945), McLeran (1973), Miller (1976a, 1977a, 1979, 1980a, 1981d, 1982a, 1986b, 1996a, 1996b), Minton (1972, 2001), Mitchell (1994), Mount (1975), Mozley (1878), Murrow (2008a, 2009a, 2009b), Olson (1977), Parker and Brown (1973), Parmelee and Fitch (1995b), Perkins (1940), Perry (1978), Peterson (1992), Platt (1984a, 1984b, 1985b, 1998, 2003), Platt and Rousell (1963), Plummer (1977b), Pope (1944b), Powell (1980a), Prior et al. (2001), Rainey (1956), Riedle (1994a), Riedle and Hynek (2002), Rossi (1992), Rossi and Rossi (1995), Roth (1959), Rues (1986), Rundquist (1996a), Schmidt (1953), Schmidt and Davis (1941), Schmidt and Murrow (2006, 2007c), Schultz (1996), Sexton et al. (1976), Shirer and Fitch (1970), Shoup (1996b), Simmons (1997), Simon and Dorlac (1990), Smith (1950, 1956a), Smits and Lillywhite (1985), Stickel et al. (1980), Suleiman (2005), Taggart (2000b, 2001a, 2002d, 2003g, 2005), Taylor (1929a), Tennant and Bartlett (1999), Terman (1988), Tihen (1937), Trauth et al. (1994), Trauth et al. (2004), Trott (1977), Turner (1977), Viets (1993), Wagner (1984), Warner and Wencel (1978), Wilgers et al. (2006), Wolfenbarger (1951, 1952), Wright and Wright (1952, 1957), and Yarrow (1883).

Ground Snake
Sonora semiannulata BAIRD AND GIRARD, 1853

Identification: Smooth body scales; divided anal scale; head, body, and tail either uniform gray or light reddish brown with no pattern, or may have anywhere from 1 to 25 black crossbands; belly cream or white; adult males have longer tails than females.

Unbanded adult: Clark County, Kansas.   Banded adult: Sumner County, Kansas.

Two adults showing variation in color and
pattern typical of species:
Clark County, Kansas.

Partially-banded adult:
Comanche County, Kansas.

Size: Adults normally grow 215–306 mm (8½–12 inches) in TL; largest specimen from Kansas: female (KU 83179) from Cowley County with TL of 370 mm (14⅝ inches) collected by Charles E. Burt between 1938 and 1941; maximum length throughout range: 18⅞ inches (Conant and Collins, 1998).

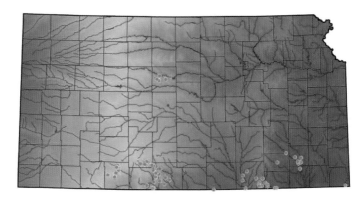

The small **Ground Snake** (*Sonora semiannulata*) is found in localized populations in the southern part of the state, from Cherokee County in extreme southeastern Kansas west along the Oklahoma border to Clark County; to the north, an isolated colony exists in Russell County. Red dots indicate records backed by voucher specimens or images.

Natural History: Inhabits dry, rocky, prairie hillsides and in eastern Kansas may occur in open areas along forest edge; recorded active annually in Kansas from late

March to early October, with a peak from late March through April; little known of daily activity; spends part of day beneath flat rocks or in cracks in dry soil; may be nocturnal; during winter, burrows deep beneath ground and emerges in spring; mates in May or June and occasionally during fall; courtship consists of male rubbing his chin on female's back just behind her head and occasionally biting her; male then curls tail beneath hers until their cloacal openings meet, and copulation occurs; egg clutches 4–6, laid in late June or July and hatch in about 2 months; feeds on spiders, scorpions, centipedes, and various insects; predators include larger snakes, mammals, and birds.

Pertinent References: Anderson (1965), Ball (1992a), Ballinger and Lynch (1983), Bartlett and Bartlett (2006c), Bartlett and Tennant (2000), Behler and King (1979), Branson (1904), Brennan (1934), Burt (1933, 1935), Burt and Hoyle (1934), Capron (1978b), Clark (1966b), Clarke (1980b), Collins (1959, 1974b, 1982a, 1990b, 2006b), Collins and Collins (1993a, 2006b, 2009a), Collins et al. (1991), Conant (1958, 1975), Conant and Collins (1991, 1998), Cragin (1894), Cross (1971), Ditmars (1907, 1939), Ellis and Henderson (1913), Ernst and Ernst (2003), Fitch (1993b), Frost (1983), Gish (1962), Grow (1977), Hall and Smith (1947), Hammerson (1999), Hoyle (1936), Johnson (2000), Jordan (1929), Karns et al. (1974), Kingsbury and Gibson (2002), Knight (1935), Leviton (1971), Miller (1981a, 1982a, 1982b), Perry (1974a), Pope (1937), Reagan (1974b), Rossi and Rossi (1995), Rundquist (1992), Schmidt (1953), Schmidt and Davis (1941), Simmons (1987c), Smith (1950, 1956a), Stebbins (1954, 1966, 1985, 2003), Stejneger and Barbour (1917, 1923, 1933, 1939, 1943), Stickel (1938, 1943), Taylor (1929a), Tennant and Bartlett (1999), Toepfer (1993a), Trauth et al. (2004), Trott (1983), and Wright and Wright (1952, 1957).

Flathead Snake

Tantilla gracilis BAIRD AND GIRARD, 1853

Adult: Chase County, Kansas.   Adult: Cherokee County, Kansas.

Adult: Cherokee County, Kansas.   Adult: Crawford County, Kansas.

Identification: Smallest snake in Kansas; smooth body scales; divided anal scale; unpatterned, light brown body and tail, which becomes gradually darker toward head; dark brown or gray head; 1 scale directly behind each eye; 6 scales on each upper lip; belly salmon pink; adult males have slightly longer tails than females; females grow larger than males.

Size: Adults normally grow 180–203 mm (7–8 inches) in TL; largest specimen from Kansas: female (KU 83480) from Cowley County with TL of 244 mm (9⅝ inches) collected by Charles E. Burt between 1938 and 1941; maximum length throughout range: 9⅝ inches (Conant and Collins, 1998).

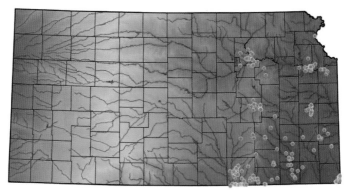

The **Flathead Snake** (*Tantilla gracilis*) is a small reptile found in localized populations throughout the eastern third of the state, from Atchison and Riley counties in the north to Cherokee and Cowley counties in the south. Red dots indicate records backed by voucher specimens or images.

Natural History: Inhabits rocky hillsides of open prairie and woodland; recorded active annually in Kansas from early March to mid-October, with a peak from mid-April to late May; very secretive, spending day beneath flat rocks; may be nocturnal (little known of daily activity cycle); discovered most frequently under rocks after spring rains; during winter, retires beneath ground to avoid cold temperatures; mating occurs during early May, but courtship unknown; females deposit 1–3 eggs beneath rocks in June; eggs hatch in about 7 weeks; young are 3–4 inches long at hatching; feeds primarily on centipedes and soft-bodied insect larvae; predators include birds, small mammals, lizards, and snakes.

Pertinent References: Anderson (1965), Ballinger and Lynch (1983), Bartlett and Bartlett (2006c), Behler and King (1979), Blanchard (1938), Branson (1904), Breukelman and Clarke (1951), Breukleman and Downs (1936), Burt (1933, 1935, 1936b), Burt and Burt (1929b), Burt and Hoyle (1934), Busby and Parmelee (1996), Busby and Pisani (2007), Busby et al. (1994, 1996, 2005), Capron (1978b), Capron et al. (1982), Clark (1966b), Clarke (1980b), Clarke et al. (1958), Cobb (1990, 2004), Cochran and Goin (1970), Cole and Hardy (1981), Collins (1959, 1974b, 1982a, 1990b), Collins and Collins (1993a, 2006b, 2009a), Conant (1958, 1975), Conant and Collins (1991, 1998), Cragin (1885a, 1885d), Cross (1971), Davenport (1943), Ditmars (1907, 1939), Easter (1996), Elick and Sealander (1972), Ernst and Ernst (2003), Fitch (1958b, 1985b, 1991, 1992, 1993b, 1999), Force (1935), Fraser (1983a, 1983b), Garman (1883, 1884), Gier (1967), Hall and Smith (1947), Hallowell (1857a), Hardy and Cole (1968), Heinrich and Kaufman (1985), Hoyle (1936), Johnson (2000), Jordan (1929), Karns et al. (1974), Kingsbury and Gibson (2002), Kirn et al. (1949), Knight (1935), Lardie and Black (1981), Leviton (1971), Loomis (1956), Miller (1976a, 1983a, 1996a), Murrow (2009a, 2009b), Perkins (1940), Perry (1975), Pope (1937), Reichman (1987), Rossi and Rossi (1995), Roth (1959), Schmaus (1959), Schmidt (1953), Schmidt and Davis (1941), Smith (1950, 1956a), Stejneger and Barbour (1939, 1943), Suleiman (2005), Taggart (2000b, 2005, 2008), Taylor (1929a, 1937), Tennant and Bartlett (1999), Tihen (1937), Trauth et al. (1994, 2004), Viets (1993), Wilgers et al. (2006), and Wright and Wright (1952, 1957).

Plains Blackhead Snake
Tantilla nigriceps KENNICOTT, 1860

Identification: Smooth body scales; divided anal scale; unpatterned light yellowish brown body and tail; black head; 2 scales behind each eye; 7 scales on each upper lip; belly pink-red; adult males have slightly longer tails than females.

Adult: Comanche County, Kansas.

Adult: Clark County, Kansas.

Adult: Comanche County, Kansas.

Adult showing coral-pink belly coloration: Comanche County, Kansas.

Juvenile: southwestern Kansas.

Adult: Sumner County, Kansas.

Size: Adults normally grow 180–254 mm (7–10 inches) in TL; largest specimen from Kansas: female (KU 203313) from Phillips County with TL of 375 mm (14⅝ inches) collected by Larry Miller on 18 May 1985; maximum length throughout range: 15¹⁄₁₆ inches (Conant and Collins, 1998).

Natural History: Found on rocky hillsides of grassland prairies and along prairie streams; recorded active annually in Kansas from mid-January through September, with a peak from late April to early June; secretive, spending most of day beneath flat rocks; may be nocturnal (little known of daily activity); during winter, burrows to great depths (up to 8 feet) beneath ground to avoid cold temperatures; nothing known of breeding habits; presumably mates during spring, females deposit egg clutches during summer, and eggs hatch in late summer or fall; eats centipedes and probably soft-bodied grubs; predators include birds, small mammals, lizards, and larger snakes.

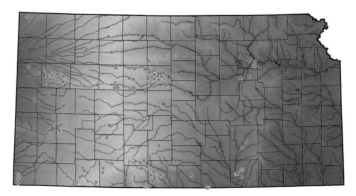

The **Plains Blackhead Snake** (*Tantilla nigriceps*) is a small serpent found in the western three-fourths of the state, from Riley County in the north to Cowley County in the south, and thence westward to the Colorado border. Red dots indicate records backed by voucher specimens or images.

Pertinent References: Ball (1992a), Ballinger and Lynch (1983), Bartlett and Bartlett (2006c), Bartlett and Tennant (2000), Behler and King (1979), Blanchard (1938), Branson (1904), Brennan (1934, 1937), Breukelman and Smith (1946), Bugbee (1945), Burt (1933, 1936b), Burt and Hoyle (1934), Busby and Parmelee (1996), Busby et al. (1994, 1996, 2005), Caldwell and Collins (1977), Carpenter (1940), Clarke (1980a, 1980b), Cochran and Goin (1970), Cole and Hardy (1981), Collins (1959, 1974b, 1979, 1982a, 1985, 1990b, 2006b), Collins and Collins (1991, 1993a, 2006b, 2009a), Conant (1958, 1975), Conant and Collins (1991, 1998), Cope (1900), Cragin (1881), Crampton (1983), Ditmars (1907, 1939), Ellis and Henderson (1913), Ernst and Ernst (2003), Fitch (1993b), Gier (1967), Gish (1962), Grow (1977), Hammerson (1999), Hudson (1942), Jordan (1929), Karns et al. (1974), Kingsbury and Gibson (2002), Knight and Collins (1977), Kretzer and Cully (2001), Lardie and Black (1981), Miller (1980a, 1981a, 1982a, 1985b), Murrow (2008a), Perkins (1940), Pope (1937), Porter (1972), Rossi and Rossi (1995), Royal (1982), Rush (1981), Schmaus (1959), Schmidt (1953, 2004l), Schmidt and Davis (1941), Schmidt and Murrow (2007b), Schwarting (1984), Schilling and Schaefer (1983), Simmons (1987c), Smith (1931, 1950, 1956a), Stebbins (1954, 1966, 1985, 2003), Suleiman (2005), Taggart (1992b, 2000a, 2001b, 2006b, 2006c, 2007c), Taggart et al. (2007), Taylor (1929a, 1929b, 1937), Tennant and Bartlett (1999), Tihen (1937), Toepfer (1993a), Trott (1983), Wright and Wright (1952, 1957), and Yarrow (1883).

HARMLESS REAR-FANGED SNAKES
FAMILY DIPSADIDAE BONAPARTE, 1838

The family Dipsadidae (Harmless Rear-fanged Snakes) is an assemblage that consists of 9 genera and 15 species in the United States (Collins and Taggart, 2009), with updates from the CNAH web site. Five species belonging to 4 genera represent this family in Kansas. The Kansas genera are: *Carphophis*, *Diadophis*, *Heterodon*, and *Hypsiglena*. This family is generally characterized as being fossorial or semi-fossorial, having a smaller number of dorsal scales through reduction of the rows on the back, a few posterior teeth enlarged into fangs (some grooved), and an egg-laying reproductive mode. In Kansas, species in this family are variable in size (up to 45½ inches) and range from small and slender to large and robust.

Western Worm Snake
Carphophis vermis (KENNICOTT, 1859)

Identification: Smooth body scales; divided anal plate; head, body, and tail black or gray-black and contrast sharply with belly color of bright reddish pink, which extends short distance onto sides; males have longer tails than females and keels or ridges on scale rows above cloacal opening; females grow larger than males.

Adult: Chase County, Kansas.

Adult showing coral-red belly coloration: Douglas County, Kansas.

Adult: Franklin County, Kansas.

Adult: Douglas County, Kansas.

Size: Adults normally grow 190–280 mm (7½–11 inches) in TL; largest specimen from Kansas: female (KU 206169) from Doniphan County with TL of 366 mm (14⅜ inches) collected by Robert Powell on 2 May 1986; maximum length throughout range: 15⅜ inches (Conant and Collins, 1998); maximum weight for Kansas specimen: 14.5 grams (½ ounce).

Natural History: Secretive; found beneath rocks or in loose, damp soil of wooded or partly wooded hillsides; sometimes found at woodland edge; recorded active annually in Kansas from mid-March to late October, with peak from mid-April through May; few individuals seen during August because high temperatures and dry soil ctreate inhospitable environment for preferred food, earthworms; during winter, retreats beneath soil to depths of at least 2 feet to avoid cold temperatures; active day and night, but daily activity more dependent on soil temperature; home ranges vary from 700 to 7,500 square feet; males have much larger home ranges than females; population density in northeastern Kansas estimated to be 150–300 snakes per acre within suitable habitat; when burrowing through soil, pushes and rotates head into areas of least resistance, such as crevices and cracks; mates during spring (April and May) and again in fall (September and October); courtship unknown; females normally lay 1–5 eggs per clutch from mid-June to early July;

REPTILES

141

eggs, presumably laid in burrows beneath soil or under rocks, known to hatch in August; feeds exclusively on earthworms; other snakes and moles main predators.

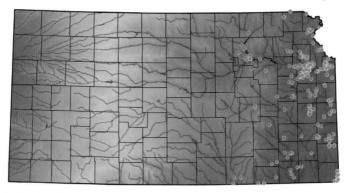

The distribution of the semi-fossorial **Western Worm Snake** (*Carphophis vermis*) in Kansas is restricted to the eastern third of the state, east of a line from Riley County in the north to Cowley County in the south. Red dots indicate records backed by voucher specimens or images.

Pertinent References: Aldridge and Metter (1973), Anderson (1965), Ballinger and Lynch (1983), Bartlett and Bartlett (2006c), Behler and King (1979), Branson (1904), Breukelman and Clarke (1951), Breukleman and Downs (1936), Burt (1933), Burt and Hoyle (1934), Busby and Parmelee (1996), Busby and Pisani (2007), Busby et al. (1994, 1996, 2005), Caldwell and Glass (1976), Clark (1966b, 1967, 1968, 1970b), Clarke (1958, 1980b), Clarke et al. (1958), Coleman (1987), Collins (1959, 1974b, 1982a, 1985, 1990b), Collins and Collins (1993a, 2006b, 2009a), Conant (1958, 1975), Conant and Collins (1991, 1998), Cope (1900), Cragin (1881), Cross (1971), Dice (1923), Ditmars (1907, 1915, 1939), Dixon and Werler (2005), Dloogatch (1978), Eddy and Babcock (1973), Elick and Sealander (1972), Ernst and Ernst (2003), Ernst et al. (2003), Fitch (1956e, 1958b, 1970, 1982, 1985b, 1987, 1991, 1992, 1993b, 1999, 2006a), Fitch and Echelle (2006), Fraser (1983a), Garman (1883), Gier (1967), Gloyd (1928, 1932), Guarisco (1982a), Hay (1892), Heinrich and Kaufman (1985), Henderson (1974), Hudson (1942), Hurter (1911), Johnson (2000), Karns et al. (1974), Kessler (1998), Kingsbury and Gibson (2002), Knight (1935), Linsdale (1925, 1927), Linzey and Clifford (1981), Lokke (2003), Loomis (1956), Mitchell (1994), Mount (1975), Mozley (1878), Murrow (2008a, 2009a, 2009b), Orr (2007), Pearce (1991), Perry (1978), Pisani (2007, 2009b), Platt (1984b), Pope (1937), Riedle (1994a), Riedle and Hynek (2002), Rossi (1992), Schmidt (1953), Schmidt and Davis (1941), Schmidt and Murrow (2007c), Schoener (1977), Shepherd (1973), Smith (1950, 1956a, 1957), Stejneger and Barbour (1917, 1923, 1933, 1939, 1943), Suleiman (2005), Taggart (2001a, 2005, 2008), Taylor (1929a), Tennant and Bartlett (1999), Tihen (1937), Trauth et al. (1994, 2004), Turner (1977), von Frese (1973), Wilgers and Horne (2006), Wright and Wright (1952, 1957), and Yarrow (1883).

Ringneck Snake
Diadophis punctatus (Linnaeus, 1766)

Identification: Smooth body scales; divided anal scale; bright yellow or orange-yellow ring around neck; head, body, and tail shiny gray-black or blue-black except for conspicuous neck ring; belly yellow, with numerous scattered black dots; underside of tail bright red; adult males have longer tails than females and have keels or ridges on scale rows above cloaca; females grow larger than males.

Adult: Osage County, Kansas.

Adult displaying profusely spotted belly
typical of this snake in Kansas:
Osage County, Kansas.

Adult: Rooks County, Kansas.

Normal colored adult (**left**): Logan County,
Kansas. Albino adult (**right**):
Johnson County, Kansas.

Adult beginning defensive tail-curling
display: Clark County, Kansas.

Adult in full tail-curling defensive display:
Clark County, Kansas.

Size: Adults normally grow 254–380 mm (10–15 inches) in TL; largest specimen from Kansas: female (KU 216516) from Shawnee County with TL of 456 mm (17⅞ inches) collected by Mark Ellis on 3 July 1990; maximum length throughout range: 27¹¹⁄₁₆ inches (Conant and Collins, 1998); maximum weight for Kansas specimen: 15.2 grams (slightly over ½ ounce).

Natural History: Inhabits rocky hillsides of open woodlands; recorded active annually in Kansas from early March to early November, with a peak from mid-April to mid-July, but becomes scarce during late July and August, probably due to high temperatures and consequent lack of moist surface soil; apparently prowls at night and spends day resting on moist soil beneath large rocks; rarely basks in sun, preferring to lie beneath sun-warmed rocks to maintain an optimal body temperature; large numbers will congregate beneath single rock; probably uses smell to follow others of its species to the same rock; during winter, retreats well below ground

REPTILES

surface to avoid cold temperatures; in Douglas County, considered to be most abundant vertebrate in upland areas with estimates of up to 730 per acre; mating occurs during spring; courtship unknown; egg clutches 1–10, with an average of 4; eggs laid in late June or early July, probably deep beneath ground, and hatch in late August or early September; males become sexually mature during second season, whereas females require 3 seasons to reach reproductive maturity; number of eggs per clutch increased during breeding seasons when there was ample rain from previous autumn, apparently because earthworms (main diet in Kansas) more available for females to eat; feeds primarily on earthworms, small frogs, insect larvae and occasionally small lizards; estimated to consume 3 times its body weight in earthworms per year; appears to rely heavily on sense of smell to locate prey; when molested or handled, may tightly coil and elevate tail, exposing bright red undersurface; large adult males exhibit tail-coiling behavior less often than smaller individuals and females; this behavior may provide defense mechanism for slower-moving juveniles and gravid females; tail-coiling behavior is defensive and, when accompanied by cloacal discharge, may discourage predators, particularly birds; predators include snakes and raptorial birds; some individuals survive to an age of 15 years.

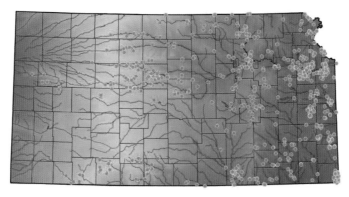

The semi-fossorial **Ringneck Snake** (*Diadophis punctatus*) is abundant in the eastern two-thirds of the state. It ranges west along the Kansas, Arkansas, and Cimarron river systems, but is apparently absent from the northwestern and southwestern corners of Kansas, and much of the upper Arkansas River drainage. Red dots indicate records backed by voucher specimens or images.

Pertinent References: Ball (1992a), Ballinger and Lynch (1983), Bartlett and Bartlett (2006c), Bartlett and Tennant (2000), Behler and King (1979), Blanchard (1942), Boyd (1991), Branson (1904), Brennan (1934, 1937), Breukelman and Clarke (1951), Breukleman and Downs (1936), Buikema and Armitage (1969), Burt (1927b, 1933, 1935), Burt and Hoyle (1934), Busby and Parmelee (1996), Busby and Pisani (2007), Busby et al. (1994, 1996, 2005), Caldwell and Glass (1976), Capron (1978b, 1985c), Capron et al. (1982), Carl (1978), Clark (1966b, 1967), Clarke (1958, 1980a, 1980b), Clarke et al. (1958), Coleman (1987), Collins (1959, 1974b, 1981, 1982a, 1985, 1990b, 2006b, 2007b, 2008d), Collins and Collins (1993a, 2006b, 2009a), Collins et al. (1991), Conant (1958, 1975), Conant and Collins (1991, 1998), Cope (1900), Cragin (1881), Cross (1971), Dice (1923), Ditmars (1915, 1939), Dundee and Miller (1968), Earle (1958), Eddy and Babcock (1973), Elick and Sealander (1972), Ellis (2002), Ellis and Henderson (1913), Ernst and Ernst (2003), Fitch (1956e, 1958b, 1960a, 1965b, 1970, 1975, 1982,

1985b, 1987, 1991, 1992, 1993b, 1999, 2000b, 2004a, 2004b, 2004c, 2006a), Fitch and Echelle (2006), Fitch and Fitch (1967), Fitch and Hall (1978), Fitch et al. (2003), Fontanella et al. (2008), Fraser (1983a, 1983b), Garman (1883), Garrigues (1962), Gehlbach (1974), Gish (1962), Gloyd (1928, 1932), Grow (1977), Guarisco (1980b, 1981c, 1982a, 2001), Hall and Smith (1947), Hammerson (1999), Heinrich and Kaufman (1985), Henderson (1970, 1974), Hensley (1959), Hoyle (1936), Hudson (1942), Hunter et al. (1992), Hurter (1911), Irwin (1983), Iverson (1976), Johnson (1987, 2000), Karns et al. (1974), Kessler (1998), Kingsbury and Gibson (2002), Knight (1935), Linsdale (1925, 1927), Loomis (1956), Maslin (1959), Mc-Nearney (2004), Meshaka (2008), Miller (1976a, 1976d, 1979, 1980a, 1981a, 1982a, 1985b, 1988, 1996a, 1996b, 2006a), Minton (2001), Mitchell (1994), Mount (1975), Mozley (1878), Murrell (1993), Murrow (2008a, 2008b, 2009a, 2009b), Parmelee and Fitch (1995b), Perry (1974a, 1975, 1978), Peterson (1992), Pisani (1974, 2009b), Platt (1984b, 2003), Pope (1937), Reichman (1987), Riedle (1994a), Rossi and Rossi (1995), Roth (1959), Rundquist (1992), Rush and Fleharty (1981), Schmidt (1953, 2004l), Schmidt and Murrow (2006, 2007b, 2007c), Schwarting (1984), Seigel and Fitch (1985), Shirer and Fitch (1970), Simmons (1987c, 1998), Skie and Bickford (1978), Smith (1950, 1956a, 1961, 1976), Sprackland (1985), Stebbins (1954, 1966, 1985, 2003), Stejneger and Barbour (1917, 1923, 1933, 1939, 1943), Suleiman (2005), Taggart (1992b, 2000a, 2000b, 2001a, 2002d, 2003g, 2004h, 2005, 2008), Taggart and Schmidt (2005f), Taggart et al. (2007), Taylor (1892, 1929a), Tennant and Bartlett (1999), Tihen (1937), Tinkle and Gibbons (1977), Toepfer (1993a), Trauth et al. (1994, 2004), Trott (1977, 1983), Viets (1993), Von Achen and Rakestraw (1984), Warner and Wencel (1978), Wilgers and Horne (2006), Wilgers et al. (2006), Wright and Wright (1952, 1957), and Yarrow (1883).

Western Hognose Snake

Heterodon nasicus BAIRD AND GIRARD, 1852

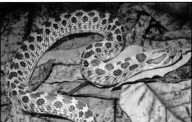

Young adult curled in defensive bluff display: Morton County, Kansas.

Adult: Pawnee County, Kansas.

Identification: Strongly keeled body scales; divided anal scale; sharply turned-up snout; belly and underside of tail extensively colored jet black; dorsal surface of body, head, and tail vary from gray to yellow or light brown; 23–50 dark brown blotches down back, with rows of smaller, similarly colored spots alternating on sides; jet black areas on belly and underside of tail may be edged with yellow; adult males have fewer blotches on back and longer tails than females; females grow larger than males.

Size: Adults normally grow 380–635 mm (15–25 inches) in TL; largest specimen from Kansas: female (KU 218806) from Morton County with TL of 918 mm (36 inches) collected by Robert L. Ball on 20 May 1991; maximum length throughout range: 39½ inches (Conant and Collins, 1998).

Adult: Finney County, Kansas.

Ventral surface of a Western Hognose Snake
(**left**) and an Eastern Hognose Snake (**right**):
Trego County, Kansas.
Note difference in pattern and color.

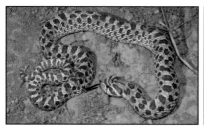

Adult: Hamilton County, Kansas.

Adult: Sumner County, Kansas.

Natural History: Generally found in grassland or sand prairie in western Kansas; most records in eastern part of state probably from isolated populations in sandy areas; recorded active annually in Kansas from April to October with a peak of activity in June and July; diurnal, active primarily in morning and late afternoon; when not active, burrows beneath sandy loose soil (rarely found beneath cover like other snakes); home range quite variable, depending on available habitat and food; not territorial; population density may be as much as 1–3 snakes per acre; during winter, burrows deep beneath ground to avoid cold temperatures; usually mates during May after emerging from winter inactivity, but few matings occur in fall; egg clutches 4–23, with an average of 9; eggs laid in July in nests a few inches below soil; females evidently deposit a clutch every other year; incubation time for eggs is 50–60 days; courtship not observed; detects prey by smell, digging a food item from beneath soil; upturned snout used to dig up and eat toads, reptile eggs, small lizards, and snakes; also consumes rodents and birds; species in need of conservation (protected by state law); predators not well known; visibly enlarged tooth on each side of upper jaw at rear of mouth, presumably an adaptation to puncture toads in order to swallow them (toads typically swell their bodies with air when grabbed); exhibits interesting defensive behavior; when approached, generally attempts to escape by crawling clumsily away; when more closely threatened, may attempt to conceal head beneath coils; if this does not dissuade intruder, spreads a "hood" by flattening neck and hisses loudly; occasionally may "strike" at intruder, but in all instances strike short and done with mouth closed; if this fails to frighten intruder, will writhe and contort, disgorge recently eaten food, roll over on back, and "play dead;" may remain "dead" for up to 5 minutes; if left alone, rolls over on belly and crawls away; maximum captive longevity: 19 years, 10 month, and 28 days (Snider and Bowler, 1992).

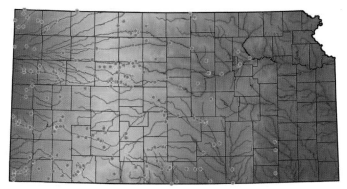

The **Western Hognose Snake** (*Heterodon nasicus*) occurs throughout the western three-quarters of Kansas. In the northern part of the state, its range extends east to Riley County and thence southeast to the eastern borders of Chautauqua, Elk, and Greenwood counties. Red dots indicate records backed by voucher specimens or images.

Pertinent References: Anderson (1945, 1965), Ball (1992a), Ballinger and Lynch (1983), Barten (1980), Bartlett and Bartlett (2006c), Bartlett and Tennant (2000), Baxter and Stone (1980, 1985), Behler and King (1979), Bender and Stark (2007), Berglund (1967), Branson (1904), Brennan (1934, 1937), Breukelman and Clarke (1951), Bugbee (1945), Burt (1933, 1935), Burt and Burt (1929b, 1929c), Burt and Hoyle (1934), Busby and Parmelee (1996), Busby et al. (1994, 1996, 2005), Capron (1978b), Carpenter (1940), Clarke (1980a, 1980b), Cochran and Goin (1970), Collins (1959, 1974b, 1982a, 1984a, 1985, 1990b), Collins and Collins (1991, 1993a, 2006b, 2009a), Collins et al. (1991), Conant (1958, 1975, 1978a), Conant and Collins (1991, 1998), Cope (1900), Cragin (1881), Diener (1956, 1957a), Ditmars (1907), Edgren (1952b, 2001, 2004), Ellis (2002), Ellis and Henderson (1913), Ernst and Ernst (2003), Falls (1933), Fitch (1970, 1985b, 1993b), Garrigues (1962), Gish (1962), Gloyd (1929), Grant (1937), Gress (2003), Guarisco (1981c), Hammerson (1982, 1999), Hay (1892), Hudson (1942), Jan and Sordelli (1865), Johnson (1987, 2000), Karns et al. (1974), Kingsbury and Gibson (2002), Knight (1935), Knight and Collins (1977), Layher et al. (1986), Levell (1997), Marr (1944), Miller (1980a), Monro (1949b, 1949c, 1949d), Mozley (1878), Platt (1969, 1975, 1983, 1984a, 1984b, 1985b, 1989, 1998, 2003), Preston (1982), Rossi and Rossi (1995), Royal (1982), Rush (1981), Schmidt (2001, 2004l), Schmidt and Davis (1941), Schoener (1977), Schwarting (1984), Schilling and Schaefer (1983), Simmons (1987c, 1989), Smith (1950, 1956a, 1957), Smith and Smith (1962), Smith et al. (2003), Stebbins (1954, 1966, 1985, 2003), Stejneger and Barbour (1917, 1923, 1933, 1939, 1943), Taggart (1992a, 2000a, 2001b, 2002c, 2004h, 2006b, 2006c, 2007c), Taggart et al. (2007), Taylor (1929a, 1929b), Tennant and Bartlett (1999), Tihen (1937), Tihen and Sprague (1939), Tinkle and Gibbons (1977), Toepfer (1993a), Turner (1977), Van Doren and Schmidt (2000), Walley and Eckerman (1999), Werth (1969), Wolfenbarger (1951, 1952), Wooster (1925), Wright and Wright (1952, 1957), and Yarrow (1883).

Eastern Hognose Snake

Heterodon platirhinos LATREILLE, 1801

Identification: Strongly keeled body scales; divided anal scale; sharply upturned snout; dark-colored belly with underside of tail much lighter; highly variable in dorsal color; back, head, and tail may be yellow, brown, reddish, olive, or gray, with series of 20–30 dark brown or black blotches on back and similarly colored bands

REPTILES

147

on tail; sides of body have 2–3 series of small, dark spots alternating with blotches on back; belly yellowish, gray, olive, or reddish and becomes darker toward cloaca; underside of tail and chin usually much lighter than belly; adult males have longer tails and fewer blotches on back than females; females grow larger than males.

Adult gaping its mouth in threat display that is all bluff: Gove County, Kansas. Note small fangs, one each on rear of upper jaw.

Juvenile: Riley County, Kansas.

Adult: Ellis County, Kansas.

Adult spreading skin on neck to appear larger and more threatening in this defensive display: Gove County, Kansas.

Adult: Logan County, Kansas.

Adult: Barber County, Kansas.

Size: Adults normally attain 510–760 mm (20–30 inches) in TL; largest specimen from Kansas: female (KU 221200) from Harper County with TL of 1,097 mm (43⅛ inches) collected by Kevin Albright on 23 August 1992; maximum length throughout range: 45½ inches (Conant and Collins, 1998).

Natural History: Lives in widely varying habitats, from forested areas of eastern Kansas west along prairie rivers and streams to Colorado border; prefers sandy floodplains; most common in valleys of major rivers; recorded active annually in Kansas from April to October with a peak from May to mid-July; daily activity period and habits correspond to those of the Western Hognose Snake; home ranges larger than those of the Western Hognose Snake but poorly defined; population density no more than 1 per acre; during winter months, burrows deep into loose soil or sand to avoid cold temperatures; mating occurs during April and May after emergence from winter inactivity; courtship unknown; females lay single clutch of eggs each year; number of eggs 4–61, with an average of 22; eggs deposited in late June or July in nests burrowed out several inches below soil or sand by female; incubation requires 50–65 days; feeds primarily on frogs and toads, although recorded eating salamanders; upturned snout useful in digging for toads; visibly enlarged tooth on each side of upper jaw at rear of mouth, presumably an adaptation that permits snake to puncture toads in order to swallow them (toads typically swell their bodies with air when grabbed); species in need of conservation (protected by state law); defensive behavior resembles that of the Western Hognose Snake except that immediate attempts to escape by crawling away or hiding head beneath body not as frequent; instead, more readily spreads "hood," hisses and "strikes;" apparently, engages in "playing dead" for much longer periods than the Western Hognose Snake; maximum captive longevity: 7 years, 8 month, and 15 days (Snider and Bowler, 1992).

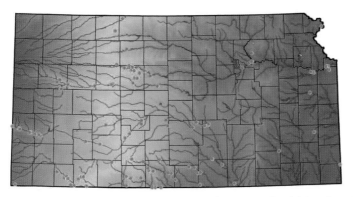

The **Eastern Hognose Snake** (*Heterodon platirhinos*) occurs on floodplains along river drainages throughout most of the state. It is apparently absent from many of the northern border counties and the northwestern corner of Kansas. The distribution of this reptile in the eastern part of the state is sporadic. Red dots indicate records backed by voucher specimens or images.

Pertinent References: Axtell (1983), Ball (1992a), Ballinger and Lynch (1983), Bartlett and Bartlett (2006c), Behler and King (1979), Blem (1981), Branson (1904), Brennan (1934), Breukelman and Clarke (1951), Bugbee (1945), Burt (1933, 1935), Burt and Burt (1929b), Burt and Hoyle (1934), Busby and Parmelee (1996), Busby et al. (1994, 1996, 2005), Clark (1966b), Clarke (1980a, 1980b), Clarke et al. (1958), Cochran and Goin (1970), Collins (1959, 1974b, 1982a, 1985, 1990b, 1994b), Collins and Collins (1991, 1993a, 1996, 2001, 2006b, 2009a), Collins et al. (1991), Conant (1958, 1975), Conant and Collins (1991, 1998), Cope (1900), Cragin (1881), Cross (1971), Ditmars (1907, 1939), Edgren (1952a, 1952b, 1957, 1961, 2001), Ernst and Ernst (2003), Fitch (1970, 1985b, 1993b), Fitzgerald (1994), Fitzgerald and Nilon (1994), Freeman and Kindscher (1992), Garrigues (1962), Gish (1962), Gloyd (1928, 1932), Grant (1937), Hall and Smith (1947), Hammerson (1982, 1999), Irwin (1983), Johnson (1987), Karns et al. (1974), Kingsbury and Gibson (2002), Klemens (1993), Knight (1935), Layher et al. (1986), Levell (1997), Linzey and Clifford (1981), Loomis (1956), Miller (1976a, 1980a), Minton (2001), Mitchell (1994), Mount (1975), Mozley (1878), Nulton and Rush (1988), Palmer and Snelson (1966), Perry (1974a), Platt (1969, 1975, 1983, 1984b, 1985b,

1989, 1998, 2003), Pope (1937), Rossi (1992), Rossi and Rossi (1995), Royal (1982), Rush and Fleharty (1981), Schmidt (2001, 2004l), Schmidt and Murrow (2007b), Schoener (1977), Schwarting (1984), Simmons (1989), Smith (1950, 1956a, 1961), Stebbins (1954), Taggart (1992a, 1992g, 2000a, 2001b, 2006b, 2007b, 2008), Taggart et al. (2007), Taylor (1929a, 1929b), Tennant and Bartlett (1999), Tihen (1937), Toepfer (1993a), Trauth et al. (2004), Turner (1977), Uhler and Warren (1929, 2008), Van Doren and Schmidt (2000), Wagner (1984), Willimon and Collins (1996), Wooster (1925), Wright and Wright (1952, 1957), and Yarrow (1883).

Chihuahuan Night Snake
Hypsiglena jani (DUGÈS ,1865)

Identification: Smooth body scales; divided anal scale; pupils that close to vertical slits when exposed to strong light (a character also present in venomous snakes in Kansas); body and tail gray or grayish yellow with 50–70 dark brown blotches on back; head gray or brown and neck has 3 elongate large brown blotches; sides of body have 2–3 rows of smaller brown spots, which alternate with blotches on back; external sexual differences unknown.

Adult: Comanche County, Kansas.   Juvenile: Barber County, Kansas.

Adult: Barber County, Kansas.   Adult: Barber County, Kansas.

Size: Adults normally grow 360–410 mm (14–16 inches) in TL; largest specimen from Kansas: female (KU 193259) from Clark County with TL of 412 mm (16³⁄₁₆ inches) collected by John Tollefson, Martin Capron, and Chris Stammler on 29 May 1983; maximum length throughout range: 22⅞ inches (Boundy, 1995).

Natural History: Found beneath rocks on hillside slopes and along canyon rims; recorded active annually in Kansas from late May to late October, with a peak from mid-April to mid-May; nocturnal; vertically slit eyes are an adaptation for prowling at night; little known of breeding cycle in Kansas; presumably, mating occurs shortly after emergence from winter inactivity; females probably produce 1 clutch of 2–6 eggs per summer; eggs hatch within 2 months; may have lengthy breeding season; feeds principally on small lizards and other reptile eggs, as well as other small snakes; predators probably consist of larger snakes; maximum captive longevity: 9 years and 3 months (Snider and Bowler, 1992).

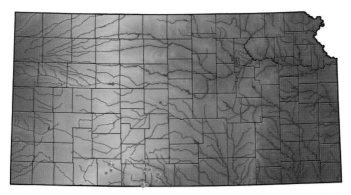

The distribution of the semi-fossorial **Chihuahuan Night Snake** (*Hypsiglena jani*) in Kansas is restricted to the south-central part of the state. The population in northern Clark County is isolated from the Barber and Comanche county populations, which extend into Oklahoma. Red dots indicate records backed by voucher specimens or images.

Pertinent References: Ball (1992a), Ballinger and Lynch (1983), Bartlett and Bartlett (2006c), Bartlett and Tennant (2000), Behler and King (1979), Clarke (1980b), Cochran and Goin (1970), Collins (1959, 1974b, 1982a, 1985, 1990b), Collins and Collins (1993a, 2006b, 2009a), Collins et al. (1995), Conant (1958, 1975), Conant and Collins (1991, 1998), Davenport (1943), Ditmars (1939), Ernst and Ernst (2003), Fitch (1970, 1993b), Grow (1977), Guarisco et al. (1982), Hammerson (1999), Hibbard (1937), Johnson (1974), Karns et al. (1974), Kingsbury and Gibson (2002), Lardie (1990), Layher et al. (1986), Levell (1997), Leviton (1971), Miller (1983c, 1987a), Mulcahy (2008), Murrell (1993), Perkins (1940), Platt et al. (1974), Rossi and Rossi (1995), Rundquist (1992, 2002), Schmaus (1959), Schmidt (1953), Schmidt and Davis (1941), Schilling and Nilon (1988), Scott (1996), Shaw and Campbell (1974), Simmons (1989), Smith (1950, 1956a), Stebbins (1954, 1966, 1985, 2003), Stejneger and Barbour (1939, 1943), Taggart (2006b), Tanner (1946), Tennant and Bartlett (1999), Tihen (1937), Vitt (1975), and Wright and Wright (1952, 1957).

HARMLESS LIVE-BEARING SNAKES
FAMILY NATRICIDAE BONAPARTE, 1840

The family Natricidae (Harmless Live-bearing Snakes) consists of 8 genera and 38 species in the United States (Collins and Taggart, 2009). In Kansas, this family is represented by 13 species in 6 genera, *Nerodia*, *Regina*, *Storeria*, *Thamnophis*, *Tropidoclonion*, and *Virginia*. Its members are semiaquatic or semi-fossorial, feed primarily on cold-blooded prey, and have (in North America) a live-bearing reproductive mode.

REPTILES

151

Plainbelly Water Snake
Nerodia erythrogaster (FORSTER, 1771)

Identification: Keeled body scales; divided anal scale; uniform cream or yellowish belly with indistinct dark shading on edges of scales; uniform cream or yellowish unpatterned underside of tail; head, body, and tail either uniform dark gray, olive, or brown (old adults) or with 30–40 blotches of similar color separated by very indistinct light bands (young adults); in contrast to adults, young individuals strikingly marked with dark brown bands and/or blotches separated by light bands; adult males have longer tails than females; females grow larger than males.

Adult: Crawford County, Kansas.
Note blue color of eye, characteristic of
a snake preparing to shed its skin.

Adult beginning to shed skin:
Barber County, Kansas.

Adult: Butler County, Kansas.

Adult: Linn County, Kansas.

Adults in mating embrace:
Linn County, Kansas.

Juvenile Plainbelly Water Snake (**right**);
juvenile Northern Water Snake (**left**):
Crawford County, Kansas. Note
differences in color, pattern, and eye size.

Size: Adults normally grow 760–1,220 mm (30–48 inches) in TL; largest specimen from Kansas: female (KU 203662) from Pratt County with TL of 1,415 mm (55½ inches) collected by Tom Dillenbeck on 9 March 1985; maximum length throughout range: 62 inches (Conant and Collins, 1998).

Natural History: Inhabits swamps, marshes, ponds, and slow-moving portions of streams; least aquatic of Kansas water snakes; wanders great distances from water during summer; recorded active annually in Kansas from early March to early November, with a peak from late April to early July; during day, basks near water on driftwood, low-hanging branches, stumps, or brush; at night, forages for food, sometimes remaining active until after midnight; seeks cover during cooler temperatures of fall beneath large brush or rock piles or in underground burrows of other animals, remaining inactive until spring; mating occurs during April and May; courtship not reported; young born in late July, August, and September; number of young per brood 4–30; feeds primarily on Leopard Frogs and Bullfrogs, but young apparently prefer fishes; when hunting for fishes trapped in drying pools, large adults demonstrate an "entrapment" behavior by whipping coils of body in C-shaped curves, using head to herd fishes while encircling them, and then lashing head about with mouth open inside circle of body, seizing any fish on contact; this behavior reported in Diamondback Water Snakes and Northern Water Snakes; Common Snapping Turtles prey on young and people appear to be chief predators of adults; see account of Northern Water Snake for comments on function of contrasting pattern in young; maximum captive longevity: 14 years, 11 months and 17 days (Snider and Bowler, 1992).

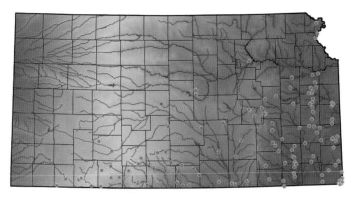

The semi-aquatic **Plainbelly Water Snake** (*Nerodia erythrogaster*) is found in the southeastern half of the state, generally southeast of line from Doniphan County in the northeastern corner of Kansas to Seward County in the southwestern corner. Red dots indicate records backed by voucher specimens or images.

Pertinent References: Anderson (1965), Anderson et al. (1995), Ball (1992a), Ballinger and Lynch (1983), Bartlett and Bartlett (2006c), Bartlett and Tennant (2000), Behler and King (1979), Boyd (1991), Branson (1904), Breukelman and Clarke (1951), Breukleman and Downs (1936), Burt (1933, 1935), Burt and Burt (1929b), Burt and Hoyle (1934), Busby and Parmelee (1996), Busby and Pisani (2007), Busby et al. (1994, 1996, 2005), Caldwell and Glass (1976), Capron (1991), Clark (1966b), Clarke (1958, 1980a, 1980b), Clarke et al. (1958), Clay (1936, 1938), Cochran and Goin (1970), Collins (1959, 1974b, 1982a, 1990b, 2006b, 2007b), Collins and Collins (1993a, 1996, 2001, 2006b, 2009a), Collins et al. (1994), Conant (1958, 1975, 1978a, 1978b), Conant and Collins (1991, 1998), Cragin (1881), Cross (1971), Davenport (1943), Diener (1957b), Ditmars (1907, 1939), Dolman (1929), Ernst and Ernst (2003), Fitch (1985b, 1993b), Garrigues (1962), Gibbons and Dorcas (2004), Gier (1967), Gloyd (1928, 1932), Grow (1976b), Hall and Smith (1947), Hoyle (1936), Hurter (1911), Irwin (1983), Johnson (1987, 2000), Karns et al. (1974), Kingsbury and Gibson (2002), Knight (1935), Loomis

(1956), Mara (1995), Marr (1944), McCauley (1945), Miller (1976a, 1976d, 1980a, 1982a, 1988), Mitchell (1994), Mount (1975), Murrow (2008a, 2009a, 2009b), Perkins (1940), Perry (1978), Platt (1975, 1984a, 1985b, 2003), Pope (1937), Riedle (1994a), Riedle and Hynek (2002), Rossi (1992), Rundquist (1992), Schmidt (1953, 2003a), Schmidt and Davis (1941), Schmidt and Murrow (2007b, 2007c), Schmidt and Taggart (2005b), Smith (1950, 1956a, 1983), Stebbins (1954, 1966, 1985, 2003), Stejneger and Barbour (1933, 1939, 1943), Suleiman (2005), Taggart (2000b, 2001a, 2005), Taylor (1929a), Tennant and Bartlett (1999), Tihen (1937), Tihen and Sprague (1939), Toland and Dehne (1960), Trauth et al. (2004), Van Doren and Schmidt (2000), Viets (1993), Wagner (1984), Warner and Wencel (1978), Wolfenbarger (1951, 1952) and Wright and Wright (1952, 1957).

Diamondback Water Snake
Nerodia rhombifer (HALLOWELL, 1852)

Identification: Keeled body scales; divided anal scale; dark half-moon shaped spots scattered irregularly on yellow belly; 30–65 narrow dark brown, black, or gray bands on light gray or yellowish gray body; young individuals have same patterns and colors as adults; adult males have numerous tiny raised bumps on chin; females have shorter tails than males and grow slightly larger.

Young adult: Elk County, Kansas.

Litter: Cherokee County, Kansas.

Adult: Sedgwick County, Kansas.

Adult: Wabaunsee County, Kansas.

Size: Adults normally attain 760–1,220 mm (30–48 inches) in TL; largest specimen from Kansas: female (KU 204483) from Douglas county with TL of 1,415 mm (55½ inches) collected by Gregory Beilfuss on 25 March 1986; heaviest example from state: female (KU 207187) from Lyon County that weighed 1,750 grams (3 pounds, 14 ounces) collected by B. Stapp on 18 May 1987; maximum length throughout range: 69 inches (Boundy, 1995).

Natural History: Usually inhabits permanent lakes, marshes, and swamps, and backwaters of rivers, although may have preference for fast-flowing water when available; recorded active annually in Kansas from early March to late October, with a peak from late April to mid-June; basks during day on brush, logs, and grassy banks along water edge; in summer, searches for food at night; like all members of genus, quite irritable and will not hesitate to bite; large individuals can inflict a painful wound; during winter, retreats into muskrat dens and other burrows around water to avoid cold temperatures; mates in spring after emergence from winter inactivity; although courtship not observed, probably involves male rubbing female's back with his bumpy chin; litters large, number of young per litter 13–62; young usually born from August to early October; feeds primarily on slow-moving and dead fishes; uses an "entrapment" behavior when preying on fishes in drying pools (see account of Plainbelly Water Snake for details); predators include mammals (especially people) and large turtles; maximum captive longevity: 4 years and 11 months (Snider and Bowler, 1992).

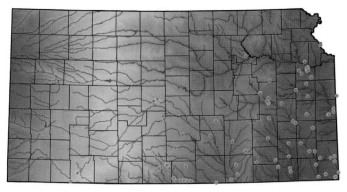

The large, semi-aquatic **Diamondback Water Snake** (*Nerodia rhombifer*) is found in the southeastern half of the state, generally southeast of line from northern Leavenworth County in northeastern Kansas southwestward to Cheyenne Bottoms in Barton County and thence southwestward to Seward County in the southwestern corner of the state. Red dots indicate records backed by voucher specimens or images.

Pertinent References: Anderson (1965), Ballinger and Lynch (1983), Bartlett and Bartlett (2006c), Behler and King (1979), Berglund (1967), Boyd (1991), Branson (1904), Breuklelman and Clarke (1951), Breuklelman and Downs (1936), Burt and Hoyle (1934), Busby and Parmelee (1996), Busby et al. (1994, 1996, 2005), Caldwell and Glass (1976), Capron (1991), Capron and Perry (1976), Clark (1966b), Clarke (1980a, 1980b), Clarke et al. (1958), Cochran and Goin (1970), Collins (1959, 1974b, 1982a, 1990b, 2001c), Collins and Collins (1993a, 1993b, 1996, 2001, 2006a, 2006b, 2009a), Collins et al. (1994), Conant (1958, 1975, 1978b), Conant and Collins (1991, 1998), Cragin (1881), Cross (1971), Crow (1913), Diener (1956, 1957b), Ditmars (1939), Dolman (1929), Ernst and Ernst (2003), Fitch (1985b, 1993b), Fleharty and Ittner (1967), Gibbons and Dorcas (2004), Gier (1967), Gloyd (1928, 1932), Grant (1937), Grow (1976b), Hall and Smith (1947), Hurter (1911), Irwin and Collins (1987), Karns et al. (1974), Kingsbury and Gibson (2002), Knight (1935), Loomis (1956), Mara (1995), Miller (1976a, 1979, 1982a, 1983a), Minton (2001), Mount (1975), Mozley (1878), Murrow (2009a), Perkins (1940), Platt (1975, 1985b, 1998, 2003), Rossi (1992), Rossi and Rossi (1995), Schmidt (1998), Schmidt and Davis (1941), Smith (1950, 1956a), Taggart (2000b, 2001a), Taylor (1929a), Tennant and Bartlett (1999), Tihen (1937), Trauth et al. (2004), Van Doren and Schmidt (2000), Wagner (1984), Wiley and Mayden (1985), Wright and Wright (1952, 1957), and Zimmerman (1990).

Northern Water Snake
Nerodia sipedon (LINNAEUS, 1758)

Identification: Keeled body scales; divided anal scale; dark brown, orange, yellow, red, and gray half-moon and speckled pattern widely spaced and scattered on front part of gray or whitish belly (markings closer together or fused on rear portion of belly, making it appear darker); dark, complete bands on front part of body; alternating rows of dark blotches on back and sides of rear portion of body; may be gray or light brown, with dark gray or brown bands and blotches (young and young adults), or uniform dark gray with no pattern (old adults); young individuals more brightly colored and have more contrasting pattern than adults; adult males have longer tails than females; females grow larger than males.

Size: Adults attain 560–1,067 mm (22–42 inches) in TL; largest specimen from Kansas: sex undetermined (KU 288637) from Jackson County with TL of 1,208 mm (47½ inches) collected by James Gubanyi on 21 October 1998; maximum length throughout range: 59 inches (Conant and Collins, 1998); maximum weight for Kansas specimen: 480 grams (1 pound, 1 ounce).

Adult: Sumner County, Kansas.

Adult.

Adult: Linn County, Kansas.

Adult: Johnson County, Kansas.

Three juveniles: Douglas County, Kansas.

Habitat: Montgomery County, Kansas.

Natural History: Found in almost any aquatic situation, from fast-flowing rocky streams and rivers to swamps, lakes, and marshes; appears to become active at lower temperature than most other water snakes; recorded active annually in

Kansas from late January through October, with a peak from early May to late July; active during day in spring and fall; becomes more nocturnal in summer; spends much time basking in sun on branches and logs that overhang, or are near, water; can be discovered easily by looking under rocks and other cover along streams; spends cold winter months in variety of retreats, ranging from deep, dry crevices and holes on rocky, wooded hillsides to below water level in crayfish burrows in lowland areas; mating occurs during spring, and courtship involves male positioning himself alongside female and rubbing chin on her neck while he spasmodically jerks his body to stimulate copulation; more than 1 male may copulate with single female; young born from August to October; number of young per litter 6–66, with an average of 20–25; some females may produce young only every other year; fishes make up over 75% of diet; also eats salamanders, frogs, and toads; uses an "entrapment" behavior when preying on fishes in drying pools (see account of Plainbelly Water Snake for details); change from contrasting bright body color and pattern of young to more uniform darker pattern of old adults related to change in habitat and pressure from predators; young individuals usually inhabit small, rocky, fast-flowing streams, and their color and pattern blend readily with surrounding debris; adults tend to migrate to larger, more open bodies of water where contrasting pattern and bright colors would be too conspicuous (reason for change in color pattern with age undoubtedly holds for Plainbelly Water Snake as well); maximum captive longevity: 9 years, 7 months, and 24 days (Snider and Bowler, 1992).

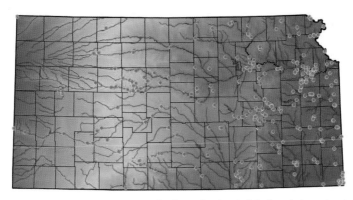

The semi-aquatic **Northern Water Snake** (*Nerodia sipedon*) is found throughout most Kansas. It is least abundant in the western fourth of the state and apparently is absent from southwestern Kansas. This reptile has the most northern distribution in North America of any member of its genus, hence the common name, Northern Water Snake. Red dots indicate records backed by voucher specimens or images.

Pertinent References: Anderson (1965), Anderson et al. (1995), Ballinger and Lynch (1983), Bartlett and Bartlett (2006c), Bartlett and Tennant (2000), Beatson (1976), Behler and King (1979), Berglund (1967), Branson (1904), Brennan (1934, 1937), Breukelman and Clarke (1951), Breukleman and Downs (1936), Burkett (1984), Burt (1933, 1935), Burt and Burt (1929b), Burt and Hoyle (1934), Busby and Parmelee (1996), Busby et al. (1994, 1996, 2005), Capron (1975a, 1991), Capron and Perry (1976), Capron et al. (1982), Clark (1966b), Clarke (1980a, 1980b), Clarke et al. (1958), Clay (1936), Coleman (1987), Collins (1959, 1974b, 1982a, 1990b, 2001c, 2007b), Collins and Caldwell (1978), Collins and Collins (1993a, 1993b, 1996, 2001, 2006a, 2006b, 2009a), Collins et al. (1991, 1994), Conant (1958, 1963, 1975), Conant and Collins (1991, 1998), Cook (1954), Cope (1900), Cragin (1881), Cross (1971), Dice

(1923), Diener (1957b), Ditmars (1915, 1939), Dolman (1929), Ellis (2002), Ellis and Henderson (1913), Ernst and Ernst (2003), Falls (1933), Fitch (1951, 1956e, 1958b, 1982, 1985b, 1991, 1992, 1999, 2000b, 2006a), Fitch and Echelle (2006), Fitch and Shirer (1971), Fitch et al. (2003), Garrigues (1962), Gibbons and Dorcas (2004), Gish (1962), Gloyd (1928, 1932), Grant (1937), Gray (1979), Grow (1976b), Guarisco (1981c), Hall and Smith (1947), Hammerson (1999), Heinrich and Kaufman (1985), Henderson (1974), Hoyle (1936), Hunter et al. (1992), Irwin and Collins (1987), Johnson (2000), Karns et al. (1974), Kingsbury and Gibson (2002), Klemens (1993), Knight (1935), LaForce (2004), Lardie and Black (1981), Linsdale (1925, 1927), Loomis (1956), Loraine (1980), Mara (1995), Miller (1977c, 1979, 1980a, 1981c, 1981d, 1982a, 1983a, 1996a, 1996b), Minton (2001), Mitchell (1994), Mount (1975), Mozley (1878), Murrow (2008a, 2009a, 2009b), Nulton and Rush (1988), Ortenburger (1926b), Peterson (1992), Pilch (1982), Platt (1975, 2003), Pope (1937, 1944b), Riedle (1994a, 1998a), Riedle and Hynek (2002), Rossi (1992), Rossi and Rossi (1995), Rush and Fleharty (1981), Savage (1959), Schmaus (1959), Schmidt (1953, 2000, 2009a), Schmidt and Davis (1941), Schmidt and Murrow (2006, 2007c), Shirer and Fitch (1970), Shoup (1996b), Skie and Bickford (1978), Smith (1950, 1956a, 1961), Stebbins (1954, 1966, 1985, 2003), Stejneger and Barbour (1943), Suleiman (2005), Taggart (1992a, 1992b, 1992j, 2000a, 2000b, 2001a, 2002d, 2003g, 2005), Taylor (1929a), Taylor et al. (1975), Tennant and Bartlett (1999), Tihen (1937), Tihen and Sprague (1939), Toepfer (1993a), Trauth et al. (1994, 2004), Trott (1977), Van Doren and Schmidt (2000), Viets (1993), Von Achen and Rakestraw (1984), Wagner (1984), Warner and Wencel (1978), Werth (1969), Wilgers et al. (2006), Woodman (1959), Wright and Wright (1952, 1957), Yarrow (1883), Zerwekh (2003), and Zimmerman (1990).

Graham's Crayfish Snake

Regina grahamii BAIRD AND GIRARD, 1853

Adult: Douglas County, Kansas.

Adult: Douglas County, Kansas

Adult showing belly color and pattern:
Douglas County, Kansas.

Adult: Linn County, Kansas.

Identification: Keeled body scales; divided anal scale; distinct color and pattern, as follows: head, body, and tail uniform brown or gray, with wide yellow or cream stripe running along each side near belly; belly cream or white with series of small dark spots; adult males have slightly longer tails than females; females grow larger than males.

Size: Adults normally grow 457–710 mm (18–28 inches) in TL; largest specimen from Kansas: female (KU 188175) from Lyon County with TL of 1,029 mm (40⅝ inches) collected by Robert F. Clarke on 11 June 1960; maximum length throughout range: 47 inches (Conant and Collins, 1998).

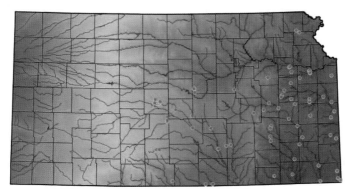

Graham's Crayfish Snake (*Regina grahamii*) is a semi-aquatic species that is generally found in the eastern half of the state, most abundantly from the Kansas River system south to the Oklahoma border. Red dots indicate records backed by voucher specimens or images.

Natural History: Lives near ponds and sluggish streams of prairies, wet meadows, and river valleys; recorded active annually in Kansas from early February to late October, with a peak from late April to early June; like many other snakes, active during day in spring and fall and becomes nocturnal during hot summer months; when not foraging for food, retires into crayfish burrows to rest; during winter, retreats into burrows to avoid cold temperatures; probably breeds during April and May; courtship occurs in water and consists of preliminary frenzied swimming by several males, probably trying to locate females; eventually, males discover female and all may entwine with her, forming fairly compact mass; more than 1 male may copulate with single female; females give birth in late July, August, or September; number of young per litter 4–39, with an average of 16; males reach sexual maturity after 1 season, but females do not breed until their third year; feeds almost exclusively on crayfishes; predators include large birds, mammals, and other snakes.

Pertinent References: Anderson (1965), Ballinger and Lynch (1983), Bartlett and Bartlett (2006c), Behler and King (1979), Berglund (1967), Boyd (1988, 1991), Branson (1904), Breukelman and Clarke (1951), Breukelman and Downs (1936), Burt (1933), Burt and Burt (1929b), Burt and Hoyle (1934), Busby and Parmelee (1996), Busby and Pisani (2007), Busby et al. (1994, 1996, 2005), Capron and Perry (1976), Cink (1995b), Clark (1966b), Clarke (1958, 1980a, 1980b), Clarke et al. (1958), Clay (1938), Cleveland (1986), Collins (1959, 1974b, 1982a, 1990b, 2001c, 2007b), Collins and Collins (1993a, 1993b, 2006a, 2006b, 2009a, 2009b), Collins et al. (1994), Conant (1958, 1975), Conant and Collins (1991, 1998), Cook (1954), Cragin (1881, 1885a, 1885d), Cross (1971), Davenport (1943), Ditmars (1939), Dolman (1929), Eddy and Hodson (1961), Ernst and Ernst (2003), Ernst et al. (2002), Fitch (1985b, 1993b), Garrigues (1962), Gibbons and Dorcas (2004), Gloyd (1928, 1932), Grant (1937), Grow (1976b), Hall (1969), Hall and Smith (1947), Hudson (1942), Irwin and Collins (1987),

REPTILES

Johnson (2000), Jordan (1929), Karns et al. (1974), Kingsbury and Gibson (2002), Knight (1935), Loomis (1956), Loraine (1980), Mozley (1878), Murrow (2008a, 2009a, 2009b), Platt (2003), Pope (1937), Riedle (1994a), Riedle and Hynek (2002), Rossi and Rossi (1995), Schmidt (1953, 2003b), Smith (1950, 1956a), Stejneger and Barbour (1939, 1943), Taggart (2001a, 2008), Taylor (1929a), Tennant and Bartlett (1999), Tihen (1937), Trauth et al. (1994, 2004), Uhler and Warren (1929, 2008), Viets (1993), Wagner (1984), Wright and Wright (1952, 1957), Young (1973), and Zimmerman (1990).

Brown Snake
Storeria dekayi (HOLBROOK, 1836)

Adult: Miami County, Kansas.

Adult: Linn County, Kansas.

Adult: Sumner County, Kansas.

Adult: Anderson County, Kansas.

Adult: Montgomery County, Kansas.

Adult Brown (**upper**) and Redbelly Snakes (**lower**): Miami County, Kansas. Note light orange spot on neck of Redbelly Snake, characteristic of the species.

Identification: Keeled body scales; divided anal scale; 7 scales on each upper lip; 2 rows of small, dark gray, brown, or black spots on either side of wide, indistinct gray stripe running down back; sides brownish or gray; black spot on each side of neck and below each eye; belly whitish; adult males have longer tails than females; females grow larger than males.

Size: Adults normally grow 230–330 mm (9–13 inches) in TL; largest specimen from Kansas: female (KU 154516) from Douglas County with TL of 405 mm (16 inches) collected by Joseph T. Collins on 6 March 1974; maximum length throughout range: 20⅝ inches (Conant and Collins, 1998); maximum weight for Kansas specimen: 12.1 grams (slightly under ½ ounce).

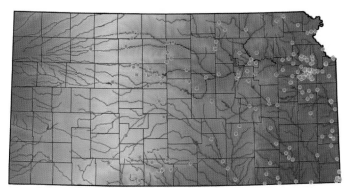

The small **Brown Snake** (*Storeria dekayi*) is abundant in the eastern part of the state, but less common in the west where it is confined to lowland riparian corridors. It is absent from the western fourth of Kansas. Red dots indicate records backed by voucher specimens or images.

Natural History: Generally lives near moist situations in woodland and along woodland edge; recorded active annually in Kansas from early January to late November, with a peak from mid-April to mid-July; may be found crawling on ground or beneath rocks during winter if warm temperatures occur; active during day in spring and fall; becomes nocturnal through hot summer months; frequently found near water; in winter, retreats as much as 2 feet beneath ground and remains inactive; mating occurs in both spring and fall; courtship unknown; females give birth to young from late July to September; number of young per litter 3–31, with an average of 12; prefers diet of earthworms; predators include birds, mammals, and larger snakes; when disturbed, emits smelly musk from cloaca; maximum captive longevity: 7 years and 13 days (Snider and Bowler, 1992).

Pertinent References: Anderson (1965), Ballinger and Lynch (1983), Bartlett and Bartlett (2006c), Behler and King (1979), Boyd (1991), Branson (1904), Breukelman and Clarke (1951), Burt (1933, 1935), Burt and Hoyle (1934), Busby and Parmelee (1996), Busby et al. (1994, 1996, 2005), Cavitt (2000a), Christman (1982), Clark (1966b), Clarke (1956c, 1958, 1980b), Clarke et al. (1958), Collins (1959, 1974b, 1982a, 1990b, 2004c, 2006b, 2007b), Collins and Collins (1993a, 2006b, 2009a), Conant (1958, 1975), Conant and Collins (1991, 1998), Cope (1900), Cragin (1881), Cross (1971), Davenport (1943), Ditmars (1905b, 1939), Elick and Sealander (1972), Ellis (2002), Ellis and Henderson (1913), Ernst and Ernst (2003), Fitch (1956e, 1958b, 1982, 1985b, 1991, 1992, 1993b, 1999, 2006a), Fitch and Echelle (2006), Fleharty and Ittner (1967), Garrigues (1962), Gloyd (1928, 1932), Gray (1979), Hall and Smith (1947), Hammerson (1982, 1999), Henderson (1974), Hudson (1942), Hunter et al. (1992), Hurter (1911), Jordan (1929), Karns et al. (1974), Kingsbury and Gibson (2002), Knight (1935), Lamson (1935), Linsdale (1925, 1927), Linzey and Clifford (1981), Loomis (1956), Lueth (1941), Mansueti (1941), Maslin (1959), McCauley (1945), Miller (1979, 1983a), Minton (2001), Mitchell (1994), Mount (1975), Murrow (2008a, 2008b, 2009a, 2009b), Parmelee and Fitch (1995b), Perkins (1940), Perry (1978), Pisani (2009a, 2009b), Riedle (1994a), Riedle and Hynek (2002), Rossi (1992), Rossi and Rossi (1995), Rundquist (1975, 1992), Schmidt (1953, 2004j, 2006), Schmidt and Davis (1941), Schmidt and Murrow (2006, 2007c), Simmons (1987c), Smith (1950, 1956a), Stejneger and Barbour (1917, 1923, 1933, 1939, 1943), Suleiman (2005), Taggart (1992b, 2000a, 2000b, 2001a, 2003g, 2006d, 2008), Taylor (1929a), Taylor et al. (1975), Tennant and Bartlett (1999), Tihen (1937), Toal (1992), Toepfer (1993a), Trapido (1944), Trauth et al. (2004), Trott (1983), Viets (1993), Von Achen and Rakestraw (1984), Wagner (1984), Warner and Wencel (1978), Wilgers and Horne (2006), Wilgers et al. (2006), Wright and Wright (1952, 1957), and Yarrow (1883).

Redbelly Snake
Storeria occipitomaculata (STORER, 1839)

Adult: Jefferson County, Kansas.

Adult: Cherokee County, Kansas.

Two adults showing gray and brown color variants typical of this species in Kansas: Miami County, Kansas.

Adult: Douglas County, Kansas.

Adults showing brilliantly-colored belly that is it's namesake: Jefferson County, Kansas (**left**); Cherokee County, Kansas (**right**).

Identification: Keeled body scales; divided anal scale; distinctive color and pattern, as follows: head, body, and tail may be slate gray with 2 thin darker stripes on each side of back, or reddish brown with indistinct darker stripes; belly may be bright orange-red or jet black (any combinations of dorsal and belly colors occur); 3 light spots on neck; sometimes a faint light stripe down middle of back; adult males have longer tails than females; females grow larger than males.

Size: Adults normally attain 203–254 mm (8–10 inches) in TL; largest specimen from Kansas: female (KU 28750) from Miami County with TL of 305 mm (12 inches) collected by A. Byron Leonard in 1950; maximum length throughout range: 16 inches (Conant and Collins, 1998).

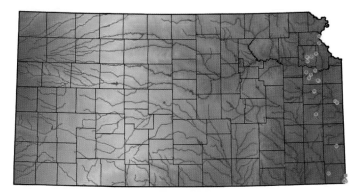

The range of the small and secretive **Redbelly Snake** (*Storeria occipitomaculata*) in Kansas is restricted to the two tiers of counties along the eastern border of the state, from Atchison County in the north to Cherokee County in the south. The recent discovery of this species along the Platte River in south-central Nebraska has initiated speculation that a single old (1884) record from Long Island (Republican River drainage in Phillips County) in north-central Kansas, might be valid; however, until voucher specimens are obtained to substantiate the Long Island record, we have not mapped it for Kansas. Red dots indicate records backed by voucher specimens or images.

Natural History: Extremely secretive; inhabits woodlands with dense leaf litter, found beneath rocks on wooded hillsides, beneath rotten logs at forest edge, under boards in sandstone woodlands, and in dry stream beds in woodlands; recorded active annually in Kansas from mid-March to early November, with a peak from late March to early May; little known of daily habits; may be nocturnal; daily cycle probably shifts with seasonal temperature (as with Brown Snake); during winter, retires beneath ground and remains inactive until spring emergence; mating occurs both in spring and fall; when fall matings occur, sperm can be retained by female through winter; courtship unknown; females give birth during late July, August, or early September; number of young per litter 1–18, with an average of 8; prefers small forest slugs, but will also eat terrestrial forest snails and earthworms; threatened species (protected by state law); predators include birds, mammals, and other snakes; maximum captive longevity: 4 years and 7 months (Snider and Bowler, 1992).

Pertinent References: Anderson (1965), Ballinger and Lynch (1983), Bartlett and Bartlett (2006c), Baxter and Stone (1980, 1985), Behler and King (1979), Branson (1904), Burt and Hoyle (1934), Choate (1967), Clark (1966b), Clarke (1980b), Collins (1959, 1974b, 1982a, 1982b, 1985, 1990b, 1994b), Collins and Collins (1993a, 2004, 2006b, 2009a), Collins et al. (1995), Conant (1958, 1975), Conant and Collins (1991, 1998), Cook (1954), Cope (1900), Cragin (1881), Cross (1971), Ditmars (1905b), Elick and Sealander (1972), Ernst (2002), Ernst and Ernst (2003), Fitch (1991, 1999), Fitzgerald (1994), Fitzgerald and Nilon (1994), Freeman and Kindscher (1992), Gloyd (1928, 1932), Guarisco et al. (1982), Hay (1972), Hunter et al. (1992), Irwin (1985b), Johnson (1974), Jordan (1929), Karns et al. (1974), Kingsbury and Gibson (2002), Klemens (1993), Knight (1935), Lamson (1935), Lardie (1990), Layher et al. (1986),

REPTILES

163

Levell (1997), Linzey and Clifford (1981), Lueth (1941), Lynch (1985), Mansueti (1941), McCauley (1945), Miller and Collins (1993), Miller and Gress (2005), Minton (2001), Mitchell (1994), Mosier and Collins (1998), Mount (1975), Peyton (1989), Platt et al. (1974), Rossi and Rossi (1995), Schmidt (1953), Schmidt and Murrow (2007c), Schilling and Schaefer (1983), Scott (1996), Simon (1973), Simmons (1989), Smith (1950, 1956a), Stebbins (1966, 1985, 2003), Stejneger and Barbour (1917, 1923, 1933, 1939, 1943), Taggart (2006b, 2007b, 2008), Taylor (1892, 1929a), Tennant and Bartlett (1999), Trapido (1944), Trauth et al. (2004), Wright and Wright (1952, 1957), and Yarrow (1883).

Checkered Garter Snake
Thamnophis marcianus (BAIRD AND GIRARD, 1853)

Identification: Keeled body scales; single anal scale; 3 longitudinal stripes on body (1 down middle of back and 1 on each side); stripe on each side of body located on second and third scale rows (counting from belly up); large, yellow or cream crescent-shaped mark on each side of head behind angle of jaws; in addition to narrow yellowish stripe on each side of body, similarly colored middorsal stripe runs down middle of back; all 3 stripes start at head and run down body onto tail; area between and below stripes brownish yellow with bold checkered pattern of dark brown or black spots; spots often encroach on middorsal stripe, creating zig-zag appearance; belly yellowish with no pattern; adult males have longer tails than females; females grow larger than males.

Gravid female.

Adult female: Comanche County, Kansas.

Head of adult: Sumner County, Kansas.
Note dark-bordered light crescent mark on side of neck, typical of this species.

Adult: Barber County, Kansas.
Note how dark spots encroach on yellow middorsal stripe, giving the stripe an irregular, wavy appearance unlike any other species of this genus in Kansas.

©Larry L. Miller

REPTILES

164

Size: Adults grow 457–610 mm (18–24 inches) in TL; largest specimen from Kansas: female (KU 158493) from Comanche County with TL of 846 mm (33¼ inches) collected by Stanley Roth, William Bradley, and Ray E. Ashton on 10 May 1975; maximum length throughout range: 42¾ inches (Boundy, 1995).

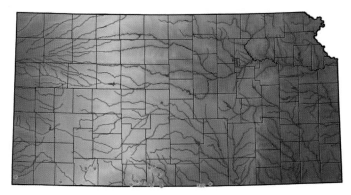

The distribution of the **Checkered Garter Snake** (*Thamnophis marcianus*) in Kansas is restricted to a single row of counties along the southern border with Oklahoma, from Sumner County in the east to Morton County in the southwestern corner. Red dots indicate records backed by voucher specimens or images.

Natural History: Generally found around lakes or near springs; often observed following heavy rains; sometimes found in twilight zones of caves; recorded active annually in Kansas from mid-February to late September, with a peak from mid-April to mid-May; forages for food by day around bodies of water, but may become nocturnal during extremely hot weather; during winter, retreats into small mammal burrows or deep in crevices on rocky hillsides to avoid cold temperatures; mating occurs during spring after emergence from winter inactivity; courtship not known; females may produce 6–18 young during June, July, or August; feeds on earthworms, frogs, toads, fishes, and possibly small rodents; threatened species (protected by state law); predators include large birds, mammals, and other snakes; maximum captive longevity: 7 years, 6 months, and 26 days (Snider and Bowler, 1992).

Pertinent References: Ball (1992a), Ballinger and Lynch (1983), Bartlett and Bartlett (2006c), Bartlett and Tennant (2000), Behler and King (1979), Burt (1935), Clarke (1980b), Cochran and Goin (1970), Collins (1959, 1974b, 1982a, 1985, 1990b), Collins and Caldwell (1978), Collins and Collins (1991, 1993a, 2006b, 2009a), Collins et al. (1991, 1994, 1995), Conant (1958, 1975), Conant and Collins (1991, 1998), Cragin (1881), Davenport (1943), Ernst and Ernst (2003), Fitch (1993b), Fleharty and Ittner (1967), Ford and Schofield (1984), Garman (1883, 1884), Guarisco et al. (1982), Johnson (1974), Karns et al. (1974), Kingsbury and Gibson (2002), Lardie (1976, 1981), Layher et al. (1986), Levell (1997), Leviton (1971), Mara (1994), Marr (1944), Miller (1978, 1983c, 1987a), Miller and Gress (2005), Mittleman (1949), Mozley (1878), Patton (1978), Perkins (1940), Platt et al. (1974), Pope (1937), Rossi and Rossi (1995), Rossman et al. (1996), Ruthven (1908), Schmidt (1953), Schmidt and Davis (1941), Schilling and Nilon (1988), Scott (1996), Shaw and Campbell (1974), Simmons (1989), Smith (1946d, 1950, 1956a), Stebbins (1954, 1966, 1985, 2003), Stejneger and Barbour (1939, 1943), Sweeney (1992), Taggart (2006b), Taylor (1929a, 1929b), Tennant and Bartlett (1999), Tihen (1937), Tihen and Sprague (1939), Webb (1970), and Wright and Wright (1952, 1957).

Western Ribbon Snake
Thamnophis proximus (SAY, 1823)

Identification: Very slender; keeled body scales; single anal scale; 3 longitudinal stripes on body (1 down middle of back and 1 on each side); stripe on each side of body situated on third and fourth scale rows (counting from belly up); upper lips without dark vertical bars; uniform olive-gray, dark green, or black skin color between stripes; stripe down middle of back orange and varies in width (narrow throughout most of state, but wider in southwestern edge of range in Kansas); stripes on each side of body light yellow or cream; belly white, cream, or gray with some dark coloration along edges; females grow slightly larger than males.

Adult: Douglas County, Kansas.

Adult: Allen County, Kansas.

Adult: Labette County, Kansas.

Adult: Meade County, Kansas.

Adult: Meade County, Kansas.

Adult: Douglas County, Kansas.

Size: Adults normally attain 510–760 mm (20–30 inches) in TL; largest specimen from Kansas: female (MHP 7243) from Edwards County with TL of 1,013 mm (39½ inches) collected by James Gubanyi and Keith Coleman on 2 June 2002; maximum length throughout range: 49⅞ inches (Boundy, 1995).

Natural History: Frequents edges of swamps, marshes, lakes, streams, and rivers; recorded active annually in Kansas from early March to mid-November, with a peak from early May to early August; both diurnal and nocturnal depending upon prevailing temperatures; spends day foraging along water-edge vegetation or basking in sun on matted reeds and grasses; when frightened, glides swiftly across water to escape; during winter, retreats to burrows beneath ground to avoid cold temperatures; mates during April and May; courtship unknown; females give birth to young from late June to September; number of young per litter 4–28, with an average of 12–13; feeds on small frogs, toads, salamanders, and fishes; predators of adults include birds, mammals, and larger snakes; young also eaten by large frogs and fishes; maximum captive longevity: 3 years, 7 months, and 1 day (Snider and Bowler, 1992).

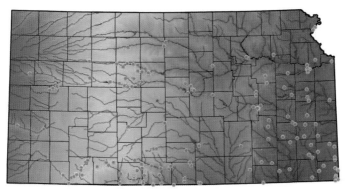

The slender **Western Ribbon Snake** (*Thamnophis proximus*) is found throughout most of Kansas, but is absent from the northern border counties and the northwestern and southwestern corners of the state. It is restricted in western Kansas to major river corridors. Red dots indicate records backed by voucher specimens or images.

Pertinent References: Ballinger and Lynch (1983), Bartlett and Bartlett (2006c), Behler and King (1979), Boyd (1988, 1991), Branson (1904), Brennan (1934, 1937), Breukelman and Clarke (1951), Breukleman and Downs (1936), Breukelman et al. (1961), Burt (1933, 1935), Burt and Burt (1929b), Burt and Hoyle (1934), Busby and Parmelee (1996), Busby and Pisani (2007), Busby et al. (1994, 1996, 2005), Caldwell and Glass (1976), Capron et al. (1982), Cink (1995b), Clarke (1958, 1980a, 1980b), Clarke et al. (1958), Collins (1959, 1974b, 1982a, 1990b, 2006b, 2007b, 2009a), Collins and Collins (1993a, 2006b, 2009a), Collins et al. (1991, 1994), Conant (1958, 1975), Conant and Collins (1991, 1998), Cope (1900), Cragin (1881, 1885a, 1885d), Cross (1971), Ditmars (1939), Ellis (2002), Ernst and Ernst (2003), Falls (1933), Fitch (1985b, 1993b), Garrigues (1962), Gish (1962), Gloyd (1928, 1932), Gomez (2008b), Grant (1937), Grow (1976b), Guarisco et al. (1982), Gubanyi (2003c), Gubanyi and Coleman (2002), Hall and Smith (1947), Hammerson (1999), Irwin (1983), Johnson (2000), Karns et al. (1974), Kingsbury and Gibson (2002), Knight (1935), Linsdale (1925, 1927), Loomis (1956), Lund (1974), Marr (1944), Miller (1976d, 1988, 1996a), Minton (1972, 2001), Mozley (1878), Murrow (2004, 2008a, 2009a, 2009b), Perry (1978), Pope (1944b), Riedle and Hynek (2002), Rossi and Rossi (1995), Rossman (1963, 1970), Rossman et al. (1996), Rundquist (1992), Rush and Fleharty (1981), Ruthven (1908), Schmidt and Murrow (2007b, 2007c), Smith (1950, 1956a), Stebbins (1954, 1966, 1985, 2003), Suleiman (2005), Sweeney (1992), Taggart (1992a, 2000a, 2000b, 2001a, 2003g, 2005), Taylor (1929a), Tennant and Bartlett (1999), Tihen (1937), Tihen and Sprague (1939), Toepfer (1993a), Trauth et al. (1994, 2004), Van Doren and Schmidt (2000), Viets (1993), Wagner (1984), Wilgers et al. (2006), Wolfenbarger (1951, 1952), Wright and Wright (1952, 1957), and Yarrow (1883).

REPTILES

Plains Garter Snake

Thamnophis radix (BAIRD AND GIRARD, 1853)

Adult: Ellsworth County, Kansas.

Adult: Hamilton County, Kansas.

Adult: Logan County, Kansas.

Adult: Barton County, Kansas.

Adult: Ellis County, Kansas.

Adult: Logan County, Kansas.

Identification: Keeled body scales; single anal scale; 3 longitudinal stripes on body (1 down middle of back and 1 on each side); stripe on each side of body situated on third and fourth scale rows (counting from belly up); dark vertical bars on upper lips; alternating rows of black spots between stripes on back; body greenish gray, light olive, or tan, with black spots between 3 stripes on body; stripe running down middle of back may be bright yellow or orange; stripes on sides are normally yellow; belly white, grayish, or greenish, with row of black spots down each side; some individuals from Barton County, in central Kansas, exhibit bright red color on sides of bodies; adult males have longer tails than females; females grow slightly larger than males.

Size: Adults normally grow to 380–710 mm (15–28 inches) in TL; largest specimen from Kansas: female (KU 221508) from Ellis County with TL of 1,044 mm (41⅛ inches) collected by Mark Van Doren, Curtis J. Schmidt, and Russell Toepfer in September 1993; maximum length throughout range: 43 inches (Conant and Collins, 1998).

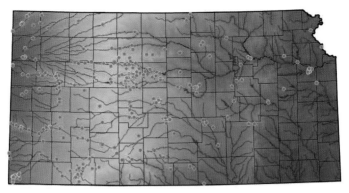

The **Plains Garter Snake** (*Thamnophis radix*) is found throughout the western two-thirds of the state and ranges east along the Kansas River corridor to Douglas County. It is also found along the Missouri River floodplain in Atchison and Leavenworth counties, in the upper reaches of the Marais des Cygnes River in Franklin and Osage counties, and in the upper reaches of the Neosho River in Lyon County. This species is absent from southeastern Kansas. Red dots indicate records backed by voucher specimens or images.

Natural History: Prefers open grassy prairies or sand prairies, particularly along edges of streams, stillwater ditches, marshes, and lakes (particularly in western Kansas); recorded active annually in Kansas from early March to late December, with a peak from mid-May to mid-August; during warm spells in winter will emerge from underground retreats; usually active during day, basking in sun or foraging for food; commonly basks in sun on dirt or gravel roads; frequently killed by automobiles for doing so; mates during April and May, and sometimes in fall; courtship involves 1 or more males crawling over and alongside female with jerking, writhing movements; successful male curls tail beneath female's until their cloacal openings meet and copulation occurs; more than 1 male may mate with female; during late July, August, or September, females give birth to litters of 5–60 young, with an average of 20; feeds on earthworms, toads, frogs, salamanders, fishes, and small rodents; predators include large birds, mammals, and other snakes; maximum captive longevity: 8 years, 5 months, and 3 days (Snider and Bowler, 1992).

Pertinent References: Anderson (1965), Anderson et al. (1995), Baker et al. (1972), Ball (1992a), Ballinger and Lynch (1983), Bartlett and Bartlett (2006c), Bartlett and Tennant (2000), Behler and King (1979), Berglund (1967), Boyd (1991), Branson (1904), Brennan (1934, 1937), Breukleman and Clarke (1951), Breukleman and Downs (1936), Breukleman et al. (1961), Brons (1882), Burt (1933, 1935), Burt and Burt (1929b), Burt and Hoyle (1934), Busby and Parmelee (1996), Busby et al. (1994, 1996, 2005), Capron and Perry (1976), Carpenter (1940), Clark (1966b), Clarke (1980a, 1980b), Clarke et al. (1958), Coleman (1987), Collins (1959, 1974b, 1981, 1982a, 1985, 1990b, 2001c, 2002, 2006b, 2007b), Collins and Collins (1991, 1993a, 1993b, 1996, 2001, 2006a, 2006b, 2009a), Conant (1958, 1975), Conant and Collins (1991, 1998), Cragin (1881), Cross (1971), Ditmars (1939), Ellis and Henderson (1913), Ernst and Ernst (2003), Fitch (1985b, 1993b), Ford (1976), Garrigues (1962), Gish (1962), Gloyd (1928), Grant (1937), Gray (1980), Gray and Douglas (1989), Gress (2003), Hammerson (1999), Hudson (1942), Hurter (1911), Irwin (1983), Irwin and Collins (1987, 2000), Karns et al. (1974), Kingsbury and Gibson (2002), Knight (1935), Knight and Collins (1977), Lardie (1981), Lardie and Black (1981), Loomis (1956), Lueth (1941), Marr (1944), Martin (1960), Miller (1988), Minton (2001), Murrow (2009a), Platt (1975, 1984a, 1985b, 1989, 1998, 2003), Pope (1944b), Rossi and Rossi (1995), Rossman et al. (1996), Royal (1982), Rundquist

(2000a), Russell and Bauer (1993), Ruthven (1908), Schmidt (2001, 2004l), Schmidt and Davis (1941), Schmidt and Murrow (2007b), Schilling and Schaefer (1983), Seigel and Fitch (1985), Smith (1946d, 1949, 1950, 1956a, 1957), Stebbins (1954, 1966, 1985, 2003), Stejneger and Barbour (1917, 1923, 1933, 1939, 1943), Sweeney (1992), Taggart (1992a, 1992b, 2000a, 2002c, 2002d, 2003b, 2004h), Taggart et al. (2007), Taylor (1929a, 1929b), Tennant and Bartlett (1999), Tihen (1937), Tihen and Sprague (1939), Toepfer (1993a), Uhler and Warren (1929, 2008), Van Doren and Schmidt (2000), Von Achen and Rakestraw (1984), Walley et al. (2003), Webb (1970), Werth (1969), Wilgers et al. (2006), Wolfenbarger (1951, 1952), Wright and Wright (1952, 1957), and Zimmerman (1990).

Common Garter Snake
Thamnophis sirtalis (Linnaeus, 1758)

Adult: Lyon County, Kansas.

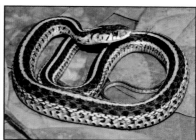

Adult: Sumner County, Kansas.

Adult: Cherokee County, Kansas. Note lack of red skin color between the stripes, an uncommon color variant found in examples of this species in extreme southeastern Kansas.

Adult: Seward County, Kansas. Note lack of red skin color between the stripes, typical of southwestern populations of this species in Kansas.

Adult: Osage County, Kansas.

Adult eating a Plains Leopard Frog: Comanche County, Kansas.

Identification: Keeled body scales; single anal scale; 3 longitudinal stripes on body (1 down middle of back and 1 on each side); stripe on each side of body situated on second and third scale rows (counting from belly up); absence of crescent-shaped mark behind angle of jaw; throughout most of Kansas range, pattern of black spots on brick red background between stripes on back with 3 yellow stripes on body (however, examples from extreme southeastern and southwestern Kansas lack red skin and spots on body between stripes and have orange middorsal stripe); belly white, greenish, or gray, with row of small dark spots along edges; adult males have longer tails than females; females larger than males.

Size: Adults normally attain 410–710 mm (16–28 inches) in TL; largest specimen from Kansas: female (KU 189179) from Reno County with TL of 1,130 mm (44½ inches) collected by George Ratzlaff on 5 March 1981; maximum length throughout range: 52 inches (Boundy, 1995); maximum weight for Kansas specimen: 410 grams (14½ ounces).

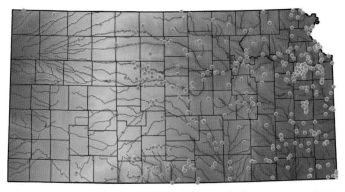

The **Common Garter Snake** (*Thamnophis sirtalis*) occurs in the eastern two-thirds of the state, ranges west along the northern border as far as Cheyenne County and ranges west along the southern border to Seward County via the Cimarron River drainage. This species is absent from west-central Kansas. Red dots indicate records backed by voucher specimens or images.

Natural History: Found in wide variety of habitats, including marshes and wet meadows, margins of ponds, woodland and woodland edge, floodplains, and cultivated fields; prefers moist areas with moderate vegetation, but also found in sand prairie; recorded active annually in Kansas from early February to mid-November, with a peak from early May to late August; during warm periods in December and January, frequently emerges from underground winter retreats; active during day; home range 22–35 acres; population density estimated as high as 3 per acre in northeastern Kansas; mates during early spring and occasionally in fall; on first warm spring days, males search actively for females, locating them by scent; several males may court same female; during courtship, males lie alongside female, their bodies rhythmically writhing; successful male curls tail beneath that of passive female until his cloaca meets hers; copulation occurs, and female may drag male along as she crawls away; females give birth in late summer or fall; number of young per litter 4–85, with an average of 20; feeds chiefly on frogs and earthworms but occasionally eats toads, small mice, and other small snakes; predators include large spiders, hawks, large snakes, and mammals; during agricultural harvests, swathing (mowing) machines kill many of these snakes; maximum captive longevity: 14 years (Snider and Bowler, 1992).

REPTILES

Pertinent References: Anderson et al. (1995), Ballinger and Lynch (1983), Bartlett and Bartlett (2006c), Bartlett and Tennant (2000), Baxter and Stone (1980, 1985), Behler and King (1979), Benton (1980), Berglund (1967), Boyd (1991), Branson (1904), Brennan (1937), Breukelman and Clarke (1951), Breukelman and Downs (1936), Breukelman et al. (1961), Burkett (1984), Burt (1931a, 1933, 1935, 1949), Burt and Burt (1929b), Burt and Hoyle (1934), Busby and Parmelee (1996), Busby and Pisani (2007), Busby et al. (1994, 1996, 2005), Caldwell and Glass (1976), Capron (1978b, 1985a, 1991), Capron and Perry (1976), Capron et al. (1982), Carpenter (1940), Cavitt (2000a, 2000b), Cink (1995b), Clark (1966b), Clarke (1958, 1980a, 1980b), Clarke et al. (1958), Collins (1959, 1974b, 1982a, 1984a, 1990b, 2001c, 2006b, 2007b), Collins and Collins (1993a, 1993b, 1996, 2001, 2006a, 2006b, 2009a), Collins et al. (1991, 1994), Conant (1958, 1975, 1978a, 1978b), Conant and Collins (1991, 1998), Cook (1954), Cope (1900), Cragin (1881, 1885a, 1885d), Cross (1971), Dillenbeck (1986, 1988), Ellis (2001, 2002), Ellis and Henderson (1913), Ernst and Ernst (2003), Fitch (1951, 1956e, 1958b, 1965a, 1970, 1980a, 1980b, 1982, 1985b, 1987, 1991, 1992, 1993b, 1999, 2000b, 2001b, 2004a, 2004b, 2005, 2006a), Fitch and Echelle (2006), Fitch and Hall (1978), Fitch and Maslin (1961), Fitch and Shirer (1971), Fitch et al. (2003), Ford (1976), Fraser (1983a), Garrigues (1962), Gish (1962), Gloyd (1928, 1932), Grant (1937), Gray and Stegall (1979), Gress (2003), Grow (1976a), Guarisco (1981c, 2001), Hall and Smith (1947), Hallowell (1857a), Hammerson (1999), Heinrich and Kaufman (1985), Henderson (1974), Hoyle (1936), Hunter et al. (1992), Hurter (1911), Irwin (1992b), Irwin and Collins (1987), Iverson (1978), Karns et al. (1974), Kingsbury and Gibson (2002), Knight (1935), Lardie (1975b, 1990, 2001), Li (1978a, 1978b), Linsdale (1925, 1927), Loomis (1956), Loraine (1980), Low (2005), Miller (1976a, 1979, 1980a, 1981d, 1982a, 1983a, 1988, 1996a, 1996b, 2000a), Minton (2001), Mitchell (1979, 1994), Minton (2001), Mozley (1878), Murrow (2008a, 2008b, 2009a, 2009b), Parker and Brown (1973), Parmelee and Fitch (1995b), Perry (1978), Peterson (1992), Pisani (1976, 2009a, 2009b), Platt (1975, 1984a, 1984b, 1985b, 1989, 1998, 2003), Preston (1982), Rehmeier and Matlack (2004), Riedle (1994a), Riedle and Hynek (2002), Rossman et al. (1996), Rues (1986), Rundquist (1975, 1992, 1996a, 2000a), Ruthven (1908), Schmidt (2004k), Schmidt and Murrow (2006, 2007b, 2007c), Schmidt and Phelps (2007), Schoener (1977), Seigel and Fitch (1985), Shine and Crews (1988), Shirer and Fitch (1970), Simmons (1987c), Simon and Dorlac (1990), Smith (1950, 1956a, 1961), Smith et al. (1983), Stebbins (1954, 1966, 1985, 2003), Stewart (1960), Suleiman (2005), Sweeney (1992), Taggart (1992k, 2000a, 2000b, 2001a, 2002d, 2003g, 2005), Tanner (1988), Taylor (1892, 1929a), Tennant and Bartlett (1999), Tihen (1937), Tihen and Sprague (1939), Toepfer (1993a), Trauth et al. (1994, 2004), Trott (1977), Turner (1977), Van Doren and Schmidt (2000), Viets (1993), Von Achen and Rakestraw (1984), Walley et al. (2005), Warner and Wencel (1978), Webb (1970), Wilgers and Horne (2006), Wilgers et al. (2006), Wolfenbarger (1951, 1952), Wooster (1925), Wright and Wright (1952, 1957), and Yarrow (1883).

Lined Snake

Tropidoclonion lineatum (HALLOWELL, 1856)

Identification: Keeled body scales; single anal scale; 3 longitudinal, gray-white to yellowish stripes on body (1 down middle of back and 1 on each side); 6 or fewer scales on each upper lip; white or yellowish belly with 2 rows of black spots down middle; head, body, and tail gray or brown; adult males have longer tails than females.

Size: Adults normally grow 224–380 mm (8⅝–15 inches) in TL; largest specimen from Kansas: female (KU 208123) from Sedgwick County with TL of 446 mm (17½ inches) collected by Jack Shumard on 15 June 1987; maximum length throughout range: 22½ inches (Boundy, 1995).

Adult: Ottawa County, Kansas.   Adult: Morris County, Kansas.

Adult: Jefferson County, Kansas.   Adult: Morris County, Kansas.

Natural History: Secretive; inhabits hillsides of open prairies and woodland edge; often found inside towns and cities beneath debris in vacant lots; habitat disturbance by human activity seems to have little effect on populations; recorded active annually in Kansas from late February to mid-November, with a peak from late April to mid-August; nocturnal during hot weather; spends day hiding beneath rocks and debris, including dried cow dung; during winter, retires beneath ground to avoid cold temperatures; mates in both spring and fall; courtship not recorded; females give birth during August and number of young per litter 2–12, with an average of 6–7; young attain sexual maturity during their second season; feeds exclusively on earthworms; predators include larger snakes, birds, and mammals.

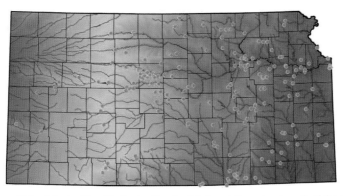

The small **Lined Snake** (*Tropidoclonion lineatum*) is found throughout much of the state. In western Kansas, it occurs along major river corridors. This species is absent from much of the northwestern, southwestern, and extreme southeastern corners of Kansas. Red dots indicate records backed by voucher specimens or images.

Pertinent References: Anderson (1965), Ballinger and Lynch (1983), Bartlett and Bartlett (2006c), Behler and King (1979), Boyd (1991), Branson (1904), Brennan (1934, 1937), Breukelman and Clarke (1951), Breukleman and Downs (1936), Breukelman et al. (1961), Brumwell (1951), Burt (1933, 1935), Burt and Hoyle (1934), Busby and Parmelee (1996), Busby and Pisani (2007), Busby et al. (1994, 1996, 2005), Caldwell and Collins (1977), Cavitt (2000a), Clark (1966b, 1967), Clarke (1956c, 1958, 1980b), Clarke et al. (1958), Cochran and Goin (1970), Collins (1959, 1974a, 1974b, 1981, 1982a, 1990b, 2001c, 2006b, 2007b), Collins and Collins (1993a, 1993b, 2006a, 2006b, 2009a), Conant (1958, 1975, 1978a), Conant and Collins (1991, 1998), Cope (1900), Cragin (1881, 1885a, 1885d), Cross (1971), Davenport (1943), Dice (1923), Ditmars (1939), Dixon and Werler (2005), Dunn (1932), Ellis (2002), Ernst and Ernst (2003), Fitch (1956e, 1985b, 1991, 1992, 1993b, 1999), Fraser (1983a), Garman (1883, 1884, 1892), Garrigues (1962), Gish (1962), Gloyd (1928, 1932), Grant (1937), Guarisco (1981c), Hammerson (1999), Hay (1892), Heinrich and Kaufman (1985), Hoyle (1936), Hudson (1942), Hurter (1911), Hurter and Strecker (1909), Irwin and Collins (1987), Jordan (1929), Karns et al. (1974), Kingsbury and Gibson (2002), Knight (1935), Knight et al. (1973), Lardie and Black (1981), Legler (1960b), Loomis (1956), Lueth (1941), Maslin (1959), Miller (1976a, 1980a, 1996a), Murrow (2008a, 2008b, 2009a, 2009b), Nulton and Rush (1988), Parmelee and Fitch (1995b), Perkins (1940), Platt (2003), Ramsey (1953), Rossi and Rossi (1995), Rush and Fleharty (1981), Schmidt (1953, 2009b), Schmidt and Davis (1941), Schmidt and Murrow (2006), Simmons (1987c), Smith (1950, 1956a, 1957), Smith and Chiszar (1994), Smith and Smith (1962, 1963), Stejneger (1891), Stejneger and Barbour (1917, 1923, 1933, 1939, 1943), Suleiman (2005), Taggart (1992a, 2000a, 2002d, 2004h, 2005, 2008), Taggart et al. (2005h, 2007), Taylor (1892, 1929a), Tennant and Bartlett (1999), Tihen (1937), Toepfer (1993a), Viets (1993), Wilgers and Horne (2006), Wilgers et al. (2006), Wright and Wright (1952, 1957), and Yarrow (1883).

Rough Earth Snake
Virginia striatula (LINNAEUS, 1766)

Adult: Chautauqua County, Kansas.

Adult: Chautauqua County, Kansas.

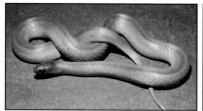
Adult: Cherokee County, Kansas.

Adult: Chautauqua County, Kansas.

Identification: Body scales generally smooth, but some weakly keeled body scales occur on middle of back and over upper rear portion of body; uniform gray or brown head, body, and tail; divided anal scale; 5 scales on each upper lip; a single scale directly behind each eye; belly white or grayish; newborns may have yellow area on middle of head just behind eyes; adult males have longer tails than females; females attain a greater total length than males.

Size: Adults normally attain 180–254 mm (7–10 inches) in TL; largest specimen from Kansas: female (KU 192191) from Cherokee County with TL of 290 mm (11⅜ inches) collected by Kelly J. Irwin, Larry Miller, and Joseph T. Collins on 10 April 1982; maximum length throughout range: 12⅝ inches (Conant and Collins, 1998).

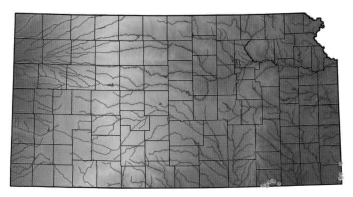

The **Rough Earth Snake** (*Virginia striatula*) is a small serpent that in Kansas occurs as two separate populations in the southeastern part of the state, one in Cherokee and Crawford counties and another to the west in Chautauqua County. Red dots indicate records backed by voucher specimens or images.

Natural History: Secretive; lives beneath flat rocks near tops of hillsides in dry open woodland and woodland edge; also found beneath sandstone slabs and discarded asphalt slabs; recorded active annually in Kansas from late March to early November, with a peak from late March to early May; probably nocturnal; spends day beneath rocks and logs or in leaf litter; during winter, burrows deep beneath ground to avoid cold temperatures; mates during April and May after emergence from winter inactivity; courtship unknown; females give birth anytime from July to September; number of young per litter 2–9, with an average of 5; prefers diet of earthworms; also eats slugs and snails; species in need of conservation (protected by state law); predators similar to those of Smooth Earth Snake; maximum captive longevity: 7 years, 3 months, and 8 days (Snider and Bowler, 1992).

Pertinent References: Anderson (1965), Ballinger and Lynch (1983), Bartlett and Bartlett (2006c), Behler and King (1979), Breukleman and Downs (1940), Breukelman and Smith (1946), Clark (1966b), Clarke (1980b), Cochran and Goin (1970), Collins (1959, 1974b, 1982a, 1982b, 1990b), Collins and Collins (1993a, 2006b, 2009a), Conant (1958, 1975), Conant and Collins (1991, 1998), Cook (1954), Cross (1971), Ernst and Ernst (2003), Fleharty and Johnson (1975), Guarisco et al. (1982), Hall and Smith (1947), Johnson (1974, 2000), Karns et al. (1974), Kingsbury and Gibson (2002), Lardie (1990), Layher et al. (1986), Levell (1997), Linzey and Clifford (1981), Mitchell (1994), Minton (2001), Perkins (1940), Platt et al. (1974), Powell et al. (1994), Rossi and Rossi (1995), Schmidt (1953), Schilling and Schaefer (1983), Simmons (1989), Smith (1950, 1956a), Stejneger and Barbour (1943), Taggart (2006b), Tennant and Bartlett (1999), Trauth et al. (2004), and Wright and Wright (1952, 1957).

Smooth Earth Snake
Virginia valeriae Baird and Girard, 1853

Adult: Jefferson County, Kansas.

Adult: Shawnee County, Kansas.

Adult: Jefferson County, Kansas.

Adult: Franklin County, Kansas.

Identification: Smooth dorsal scales on front part and weakly keeled scales on rear part of body; divided anal scale; gray or brown unpatterned head, body, and tail; 6 scales on upper lip; 2 or more scales bordering rear edge of eye; belly white; adult males have longer tails than females; females grow to a greater TL.

Size: Adults normally grow 180–254 mm (7–10 inches) in TL; largest specimen from Kansas: female (KU 177023) from Jefferson County with TL of 305 mm (12 inches) collected by Jill Krebs in April 1978; maximum length throughout range: 15⅜ inches (Conant and Collins, 1998).

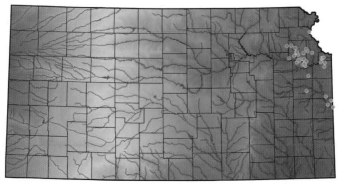

Distribution of the **Smooth Earth Snake** (*Virginia valeriae*) in Kansas. This small reptile is restricted to the forested hills and valleys of the northeastern part of the state in the Missouri, Kansas, and Marais des Cygnes river drainages, from Atchison County in the north to Anderson and Linn Counties in the south, and thence westward to Jackson and Shawnee counties. Red dots indicate records backed by voucher specimens or images.

Natural History: Secretive; lives seasonally in a variety of habitats, such as rocky hillsides in moist woodlands, woodland edge, and meadows, pastures and grasslands with abundant thatch; may hibernate in habitats not used for feeding and reproduction; recorded active annually in Kansas from late March to late-October or even November (weather-dependent) with peaks from mid-April to late May and then in September–October; forages for food in moist areas that support good earthworm populations; avoids hot and/or dry habitats; during winter, crawls deep into crevices on rocky hillsides to avoid cold temperatures; sometimes uses ant hills to overwinter; mates in spring after emergence from winter inactivity and also may mate in fall; courtship unknown; 11–14 weeks after mating, females give birth; number of young per litter 2–14, with an average of 6–7; birth generally occurs during August or September; feeds exclusively on earthworms; threatened species (protected by state law); predators include birds, mammals, and other snakes; maximum captive longevity: 6 years, 1 month, and 7 days (Snider and Bowler, 1992).

Pertinent References: Anderson (1965), Ballinger and Lynch (1983), Bartlett and Bartlett (2006c), Behler and King (1979), Choate (1967), Clark (1966b), Clarke (1980b), Cochran and Goin (1970), Collins (1959, 1974b, 1982a, 1983, 1990b), Collins and Collins (1993a, 2006b, 2009a), Collins et al. (1995), Conant (1958, 1975), Conant and Collins (1991, 1998), Ditmars (1939), Elick and Sealander (1972), Ernst and Ernst (2003), Fitch (1956e, 1958b, 1991, 1993b, 1999), Fitzgerald (1994), Fitzgerald and Nilon (1994), Gloyd (1928), Hay (1972), Henderson (1974), Irwin and Collins (1994, 2005), Johnson (1974, 2000), Karns et al. (1974), Kingsbury and Gibson (2002), Lardie (1990), Layher et al. (1986), Levell (1997), Leviton (1971), Linzey and Clifford (1981), Loomis (1956), Miller and Collins (1993), Miller and Gress (2005), Minton (1972, 2001), Mitchell (1994), Minton (2001), Perkins (1940), Pisani (2005, 2007, 2009a, 2009b), Platt et al. (1974), Powell et al. (1992), Rossi (1992), Rossi and Rossi (1995), Schmidt and Davis (1941), Scott (1996), Simmons (1989), Smith (1950, 1956a), Taggart (2006b), Taylor (1929a), Tennant and Bartlett (1999), Tihen (1937), Trauth et al. (2004), and Wright and Wright (1952, 1957).

PITVIPERS
FAMILY CROTALIDAE OPPEL, 1811

The family Crotalidae (Pitvipers) consists of 3 genera and 24 species in the United States (Collins and Taggart, 2009). Five species assigned to 3 genera represent this family in Kansas: they are the Copperhead (*Agkistrodon contortrix*), Cottonmouth (*Agkistrodon piscivorus*), Timber Rattlesnake (*Crotalus horridus*), Prairie Rattlesnake (*Crotalus viridis*), and Massasauga (*Sistrurus catenatus*). In Kansas, this family is composed of venomous species with a pair of long fangs (1 each on each side of front of upper jaw), a vertically slit pupil (when exposed to light), and a live-bearing reproductive mode.

Copperhead
Agkistrodon contortrix (LINNAEUS, 1766)

Identification: **VENOMOUS.** Distinctly banded snake; no rattle on tail; small pit on each side of head between and slightly below eye and nostril; keeled body scales;

pattern of 7–20 light-edged dorsal crossbands on body; bands narrow on middle of back and wide on sides (in individuals from Cowley County area, crossbands may be nearly equal in width on back and on sides); varies in color from gray to light brown, with dark gray or brown crossbands; head may be gray, brown, or reddish; belly white with large dark gray, brown, or black blotches on edges extending short distance up onto sides of body; young have lemon-yellow tails; adult males grow much longer than females; in northeastern Kansas, several males were discovered with no pattern of crossbands (uniform dull brown).

Adult: Chautauqua County, Kansas.

Adult: Miami County, Kansas.

Adult: Johnson County, Kansas.

Adult: Douglas County, Kansas.
Note flattened body, a defensive threat
posture to make the snake appear larger.

Juvenile: Douglas County, Kansas.
Note sulphur-yellow tail, typical
of newborn and young.

Patternless adult: Jefferson County, Kansas.

Size: Adults normally grow 560–915 mm (22–36 inches) in TL; largest specimen from Kansas: male (KU 196643) from Jefferson County with TL of 1,020 mm (40 inches) collected by Henry S. Fitch on 13 July 1984; maximum length throughout range: 53 inches (Conant and Collins, 1998); maximum weight for Kansas specimen is about 400 grams (14 ounces).

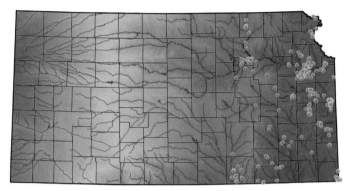

The **Copperhead** (*Agkistrodon* contortrix) is a venomous reptile with a distribution in Kansas restricted to eastern third of the state, from Marshall County in the north to Cowley County in the south. Red dots indicate records backed by voucher specimens or images.

Natural History: Found in open rocky woodland, woodland edge, and meadows with clumps of brush adjacent to woodland; body pattern and color blend perfectly with forest leaf litter; recorded active annually in Kansas from late March to November with a peak from late April to late July; during cool spring and fall months, active only during day and normally found coiled on rocky hillsides or forest floors where it awaits approach of prey; in summer, becomes nocturnal and prowls for food, aided by night vision and 2 sensory pits between eyes and nostrils; these pits detect body heat of warm-blooded prey such as mice; home ranges vary from 8–25 acres (males tend to wander in larger areas than females); because of shy, retiring disposition and camouflage pattern, easily exists in reasonably large numbers near areas of heavy human population; estimates indicate population density as high as 3–4 per acre in northeastern Kansas (people annually bitten by these snakes are few and normally victims of accidents, definitely not victims of aggressive, dangerous animals); during fall, returns to wooded hilltop rock outcrops with southern exposure used in previous years and retires deep beneath ground to avoid cold temperatures of winter; breeds from April to August, with peak of activity in April and May just after spring emergence from winter inactivity; many females bear young every other year; courtship and mating not observed adequately and presumably occur at night; young born venomous and number 1–14 per litter, with an average of about 5; females give birth in August, September, and October; feeds on insects (particularly cicadas), frogs, toads, lizards, small birds, and other snakes; particularly fond of rodents; young may use their lemon-yellow tails as lures to attract prey such as small frogs and toads; predators include large birds, mammals, and larger snakes, but people chief enemy; adult males among snakes that engage in combat dances; typical combat dance involves 2 males facing each other with heads and front of bodies raised off ground; they come together with bellies firmly adpressed and slowly intertwine their necks with writhing movements, until suddenly 1 male hurls other to the ground with quick body contraction; combat dance evidently establishes dominance of 1 male over another; estimated natural longevity of up to 15 years; maximum captive longevity: 29 years, 10 months, and 6 days (Snider and Bowler, 1992).

Pertinent References: Anderson (1942, 1965), Ashton et al. (1974a), Ballinger and Lynch (1983), Bartlett and Bartlett (2006c), Barton (1948), Behler and King (1979), Branson (1904),

Breukelman and Clarke (1951), Breukleman and Downs (1936), Brown (1973), Burt (1933, 1935), Burt and Hoyle (1934), Busby and Parmelee (1996), Busby and Pisani (2007), Busby et al. (1994, 1996, 2005), Caldwell and Collins (1977), Caldwell and Glass (1976), Campbell and Lamar (2004), Capron (1978a, 1983, 1985e), Caras (1974), Clarke (1958, 1959, 1980b), Clarke et al. (1958), Cochran and Goin (1970), Coleman (1987), Collins (1959, 1974b, 1978b, 1979, 1980a, 1982a, 1985, 1990b), Collins and Collins (1993a, 1996, 2001, 2006b, 2009a), Collins et al. (1991), Colt (1862), Conant (1958, 1975, 1978a, 1992, 1995), Conant and Collins (1991, 1998), Cope (1859, 1900), Costello (1969), Cragin (1881), Cross (1971), Crow (1913), Davenport (1943), deWit (1982a, 1982b), Dice (1923), Ditmars (1939), Do Amaral (1927), Eddy and Babcock (1973), Eddy and Hodson (1961), Ernst (1992), Ernst and Ernst (2003), Fitch (1951, 1956e, 1958b, 1959, 1960a, 1960b, 1965b, 1970, 1982, 1985b, 1987, 1991, 1992, 1993b, 1999, 2000b, 2004a, 2004b, 2004d, 2005, 2006a), Fitch and Collins (1985), Fitch and Echelle (2006), Fitch and Hall (1978), Fitch and Shirer (1971), Fitch et al. (2003), Foster and Caras (1994), Garrigues (1962), Gibbons et al. (1990), Gier (1967), Gloyd (1928, 1932, 1934, 1938, 1947, 1958, 1969), Gloyd and Conant (1938, 1943, 1990), Guarisco (1980b, 2001), Guiher and Burbrink (2008), Hall and Smith (1947), Harding and Welch (1980), Hay (1892), Heinrich and Kaufman (1985), Henderson (1974), Henderson and Lee (1992), Horr (1949), Hoyle (1936), Hudson (1942), Hurter (1911), Johnson (1987, 2000), Jones (1976), Jones and Burchfield (1971), Jordan (1929), Karns et al. (1974), Kingsbury and Gibson (2002), Knight (1935, 1992), Knight et al. (1992), Lardie (1990), Lardie and Black (1981), Legler (1960b), Linsdale (1925, 1927), Linzey and Clifford (1981), Lokke (1981), Loomis (1956), Maher and Sievert (2006), Mansueti (1941), Martin (1956), McCauley (1945), McLeran (1972a, 1974a), Merriman (1975), Miller (1980c, 1983a, 1987c), Minton (1957, 1971, 2001), Mitchell (1994), Monro (1949a), Mount (1975), Mozley (1878), Murphy (1964), Murphy and Barten (1979), Murrell (1993), Murrow (2009a, 2009b), Parmelee and Fitch (1995b), Parrish (1980), Perkins (1940), Perry (1978), Pinney (1981), Pisani (2009b), Pope (1937, 1944a), Rainey (1956), Reichman (1987), Reiserer (2002), Richardson (1972), Richardson and Phillips (1997), Riedle (1994a), Rossi (1992), Rossi and Rossi (1995), Roth (1959), Rues (1986), Russell (1980), Schmaus (1959), Schmidt (1953, 1998), Schmidt and Davis (1941), Schmidt and Inger (1957), Schmidt and Murrow (2007c), Schoener (1977), Schuett (1992), Seigel and Fitch (1985), Shirer and Fitch (1970), Smith (1940, 1950, 1956a, 1978, 1980), Smith et al. (1983), Snelson ands Palmer (1966b), Spanbauer (1984), Stejneger and Barbour (1939, 1943), Suleiman (2005), Taggart (2000b, 2003g, 2008), Taylor (1929a), Tennant and Bartlett (1999), Tihen (1937), Trauth et al. (2004), Turner (1977), Vial (1974), Vial et al. (1977), Von Achen and Rakestraw (1984), von Frese (1973), Welch (1994a), Wilgers et al. (2006), Wolfenbarger (1951, 1952), Wooster (1925), Wright and Wright (1952, 1957), and Yarrow (1883).

Cottonmouth
Agkistrodon piscivorus (LACÉPÈDE, 1789)

Adult.   Juvenile. Note sulphur-yellow tail color.

Identification: **VENOMOUS**. Adults uniformly gray or black; no rattle on tail; small pit on each side of head between and slightly below eye and nostril; keeled body scales; sometimes faint banded pattern present on back and sides, but normally seen clearly only when snake is submerged in water; belly dark; young adults may be banded; juveniles have lemon-yellow tails and resemble young Copperheads; tails of males slightly longer than those of females.

Adult captured in mid-1970s from an introduced colony in Montgomery County, Kansas.

Adult.

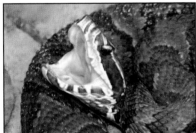

Adult displaying a gaping mouth, behavior typical of this species.

Mud-covered juvenile recently emerged from winter brumation.

Size: Adults normally grow 760–1,220 mm (30–48 inches) in TL; largest specimen from Kansas: male (KU 218677) from Cherokee County with a total length of 915 mm (36 inches) collected by Shane Eckhardt on 14 September 1991; maximum length throughout range: 74½ inches (Conant and Collins, 1998).

Natural History: Little known; probably active from April to early October; during cool spring and fall months, probably active only during day; normally found coiled along small streams and backwaters awaiting approach of prey; in summer months, becomes nocturnal and prowls oxbows and other waterways for food, aided by night vision and 2 sensory pits between eyes and nostrils; pits detect body heat of warm-blooded prey; because of irritable disposition and proximity to aquatic areas used by people for recreation, often killed; people bitten annually by these snakes are few and normally victims of accidents (definitely not victims of aggressive, dangerous animals); during fall, leaves aquatic environs, travels to upland den site used in previous years, and retires deep beneath ground to avoid cold temperatures of winter; breeds in spring, just after emergence from winter inactivity; females may bear young only every other year; courtship and mating not observed in Kansas; young born venomous and number 2–15 per litter, with an average of 6–7;

females give birth in August and September; feeds on insects, fishes, salamanders, frogs, turtles, lizards, other snakes (including smaller Cottonmouths), birds, and small mammals; predators include large birds, mammals, and larger snakes, but people chief enemies; adult males engage in combat dances (refer to Copperhead account for more detailed description of behavior); maximum captive longevity: 21 years, 4 months, and 10 days (Snider and Bowler, 1992).

The venomous **Cottonmouth** (*Agkistrodon piscivorus*) is rarely encountered in Kansas. Its natural range is restricted to the Spring River drainage in eastern Cherokee County in the extreme southeastern corner of the state. However, a number of these snakes apparently were collected outside of Kansas in the mid-1970s and released in the Verdigris River between Independence and Coffeyville, in Montgomery County; these introduced snakes may have since perished because of lack of habitat and inhospitable weather (none has been seen since 1977). An old record for this species from the Neosho River at Chetopa in Labette County has been discredited and is not mapped. Red dots indicate records considered to be native specimens by the authors and are backed by voucher specimens.

Pertinent References: Baker (1983), Bartlett and Bartlett (2006c), Brown (1973), Burkett (1966), Campbell and Lamar (2004), Capron (1984), Clarke (1959, 1980b), Clarke and Boles (1980), Cochran and Goin (1970), Collins (1959, 1974b, 1978a, 1979, 1982a, 1985, 1992a), Collins and Collins (1993a, 1996, 2001, 2006b, 2009a), Conant (1958), Cross (1971), Ernst and Ernst (2003), Fitch (1985b), Gloyd and Conant (1990), Guarisco (1979b), Guiher and Burbrink (2008), Hall and Smith (1947), Henderson and Lee (1992), Jones-Burdick (1963), Kingsbury and Gibson (2002), Knight (1935), Kraus (2009), McLeran (1972c), Miller (1987c), Minton (2001), Mount (1975), Murrell (1993), Parrish (1980), Perkins (1940), Perry (1977b, 1977d), Platt et al. (1974), Powell and Gregory (1978), Rundquist and Triplett (1993), Russell (1980), Schmaus (1959), Shoup (1992a, 1996b), Simmons (1983), Smith (1950, 1956a, 1957), Strimple (1986), Taggart (2006b), Taylor (1929a), Tennant and Bartlett (1999), Trauth et al. (2004), Wooster (1925), and Wright and Wright (1952, 1957).

Timber Rattlesnake
Crotalus horridus LINNAEUS, 1758

Identification: **VENOMOUS**. Pit on each side of head between and slightly below eye and nostril; rattle on tail; strongly keeled body scales; small scales cover top of head except for 1 large scale over each eye; pattern of dark bands or chevrons on back; tail uniformly black; head and body vary from pinkish gray to yellowish brown; back has 18–33 dark brown to black bands or chevrons; rusty, reddish stripe often runs down middle of back; belly grayish white; adult males have longer and thicker tails and grow larger than females.

Adult: Chautauqua County, Kansas.

Adult: Crawford County, Kansas.

Adult: Jefferson County, Kansas.

Adult: Douglas County, Kansas.

Young adult: Chautauqua County, Kansas.

Adult showing how effectively cryptic coloration and pattern make a snake difficult to detect: Atchison County, Kansas.

Size: Adults normally grow 900–1,520 mm (36–60 inches) in TL; largest specimen from Kansas: female (KU 1645) from Douglas County with a total length of 1,613 mm (63½ inches, including rattle) collected by Charles D. Bunker and G. I. Adams in June 1899; maximum length throughout range: 74½ inches (Conant and Collins, 1998); maximum weight for Kansas specimen: 2,386 grams (5 pounds, 4 ounces).

Natural History: Largest venomous snake in Kansas; found in rugged terrain along heavily vegetated, rocky outcrops on partially forested hillsides; recorded active annually in Kansas from January to October with a peak from mid-April to early June; diurnal during spring and fall, but prowls at night during summer months to avoid higher daytime temperatures; although may travel over 30 yards a day to seek areas of abundant food, frequently spends long periods of time coiled and immobile, patiently waiting for prey to approach; females apparently wander less than males

REPTILES

183

and generally do not feed during pregnancy, relying on stored fat to maintain them until birth of young; during winter, retreats deep into burrows and crevices of rocky outcrops to avoid cold weather; mates during spring soon after emergence from winter inactivity; courtship poorly documented, but involves male positioning himself alongside female and stimulating her with quick, rapid jerks of head and body; male curls tail beneath female's until their cloacal openings meet and copulation occurs; females may produce litters only every other year; young born during August, September, or October; litters 5–14, with an average of 8–9; young venomous at birth and have single button on tail; additional segments are added to rattle each time shedding of skin occurs, sometimes producing a rattle with up to 15 segments; females normally reach maturity in fourth year; feeds on mice, rats, squirrels, rabbits, bats, and other small mammals, as well as on smaller snakes; species in need of conservation (protected by state law); has fairly mild disposition (compared to Prairie Rattlesnake or Massasauga); when approached, frequently remains motionless and quiet in order to avoid being seen (no one should rely on any rattlesnake to "warn" them by rattling, since many rattlesnakes never rattle until stepped on or otherwise molested); males may engage in combat dances, as do Prairie Rattlesnakes (see account on Copperheads for description of this activity); maximum captive longevity: 30 years, 2 months, and 1 day (Snider and Bowler, 1992).

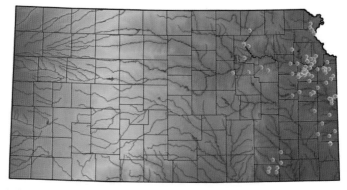

The **Timber Rattlesnake** (*Crotalus horridus*) is a large, venomous serpent with a distribution in Kansas restricted to the eastern third of the state, mostly along major river drainages. There are two disjunct populations in Kansas, one north of the Neosho River in northeastern Kansas and a second southwest of the Verdigris River. This species has not been discovered in the Neosho River basin. Red dots indicate records backed by voucher specimens or images.

Pertinent References: Anderson (1965), Ashton et al. (1974a), Ballinger and Lynch (1983), Bartlett and Bartlett (2006c), Behler and King (1979), Branson (1904), Brown (1973, 1987, 1993), Burt (1933), Busby and Parmelee (1996), Busby et al. (1994, 1996, 2005), Campbell and Lamar (2004), Caras (1974), Clarke (1958, 1959, 1980b), Cochran and Goin (1970), Coleman (1987), Collins (1959, 1974b, 1979, 1982a, 1985, 1990b, 1994b), Collins and Collins (1993a, 1996, 2001, 2006b, 2009a), Collins and Knight (1980), Colt (1862), Conant (1958, 1975, 1978a), Conant and Collins (1991, 1998), Cope (1859, 1900), Cragin (1881), Cross (1971), Ditmars (1907, 1915, 1939), Eddy and Babcock (1973), Ernst (1992), Ernst and Ernst (2003), Fitch (1951, 1956e, 1958b, 1970, 1982, 1985a, 1985b, 1991, 1992, 1993b, 1999, 2000b, 2006a), Fitch and Echelle (2006), Fitch and Pisani (2002, 2005, 2006), Fitch and Shirer (1971), Fitch et al. (2003, 2004), Foster and Caras (1994), Galligan and Dunson (1979), Gibbons et al. (1990), Gier (1967), Glenn and Straight (1982), Gloyd (1928, 1932, 1940), Gress and Potts (1993), Guarisco (1980b, 2001), Hall (1953), Hall and Smith (1947), Harris and Simmons (1978), Hay (1892), Henderson (1974), Henderson and Lee (1992), Hubbs and O'Connor (2009), Hudson (1942), Hunter et al. (1992), Hurter (1911), Johnson (2000), Karns et al. (1974), Kauffeld (1942), Kingsbury and Gibson (2002), Klauber (1972), Klemens (1993), Knight (1935, 1992), Levell (1997), Linsdale (1925, 1927), Linzey and Clifford (1981), Lokke (1983a), Loomis (1956), Loraine (1980), Lynch (1985), Martin (1956, 1992, 2002), McCauley (1945), McCoy (1975c), McLeran (1972b), Miller (1987c, 1996a), Minton (1957, 2001), Mitchell (1994), Monro (1947), Mount (1975), Mozley (1878), Murphy and Armstrong (1978), Murrell (1993), Parmelee and Fitch (1995b), Parrish (1980), Pearce (1991), Pisani and Fitch (2006), Pisani et al. (1973), Pope (1944a), Rainey (1956), Richardson and Phillips (1997), Riedle (1994b), Rossi (1992), Rues (1986), Russell (1980), Schmaus (1959), Schmidt (1953), Schmidt and Davis (1941), Schmidt and Inger (1957), Scott (1996), Shirer and Fitch (1970), Shoup (1996b), Simon and Dorlac (1990), Smith (1950, 1956a, 1957), Smith et al. (1983), Spanbauer (1984), Stejneger and Barbour (1943), Taggart (2005, 2006b, 2008), Taggart and Schmidt (2005g), Taylor (1929a), Tennant and Bartlett (1999), Tihen (1937), Trauth et al. (2004), von Frese (1973), Walker et al. (2008), Wolfenbarger (1951, 1952), Wright and Wright (1952, 1957), and Yarrow (1883).

Prairie Rattlesnake
Crotalus viridis (RAFINESQUE, 1818)

Adult: Barber County, Kansas.

Adult: Comanche County, Kansas.

Adult: Hamilton County, Kansas.

Adult: Comanche County, Kansas.

Adult: Graham County, Kansas.

Adult consuming an Ord's Kangaroo Rat: Finney County, Kansas.

REPTILES

185

Identification: **VENOMOUS**. Pit on each side of head between and slightly below eye and nostril; rattle on tail; strongly keeled body scales; small scales cover top of head except for 1 large scale over each eye; head, body, and tail greenish gray to brown; 30–55 dark gray or brown blotches on back; tail bands similar in color to body blotches; belly grayish or white; adult males have longer and thicker tails than females.

Size: Adults normally grow 890–1,140 mm (35–45 inches) in TL; largest specimen from Kansas: male (MHP 8564) from Hamilton County with TL of 1,454 mm (57⅛ inches, including rattle) collected by Dick Grusing on 28 April 2004; exceeds maximum length throughout range, as reported in Conant and Collins (1998).

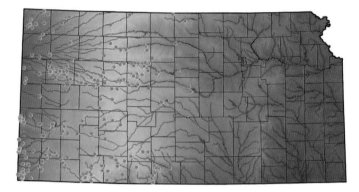

The **Prairie Rattlesnake** (*Crotalus* viridis) is a venomous reptile found in the western half of the state, from Jewell County in the north to Barber County in the south. Populations on the eastern edge of the range are isolated. Old records for Riley County have been discredited and are not mapped. Red dots indicate records backed by voucher specimens or images.

Natural History: Prefers rocky canyons or open prairies with abundance of small burrows, particularly those of prairie dogs; recorded active annually in Kansas from mid-February to mid-December, with a peak from mid-May to early August; has daily activity cycle similar to that of Timber Rattlesnake; little known of home range or habits in Kansas; during extremely hot daytime temperatures, retreats into small mammal burrows; during winter, uses these burrows to avoid extreme cold; on warm spring days, can be found sunning at edge of den sites on rocky southfacing hillside; as spring progresses, moves away from den sites in favor of open prairies; can be found under large objects such as scrap piles, in late afternoon; on hot summer days, can be located by probing with stick in shady retreats under rock overhangs at base of an outcrop; also often found by roadcruising, both during day on or above river floodplains in late spring and early summer, and during summer at night near prairies in association with rocky areas; after first or second cold snap in fall, can be found at denning sites such as hillsides with massive rocks, caves, wells, and mammal burrows; mates in early spring after emergence from winter inactivity or in fall; females produce litters only every other year; courtship resembles that of Timber Rattlesnake; young born in spring, summer, or fall (depending on time of mating) and venomous at birth; litters 5–18, with an average of 10; each newborn has single button and gains an additional segment each time it sheds skin; females normally reach maturity in third year; feeds on lizards, rats, mice, gophers, and young prairie dogs; males engage in combat dances (see account of Copperhead for details); unlike Timber Rattlesnake, quite aggressive and irritable; invariably rattles when approached too closely and should be avoided; people are the main enemies of this serpent; maximum captive longevity: 27 years and 9 months, based on specimen from Kearny County (Bailey et al., 1989).

Pertinent References: Aldridge (1993), Ashton et al. (1974a), Bailey et al. (1989), Ball (1992a), Ballinger and Lynch (1983), Bammes (1995), Bartlett and Bartlett (2006c), Bartlett and Tennant (2000), Baxter and Stone (1980, 1985), Behler and King (1979), Bennett (2004), Branson (1904), Brennan (1934, 1937), Brewer (1897), Brons (1882), Brown (1973), Burt (1933, 1935), Burt and Hoyle (1934), Busby and Parmelee (1996), Busby et al. (1994), Campbell and Lamar (2004), Capron (1985c), Caras (1974), Carpenter (1940), Clarke (1959, 1980a, 1980b), Cochran and Goin (1970), Collins (1959, 1974b, 1982a, 1985, 1990b, 2001c), Collins and Collins (1991, 1993a, 1996, 2001, 2006b, 2009a), Collins et al. (1991), Conant (1958, 1975, 1978a), Conant and Collins (1991, 1998), Cope (1859, 1866, 1900), Cragin (1881), Diener (1956), Ditmars (1939), Do Amaral (1927, 1929), Douglas et al. (2002), Dyche (1908), Dyrkacz (1981), Edds (1992c, 1993a, 1993b), Ellis and Henderson (1913), Ernst (1992), Ernst and Ernst (2003), Fitch (1985a, 1985b, 1993b, 1995a, 1995b, 2000b, 2003), Fleharty and Hulett (1988), Foster and Caras (1994), Garman (1883), Garrigues (1962), Gier (1967), Gish (1962), Glenn and Stinson (1988), Glenn and Straight (1982), Gloyd (1935, 1940), Gray and Stegall (1979), Gress (2003), Hallowell (1857a), Hammerson (1999), Harris and Simmons (1978), Henderson and Lee (1992), Hensley (1959), Holycross (1995), Hubbs and O'Connor (2009), Irwin (1979, 1983), Karns et al. (1974), Kingsbury and Gibson (2002), Knight (1935), Knight and Collins (1977), Kraus (2009), Kretzer and Cully (2001), Kauffeld (1942), Klauber (1930, 1935, 1972), Lardie (1990), Levell (1995, 1997), Loomis (1956), Marr (1944), Martin (1960), McLeran (1972b, 1974b, 1975a), Miller (1981a, 1987c), Minton (1957), Mozley (1878), Murrell (1993), Nulton and Rush (1988), Parrish (1980), Perry (1974a), Pinney (1981), Pisani and Fitch (1993), Pope (1937), (1977), Reber and Smith-Reber (1994), Reiserer and Reber (1992), Rossi and Rossi (1995), Royal (1982), Russell (1980), Savage (1877), Schmaus (1959), Schmidt (1953, 2000, 2001, 2002a, 2002b, 2004l), Schmidt and Davis (1941), Schmidt and Inger (1957), Schmidt and Murrow (2007b), Schmidt and Stark (2002), Schwarting (1984), Schilling and Schaefer (1983), Smith (1950, 1956a), Smith and Kohler (1977), Spanbauer (1984), Stebbins (1954, 1966, 1985, 2003), Taggart (1992a, 1992b, 1992d, 2000a, 2001b, 2002c, 2004h, 2006b, 2006c, 2007c), Taggart and Schmidt (2004), Taggart et al. (2007), Taylor (1929a, 1929b), Tennant and Bartlett (1999), Tihen (1937), Tihen and Sprague (1939), Toepfer (1993a), Van Doren and Schmidt (2000), Von Achen and Rakestraw (1984), Wooster (1925), and Wright and Wright (1952, 1957).

Massasauga
Sistrurus catenatus (RAFINESQUE, 1818)

Identification: **VENOMOUS.** Small pit on each side of head between and slightly below eye and nostril; small rattle on tail; strongly keeled body scales; 9 large scales on top of head; head, body, and tail gray or light brown, with 20–50 dark gray or brown blotches on back and smaller blotches on tail; belly may be mottled or blotched, or light with indistinct pattern; adult males have longer and thicker tails and reach greater length than females.

Size: Adults normally grow 457–760 mm (18–30 inches) in TL; largest specimen from Kansas: male (KU 197200) from Coffey County with a TL of 850 mm (33½ inches, including rattle) collected by G. Lairson on 14 June 1984; maximum length throughout range: 39½ inches (Conant and Collins, 1998).

Adult: Ellsworth County, Kansas.

Adult: Barton County, Kansas.

Adult: Comanche County, Kansas.

Juvenile: Russell County, Kansas. Note the orange-colored tail.

Adult: Comanche County, Kansas.

Adult. Note nine large scales on top of the head, typical of this genus of snakes.

Natural History: Smallest rattlesnake in Kansas; found in wide variety of habitats ranging from arid open sagebrush prairie and rocky, prairie hillsides to open wetlands; seems to reach peak of abundance in grassy wetlands such as Cheyenne Bottoms in Barton County; recorded active annually in Kansas from late March to early December, with a peak from early May to late July; diurnal during spring and fall; prowls at night during hot summer months; much time spent basking in sun and waiting for food; during winter, crawls deep into rock crevices or down rodent burrows to avoid cold; mates in both spring and fall; courtship involves male crawling beside female with quick jerking movements of his body; bends his tail beneath hers until cloacal openings meet and copulation occurs; young born in July and August; venomous at birth; number of young per litter 3–13, with an average of 6; eats frogs, lizards, other snakes, bird eggs, and rodents; because of small size, difficult to hear when rattling; apparently, males engage in combat dances typical of other venomous species found in Kansas; predators unknown, but may include large water birds in areas such as Cheyenne Bottoms; maximum captive longevity: 20 years and 5 days (Snider and Bowler, 1992).

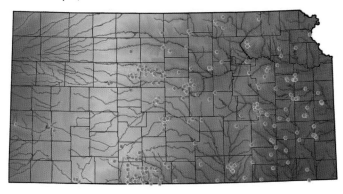

The **Massasauga** (*Sistrurus catenatus*) is a small, venomous rattlesnake that is generally found throughout the eastern two-thirds of the state, except for the southeastern corner. It is found west to the Colorado border along the Arkansas River floodplain in Kansas. Red dots indicate records backed by voucher specimens or images.

Pertinent References: Anderson (1965), Ashton et al. (1974a), Ballinger and Lynch (1983), Bartlett and Bartlett (2006c), Behler and King (1979), Beltz (1993), Bender (2006), Bender and Stark (2007), Berglund (1967), Blythe (1979), Branson (1904), Breukelman and Clarke (1951), Breukleman and Downs (1936), Brush and Ferguson (1986), Burt (1933, 1935), Burt and Burt (1929b), Burt and Hoyle (1934), Busby and Pisani (2007), Busby et al. (1996, 2005), Campbell and Lamar (2004), Capron (1978b, 1985c, 1986b), Capron and Perry (1976), Caras (1974), Clarke (1956c, 1959, 1980a, 1980b), Clarke et al. (1958), Cochran and Goin (1970), Collins (1959, 1974b, 1982a, 1985, 1990b, 2001c, 2006b, 2007b), Collins and Collins (1993a, 1993b, 1996, 2001, 2006a, 2006b, 2009a), Collins et al. (1991, 1994), Conant (1958, 1975, 1978a), Conant and Collins (1991, 1998), Cope (1859), Cragin (1881, 1885a, 1885d), Cross (1971), Diener (1956), Ditmars (1907, 1915, 1930, 1939), Do Amaral (1927), Dyrkacz (1981), Ellis (2002), Ellis and Henderson (1913), Ernst (1992), Ernst and Ernst (2003), Evans and Gloyd (1948), Fitch (1985b, 1993b), Foster and Caras (1994), Garrigues (1962), Gish (1962), Glenn and Straight (1982), Gloyd (1928, 1940, 1955), Greene and Oliver (1965), Gress (2003), Guar-isco (1980d, 1981c), Hall and Smith (1947), Haltom (1931), Hammerson (1999), Harris and Simmons (1978), Henderson and Lee (1992), Hobert (1997), Holycross and Mackessy (2002), Holycross et al. (2008), Hoyle (1936), Hubbs and O'Connor (2009), Hudson (1942), Hurter (1911), Irwin (1979), Irwin and Collins (1987), Jantzen (1960), Karns et al. (1974), Kauffeld (1942), Kingsbury and Gibson (2002), Klauber (1972), Knight (1935), Knight et al. (1973), Jordan (1929), Lardie (1990), Lardie and Black (1981), Lokke (1989), Loomis (1956), Lueth (1941), Maslin (1959), McCoy (1975a), McLeran (1972b), Miller (1980a, 1983a, 1986a, 1987c, 1988), Minton (1983), Mozley (1878), Murrell (1993), Murrow (2008a, 2009a, 2009b), Parrish (1980), Patten et al. (2009), Perkins (1940), Peterson (1992), Platt (1975, 2003), Pope (1937, 1944b), (1977), Porter (1972), Powell and Gregory (1978), Reinert (1978), Reiserer (2002), Richardson and Phillips (1997), Rossi (1992), Rossi and Rossi (1995), Royal (1982), Rues (1986), Rush (1981), Rush and Fleharty (1981), Rush et al. (1982), Russell (1980), Schmaus (1959), Schmidt (1938b, 1953, 1998), Schmidt and Davis (1941), Schmidt and Inger (1957), Schmidt and Murrow (2006, 2007c), Schilling (1985), Simmons (1987c), Smith (1931, 1950, 1956a), Spanbauer (1984), Stebbins (1954, 1966, 1985, 2003), Stejneger and Barbour (1917, 1923, 1933, 1939, 1943), Taggart (2000a, 2002d, 2008), Taylor (1929a), Tennant and Bartlett (1999), Tihen (1937), Tihen and Sprague (1939), Toepfer (1993a), Tucker (1912), Uhler and Warren (1929, 2008), Van Doren and Schmidt (2000), Viets (1993), Von Achen and Rakestraw (1984), von Frese (1973), Washburne (2003a), Woodburne (1956), Wright (1941), Wright and Wright (1952, 1957), and Zimmerman (1990).

TURTLES
CLASS CHELONIA MACARTNEY (IN ROSS), 1802

Turtles are members of the Class Chelonia, which consists in Kansas of 4 families, 10 genera, and 14 species. Seven families are represented in the United States and Canada by 22 genera and 58 species (Collins and Taggart, 2009, with updates from the CNAH web site).

As a vertebrate group, turtles are easily distinguished from amphibians and reptiles. Turtles, unlike amphibians and reptiles, have an upper (carapace) and lower (plastron) shell and, unlike amphibians and some reptiles (snakes), they have claws on their toes and fingers. As a group, turtles are terrestrial or semiaquatic vertebrates that are most closely related to birds and crocodilians. In Kansas, 11 kinds live in or near water and 2 kinds (Box Turtles) lead a terrestrial existence. Put another way, about 15% of the species of turtles in Kansas live on land, whereas more than 92% of the species of reptiles are primarily terrestrial.

All turtles can spend time out of water because their skin is better able to contain body moisture; thus they are much less likely to dry up and die than amphibians. Unlike amphibians in Kansas, male turtles fertilize females internally with a single copulatory organ during mating (reptiles, such as lizards and snakes, have a paired copulatory organ). Females dig underground nests and lay their eggs on land.

Young turtles hatch as small duplicates of adults. Some may have a different pattern and color, but all generally look like their parents. Turtles, like reptiles, do not have a larval or tadpole stage.

STRAIGHTNECK TURTLES
ORDER CRYPTODIRA COPE, 1868

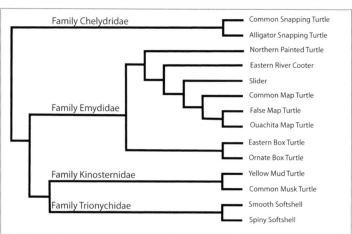

A phylogeny of Kansas Turtles (Order Cryptodira) that is consistent with evolutionary history based on current scientific evidence.

Turtles in Kansas (and the United States) all belong to the Order Cryptodira, the Straightneck Turtles. Straightneck Turtles have been moderately successful in adapting to the arid landscape of Kansas. Four species (one terrestrial) are found throughout the state. One kind, the Yellow Mud Turtle, is primarily found in the western two-thirds of Kansas.

SNAPPING TURTLES
FAMILY CHELYDRIDAE GRAY, 1870

The family Chelydridae (Snapping Turtles) is found only in the western hemisphere and consists of 2 genera, each with a single species (Collins and Taggart, 2009) and both of which occur in Kansas. These are the Common Snapping Turtle (*Chelydra serpentina*) and Alligator Snapping Turtle (*Macrochelys temminckii*). The family is distinguished by a large head, a strong, hooked beak, a long tail, a small plastron, an overall dull uniform color, and large size (up to 132½ pounds in Kansas).

Common Snapping Turtle
Chelydra serpentina (LINNAEUS, 1758)

Adult: Douglas County, Kansas.

Adult in defensive posture, with rear of body elevated to increase thrust when it lunges: Linn County, Kansas. Note leech on hind leg, a common occurrence with these highly aquatic chelonians.

© Larry L. Miller

Adult: Shawnee County, Kansas.

Subadult: Lyon County, Kansas.

TURTLES

191

Juvenile showing small plastron when compared to many other species of Kansas turtles: Douglas County, Kansas.

Adult: Barton County, Kansas.

Identification: Carapace rigid; plastron rigid and small; very long (at least half length of carapace) sawtoothed tail; large size; eyes can be seen from directly above head; relatively smooth skin on top of head; carapace tan or brown, often covered with mud and algae; plastron white or yellowish; head, limbs, and tail brown; cloacal opening of males situated on underside of tail at greater distance out from under rear edge of carapace than in females; females reach slightly larger size than males.

Size: Adults normally 203–360 mm (8–14 inches) in carapace length; largest specimen from Kansas: male (MHP 13387) from Reno County collected by Jay E. Mattison and Allen Andresen on 16 October 2006 with carapace length of 406 mm (16 inches); turtle weighed 20.5 kilograms (45 pounds), heaviest specimen recorded from state; maximum carapace length throughout range: 19⅜ inches; maximum weight known (captive specimen): 86 pounds (Conant and Collins, 1998).

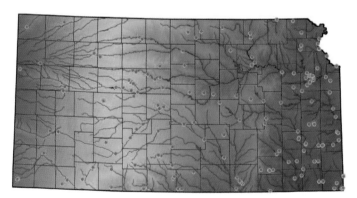

The large, aggressive, semi-aquatic **Common Snapping Turtle** (*Chelydra serpentina*) is found throughout the state, but is least abundant in the western fourth of Kansas. Red dots indicate records backed by voucher specimens or images.

Natural History: Found in every aquatic situation; prefers water with soft mud bottom, abundant pond-edge vegetation, and numerous sunken logs and branches; most active at night when foraging for food; spends time buried in mud in water about same depth as long neck, allowing it to raise head to surface and breathe with minimal movement; recorded active annually in Kansas from early March to late December (with a peak from early May to late July), digging beneath mud of ponds and lakes during coldest months; unable to retain body moisture as well as some other turtles (might explain why it does not often bask on logs); mating generally occurs between April and November; courtship involves a pair of turtles facing each other and waving their heads and necks sidewise in opposite directions; they gulp water and violently expel it through their nostrils, creating turbulence at surface; male mounts female by gripping her carapace with claws and curling his tail beneath hers until their cloacal openings in contact; during penetration female remains passive; sperm may retain potency for several years; females lay 12–87 white, round eggs and may produces two clutches per season; eggs laid on land in nests dug by female; nests 4–7 inches deep in sandy or loamy soil and may be some distance from water; eggs hatch in 55–125 days, depending on temperature and humidity; young have carapace length of 1–1½ inches at birth; eats aquatic plants, insects, crayfishes, earthworms, clams, snails, fishes, frogs, toads, salamanders, snakes, other turtles, birds, small mammals, and carrion; nasty disposition; large individuals can inflict painful bite; when disturbed, emits foul-smelling musk; excellent to eat and probably only turtle in Kansas of any commercial value due to abundance and large size; predators on eggs and young include skunks, raccoons, crows, herons, hawks, bullfrogs, large fishes, and snakes; people are the main predators of adults; maximum longevity: 38 years, 8 months, and 27 days (Snider and Bowler 1992).

Pertinent References: Anderson et al. (1995), Ballinger and Lynch (1983), Bartlett and Bartlett (2006b), Behler and King (1979), Bonin et al. (2006), Boyd (1991), Brennan (1934, 1937), Breukelman and Clarke (1951), Breukleman and Downs (1936), Breukelman et al. (1961), Burkett (1984), Burt (1931a, 1933, 1934, 1935), Burt and Burt (1929b), Burt and Hoyle (1934), Busby and Parmelee (1996), Busby and Pisani (2007), Busby et al. (1994, 1996, 2005), Capron and Perry (1976), Caldwell and Collins (1981), Caldwell and Glass (1976), Capron (1975a, 1985c, 1987b), Carr (1952), Cink (1991), Clarke (1956a, 1958, 1980a), Clarke et al. (1958), Cloutman (1982), Coleman (1987), Collins (1959, 1974b, 1982a, 1985, 1990b, 2001c, 2006b, 2007a, 2007b), Collins and Collins (1991, 1993a, 1993b, 2006a), Collins et al. (1994), Conant (1958, 1975, 1978b), Conant and Collins (1991, 1998), Cragin (1881), Cross (1971), Dice (1923), Dyche (1914), Edds (1992a), Eddy and Babcock (1973), Ellis (2002), Ellis and Henderson (1913), Ernst (2008), Ernst and Barbour (1972), Ernst & Lovich (2009), Ernst et al. (1994), Fitch (1956e, 1958b, 1991, 2006a), Fraser (1983a), Gish (1962), Gloyd (1928, 1932), Grant (1937), Grow (1976b), Guarisco (1982a), Gubanyi (2004c), Hall and Smith (1947), Heinrich and Kaufman (1985), Householder (1917), Hoyle (1936), Hunter et al. (1992), Irwin and Collins (1987), Iverson (1986, 1992), Karns et al. (1974), Kern et al. (1978), Kingsbury and Gibson (2002), Knight (1935), Knight and Collins (1977), Legler (1955, 1956), Levell (1995, 1997), Linsdale (1925, 1927), Matthews (1935), Miller (1976a, 1977a, 1980a, 1982a, 1983a, 1986a, 1988, 1996a, 1996b), Minton (2001), Mitchell (1994), Moll and Iverson (2008), Mount (1975), Murrow (2008a, 2009a, 2009b), Nadeau (1979), Peterson (1992), Pisani (2004), Platt (1975), Reichman (1987), Riedle (1994a), Riedle and Hynek (2002), Rues (1986), Rush and Fleharty (1981), Salazar (1968a), Schmidt (2004a, 2004l, 2005), Schmidt and Murrow (2007b, 2007c), Schwilling and Schaefer (1983), Simon and Dorlac (1990), Smith (1950, 1956a, 1961, 2001), Sparks et al. (1999), Stebbins (1954, 1966, 1985, 2003), Suleiman (2005), Taggart (1992b, 2001a, 2002c, 2004h, 2005, 2006b, 2006c, 2007a, 2007c), Taggart and Schmidt (2005d), Taggart et al. (2007), Taylor (1929b, 1933), Tihen (1937), Tihen and Sprague (1939), Toepfer (1993a), Trauth et al. (2004), Uhler and Warren (1929, 2008), Van Doren and Schmidt (2000), Viets (1993), Vitt (2000), Voorhees et al. (1991), Wagner (1984), Wilgers et al. (2006), Wooster (1925), Yarrow (1883), and Zimmerman (1990).

Alligator Snapping Turtle
Macrochelys temminckii (HARLAN, 1835)

Identification: Largest turtle in Kansas; sawtoothed tail longer than half length of carapace; large carapace rigid; plastron rigid and small; eyes not visible from above; large scales present on top of head; carapace, head, limbs, and tail uniform dark brown or gray; plastron brown or gray; cloacal opening on underside of tail situated out from under rear edge of carapace in males; males reach carapace length over 30 inches and total weight over 300 pounds; females do not exceed 25 inches carapace length and 62 pounds weight.

Subadult. Note pink, wormlike appendage in lower mouth, used as a lure to attract prey.

Adult female: Montgomery County, Kansas.

Skeleton of adult: Lyon County, Kansas.

Adult. Compliments of Topeka Zoo.

Largest adult Alligator Snapping Turtle (*Macroclemys temminckii*) ever found in Kansas, taken in Labette County. It was a male that weighed 132½ pounds.

Size: Adults normally 380–660 mm (15–26 inches) in carapace length; largest specimen from Kansas: sex undetermined adult (KU 204150) from Lyon County with carapace length of 558 mm (22 inches) collected by Rick Christie and J. M. McDaniel on 21 February 1967; heaviest specimen from Kansas: male (KU Color Slide 7592) from Labette County with weight of 60 kilograms (132½ pounds) collected by Jack

Gearhart, Ralph Stice, and Henry Stice in April 1938; maximum carapace length throughout range: 31½ inches; maximum weight 316 pounds (Conant and Collins, 1998).

Second adult male: Labette County, Kansas.

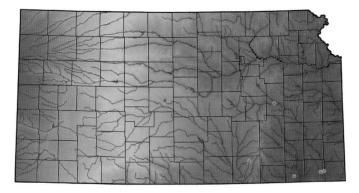

The **Alligator Snapping Turtle** (*Macrochelys temminckii*) is a very large, secretive, aggressive, semi-aquatic chelonian that is seldom seen in the state. Only five specimens are documented from Kansas, one from the Arkansas River drainage in Cowley County, one from the Verdigris River drainage in Montgomery County, and three from the Neosho River drainage in Labette and Lyon counties. This highly aquatic species is restricted to streams and rivers of southeastern Kansas; it may be more abundant than records indicate. Red dots indicate records backed by voucher specimens or images.

Natural History: Found in rivers, larger creeks, swamps, lakes, and sloughs; probably active annually from March to October; spends days under large logs or overhanging banks; actively moves about and forages for food at night, particularly between midnight and dawn; almost completely aquatic; only females known to leave water, and only for nesting; a single Kansas specimen showed movements of up to 176 yards, rested during day in concealed, shady areas beneath log jams and drifts in shallow water (no more than 30 inches deep), and migrated upstream, sometimes resting as much as 8 days between movements (movements lasted up to 3 hours and always took place between midnight and dawn); probably opportunistic breeder, but reported to mate during warm weather from late April to mid-June; courtship consists of male mounting female's carapace, sometimes accompanied by vicious neck-biting on his part; once mounted, male bends tail beneath hers until their cloacal openings meet; copulation may last nearly 30 minutes; nesting activities take place during May and June and involve nest-digging by

TURTLES

female, usually in morning, on land about 40 feet from water; females may lay single clutch per season or clutch every other year; clutches consist of 9–61 round, white eggs that hatch in approximately 3–4 months; eat anything they can stalk, overpower, and swallow; newborn or young turtles resting underwater on bottom during day use different technique for obtaining food—they sit motionless on muddy bottom and open mouth in wide gape when fishes swim nearby; attached to tongue is pink or red fleshy projection resembling a "worm," which newborn turtle wiggles to attract unwary fishes; when fish swims into or near gaping mouth to examine "worm," turtle has a meal; eats crayfishes, mollusks, fishes, salamanders, turtles, snakes, birds, and small mammals; species in need of conservation (protected by state law); people are the only predators of adults of this species; maximum longevity: 70 years, 4 months, and 26 days (Snider and Bowler, 1992)

Pertinent References: Anderson (1965), Ballinger and Lynch (1983), Bartlett and Bartlett (2006b), Behler and King (1979), Bonin et al. (2006), Caldwell and Collins (1981), Capron (1975b, 1985c, 1985d, 1986a, 1987a, 1987b), Carr (1952), Clarke (1956a, 1981), Cochran and Goin (1970), Collins (1959, 1974b, 1982a, 1984b, 1985, 1986b, 1990b, 1994b), Collins and Collins (1993a), Collins et al. (1981, 1991), Conant (1958, 1975), Conant and Collins (1991, 1998), Conant et al. (1992), Cragin (1886b), Cross (1971), Ernst (2008), Ernst and Barbour (1972), Ernst & Lovich (2009), Ernst et al. (1994), Freeman and Kindscher (1992), Fritz and Havas (2007), Gier (1967), Guarisco et al. (1982), Hall and Smith (1947), Harrison (1977), Henderson and Andelt (1982), Henderson and Schwilling (1982), Hlavachick (1978), Householder (1917), Irwin (1985a, 1992a), Iverson (1986, 1992), Johnson (1974, 2000), Karns et al. (1974), Kingsbury and Gibson (2002), Layher et al. (1986), Legler (1960b), Levell (1997), Lovich (1993), Manes (1986), Martin (1994), Minton (1972, 2001), Moss and Brunson (1981), Mount (1975), Mullen (1979), Platt (1994), Platt et al. (1974), Pope (1956), Pritchard (1967, 1979a, 1979b, 1980, 1989, 2006), Riedle et al. (2005, 2008), Salazar (1968a), Schmidt (1953), Schwilling (1981), Schwilling and Nilon (1988), Scott and Koons (1994), Shipman (1993b), Shipman and Riedle (2008), Shipman et al. (1991, 1993, 1994, 1995), Simmons (1986b, 1989), Smith (1950, 1956a), Taggart (2006b, 2008), Trauth et al. (2004), Wood (1985), and Zappalorti (1976).

MUD AND MUSK TURTLES
FAMILY KINOSTERNIDAE AGASSIZ, 1857

In Kansas, the family Kinosternidae (Mud and Musk Turtles) consists of 2 genera, each with a single species. These are the Common Musk Turtle (*Sternotherus odoratus*) and Yellow Mud Turtle (*Kinosternon flavescens*). Confined to the New World, this family is composed of 2 genera and 10 species in the United States and Canada (Collins and Taggart, 2009), and is characterized by an oval carapace, a short tail, comparatively dull coloration, a much-reduced plastron, and the presence of musk glands that can emit an offensive odor.

Yellow Mud Turtle
Kinosternon flavescens (AGASSIZ, 1857)

Identification: Short tail that ends in a horny, clawlike tip; rigid carapace; plastron with distinct movable hinge; webbed feet; carapace brown or olive-brown with dark brown margins around each scute; plastron yellowish brown with dark brown margins around each scute; head, limbs, and tail grayish, and chin yellow; adult males have longer, thicker tails and grow slightly larger than females.

Young adult: Stafford County, Kansas.

Adult: Rice County, Kansas.

Adult: Morton County, Kansas.

Adult showing plastron (lower shell) color and arrangement of scutes: Morton County, Kansas.

Adults.

Size: Adults normally 100–125 mm (4–5 inches) in carapace length; largest specimen from Kansas: sex undetermined (MHP 9661) from Rooks County with carapace length of 146 mm (5¾ inches) collected by Chad Whitney on 3 October 2004; maximum carapace length throughout range: 6⅜ inches (Conant and Collins, 1998).

Natural History: Prefers quiet water with mud or sand bottom; found in ponds, sloughs, backwaters, swamps, sinkholes, rivers, cisterns, roadside ditches, and cattle tanks; presence of aquatic vegetation preferred, but not necessary; recorded active annually in Kansas from early March to mid-October, with a peak from mid-May to mid-July; daily activity divided into 2 periods—from afternoon to dusk and from midnight to sunrise; may forage on land and frequently found crawling from one body of water to another; known to bask on brush or logs at water's edge; during winter, burrows in mud above or below water; may reside in muskrat dens or old stump holes; breeding not well documented in Kansas, but probably occurs before June; courtship involves male approaching other turtles from rear and smelling their

TURTLES

197

tails, evidently to determine sex; upon discovering female, male moves to her side and nudges underedge of her carapace with his nose; receptive female mounted immediately, but reluctant female causes male to give chase until mounting achieved; male mounts female by clasping her carapace with his clawed feet; male pinions her tail up between his rear legs and positions cloaca to hers; copulation may last up to 3 hours; during copulation, male extends head forward and rubs and bites female's head; nesting probably occurs in June, but nesting habits not observed; females lay up to 7 elongate white eggs, which hatch within 3 months; omnivorous, eating insects, crayfishes, snails, earthworms, amphibians, dead fishes, and aquatic vegetation; acute sense of underwater smell that aids in locating food; large fishes, other turtles, and snakes eat young; people are only major predators of adults; can emit foul-smelling musk when excited; maximum longevity: 10 years, 4 months, and 25 days (Snider and Bowler, 1992).

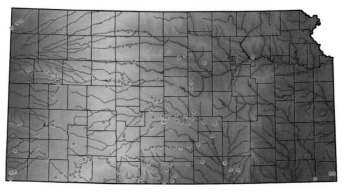

The semi-aquatic **Yellow Mud Turtle** (*Kinosternon flavescens*) is found in the western two-thirds of the state; an isolated colony is known from Cherokee and Labette counties in southeastern Kansas. The isolated record collected in 1995 in Pottawatomie County in northeastern Kansas needs corroboration. Red dots indicate records backed by voucher specimens or images.

Pertinent References: Anderson (1965), Ballinger and Lynch (1983), Bartlett and Bartlett (2006b), Behler and King (1979), Bonin et al. (2006), Brennan (1934, 1937), Breukelman and Clarke (1951), Buhlmann et al. (2008), Burt (1933, 1935), Burt and Hoyle (1934), Caldwell and Collins (1981), Caldwell and Glass (1976), Capron and Perry (1976), Carr (1952), Clarke (1956a, 1980a), Clarke et al. (1958), Cochran and Goin (1970), Collins (1959, 1974b, 1979, 1980a, 1982a, 1985, 1989c, 1990b, 1997a, 2001c), Collins and Collins (1991, 1993a, 1993b, 2006a), Collins et al. (1991, 1994), Conant (1958, 1975, 1978a), Conant and Berry (1978), Conant and Collins (1991, 1998), Cragin (1885a, 1885d, 1886a), Cross (1971), Ellis and Henderson (1913), Ernst and Barbour (1972), Ernst & Lovich (2009), Ernst et al. (1994), Fichter (1969), Gier (1967), Gish (1962), Hall and Smith (1947), Hartweg (1938), Hay (1892), Houseal et al. (1982), Householder (1917), Hoyle (1936), Irwin (1977, 1983), Irwin and Collins (1987), Iverson (1977, 1979, 1986, 1992), Karns et al. (1974), Kern et al. (1978), Kingsbury and Gibson (2002), Knight and Collins (1977), Jordan (1929), Lardie (1979), Long (1993), Matthews (1935), Miller (1976a, 1980a, 1985b, 1988), Perry (1974a), Platt (1975), Pope (1939), Pritchard (1979a), Reichard et al. (1995a), Royal (1982), Rush et al. (1982), Schmidt (1953, 2001, 2004l), Schmidt and Murrow (2007b), Schmidt and Taggart (2004a), Schwaner (1978), Smith (1950, 1951, 1956a, 1957), Sparks et al. (1999), Stebbins (1954, 1966, 1985, 2003), Stejneger and Barbour (1917, 1923, 1933), Taggart (1992b, 1992h, 2001b, 2002c, 2004b, 2004h, 2006b, 2006c, 2007c), Taggart et al. (2007), Taylor (1929b, 1933), Tihen (1937), Tihen and Sprague (1939), Toepfer (1993a),

Uhler and Warren (1929, 2008), Van Doren and Schmidt (2000), Vetter (2004), Webster (1986), Werth (1969), Whitney et al. (2004), and Zimmerman (1990).

Common Musk Turtle
Sternotherus odoratus (LATREILLE, 1802)

Identification: Short tail; rigid carapace; small plastron; two small fleshy projections or barbels on chin; carapace dark gray-brown or black; plastron yellow, gray, or brown; head, limbs, and tail dark gray or black; each side of head has pair of narrow yellow lines, except in very old adults; tail of males ends in blunt tip; females grow slightly larger than males.

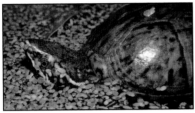

Adult: Elk County, Kansas.

Juvenile next to a U.S. quarter for size comparison: Neosho County, Kansas.

Adult: Johnson County, Kansas.

Adult: Douglas County, Kansas.

Adult: Cherokee County, Kansas.

Adult: Johnson County, Kansas.

Size: Adults normally 51–115 mm (2–4½ inches) in carapace length; largest specimen from Kansas: female (KU 45016) from Linn County with carapace length of 114 mm (4½ inches) collected by William L. Minckley on 10 September 1957; maximum carapace length throughout range: 5⅜ inches (Conant and Collins, 1998).

Natural History: Prefers still or slow-moving water of lakes, swamps, sloughs, oxbows, and rivers; recorded active annually in Kansas from late February to late

TURTLES

199

November, with a peak from mid-April to mid-July; burrows into mud a foot beneath surface and remains inactive during cold periods; stays hidden during much of day and becomes active at night, foraging for food; occasionally will bask in sun and may climb into bushes and up logs to as much as 6 feet above water; mates during May or June and sometimes again in fall; courtship similar to that described for Yellow Mud Turtle; females nest from May to August, either digging shallow nests in loose soil or laying eggs on ground in open (several females may nest together); females lay 2–7 elongate white eggs; when laid during May or June, eggs hatch in 2–3 months; when laid in fall, eggs may not hatch, and young may not emerge until following spring; newly hatched young orient toward light of large open areas (such as lakes) and thus find their way to water; an omnivorous scavenger, eating aquatic plants, insects, earthworms, snails, crayfishes, small or dead fishes, and even human garbage; gives off foul-smelling musk when excited; has minimum known life span of 15–19 years; maximum longevity: 54 years and 9 months (Snider and Bowler, 1992).

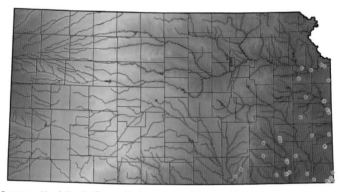

The **Common Musk Turtle** (*Sternotherus odoratus*) is a small, semi-aquatic chelonian with a distribution in Kansas restricted to river drainages in eastern part of the state, from the Kansas River drainage in Wyandotte County southwest to the Arkansas River drainage in Cowley County. Although its distribution in the state is restricted, this species has the most extensive range in North America of any member of its genus, hence the name, Common Musk Turtle. Red dots indicate records backed by voucher specimens or images.

Pertinent References: Anderson (1965), Ballinger and Lynch (1983), Bartlett and Bartlett (2006b), Behler and King (1979), Bonin et al. (2006), Burt (1933, 1935), Burt and Hoyle (1934), Caldwell and Collins (1981), Caldwell and Glass (1976), Carr (1952), Clarke (1956a, 1956c), Cochran and Goin (1970), Collins (1959, 1974b, 1981, 1982a, 1985, 1990b), Collins and Collins (1993a), Collins et al. (1994), Conant (1958, 1975), Conant and Berry (1978), Conant and Collins (1991, 1998), Cragin (1881), Cross (1971), Edds (1992a), Ernst and Barbour (1972), Ernst & Lovich (2009), Ernst et al. (1994), Fitch (1985b), Gier (1967), Gibbons (1987), Grant (1937), Grow (1976b), Hall and Smith (1947), Householder (1917), Hunter et al. (1992), Iverson (1986, 1992), Johnson (2000), Karns et al. (1974), Kingsbury and Gibson (2002), Klemens (1993), Knight (1935), Mansueti (1941), McCauley (1945), Minton (1972, 2001), Mitchell (1994), Mount (1975), Pope (1939), Reynolds and Seidel (1982, 1983), Riedle (1994a), Riedle and Hynek (2002), Sanders et al. (2005), Smith (1950, 1956a), Stejneger and Barbour (1939, 1943), Taggart (2001a, 2005, 2008), Tihen (1937), Tinkle (1961), Trauth et al. (2004), Voorhees et al. (1991), Wagner (1984), and Zappalorti (1976, 1990).

BOX AND BASKING TURTLES
FAMILY EMYDIDAE GRAY, 1825

The family Emydidae (Box and Basking Turtles) is composed of 11 genera and 34 species in the United States and Canada (Collins and Taggart, 2009). Eight species assigned to 5 genera represent this family in Kansas. They include 2 species of the Box Turtle genus *Terrapene*, 3 of the Map Turtle genus *Graptemys*, and 1 each of the basking turtle genera *Chrysemys*, *Pseudemys*, and *Trachemys*. The Emydidae is the largest family of turtles in the world. Most Kansas taxa are semiaquatic, but some are terrestrial, and all have large, well-developed lower shells.

Northern Painted Turtle
Chrysemys picta (SCHNEIDER, 1783)

Mud-covered juvenile:
Neosho County, Kansas.

Juvenile showing brightly-colored plastron
(lower shell): Neosho County, Kansas.

Adult: Lyon County, Kansas.

Adult: Douglas County, Kansas.

Adult: Douglas County, Kansas.

Adult: Chase County, Kansas.

Identification: Semiaquatic; short tail; rigid carapace and plastron; carapace with smooth rear edge; some pattern of bright red or coral red on plastron; carapace gray, brown, or green with red markings around edge; head, limbs, and tail dark gray or green with yellow lines; adult males smaller than females and have very long claws on front feet.

Size: Adults normally 90–180 mm (3½–7 inches) in carapace length; largest specimen from Kansas: sex undetermined (KU 159983) from Barton County with carapace length of 207 mm (8⅛ inches) collected by William Knighton and Natalie Fayman on 24 June 1989; maximum carapace length throughout range: 9⅞ inches (Conant and Collins, 1998).

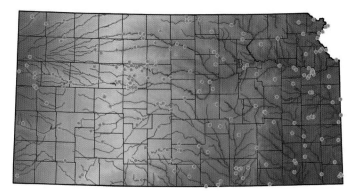

The **Northern Painted Turtle** (*Chrysemys picta*) is a semi-aquatic species that is abundant throughout the state, but appears to be least common in the southwestern corner of Kansas. Red dots indicate records backed by voucher specimens or images.

Natural History: Resides in slow-moving shallow streams and rivers and shallow ponds and lakes having soft bottoms with aquatic plants and numerous half-submerged logs and branches on which to bask in sun; recorded active annually in Kansas from mid-February to late December, with a peak from late April to late July; during cold winter months, may burrow as deep as 1½ feet beneath mud in bottoms of lakes, streams, and ponds; hatchlings only known turtles that can survive winters through natural freezing of up to 54% of their body fluids; apparently produce natural protectants to prevent cell damage during freeze and thaw process; ability to tolerate subfreezing temperatures allows hatchlings to overwinter in nest rather than face hazards of winter ponds, where they would be easy meal for predators; diurnal throughout active season, sleeping at night beneath water on submerged logs or on bottom; divides daytime hours between basking and feeding; mating normally occurs from March to June but may continue into summer; courtship starts with slow pursuit of female by male; upon catching up with her, male faces female and strokes her head and neck with backs of long claws on front feet; male periodically swims away as if trying to entice female to follow; eventually, female sinks to bottom and male swims down and mounts her carapace; male secures himself on female carapace with his claws, and curls tail down under hers until their cloacal openings meet, and copulation occurs; females lay up to 23 elongate white eggs in nests from May to July; nests are dug in soft soil by female, who uses her hind limbs to excavate earth to a depth of 4 inches; female may release fluid from her cloaca to dampen soil; eggs generally hatch in 2–2½ months, and young turtles dig free of nest and rapidly crawl to water; sometimes eggs may not hatch, or if they do, young may not emerge until late fall during cold weather; in latter case, they remain in nest during winter and hatch or emerge in following spring; omnivorous, but young are more carnivorous than adults; eats plant and animal matter, dead or alive; predators of adults primarily people and their pesticides; eggs and young eaten by skunks, badgers, raccoons, muskrats, crows, snakes, other turtles, bullfrogs, and large fishes; maximum longevity: 20 years, 6 months, and 24 days (Snider and Bowler, 1992).

Pertinent References: Ballinger and Lynch (1983), Bartlett and Bartlett (2006b), Behler and King (1979), Bonin et al. (2006), Boyd (1991), Brennan (1934, 1937), Breukelman and Clarke (1951), Breukleman and Downs (1936), Burt (1933, 1935), Burt and Hoyle (1934), Busby and Parmelee (1996), Busby and Pisani (2007), Busby et al. (1994, 1996, 2005), Caldwell and Collins (1981), Caldwell and Glass (1976), Capron and Perry (1976), Carr (1952), Clarke (1956a, 1958, 1980a), Clarke et al. (1958), Cloutman (1982), Cochran and Goin (1970), Coleman (1987), Collins (1959, 1974b, 1982a, 1985, 1990b, 2001c, 2006b, 2007b), Collins and Collins (1991, 1993a, 1993b, 2006a), Collins et al. (1991), Conant (1958, 1975, 1978b), Conant and Collins (1991, 1998), Cragin (1881, 1885a, 1885d), Cross (1971), Edds (1992a), Eddy and Babcock (1973), Ellis and Henderson (1913), Ernst (1971), Ernst and Barbour (1972), Ernst & Lovich (2009), Ernst et al. (1994), Falls (1933), Fitch (1958b, 1991), Gibbons (1987), Gish (1962), Gloyd (1928, 1932), Grant (1937), Gress and Potts (1993), Guarisco (1980a), Hall and Smith (1947), Hay (1892), Heinrich and Kaufman (1985), Householder (1917), Hunter et al. (1992), Irwin and Collins (1987), Iverson (1986, 1992), Karns et al. (1974), Kern et al. (1978), Kingsbury and Gibson (2002), Knight (1935), Knight and Collins (1977), Lardie (1975a), Lardie and Black (1981), Legler (1956), Maslin (1959), Maurer (1995), Miller (1976a, 1976d, 1977a, 1979, 1980a, 1983a, 1985b, 1988, 1996a, 1996b, 2000c), Miller and Lardie (1976), Minton (2001), Mitchell (1994), Moriarty (1996), Mount (1975), Murrow (2008a, 2008b, 2009a, 2009b), Nadeau (1979), Platt (1975), Porter (1972), Preston (1982), Pritchard (1979a), Riedle (1994a), Riedle and Hynek (2002), Rundquist (1996a), Salazar (1968b), Schmidt (2001, 2004b, 2004l), Schmidt and Murrow (2006, 2007b, 2007c), Schwilling and Schaefer (1983), Simmons (1987c), Smith (1950, 1956a, 1961), Starkey et al. (2003), Stebbins (1954, 1966, 1985, 2003), Suleiman (2005), Taggart (1992b, 2000b, 2001a, 2001b, 2002d, 2003g, 2004h, 2005, 2006b, 2006c, 2006d, 2007c), Taggart et al. (2005d, 2007, 2008), Taylor (1933), Thomas et al. (2008), Tihen (1937), Tihen and Sprague (1939), Toepfer (1993a), Uhler and Warren (1929, 2008), Van Doren and Schmidt (2000), Viets (1993), Vitt (2000), Voorhees et al. (1991), Wagner (1984), Webb (1970), Werth (1969), Wilgers et al. (2006), Wooster (1925), Yarrow (1883), and Zimmerman (1990).

Common Map Turtle
Graptemys geographica (LeSueur, 1817)

Identification: Semiaquatic; short tail; rigid carapace and plastron; carapace with a roughly jagged rear edge; small yellow spot behind each eye; dark-colored seams between large scutes on plastron; carapace dull olive-gray with yellowish lines and circles; plastron gray and lacks pattern, but seams between each scute darker than plastron color; head, limbs, and tail olive or brownish with yellow stripes; males have longer tails than females; females have broader heads and reach much larger size than males.

Size: Adults normally 90–253 mm (3½–10 inches) in carapace length; largest specimen from Kansas: female (KU Color Slide 8872) from Osage County with carapace length of 226 mm (8⅞ inches) collected by Lenn Shipman and Warren Voorhees on 3 July 1990; maximum carapace length throughout range: 10⅝ inches (Conant and Collins, 1998).

Adult male: Osage County, Kansas.

Plastron (lower shell) of adult: Osage County, Kansas.

Adult: Osage County, Kansas.

Adult: Osage County, Kansas.

Adult female: Allen County, Kansas.

Subadult.

Natural History: Prefers slow-moving or still bodies of water ranging in size from small stream to rivers, including river oxbows and lakes; reported in Kansas streams 15–60 feet wide with shorelines shaded by trees, little vegetation, and few basking sites; recorded active annually in Kansas from early April to mid-December, with a peak from early April to late June; becomes less active during winter, but may not actually burrow in mud like some other semiaquatic turtles; spends most of day basking and sleeping in sun on logs or other suitable perches; forages at twilight and after dark; gregarious, congregating in large numbers at optimal sites; extremely wary when approached; may mate twice each season but generally initiates breeding in spring; nothing known of courtship; females dig nests from May to July along water's edge in soft soil or sand but normally not on beaches; females lay

10–16 elongate dull white eggs; if laid in early spring, hatching occurs in early fall; eggs laid during summer or fall may not hatch until following spring; feeds on crayfishes and freshwater mussels; apparently does not prey on fishes; threatened species (protected by state law); maximum longevity: 18 years and 21 days (Snider and Bowler, 1992).

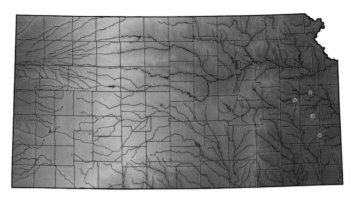

The secretive, semi-aquatic **Common Map Turtle** (*Graptemys geographica*) inhabits the Marais des Cygnes, Little Osage, and Marmaton river drainages in the east-central part of the state. Records from the Verdigris and Neosho (Grand) rivers in Oklahoma indicate that this species may eventually be discovered in those drainages in Kansas. Although its distribution is restricted in Kansas, this species has the most extensive range in North America of any member of its genus, hence the name, Common Map Turtle. Red dots indicate records backed by voucher specimens or images.

Pertinent References: Anderson (1965), Ballinger and Lynch (1983), Bartlett and Bartlett (2006b), Behler and King (1979), Bonin et al. (2006), Burt (1933), Caldwell and Collins (1981), Carr (1952), Clarke (1953, 1956a, 1958), Cochran and Goin (1970), Collins (1959, 1974b, 1982a, 1990b, 1994b), Collins and Collins (1993a), Collins et al. (1991, 1995), Conant (1958, 1975), Conant and Collins (1991, 1998), Cragin (1881), Cross (1971), Edds (1991), Edds et al. (1990), Ernst and Barbour (1972), Ernst & Lovich (2009), Ernst et al. (1994), Freeman and Kindscher (1992), Fuselier and Edds (1994), Gier (1967), Gloyd (1928), Guarisco et al. (1982), Hall and Smith (1947), Householder (1917), Iverson (1986, 1992), Karns et al. (1974), Kingsbury and Gibson (2002), Kirkpatrick (1993), Lamb et al. (1994), Levell (1997), McCauley (1945), McCoy and Vogt (1994), Miller and Gress (2005), Minton (1972, 2001), Mitchell (1994), Mount (1975), Oldfield and Moriarty (1994), Pope (1939), Schmidt (1953), Smith (1950, 1956a), Stejneger and Barbour (1939, 1943), Taggart (2006b, 2008), Trauth et al. (2004), Voorhees et al. (1991), and Welch (1994b).

Ouachita Map Turtle
Graptemys ouachitensis CAGLE, 1953

False Map Turtle
Graptemys pseudogeographica (GRAY, 1831)

Identification: Semiaquatic; short tail; rigid carapace and plastron; carapace with roughly jagged rear edge; in juveniles and young adults, distinct high raised keel down middle of carapace; head markings may be long, yellow; crescent-shaped mark behind eye extending onto side of head or thin bar (or rectangular blotch) behind eye on top of head; carapace olive or brownish with numerous light lines and dark brown keel down middle; plastron cream, yellow, or greenish yellow with faint pattern of dark lines or often smudged with dark gray but lacking pattern; head, limbs, and tail olive-green with yellow stripes; females much larger and have larger heads than males; males have long claws on front feet and longer tails than females.

Adult male: Riley County, Kansas.

Adult female: Wilson County, Kansas.

Adult: Linn County, Kansas.

Adult: Chase County, Kansas.

Adult: Geary County, Kansas.

Juvenile: Wyandotte County, Kansas.

Size: Adults normally 90–200 mm (3½–8 inches) in carapace length; largest specimen from Kansas: female (KU Color Slide 11236) from Riley County with a carapace length of 256 mm (10 inches) collected by Steven Seitz and Robert Seitz on 2 August 1996; maximum carapace length throughout range: 10⅝ inches (Conant and Collins, 1998).

Natural History: Large rivers, backwaters, lakes, sloughs, and ponds are favorite haunt; basking sites, vegetation, and soft bottoms required; prefers abundance of aquatic vegetation; recorded active annually in Kansas from early February to late October, with a peak from early May to early September; will emerge to bask on sunny winter days; during winter, burrows into soft mud or enters muskrat dens; during warmer months, basks for hours on perches over water in sun; populations of 100–120 per river mile have been estimated in lower Arkansas River valley; rarely ventures onto land except to nest; courtship consists of male swimming over female then abruptly facing her and stroking her head and chin with long claws on front feet; female settles to bottom, and male follows and mounts her; elongate eggs laid in shallow underground nests on land; females lay up to 3 clutches per season, with 5–13 eggs per clutch; nesting occurs periodically throughout spring and summer; young have shell lengths of 1–1¼ inches; omnivorous, feeding on aquatic plants and such animals as dead fishes, tadpoles, crayfishes, worms, and insects; maximum longevity: 35 years and 5 months (Snider and Bowler, 1992).

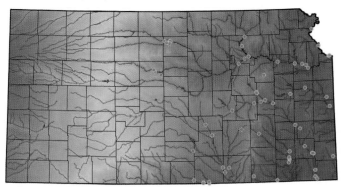

Combined distribution of the **Ouachita Map Turtle** (*Graptemys ouachitensis*) and the **False Map Turtle** (*Graptemys pseudogeographica*) in Kansas. These semi-aquatic turtles are found in the eastern two-thirds of the state in the following drainages: Kansas River, Smoky Hill River, Marais des Cygnes River, Neosho River, Verdigris River, Arkansas River, Caney, Ninnescah, and Chikaskia River. Red dots indicate records backed by voucher specimens or images.

Remarks: Vogt (1993) assessed intra- and interpopulation relationships in this complex of turtles and demonstrated that head marking patterns among individuals were highly variable and that they could be explained by changes in incubation temperatures within a single brood. In this study, the Ouachita Map Turtle (*Graptemys ouachitensis*) was recognized as a distinct species using detailed statistical analyses of skull and shell characters. Three different members of the genus were reported from Kansas, the Mississippi Map Turtle (*Graptemys kohnii*), Ouachita Map Turtle (*Graptemys ouachitensis*), and False Map Turtle (*Graptemys pseudogeographica*), although *G. kohnii* was considered a race of *G. pseudogeographica*. Most specimens from Kansas do not permit examination of head patterns; they exist as a shell only. Consequently, we cannot with any certainty define the range of any of these taxa in Kansas and have elected instead to present these turtles in a single account. We recognize that at least two (and maybe three) separate species exist in Kansas, but only intensive future collecting and analysis of molecular data will reveal their identity and where they occur in the state.

Pertinent References: Anderson (1965), Anderson et al. (1995), Ballinger and Lynch (1983), Bartlett and Bartlett (2006b), Behler and King (1979), Bonin et al. (2006), Busby and Parmelee (1996), Busby et al. (1994, 1996, 2005), Cagle (1943, 1954), Caldwell and Collins (1977, 1981), Capron (1975a, 1986d, 1987b), Carr (1952), Clarke (1956a, 1956c), Clarke et al. (1958), Cochran and Goin (1970), Collins (1959, 1974b, 1981, 1982a, 1989c, 1990b), Collins and

TURTLES

Collins (1993a), Conant (1958, 1975), Conant and Collins (1991, 1998), Cragin (1881, 1885a, 1885d), Cross (1971), Ditmars (1907), Edds (1992a), Ernst and Barbour (1972), Ernst & Lovich (2009), Ernst et al. (1994), Fitch (1985b), Fritz and Havas (2007), Fuselier and Edds (1994), Gier (1967), Guarisco et al. (1982), Hall and Smith (1947), Haltom (1931), Householder (1917), Hudson (1942), Irwin and Collins (1994, 2005), Iverson (1986, 1992), Johnson (2000), Jordan (1929), Karns et al. (1974), Kern et al. (1978), King and Burke (1989), Kingsbury and Gibson (2002), Kirkpatrick (1993), Lamb et al. (1994), McCoy and Vogt (1994), Minton (2001), Moll (1976), Mount (1975), Murrow (2008a, 2009a), Pope (1939), Pritchard (1979a), Riedle and Hynek (2002), Schmidt (1953), Schwartz and Dutcher (1961), Smith (1950, 1956a), Stejneger and Barbour (1917, 1923, 1933, 1939, 1943), Suleiman (2005), Taggart (1992f, 2006b), Trauth et al. (2004), Trott (1983), Van Doren and Schmidt (2000), Vetter (2004), Vogt (1993, 1995), Voorhees et al. (1991), Wagner (1984), Welch (1994b), and Young and Thompson (1995).

Eastern River Cooter
Pseudemys concinna (LeConte, 1830)

Plastron (lower shell) of adult: Cherokee County, Kansas.

Subadult: Crawford County, Kansas.

Adult: Lyon County, Kansas.

Adult: Cherokee County, Kansas.

Juvenile: Cowley County, Kansas.

Plastron (lower shell) of a juvenile: Cowley County, Kansas.

TURTLES

Identification: Semiaquatic; short tail; smooth, rigid carapace and plastron, carapace with roughly jagged rear edge; dark head with numerous yellow stripes; carapace dark gray or black with network of yellow lines; plastron light yellow or tan-yellow, generally unpatterned but sometimes with dark smudges or small circles, particularly in very young turtles; head, limbs, and tail dark gray or black with narrow yellow lines; adult males smaller than females and have long claws on front limbs.

Size: Adults normally 190–306 mm (7½–12 inches) in carapace length; largest specimen from Kansas: female (KU 223471) from Labette County with carapace length of 360 mm (14⅛ inches) collected by Alan Hynek on 24 April 1996; maximum carapace length throughout range: 14⅝ inches (Conant and Collins, 1998).

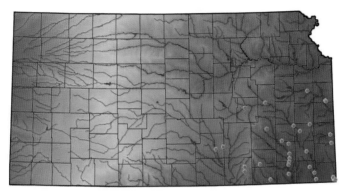

The **Eastern River Cooter** (*Pseudemys concinna*) is a semi-aquatic chelonian found in southeastern Kansas in the following river drainages: Marais des Cygnes, Little Osage, Marmaton, Spring, Neosho, Verdigris, Walnut, and Caney. The two records for the Arkansas River drainage in Sedgwick County need corroboration. Red dots indicate records backed by voucher specimens or images.

Natural History: Inhabits large streams, rivers, large ponds, and lakes; prefers still or slow-moving water with soft bottom, abundant vegetation, and plenty of logs and brush on which to bask in sun; recorded active annually in Kansas from late February to early November, with a peak from early May to late July; apparently forages for food in morning and evening, spending at least some of midday hidden under water; at night, generally sleeps beneath water on soft bottom or occasionally may search for food; more land-roving than other aquatic turtles, using sun as a compass for orientation; killed frequently while crossing highways; nothing known of courtship behavior in Kansas; nesting occurs in June, and females emerge from water to find an area of soft earth in which to lay 12–29 elongate white eggs; eggs hatch in 80–150 days; young may stay in nest during winter and emerge following spring; adults feed primarily on aquatic vegetation; young will eat meat, but diet changes to plants as they grow older; predators include hogs, raccoons, skunks, and opossums, which eat eggs, and large birds, fishes, snakes, other turtles, and mammals, which eat young; people are chief predators of adults; maximum longevity: 19 years and 5 months (Snider and Bowler, 1992).

Pertinent References: Anderson (1965), Ballinger and Lynch (1983), Bartlett and Bartlett (2006b), Behler and King (1979), Bonin et al. (2006), Breukelman and Smith (1946), Burger et al. (1949), Burt and Hoyle (1934), Caldwell and Collins (1981), Carr (1952), Clarke (1953, 1956a, 1958), Clarke et al. (1958), Cochran and Goin (1970), Collins (1959, 1974b, 1982a, 1984b, 1990b), Collins and Collins (1993a), Conant (1958, 1975, 1978b), Conant and Collins (1991, 1998), Cross (1971), Edds (1992a), Ernst and Barbour (1972), Ernst & Lovich (2009), Ernst et al. (1994), Fritz and Havas (2007), Gier (1967), Grow (1976b), Guarisco et al. (1982), Irwin and Collins (1994, 2005), Iverson (1986, 1992, 2001), Johnson (2000), Karns et al. (1974), King and Burke (1989), Kingsbury and Gibson (2002), Knight (1935), Lardie (1975a), Lindeman (1997), Miller (1976d), Minton (1972, 2001), Mitchell (1994), Mount (1975), Murrow (2009a, 2009b), Pritchard (1967, 1979a), Riedle and Hynek (2002), Schmidt (1953), Seidel (1994), Seidel and Dreslik (1996), Seidel and Ernst (1996), Smith (1950, 1956a), Stebbins (1954), Stejneger (1938), Stejneger and Barbour (1939, 1943), Taggart (2001a), Trauth et al. (2004), Voorhees et al. (1991), Ward (1984), and Welch (1994b).

Eastern Box Turtle
Terrapene carolina (LINNAEUS, 1758)

Adult male: Franklin County, Kansas.

Adult female: Linn County, Kansas.

Adult male: Neosho County, Kansas.

Adult female: Chautauqua County, Kansas.

Adult male: Crawford County, Kansas.

Juvenile: Johnson County, Kansas.

Identification: Terrestrial; short tail; rigid carapace; plastron with distinct movable hinge; 3 (sometimes 4) claws on each hind foot; uniform, patternless plastron; carapace uniform tan or olive, sometimes with faint radiating light or dark lines; plastron uniform tan or olive; limbs and tail brown, gray, or olive; head brown or olive, with small bright orange, red, or yellow spots; sometimes heads of males completely red; generally males have red eyes, whereas eyes of females yellowish brown; males have longer tails and normally grow slightly larger than females.

Size: Adults normally 113–150 mm (4½–6 inches) in carapace length; largest specimen from Kansas: female (KU 218958) from Wyandotte County with carapace length of 179 mm (7 inches), collected by Tom Sullivan and Stanley D. Roth on 1 June 1989; maximum carapace length throughout range: 8½ inches (Conant and Collins, 1998).

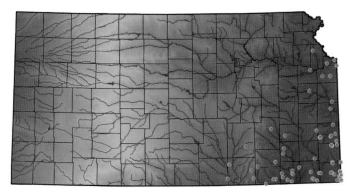

The distribution of the terrestrial **Eastern Box Turtle** (*Terrapene carolina*) in Kansas is restricted to the southeastern part of the state. Voucher specimens are known from Pottawatomie, Riley, Sedgwick, and Stafford counties, but we have not mapped them because they undoubtedly represent introductions or escaped pets; the record for Butler County (collected in 1988) needs corroboration. Red dots indicate records backed by voucher specimens or images.

Natural History: Inhabits open woodlands; occasionally found in pastures and around marshes; recorded active annually in Kansas from early March to late November, with a peak from early April to late May; spends winter months buried 2 feet deep in soil or well under leaf-covered rock overhangs to escape freezing temperatures; on warm winter days, may emerge from ground; some individuals killed by rapidly falling temperatures that prevent them from returning underground or beneath shelter; active during daylight, usually in morning or after rains; during extreme heat, retires to shaded areas; breeds primarily during spring months, but some mating may occur during summer and fall; during courtship, male approaches female with his head held high, exposing and pushing orange throat; female partially withdraws into shell, and male circles her, nipping and nudging edge of her shell (sometimes for as much as an hour) until she opens it; then male mounts her, hooking his hind toes into space between rear edges of her carapace and plastron; female responds initially by clamping shells on his toes and holding him tightly; evidently, male tickles rear inside edge of her carapace, causing her to open plastron; next, male positions hind feet near rear edge of female's plastron; male then extends head forward, exposing his bright throat again, while front feet touch her shell; finally, male slips backward with rear edge of his shell on ground and positions cloaca with hers, and copulation occurs; nesting takes place from May to July; female digs nest with hind feet at twilight and lays eggs at night; nest dug in loose sand or soil to depth of 3–4 inches on elevated patch of ground; females lay 2–8 elongate white eggs that generally hatch in 3 months; hatchlings may spend winter in nest and emerge following spring; omnivorous, eating mushrooms, berries, fruit, grass, snails, crayfishes, earthworms, numerous insects, fishes, frogs, toads, salamanders, lizards, small snakes, and carrion; maximum longevity: 26 years, 5 months, and 6 days (Snider and Bowler, 1992).

Pertinent References: Anderson (1965), Ashton et al. (1974a), Ballinger and Lynch (1983), Bartlett and Bartlett (2006b), Behler and King (1979), Bonin et al. (2006), Burt (1933), Burt and Hoyle (1934), Busby and Pisani (2007), Caldwell and Collins (1981), Capron (1987b), Carr (1952), Clarke (1956a, 1956c, 1958), Cochran and Goin (1970), Collins (1959, 1974b, 1982a, 1985, 1989c, 1990b, 1994a, 1996a), Collins and Collins (1993a), Collins et al. (1991), Conant (1958, 1975), Conant and Collins (1991, 1998), Cragin (1881), Cross (1971), Ditmars (1915), Dodd (2001), Ellis and Henderson (1913), Ernst and Barbour (1972), Ernst & Lovich (2009), Ernst et al. (1994), Fitch (1958b, 1991), Fraser (1983a), Gier (1967), Gloyd (1932), Grow (1976b), Householder (1917), Hunter et al. (1992), Iverson (1986, 1992), Johnson (2000), Karns et al. (1974), King and Burke (1989), Kingsbury and Gibson (2002), Knight (1935), Kraus (2009), Legler (1960b), Loomis (1956), Maslin (1959), Miller (1983a), Milstead (1969), Minton (2001), Minx (1996), Mitchell (1994), Mount (1975), Murphy (1964, 1976), Murrow (2008a), Perry (1978), Platt (1998), Pope (1939), Riedle (1994a), Riedle and Hynek (2002), Rodeck (1950), Rundquist (1996a), Schmaus (1959), Schmidt (1953), Schmidt and Murrow (2007c), Smith (1950, 1956a), Stejneger and Barbour (1933, 1939, 1943), Taggart (2000b, 2001a, 2005), Taylor (1895), Tihen (1937), Trauth et al. (2004), and Welch (1994b).

Ornate Box Turtle
Terrapene ornata (Agassiz, 1857)

Identification: Terrestrial; short tail; rigid carapace; distinct movable hinge on plastron; four claws on each hind foot; distinct radiating yellow or orange-yellow lines on dark plastron; carapace dark brown or reddish brown (sometimes with yellow line down middle) and covered with yellow or yellow-orange radiating lines; head and limbs dark brown, gray, or greenish, and covered with yellow or orange-yellow spots; dark tail may have yellow stripe on upper surface; adult males have red eyes whereas those of females yellowish brown; females grow slightly larger than males.

Adult male: Rooks County, Kansas.

Adult female and her hatchling: Douglas County, Kansas.

Adult male: Douglas County, Kansas.   Adult female: Sheridan County, Kansas.

Adult male: Morris County, Kansas.   Adult female: Douglas County, Kansas.
Note the bright red eye, typical of males
of this species.

Size: Adults normally 100–125 mm (4–5 inches) in carapace length; largest specimen from Kansas: female (KU 18358) from Barber County with a shell length of 154 mm (6⅛ inches), collected by Hobart M. Smith and Claude W. Hibbard on 2 September 1933; maximum carapace length throughout range: 6⅛ inches (Conant and Collins, 1998).

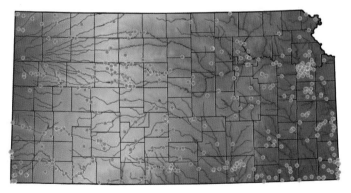

The **Ornate Box Turtle** (*Terrapene ornata*) is a terrestrial chelonian found throughout the state. Red dots indicate records backed by voucher specimens or images.

Natural History: Reaches greatest abundance on western open prairie, but equally at home along natural breaks in grassy vegetation of pastures, open woodlands, and open fields of eastern Kansas; recorded active annually in Kansas from late

TURTLES

213

February to mid-December, with a peak from mid-May to mid-August; during winter, digs beneath ground or enters dens or burrows of other animals; in open grasslands, may dig to depth of 18 inches but woodland burrows as shallow as 6 inches; emergence from burrow during spring dependent on a period of warm, moist weather; diurnal, spending daytime basking, feeding, and resting; daily, individuals may travel 200–300 feet, although pregnant females may travel somewhat greater distances; individual home range about 5 acres, but not possessive of this patch of land, several turtles utilizing same general area; shows a population density of over 1–2 per square acre; congregates around springs, seeps, intermittent pools, and still backwaters during hot, dry weather; mating occurs most commonly in spring and autumn but may also take place throughout summer; during courtship, male mounts female, hooking claws on rear edge of her plastron; female secures male's hind limbs by wrapping her own about them; male positions cloaca against that of female, and copulation occurs; when dismounting from female, male will often fall on back, but generally can right himself with little difficulty; nesting and egg deposition occur at least once a year, and probably a third of females in a population lay a second clutch in same season; nesting most common in June but may occur from May into autumn; females lay 2–8 elongate white eggs that hatch in a little over 2 months; young become sexually mature at 7–8 years of age; primarily carnivorous, feeding on beetles, caterpillars, grasshoppers, cicadas, earthworms, and dead vertebrates; also fond of berries and other fruits; was designated an official state symbol of Kansas in 1986; predators include mammals such as coyotes and skunks; infested frequently with chiggers during summer months, but apparently not harmed by these parasites; maximum recorded captive longevity: 26 years and 6 months (Snider and Bowler, 1992), but probably has a maximum age of 30–35 years in wild.

Pertinent References: Anderson (1965), Anderson et al. (1995), Ballinger and Lynch (1983), Bartlett and Bartlett (2006b), Baxter and Stone (1980, 1985), Beaver and Ingram (2008), Behler and King (1979), Blair (1976), Bonin et al. (2006), Brennan (1934, 1937), Breukelman and Clarke (1951), Breukleman and Downs (1936), Breukelman et al. (1961), Brons (1882), Brumwell (1940), Brunson (1986), Burt (1933, 1935), Burt and Burt (1929b), Burt and Hoyle (1934), Busby and Parmelee (1996), Busby and Pisani (2007), Busby et al. (1994, 1996, 2005), Carpenter (1940), Caldwell and Collins (1981), Caldwell and Glass (1976), Capron (1987b), Capron et al. (1982), Carr (1952), Clarke (1950, 1956a, 1956b, 1980a), Clarke et al. (1958), Coleman (1987), Collins (1959, 1974b, 1982a, 1985, 1990b, 2006b, 2007b, 2009a), Collins and Collins (1991, 1993a), Conant (1958, 1975, 1978a), Conant and Collins (1991, 1998), Costello (1969), Cragin (1881, 1885a, 1885d, 1894), Cross (1971), Dice (1923), Ditmars (1907, 1934), Dodd (2001), Eddy and Babcock (1973), Ellis (2002), Ellis and Henderson (1913), Ellner and Karasov (1993), Ernst and Barbour (1972), Ernst & Lovich (2009), Ernst et al. (1994), Ewert (1979), Falls (1933), Fitch (1956e, 1958b, 1965b, 1966, 1991, 2006a), Fraser (1983a), Garman (1884), Gatten (1974), Gibbons (1987), Gish (1962), Gloyd (1928, 1932), Grant (1937), Gress (2003), Grow (1976b), Guarisco (1980a, 1981c, 1983b), Hammerson (1999), Hartmann (1986), Hay (1892), Heinrich and Kaufman (1985), Householder (1917), Irwin (1977, 1983), Iverson (1982, 1986, 1992), Johnson (1987, 2000), Karns et al. (1974), Kazmaier and Robel (2201), Kern et al. (1978), Kingsbury and Gibson (2002), Knight (1935), Knight and Collins (1977), Kretzer and Cully (2001), Legler (1960a), Loomis (1956), Marr (1944), Martin (1960), Metcalf and Metcalf (1970, 1978, 1979, 1985), Miller (1976a, 1980a, 1980c, 1981c, 1981d, 1982a, 1983a, 1985b, 1988, 1996b), Miller and Collins (1995), Milstead (1969), Minton (2001), Minx (1996), Moll (1979), Murphy (1976), Murrow (2008a, 2008b, 2009a, 2009b), Nieuwolt (1996), Packard et al. (1985), Parmelee and Fitch (1995), Perry (1978),

Peterson (1992), Platt (1975, 1998), Pope (1939), Porter (1972), Rainey (1953), Reagan (1974b), Redder et al. (2006), Reichman (1987), Riedle (1994a), Riedle and Hynek (2002), Rodeck (1950), Roth (1959), Royal (1982), Rundquist (1975, 1992), Rush (1981), Rush and Fleharty (1981), Salazar (1968b), Schmidt (1938a, 1938b, 2001, 2004l), Schmidt and Murrow (2007b, 2007c), Schwarting (1984), Schwilling and Schaefer (1983), Shirer and Fitch (1970), Simmons (1986a, 1986c, 1987a, 1987c), Simon and Dorlac (1990), Skie and Bickford (1978), Smith (1950, 1956a, 1957, 1961), Sparks et al. (1999), Stebbins (1954, 1966, 1985, 2003), Switak (no date), Taggart (1992b, 2000b, 2001a, 2001b, 2002c, 2002d, 2004h, 2005, 2006b, 2006c, 2007c, 2008), Taggart et al. (2007), Taylor (1895, 1929b), Tihen (1937), Tihen and Sprague (1939), Thomasson (1980), Toepfer (1993a), Trauth et al. (2004), Trott (1983), Tyler (1979), Uhler and Warren (1929, 2008), Van Doren and Schmidt (2000), Vetter (2004), Viets (1993), Vitt (2000), Wagner (1984), Warner and Wencel (1978), Werth (1969), Wilgers and Horne (2006), Wilgers et al. (2006), Wolfenbarger (1951, 1952), Wooster (1925), Yarrow (1883), and Zug et al. (1986).

Slider
Trachemys scripta (Thunberg in Schoepff, 1792)

Melanistic adult male: Linn County, Kansas.

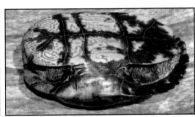

Plastron (lower shell) of melanistic adult male: Neosho County, Kansas.

Adult male: Sedgwick County, Kansas.

Adult female: Lyon County, Kansas.

Melanistic adult female with two eggs she laid in her recently dug nest: Harvey County, Kansas.

Plastron of adult male: Chase County, Kansas.

TURTLES

215

Identification: Semiaquatic; short tail; rigid carapace and plastron, carapace with roughly jagged rear edge; large dark spots on plastron; carapace dark with pattern of black and yellowish lines; plastron yellow or light brown with large dark brown or black spots; limbs and tail dark gray or greenish with yellow stripes; head same color but has broad red stripe behind each eye plus yellow stripes above and below; old adults may turn progressively darker in color until almost completely black; adult females grow much larger than males; male turtles have longer claws on front feet than females.

Size: Adults normally 125–203 mm (5–8 inches) in carapace length; largest specimen from Kansas: female (KU 189281) from Douglas County with a carapace length of 290 mm (11⅜ inches) and a weight of 4.2 kilograms (9 pounds, 3 ounces), collected by Rob Ladner on 23 June 1981; maximum carapace length throughout range: 11⅜ inches (Conant and Collins, 1998).

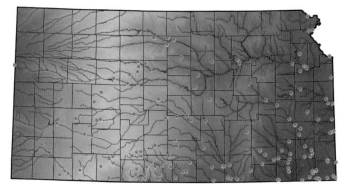

The **Slider** (*Trachemys scripta*) is a semi-aquatic turtle found throughout much of Kansas (in all major river drainages). It is absent from the northwestern portion of the state and along its northern and western borders (except for the single record from Wallace County). The range (and numbers) of this species may be increasing due to construction of new impoundments. Red dots indicate records backed by voucher specimens or images.

Backwater river-swamp in Labette County, Kansas, a favored haunt of Sliders.

Natural History: Found in nearly every permanent body of water; like most semiaquatic turtles, prefers quiet water with soft mud bottoms, plentiful aquatic vegetation, and numerous basking sites; recorded active annually in Kansas from early March to early December, with a peak from late April to late July; during winter months, burrows into bottom mud of lakes or ponds or in earth along shoreline; active during day and sleeps in water at night, either floating on surface or resting on bottom; usually breeds from March to June; fall mating also occurs; may breed twice or more each season; courtship similar to that of Northern Painted Turtle, but female more passive; females normally search for nesting sites during favorable weather, but if suitable areas not located, they can retain eggs for many weeks until proper nest found; females prepare their nest during the evening in loose, damp soil near water's edge, using hind feet to dig from 1–4 inches deep; females deposit up to 22 white eggs, elongate and granular; young turtles hatch in 2–2½ months and may emerge immediately from nest and enter water or stay in nest all winter until spring; omnivorous diet, eating both plants and animals with equal relish; known to consume insects, tadpoles, fishes, snails, and crayfishes, as well as aquatic vegetation; minimum known life span in wild over 25 years; maximum recorded longevity: 37 years, 9 months, and 10 days (Snider and Bowler, 1992).

Pertinent References: Anderson et al. (1995), Ashton et al. (1974b), Ballinger and Lynch (1983), Bartlett and Bartlett (2006b), Behler and King (1979), Bonin et al. (2006), Boyd (1991), Breukelman and Clarke (1951), Breukleman and Downs (1936), Burger (1952), Burt (1933, 1935), Burt and Burt (1929b), Burt and Hoyle (1934), Busby and Parmelee (1996), Busby and Pisani (2007), Busby et al. (1994, 1996, 2005), Cagle (1950), Caldwell and Collins (1981), Caldwell and Glass (1976), Capron (1978b), Carr (1938, 1952), Clarke (1956a, 1958, 1980a), Clarke et al. (1958), Collins (1959, 1974b, 1982a, 1984b, 1990b, 2001c, 2006b, 2007b, 2007d), Collins and Collins (1993a, 1993b, 2006a), Collins et al. (1994), Conant (1958, 1975, 1978b), Conant and Collins (1991, 1998), Cragin (1881, 1885a, 1885d), Cross (1971), Ditmars (1907, 1915), Edds (1992a), Eddy and Babcock (1973), Ernst and Barbour (1972), Ernst & Lovich (2009), Ernst et al. (1994), Gibbons (1987, 1990), Gish (1962), Gloyd (1928, 1932), Grow (1976b), Gubanyi (2005), Hall and Smith (1947), Householder (1917), Hudson (1942), Irwin (1983), Irwin and Collins (1987), Iverson (1986, 1992), Johnson (2000), Karns et al. (1974), Kern et al. (1978), Kingsbury and Gibson (2002), Knight (1935), Linsdale (1925, 1927), Mansueti (1941), McLeod (1994), Miller (1976a, 1979, 1980a, 1981a, 1982a, 1988, 1996a), Minton (2001), Mitchell (1994), Moll and Legler (1971), Mount (1975), Murrow (2008a, 2009a, 2009b), Parmelee (1994), Perry (1978), Platt (1975), Pleshkewych (1964), Pope (1939), Pritchard (1979a), Riedle (1994a), Riedle and Hynek (2002), Rush and Fleharty (1981), Schmidt (1953, 2004c), Schmidt and Murrow (2006, 2007b, 2007c), Schmidt and Schmidt (2009), Schmidt and Taggart (2002b), Seidel and Ernst (2006), Simmons (1987c), Smith (1950, 1956a), Sparks et al. (1999), Stebbins (1966, 1985, 2003), Stejneger and Barbour (1939, 1943), Suleiman (2005), Taggart (2000b, 2001a, 2006b, 2006c, 2006d, 2007c), Taylor (1933), Teller (2007), Thomas et al. (2008), Tihen (1937), Tihen and Sprague (1939), Toepfer (1993a), Trauth et al. (2004), Van Doren and Schmidt (2000), Vetter (2004), Viets (1993), Vitt (2000), Voorhees et al. (1991), Wagner (1984), Warner and Wencel (1978), Werth (1969), Wilgers et al. (2006), Yarrow (1883), Zimmerman (1990), and Zweig and Crenshaw (1957).

SOFTSHELLS
FAMILY TRIONYCHIDAE BELL, 1828

The family Trionychidae (Softshells) is composed of 1 genus and 3 species in the continental United States and Canada (Collins and Taggart, 2009). Two species belonging to the genus *Apalone* represent this turtle family in Kansas, the Smooth Softshell (*Apalone mutica*) and Spiny Softshell (*Apalone spinifera*). This family is characterized by its flexible, leathery carapace, much reduced plastron, slender elongate nose, and highly aquatic habits.

Smooth Softshell
Apalone mutica (LeSueur, 1827)

Identification: Semiaquatic; short tail; carapace with flexible, soft edges; soft plastron much smaller than carapace; uniform, patternless limbs; lack of bumps or

TURTLES

217

tubercles along front edge of carapace; carapace olive to light brown with pattern of darker dots or dashes (males) or blotches (females) and marginal dark line; plastron white or light gray; head, limbs, and tail olive to light brown above and white or light gray below; black-bordered yellow or white line extends through eye on each side of head; males have longer tails with anal opening near tip; adult females have mottled pattern on carapace and grow much larger than males.

Adult: Geary County, Kansas.

Adult: Douglas County, Kansas.

Adult: Geary County, Kansas.

Adult: Douglas County, Kansas.

Juvenile: Sumner County, Kansas.

Juvenile: Douglas County, Kansas.

Size: Adults normally 115–306 mm (4½–12 inches) in carapace length; largest specimen from Kansas: female (KU 218796) from Osage County with carapace length of 285 mm (11¼ inches) collected by John Powell and Beverly Downing on 9 June 1991; maximum carapace length throughout range: 14 inches (Conant and Collins, 1998).

Natural History: Prefers sand or mud bottoms of moderate to fast-flowing streams and rivers, rarely straying far from water except to bask and nest; recorded active

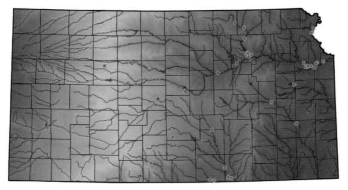

The **Smooth Softshell** (*Apalone mutica*) is a semi-aquatic chelonian found in the Missouri, Kansas, Marais des Cygnes, and Neosho river drainages in the northeastern part of the state. It is also found in the Smoky Hill, Arkansas, Chikaskia, and Medicine river drainages in central Kansas. This species is absent from the southeastern part of the state and from vast areas of the western quarter of Kansas. The single record for the Arkansas River in Kearny County (collected in 1958) needs corroboration. Red dots indicate records backed by voucher specimens or images.

annually in Kansas from late February to mid-November, with a peak from late May to mid-August, basking on sandbars, mud flats, steep mud banks, and logs along rivers; avoids open water; during warm summers, when water levels are low, may become inactive; apparently not territorial; slightest disturbance will send it scurrying into water, where it may surface and float downstream to escape enemies; during night, rests buried in mud or sand below water at depth that allows its long neck to extend to surface for air; active during daylight, basking and foraging for food; spends winter months burrowed deep in mud on bottoms of rivers and streams; population density of 1,900 turtles estimated in a one-mile stretch of Kansas River in Douglas County; population density estimated at 500–700 turtles per river mile in lower Arkansas River valley; mates during April and May and again during fall; courtship and mating take place in water, with much chasing of females by males; male maintains an extended neck while pursuing female, frequently probing with head under sides of her shell; if female is receptive, she remains passive and male mounts her; if female is not receptive, she will bite edges of male's shell; male mounts female in water but does not grasp her shell like some other turtles; instead, swims with female during copulation; nesting reaches peak during late May and June, when females emerge from water to dig holes on high open sandbars; number of eggs per clutch 3–25, with average of about 11 in Kansas; number produced by a female increases with size; round, hard-shelled eggs hatch in nest in 2–2½ months, and emerging young orient toward light of flat open areas such as rivers; adult males mature at 4 years of age, whereas females reach maturity age of 6–9 years; carnivorous diet, feeding primarily on insects, but will also eat fishes, frogs, tadpoles, salamanders, crayfishes, snails, and worms; males feed primarily on terrestrial insects, whereas females eat mostly aquatic insects; difference due to habitat preference, since males spend more time on or along shorelines than females; rapid swimmer, easily catching moving fish; raccoons and skunks relish eggs of this turtle; young preyed upon by large fishes, other turtles, snakes, large birds, and various mammals; heavy egg mortality occurs with high water levels during

TURTLES

219

flooding; in Kansas, people are chief predators of full-grown turtles; can remain submerged in water for long periods due to ability to remove oxygen from water through membranes of mouth; this adaptation to aquatic living, however, leaves it at the mercy of polluted streams and rivers; exhibits a timid and gentle disposition when handled, rarely attempting to bite; longevity unknown.

Pertinent References: Anderson et al. (1995), Ballinger and Lynch (1983), Bartlett and Bartlett (2006b), Behler and King (1979), Bonin et al. (2006), Bovee (1981), Burt (1935), Busby and Parmelee (1996), Busby et al. (1994, 1996, 2005), Caldwell and Collins (1981), Capron (1987b), Carr (1952), Clarke (1956a, 1956c, 1980a), Collins (1959, 1974b, 1982a, 1985, 1990b), Collins and Collins (1993a), Conant (1958, 1975, 1978b), Conant and Collins (1991, 1998), Cragin (1881), Dawson (1974), Dranoff (1981), Edds (1992a), Ellis (2002), Ernst and Barbour (1972), Ernst & Lovich (2009), Ernst et al. (1994), Fitch and Plummer (1975), Garrett and Barker (1987), Householder (1917), Hudson (1942), Iverson (1982, 1986, 1992), Johnson (1987, 2000), Karns et al. (1974), Kern et al. (1978), Kingsbury and Gibson (2002), Legler (1955), Levell (1995, 1997), McKenna and Seabloom (1979), Miller (1980a), Minton (2001), Moll and Moll (2004), Mount (1975), Nadeau (1979), Peterson (1992), Plummer (1975, 1976, 1977c, 1977d, 1977e), Plummer and Farrar (1981), Plummer and Shirer (1975), Plummer et al. (1994), Pope (1939), Rues (1986), Rush and Fleharty (1981), Smith (1947a, 1950, 1956a), Stebbins (1966, 1985, 2003), Stejneger (1944), Stejneger and Barbour (1933, 1939, 1943), Taggart (2006b, 2008), Taylor (1933), Toepfer (1993a), Trauth et al. (2004), Trott (1977), Wagner (1984), Warner and Wencel (1978), Webb (1962, 1973a), and Werth (1969).

Spiny Softshell
Apalone spinifera (LeSueur, 1827)

Juvenile: Ellis County, Kansas.

Adult.

Adult: Douglas County, Kansas.

Adult: Douglas County, Kansas.

Identification: Semiaquatic; short tail; carapace with flexible, soft edges, soft plastron much smaller than carapace; limbs with pattern of dark streaks and spots; small bumps or tubercles along front edge of carapace; carapace olive to light brown with patterns varying from all black-edged spots (adult males) to dark indistinct blotches (adult females); edge of carapace has marginal dark line; plastron white or yellowish; head, limbs, and tail white underneath and olive to light brown above; black-edged yellow line extends through eye on each side of head; males have longer tails than females, with anal opening near tip; females grow much larger than males.

| Adult: Linn County, Kansas. | Chikaskia River in Sumner County, Kansas, typical habitat of the Spiny Softshell. |

Size: Adults normally 125–432 mm (5–17 inches) in carapace length; largest specimen from Kansas: female (KU 197330) from Kingman County with a shell length of 523 mm (20½ inches) and a weight of 15.9 kilograms (35 pounds), collected by Richard Keller and Ralph Massoth, Jr., on 10 September 1984; maximum carapace length throughout range: 21¼ inches (Conant and Collins, 1998).

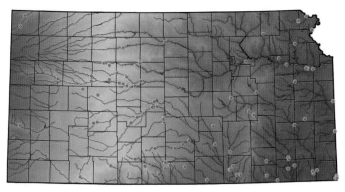

The highly aquatic **Spiny Softshell** (*Apalone spinifera*) is found throughout the state, but is least common in the western quarter of Kansas and is absent from many of the northern border counties. The single record for Hamilton County, collected in 1926 in the Arkansas River drainage, needs corroboration. Red dots indicate records backed by voucher specimens or images.

Natural History: Prefers wide variety of aquatic habitats, ranging from swift-flowing rivers and streams to still-water oxbows, lakes, and reservoirs; likes areas with sandbars or mud flats and bodies of water with soft bottoms; recorded active annually in Kansas from early March to mid-October (with a peak from mid-May to mid-July), spending cold winter months buried several inches below mud beneath water; diurnal during warmer months, basking, foraging for food, or resting buried in mud in water; occasionally active at night; population estimates of 500–700 per river mile in lower Arkansas River valley; mates during April and May; courtship unknown in Kansas, but probably resembles that of Smooth Softshell; nesting

TURTLES

occurs during day in June and possibly July; females crawl on land and dig cavity 4–10 inches deep in sand or soft soil; females lay 3–32 round, white, hard-shelled eggs, which hatch in late summer or fall; when egg-laying occurs late in season, young may stay in nest for first winter and emerge following spring; carnivorous diet, eating insects, earthworms, crayfishes, fishes, tadpoles, and frogs; can swim very fast, easily capturing small fishes; exhibits same ability to remove oxygen from water as Smooth Softshell; has an aggressive, nasty temper and will bite if not handled carefully; predators of eggs primarily mammals; young eaten by fishes, other turtles, wading birds, and snakes; people are prime predators of adults; maximum recorded longevity: 25 years, 2 months, and 17 days (Snider and Bowler, 1992).

Pertinent References: Anderson et al. (1995), Ashton et al. (1974a), Ballinger and Lynch (1983), Bartlett and Bartlett (2006b), Behler and King (1979), Bonin et al. (2006), Boyd (1991), Brennan (1934, 1937), Breukelman and Clarke (1951), Breukleman and Downs (1936), Burt (1933, 1935), Burt and Hoyle (1934), Busby and Parmelee (1996), Busby et al. (1994, 1996, 2005), Caldwell and Collins (1981), Capron (1987b, 1999), Carr (1952), Clarke (1956a, 1958, 1980a), Clarke et al. (1958), Collins (1959, 1974b, 1982a, 1984b, 1985, 1990b, 2001c), Collins and Collins (1993a, 1993b, 2006a), Conant (1958, 1975), Conant and Collins (1991, 1998), Conant and Goin (1948), Cragin (1881), Cross (1971), Dranoff (1981), Edds (1992a), Ellis and Henderson (1913), Ernst and Barbour (1972), Ernst & Lovich (2009), Ernst et al. (1994), Fitch (1985b), Garrett and Barker (1987), Gish (1962), Gloyd (1928), Graham and Graham (1997), Grow (1976b), Hall and Smith (1947), Hammerson (1999), Householder (1917), Irwin and Collins (1987), Iverson (1986, 1992), Karns et al. (1974), Kern et al. (1978), Kingsbury and Gibson (2002), Knight (1935), Knight and Collins (1977), Levell (1995, 1997), Linsdale (1925, 1927), Maslin (1959), McLeran (1975b), Miller (1976a, 1977b, 1982a), Minton (2001), Mitchell (1994), Morafka (1977), Mount (1975), Murrow (2009a), Neill (1951), Peters (1952), Pope (1939), Riedle and Hynek (2002), Rues (1986), Schmidt (2004g, 2004l), Schmidt and Murrow (2007b), Simmons (1984), Smith (1947a, 1950, 1956a), Smith et al. (1949), Stebbins (1954, 1966, 1985, 2003), Stejneger (1944), Taggart (2001a, 2008), Taggart et al. (2005e, 2007), Tihen (1937), Tihen and Sprague (1939), Toepfer (1993a), Trauth et al. (2004), Van Doren and Schmidt (2000), Vitt (2000), Voorhees et al. (1991), Wagner (1984), Webb (1962, 1973b), and Wooster (1925).

SPECIES OF POSSIBLE OCCURRENCE

This is a brief review of those species of amphibians, turtles, and reptiles that range close to Kansas but either have never been documented by voucher specimens from the state, are known from a single Kansas record taken over 50 years ago, or whose status in Kansas is questionable based on available evidence. Information contained in the Kansas Herpetofaunal Atlas (http://webcat.fhsu.edu/ksfauna/herps/) indicates that the following 10 species may yet be discovered in Kansas:

Spotted Salamander (*Ambystoma maculatum*). This amphibian is known from Delaware County, Oklahoma, 25–35 miles south of the southeastern corner of Kansas, and from Cedar and MacDonald counties, Missouri, east of the eastern Kansas border. Native Kansas populations of the Spotted Salamander may ultimately be discovered in vernal pools in Cherokee, Linn, or Miami counties. Hall and

Smith (1947) reported this species from Crawford County, but the specimen (KU 23278) upon which their record was based is apparently lost and cannot be verified. In addition, a specimen mapped by Smith (1950, 1956) for Douglas County (KU 950) was cleared and stained for use as a study specimen, and has deteriorated to the extent that it can no longer be identified.

Oklahoma Salamander (*Eurycea tynerensis*). Originally reported from Kansas as *Eurycea multiplicata* based on four larval specimens collected in 1967 in Cherokee County, three of which are preserved vouchers (KU 153033-035). However, recent research (Bonett and Chippindale, 2004) has restricted *Eurycea multiplicata* to those populations south of the Arkansas River in Arkansas and Oklahoma; all populations north of the Arkansas River were assigned to *E. tynerensis*. The Oklahoma Salamander has been recorded within 6 miles of the Kansas border in Ottawa County, Oklahoma, which adjoins the southeastern corner of Kansas to the south, and from Newton County, Missouri, which adjoins the southeastern corner of Kansas to the east. Potts and Collins (2005) excluded this species from the list of vertebrates known to occur in Kansas because no specimens had been found in the state for nearly 40 years. Cherokee County is an area where extensive field work has been conducted since the middle of the last century; this salamander has never been re-discovered there.

Western Slimy Salamander (*Plethodon albagula*). Known from Ottawa County, Oklahoma, and from Newton County in southwestern Missouri. These two counties adjoin the southeastern corner of Kansas from south and east, respectively. Although found very near the Kansas border in Missouri, no specimens of this amphibian have ever been documented from Kansas, and many voucher specimens with the erroneous locality data of "Cherokee County, Kansas" exist in university and museum collections. May eventually be discovered in southeastern Cherokee County in Kansas.

Hurter's Spadefoot (*Scaphiopus hurterii*). Hurter's Spadefoot is known from Osage County, northeastern Oklahoma, which adjoins Chautauqua County in southeastern Kansas. This amphibian should be sought on rainy spring nights in Chautauqua County.

New Mexico Spadefoot (*Spea multiplicata*). Found near Kansas in northcentral Oklahoma in Woods County, about 11 miles south of the Kansas border. Because of its close proximity to southcentral Kansas, may be discovered in Barber County. Also known from Baca County, Colorado, ca. 35 miles west of Kansas border in Arkansas River drainage. Should be sought in Kansas along Arkansas River floodplain in Hamilton County and Bear Creek drainage in Stanton County. Collins and Collins (1991) did not report it from Morton County in extreme southwestern Kansas.

Eastern Mud Turtle (*Kinosternon subrubrum*). Known from multiple records in Craig County, Oklahoma, within 14 miles of the Kansas border, lending credence to its possible occurrence in the state. This turtle may be discovered in the Neosho River drainage in Labette and Cherokee counties in southeastern Kansas.

Smooth Green Snake (*Liochlorophis vernalis*). Yarrow (1883) first reported this serpent from Kansas on the basis of a specimen (USNM 5236) collected by B. F. Goss from Neosho Falls in Woodson County sometime prior to February 1861 (the specimen was subsequently re-identified as *Opheodrys aestivus*, but cannot now be located; Addison Wynn, pers. comm.). Branson (1904) next reported this small snake from six counties in Kansas (without specific localities), but he retained no voucher specimens and his identifications cannot be verified; it is highly probable that his specimens were Rough Green Snakes (*Opheodrys aestivus*) and/or greenish variants of the Eastern Racer (*Coluber constrictor*). Smith (1931) and Hall and Smith (1947) reported single specimens from Riley and Crawford counties, respectively, but their specimens cannot be located for verification. Webb (1970) commented on the confusing situation regarding this species in Kansas. Collins (1974) originally considered this serpent a member of the Kansas fauna, but omitted it from his second edition (1982) following the recommendation of Rundquist (1979). The only fully documented Kansas specimen (UMMZ 67021), collected from the Chippewa Hills in Franklin County on 22 May 1928 by Wilbur Doudna (reported by Gloyd 1928 and considered authentic by Taylor 1929a), was not enough evidence to convince Collins to retain the species, and it was omitted from Kansas in the third edition of *Amphibians and Reptiles in Kansas* by Collins and Collins (1993) because no further evidence of its presence in the state had come to light in the intervening 65 years. In addition, the late Howard K. Gloyd, a noted herpetologist, excellent field worker, and student at Ottawa University, spent many years in Franklin County, Kansas, and never discovered this snake there, despite collecting or recording a plethora of amphibians, reptiles, and turtles. Because no other documented, extant, specimen of the Smooth Green Snake has been found in Kansas in over 80 years, it seems highly improbable that it occurs in the state. The Smooth Green Snake, however, has recently been recorded from several localities in south-central Nebraska and it is possible that populations of this diminutive species may still exist in Kansas along the northern border east of (and including) Phillips County and north and east of Morris County in association with low wet prairies.

Western Fox Snake (*Mintonius vulpinus*). Recorded from along the Missouri River in Andrew, Buchanan, and Holt counties, Missouri, opposite Atchison and Doniphan counties in northeastern Kansas. Also known from Nemaha County, Nebraska, about 30 miles north of the northeastern Kansas border (see Lokke, 1992). May eventually be found in Doniphan and Atchison counties in northeastern Kansas.

Blackneck Garter Snake (*Thamnophis cyrtopsis*). Known from Kansas based on a single specimen (KU 2088) collected in Hamilton County, Kansas, in 1903 by R. L. Moodie (specimen was initially cataloged and published as a Common Garter Snake, *Thamnophis sirtalis*); only known example of this serpent from state. Given some credence based on the existence of records from Arkansas River valley in Bent County, Colorado, less than 50 miles west of the Kansas border. This reptile may be discovered in the Arkansas River valley of Hamilton County along the western border of Kansas.

Western Terrestrial Garter Snake (*Thamnophis elegans*). The closest documented records of this species are from Cimarron County, Oklahoma, which adjoins the southwestern corner of Kansas, and from the Arkansas River drainage in Baca and Prowers counties, Colorado, which adjoins Morton County in extreme southwestern Kansas; the latter record is about 30 miles west of Kansas. Collins and Collins (1991) did not record this species from Morton County, but it may yet be found there in the Cimarron National Grasslands in southwestern Kansas.

A TECHNICAL KEY TO THE ADULT
AMPHIBIA, REPTILIA, AND CHELONIA IN KANSAS

These keys have been devised for use in the classroom with preserved specimens *of known locality in Kansas;* they may also be useful for identifying live individuals *of known locality in Kansas.* They are intended for use with adult examples. No attempt has been made to provide keys for larval amphibians or juvenile turtles or reptiles. Special attention should be given to *where* the amphibian, reptile, or turtle was found in Kansas; knowing the geographic location in Kansas where the example was found is essential in order to eliminate many species from further consideration. This key integrates external characteristics of species with geography in order to arrive at a correct identification. In addition, consult the glossary for additional information and illustrations of external characters. The line illustrations appearing in the technical key to Kansas amphibians, reptiles, and turtles, were drawn by Errol D. Hooper, Jr.

A Limbs always present; fingers and toes without claws; surface of body relatively smooth, frequently moist, and without scales.......**CLASS AMPHIBIA (AMPHIBIANS)**

B Limbs, if present, with claws on fingers and toes; body dry and covered with scales ....**CLASS REPTILIA (REPTILES)**

C With well-developed upper shell (lower shell may be less developed) and four limbs; claws present on fingers and toes ....**CLASS CHELONIA (TURTLES)**

A. CLASS AMPHIBIA

KEY TO ORDERS

A Tail present; limbs of nearly equal size ....**Order Caudata** (Salamanders)

B Tail absent; hind limbs larger than forelimbs ....**Order Anura** (Frogs and Toads)

A KEY TO SPECIES OF SALAMANDERS

1A. Four toes on each hind foot; eyelids absent; bushy external gills present (Fig. 1) throughout life; restricted to eastern third of Kansas, south of Kansas River ...**2**
1B. Five toes on each hind foot; eyelids present; gills, if present, in larval stage only **3**

2A. Belly uniform gray or pale with some spots (Fig. 1); restricted to Marais des Cygnes, Little Osage, and Marmaton river systems in Kansas...........................
....*Necturus maculosus*

2B. Center of belly grayish white with few or no spots (Fig. 1) and tinged with pink; restricted to Verdigris, Neosho, and Spring river systems in Kansas................. ....*Necturus louisianensis*

3A. No ridges on head; costal grooves distinct **4**
3B. A pair of distinct longitudinal ridges on top of head (Fig. 2); no costal (vertical) grooves on sides of body or, if present, very faint; both efts and adults often display red spots encircled by black on dorsum; found only in eastern border counties of Kansas, from Miami in north to Cherokee in south ....*Notophthalmus viridescens*

4A. Fourteen or fewer costal grooves (Fig. 3) between limbs **5**
4B. Sixteen or more costal grooves between limbs (Fig. 3); found in southeastern Cherokee County only; known only from stream-dwelling gilled larvae; adults confined to interior of caves ....*Eurycea spelaea*

5A. Slender body light colored above with small dark spots.................................**6**
5B. Stout body dark-colored above, with or without large light spots **7**

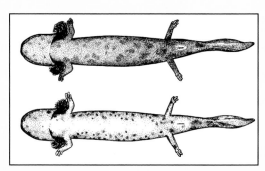

Fig. 1. Ventral view of two species of aquatic salamanders of the genus *Necturus*. **Upper:** Common Mudpuppy (*Necturus maculosus*), showing a uniformly well-patterned dark belly. **Lower:** Red River Mudpuppy (*Necturus louisianensis*), showing a lighter, less well-patterned belly. This character is highly variable in Kansas populations of both species; knowing the river system where a specimen is found is the best way to identify these two amphibians.

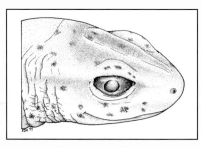

Fig. 2. Dorsolateral view of the head of an Eastern Newt (*Notophthalmus viridescens*), showing the two distinct longitudinal ridges between the eyes, typical of this salamander genus in Kansas.

Fig. 3. Drawing of a salamander showing the location of the costal grooves on the side of the body.

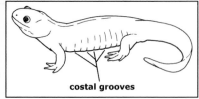

costal grooves

6A. A continuous dark streak from eye along each side of body onto tail; found in southeastern Cherokee County only.....................................*Eurycea longicauda*

6B. No dark lateral streak; entire body light with small dark spots; found in southeastern Cherokee County only ....*Eurycea lucifuga*

7A. Light spots or bars (pale yellow to bright yellow) on back and sides **8**

7B. No yellow spots on back or sides; flecking, slightly lighter than dark ground color, may be present on entire body; found in eastern Kansas only ....
....*Ambystoma texanum*

8A. Large distinct bright yellow spots or bars (numbering 6 to 36 with a mean of 17) on back and sides; found in western three-quarters of Kansas...................
....*Ambystoma mavortium*

8B. Smaller, subdued, round or irregular pale yellowish spots (numbering 15 to 58 with a mean of 30) on back and sides; found in eastern fourth of Kansas ....
....*Ambystoma tigrinum*

B KEY TO SPECIES OF FROGS AND TOADS

1A. Enlarged oval or elongate raised glands (parotoid glands) on neck, one behind each eye (Fig. 4) **2**

1B. No parotoid glands on neck..**7**

2A. Parotoid gland elongate, longer than broad (Fig. 5) **3**

2B. Parotoid gland oval, as broad as or broader than long (Fig. 5); found only in southwestern Kansas along counties bordering Oklahoma, from Barber in east to Morton in west...*Anaxyrus punctatus*

3A. Parotoid gland not extending laterally below level of middle of tympanum (Fig. 5)
....**4**

3B. Parotoid gland extending laterally below level of lower edge of tympanum (Fig. 5); found only in extreme western Kansas, from Wallace and Logan counties in north to Morton County in south..*Anaxyrus debilis*

4A. Dorsal pattern of smaller spots, usually enclosing a varying number of warts; large spots, if present, not distinctly outlined; cranial crests to not form boss (raised bump) on snout **5**

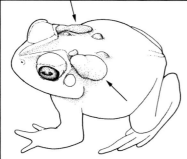

FIG. 4. Dorsolateral view of the head of a Nearctic toad (genus *Anaxyrus*) showing the location of the two large parotoid glands (arrows).

4B. Dorsal pattern of large, dark, rather closely placed, distinctly outlined, and light-edged spots or blotches, sometimes elongate so as to form reticulate pattern; ridges (cranial crests) on head between eyes unite on snout to form conspicuous boss (raised bump); found throughout western three-quarters of Kansas, with populations extending east along Kansas River valley and north along Missouri River valley in eastern part of state......................***Anaxyrus cognatus***

5A. Median anterior surface of forefoot without blackish spines; parotoid gland in contact with postorbital crest (Fig. 5); venter unspotted, or with single dark median spot on chest; small warts on dorsal surface of hind limb uniform in size and not distinctly raised; parotoid glands narrower and wider apart, usually separated by more than length of one gland **6**

5B. Median anterior surface of forefoot with blackish spines; parotoid gland separated from postorbital crest (Fig. 5); venter darkly spotted; small warts on dorsal surface of hind limb irregular in size and distinctly raised; parotoid glands broader and closer together, separated by no more than length of one gland; restricted to eastern quarter of Kansas, east of a line from Riley County in north to Cowley County in south ....***Anaxyrus americanus***

6A. Dorsal spots irregular in shape, not prominent, with each spot usually containing 1–2 warts; found statewide except southeastern part of Kansas..........
....***Anaxyrus woodhousii***

6B. Dorsal spots dark and well-defined (often 5-7 in number), with each spot containing 3 or more warts; restricted to extreme southeastern Kansas, from Bourbon County south to Cherokee County ....***Anaxyrus fowleri***

7A. Toes not webbed or only slightly so, web never extending beyond basal segment of the movable portion of longest toe on hind foot; toe discs absent or their diameter considerably less than half diameter of tympanum **8**

7B. Toes distinctly webbed, web extending beyond basal segment of movable portion of longest toe on hind foot; toe discs either absent or, if present, relatively large, and diameter of largest disc at least one-half diameter of tympanum..**12**

8A. Tympanum not visible; head much narrower than body; skin folded behind eyes to form dorsal groove (Fig. 6), which extends at least down to angle of jaw and sometimes encircles head **9**

8B. Tympanum visible and distinct, but small; head as broad as body or nearly so (except in gravid females); no skin fold behind eyes **10**

Fig. 5. Dorsolateral view of the heads of five species of Nearctic toads (genus *Anaxyrus*) showing the shape and location of their parotoid glands. **Left-to-Right:** American Toad (*Anaxyrus americanus*), Great Plains Toad (*Anaxyrus cognatus*), Green Toad (*Anaxyrus debilis*), Red-spotted Toad (*Anaxyrus punctatus*), and Woodhouse's Toad (*Anaxyrus woodhousii*).

9A. Venter heavily pigmented except in small scattered areas; found only in southeastern Cherokee County ....*Gastrophryne carolinensis*
9B. Venter without pigment, uniformly light colored; found statewide ....
....*Gastrophryne olivacea*

10A. Light line along the upper lip uniform and unbroken by dark spot beneath eye
....**11**
10B. Light line along the upper lip broken by dark spot beneath eye; found only in extreme south-central Kansas ....*Pseudacris streckeri*

11A. Dorsal pattern of irregularly arranged dark gray (green in life) spots, which are edged with black; tympanum nearly in contact with angle of jaw (Fig. 7); found in southcentral Kansas, from Haskell County in southwest to Montgomery County in southeast, and as far north as Ellis County ....*Pseudacris clarkii*
11B. Dorsal pattern of longitudinal dark gray (gray or brown in life) stripes, sometimes broken into rows of spots; tympanum distinctly separate from angle of jaw (Fig. 7); statewide except southwestern corner.......*Pseudacris maculata*

12A. No enlarged black tubercle on hind foot...**13**
12B. An enlarged black tubercle with free cutting edge present at base of hind foot (Fig. 8); found throughout western two-thirds of Kansas, but restricted to floodplains of Kansas and Missouri rivers in northeastern part of state ....
....*Spea bombifrons*

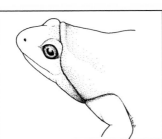

Fig. 6. Lateral view of a narrowmouth toad (genus *Gastrophryne*) showing the skin fold across the back of the head, typical of this genus.

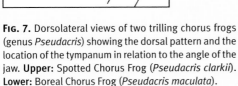

Fig. 7. Dorsolateral views of two trilling chorus frogs (genus *Pseudacris*) showing the dorsal pattern and the location of the tympanum in relation to the angle of the jaw. **Upper:** Spotted Chorus Frog (*Pseudacris clarkii*). **Lower:** Boreal Chorus Frog (*Pseudacris maculata*).

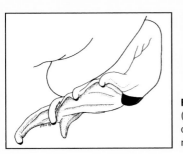

Fig. 8. Hind foot of a Plains Spadefoot (*Spea bombifrons*) showing placement of the enlarged black tubercle or spade, namesake of the family.

13A. Tips of finger and toes with enlarged discs, largest with a diameter at least half as wide as diameter of tympanum **14**

13B. No discs on tips of fingers and toes..**15**

14A. Distinct, dark, x-shaped mark on dorsum of body (Fig. 9); webs between toes of hind feet extend only to next to last joint except on longest toe; no white blotch or spot between lower edge of eye and upper lip; found only in eastern border counties of Kansas, from Miami in north to Cherokee in south............
....*Pseudacris crucifer*

14B. No x-shaped mark on dorsum of body; webs between toes of hind feet extend to terminal discs except on longest toe; color of groin area bright yellow; large white blotch or spot present between lower edge of eye and upper lip (Fig. 10); color of body may be gray or bright green; generally found in eastern third of Kansas, extending west in north to Republic County and west in south to Sumner county ....*Hyla chrysoscelis* and *Hyla versicolor*

15A. A raised, ridgelike gland, called dorsolateral fold, extends on each side of body from eye down back, sometimes reaching to the groin (Fig. 11) **16**

15B. No raised, ridgelike dorsolateral fold present down back (Fig. 11) **20**

16A. Distinctly outlined dark spots on back; dorsolateral fold extends to groin...**17**

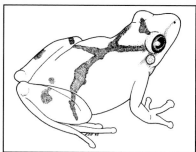

FIG. 9. Dorsolateral view of a Spring Peeper (*Pseudacris crucifer*) showing the dark X-shaped pattern on the back.

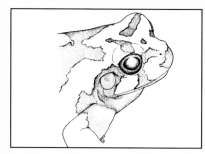

FIG. 10. Dorsolateral view of the head of a Gray Treefrog (*Hyla chrysoscelis-H. versicolor* complex) showing the light spot located between the lower edge of the eye and the upper lip.

FIG. 11. Dorsolateral view of North American true frogs (genus *Lithobates*) showing the presence (left) or absence (right) of a dorsolateral fold (arrow).

16B. No distinctly outlined dark spots on back; dorsolateral fold does not extend to the groin, but nearly so (Fig. 12); found only in southeastern Cherokee County ....*Lithobates clamitans*

17A. Spots on dorsum irregularly placed and round, not square or rectangular ..**18**
17B. Dorsal pattern of square or rectangular spots in two rows between dorsolateral folds (Fig. 13); found only in southeastern Cherokee County..................
....*Lithobates palustris*

18A. Areas between spots not reticulate or dark colored; narrow dorsolateral folds
....**19**
18B. Areas between spots distinctly reticulate; broad dorsolateral folds; found only in eastern fourth of state south of Kansas River ....*Lithobates areolatus*

19A. Dorsolateral folds broken and set inward (toward middorsum) near groin (Fig. 14); occurs statewide in every county ....*Lithobates blairi*
19B. Dorsolateral folds unbroken to groin (Fig. 14); found only in eastern Kansas south of Kansas River ....*Lithobates sphenocephalus*

20A. Skin on venter smooth; tympanum distinct, as large as eye or larger; no triangular dark mark between eyes; occurs statewide in every county ....
....*Lithobates catesbeianus*
20B. Skin on venter granular; tympanum obscure, smaller than eye; triangular mark present between eyes; found statewide, but scarce in western quarter of Kansas..*Acris blanchardi*

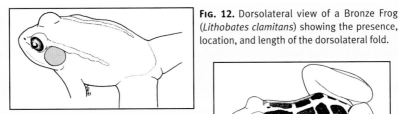

FIG. 12. Dorsolateral view of a Bronze Frog (*Lithobates clamitans*) showing the presence, location, and length of the dorsolateral fold.

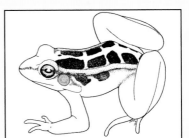

FIG. 13. Dorsolateral view of a Pickerel Frog (*Lithobates palustris*) showing the dorsal pattern of square or rectangular spots between the dorsolateral folds.

FIG. 14. Dorsolateral views of North American true frogs (genus *Lithobates*). **Left:** Dorsolateral folds are unbroken to the groin, typical of the Southern Leopard Frog (*Lithobates sphenocephalus*). **Right:** Dorsolateral folds are broken and set inward (shown as shaded) near the groin, typical of the Plains Leopard Frog (*Lithobates blairi*).

B. CLASS REPTILIA

KEY TO ORDER SQUAMATA

A Eyelids present; venter with many rows of scales, sometimes arranged in transverse series, but never fused into single transverse scales; ear openings present (except in one species); 4 limbs present (except in one limbless species) ....**Order Squamata (Lizards)**

B No eyelids (one species has an opaque scale covering eye); venter with single row of large, undivided, transverse scales (except one species); no ear openings; no limbs ....**Order Squamata (Snakes)**

A KEY TO SPECIES OF LIZARDS

1A. Body with 4 limbs..2
1B. Body without limbs; found throughout eastern two-thirds of Kansas..............
....***Ophisaurus attenuatus***

2A. Eyelids present; no toepads present on fingers and toes; pupil round when exposed to light..3
2B. Eyelids absent; toepads present on fingers and toes (Fig. 15); pupil vertical when exposed to light; known only from Johnson County, Kansas ....
....***Hemidactylus turcicus***

3A. Ear openings present on sides of head and tympanic membrane visible **4**
3B. No ear openings present; found in western two-thirds of Kansas, with isolated colonies as far east as Butler and Chase counties...........***Holbrookia maculata***

4A. Scales around middle of body numbering 35 rows or more; body scales differing markedly in size, shape, and texture..**5**
4B. Scales around middle of body numbering 30 rows or less; body scales all of uniform size and shape, absolutely smooth, shiny and overlapping.............**10**

5A. Posterior border of head without enlarged spines; no fringes of elongate scales on each side of body **6**

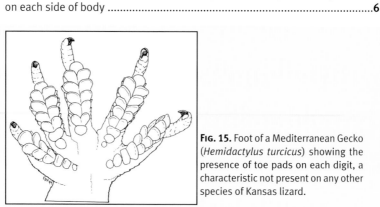

FIG. 15. Foot of a Mediterranean Gecko (*Hemidactylus turcicus*) showing the presence of toe pads on each digit, a characteristic not present on any other species of Kansas lizard.

5B. Posterior border of head with enlarged spines; 2 fringes of elongate scales on each side of body; found across southern third of Kansas, from Hamilton County in the west to Labette County in east; also found extensively in north-central part of state, from Wabaunsee County in east to Gove County in west and as far north as Phillips County ....*Phrynosoma cornutum*

6A. A granular fold of skin (called the gular fold) present across throat; scales on back small and granular **7**
6B. No fold of skin across throat; dorsal scales relatively large, keeled and strongly pointed; found throughout the western two-thirds of Kansas, with eastward extensions of its range along Kansas River valley to Johnson County and along Oklahoma border to Cherokee County ....*Sceloporus consobrinus*

7A. Scales on venter large, abruptly differentiated from small dorsal scales and in 10 rows or less **8**
7B. Scales on venter not as abruptly differentiated in size from those on dorsum; more than 15 rows of scales across venter; 2 dark patches (resembling a collar) on neck and shoulders; found in Kansas from Cherokee county in southeast to Stanton County in southwest, and as far north as Washington County; absent from northeastern Kansas and most of western third of state ....
....*Crotaphytus collaris*

8A. Ventral scales in 6 rows; single enlarged preanal scale (Fig. 16); no light stripes on dorsum, which may be a combination of green, gray-brown, and rusty brown, or uniform green...**9**
8B. Ventral scales in 8 rows; several enlarged preanal scales (Fig. 16); dorsal pattern of longitudinal light stripes on body; head, shoulders, and anterior of flanks often washed with lime green; found throughout Kansas, but scarce in the northeastern part of state...*Aspidoscelis sexlineata*

9A. Dorsum a combination of green (sometimes yellow), gray-brown, and rusty brown; SVL does not exceed 76 mm; range in Kansas restricted to within city limits of Hays (Ellis County), Lawrence (Douglas County), and Topeka (Shawnee County) ....*Podarcis siculus*
9B. Dorsum uniform green; SVL up to 105 mm; range in Kansas restricted to within city limits of Topeka (Shawnee County) ....*Lacerta bilineata*

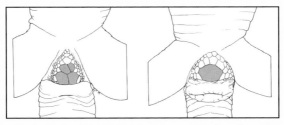

FIG. 16. Ventral view of the groin area of two species of lizards. **Left:** Six-lined Racerunner (*Aspidoscelis sexlineata*) showing the presence of several enlarged preanal scales (shaded). **Right:** A lizard of the Family Lacertidae showing the presence of a single enlarged preanal scale (shaded).

10A. Scales present on lower eyelid (when eyes are closed) (Fig. 17); frontal scale rectangular in shape with nearly parallel sides, posterior part being but little narrower than anterior part; supranasal scales present (Fig. 18)................**11**

10B. Translucent window present on lower eyelid (when eyes are closed) (Fig. 17); frontal scale V-shaped, posterior part much narrower than anterior part; supranasal scales absent (Fig. 18); found throughout southeastern Kansas, from Wyandotte County west to Riley County and thence southwest to Kiowa County ....*Scincella lateralis*

11A. All dorsal and lateral scale rows parallel with the long axis of body, none rising obliquely **12**

11B. Only 7 or fewer dorsal scale rows parallel with long axis of body; lower lateral rows oblique, rising as they continue posteriorly (Fig. 19); found throughout Kansas but scarce in far west-central part of state ....*Plestiodon obsoletus*

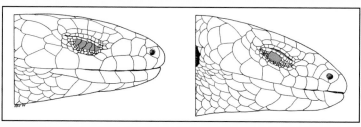

FIG. 17. Lateral view of the heads of two lizards of the family Scincidae. **Left:** The head of a lizard of the genus *Plestiodon* showing the presence of scales (shaded) on the lower eyelid when the eye is closed. **Right:** The head of a Ground Skink (*Scincella lateralis*) showing the single translucent window (shaded) on the lower eyelid when the eye is closed.

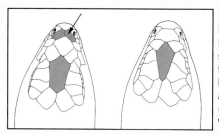

FIG. 18. Dorsal views of the heads of two lizards of the family Scincidae. **Left:** The head of a lizard of the genus *Plestiodon* showing the single rectangular-shaped frontal scale (shaded) and the presence and location of two supranasal scales (shaded, arrow). **Right:** The head of a Ground Skink (*Scincella lateralis*) showing the single V-shaped frontal scale (shaded).

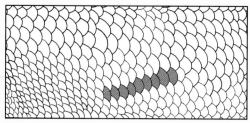

FIG. 19. Lateral view of the body scales of the Great Plains Skink (*Plestiodon obsoletus*), showing the lower lateral scale rows (shaded) as oblique, rising as they continue posteriorly (left to right), not running parallel to the long axis of the body.

12A. Two postmental scales present (Fig. 20); postnasal scale may or may not be present (see next couplet); medial light line, if present on head, may or may not be forked...**13**

12B. One postmental scale present (Fig. 20); no postnasal scale present (see next couplet); no light lines present on head; in Kansas, found south of Kansas River and east of Flints Hills, from Johnson County in north to Cherokee County in southwest and thence westward to Chautauqua County along the Oklahoma border...*Plestiodon anthracinus*

13A. Postnasal scale absent (Fig. 21); medial light line, if present on head, never forked; 7 light stripes present dorsolaterally on body, or body plain tan or brown without any stripes on back...**14**

13B. Postnasal scale present (Fig. 21); medial light line, if present on head, always forked; light dorsolateral stripes, if present on body, are 5 in number........**15**

14A. Dorsum and sides of body with 7 light stripes alternating with 6–8 dark ones; found in the eastern two-thirds of Kansas, north and east of Arkansas River, from Trego County in west to Doniphan County in northeast to Chautauqua County in southeast ....*Plestiodon septentrionalis*

14B. Dorsum of body normally plain tan or brown, sometimes with single faint middorsal stripe; some stripes may be present on sides of body; restricted to south-central Kansas, south of Arkansas River from Sumner County in east to Clark County in west ....*Plestiodon obtusirostris*

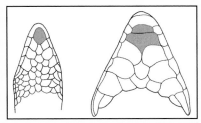

FIG. 20. Ventral view of the heads of two lizards of the genus *Plestiodon*. **Left:** One postmental scale (shaded) present, typical of the Coal Skink (*Plestiodon anthracinus*). **Right:** Two postmental scales (shaded) present, typical of other Kansas members of the genus *Plestiodon*.

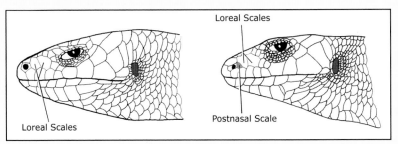

FIG. 21. Lateral view of the heads of two lizards of the genus *Plestiodon*. **Left:** No postnasal scale present in front of the loreal scales (shaded). **Right:** A single postnasal scale (shaded) present in front of the loreal scales.

15A. Generally no postlabial scales, or, if present, 1–2 of small size (Fig. 22); generally 8–9 upper labial scales (Fig. 23), sixth and seventh of which contact orbit (Fig. 22); maximum snout-vent length may exceed 85 mm (3½ inches) ; restricted to eastern border of Kansas from Franklin and Miami counties in north to Neosho and Cherokee counties in south ....*Plestiodon laticeps*

15B. Two relatively large postlabial scales present (Fig. 22); generally 7 upper labial scales (Fig. 23), fifth and sixth of which contacts orbit (Fig. 22); maximum snout-vent length does not exceed 85 mm (3½ inches); restricted to eastern fourth of Kansas..*Plestiodon fasciatus*

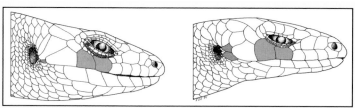

Fɪɢ. 22. Lateral view of the heads of two lizards of the genus *Plestiodon*. **Left:** Postlabial scales very small or absent and sixth and seventh upper labial scales (shaded) contact orbit of eye, typical of the Broadhead Skink (*Plestiodon laticeps*). **Right:** Two postlabial scales present (shaded just anterior to ear opening) and fifth and sixth upper labial scales (shaded) contact the orbit of the eye, typical of the Five-lined Skink (*Plestiodon fasciatus*).

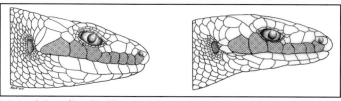

Fɪɢ. 23. Lateral view of heads of lizards of the genus *Plestiodon*. **Left.** The head of a Broadhead Skink (*Plestiodon laticeps*) showing the presence of eight upper labial scales (shaded). **Right.** The head of a Five-lined Skink (*Plestiodon fasciatus*) showing the presence of seven upper labial scales (shaded).

Ⓑ KEY TO SPECIES OF SNAKES

1A. Scales on venter wide, much wider than scales on dorsum (Fig. 24) **2**

1B. Scales on venter same size as those on dorsum (Fig. 24); found in southwestern Kansas along Oklahoma border, from Sumner county in east to Morton County in west and north into southern Kiowa County ....*Rena dissecta*

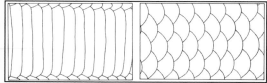

Fɪɢ. 24. Ventral view of the bodies of two snakes. **Left.** Scales on the venter much wider than those on the sides and dorsum, typical of all Kansas snakes except those in the Family Leptotyphlopidae. **Right.** Scales on venter and dorsum of uniform size, typical of the New Mexico Blind Snake (*Rena dissecta*).

2A. A conspicuous sensory pit on each side of head between and slightly below eye and nostril (Fig. 25) **3**

2B. No sensory pit between eye and nostril (Fig. 25) **7**

3A. Rattle absent **4**

3B. Rattle present..**5**

4A. Distinct brown or gray dorsal hourglasslike bands on lighter brown or gray ground color (Fig. 26); loreal scale present (Fig. 27); found in eastern third of Kansas, from Marshall County in north to Cowley County in south ....*Agkistrodon contortrix*

4B. Generally uniform dark brown, dark gray, or black; younger adults sometimes darkly banded; loreal scale absent (Fig. 27); restricted to Spring River drainage of eastern Cherokee County in southeastern Kansas.. ....*Agkistrodon piscivorus*

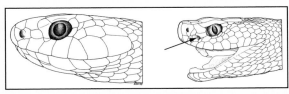

FIG. 25. Lateral view of the heads of two snakes. **Left:** The head of a harmless snake showing the lack of a sensory pit. **Right:** The head of a venomous snake showing the presence of a sensory pit (arrow) in relationship to the eye and nostril.

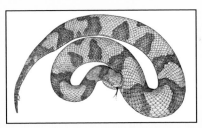

FIG. 26. Dorsalateral view of an adult Copperhead (*Agkistrodon contortrix*), showing the distinct hourglass-shaped crossbands, narrow along the top of the body and wider along the sides.

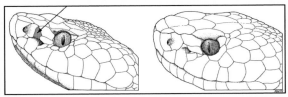

FIG. 27. Dorsalateral view of the heads of two venomous snakes of the genus *Agkistrodon*. **Left:** The head of a Copperhead (*Agkistrodon contortrix*), showing the presence of a loreal scale (shaded, arrow). **Right:** The head of a Cottonmouth (*Agkistrodon piscivorus*), showing the absence of a loreal scale.

5A. Numerous small scales on top of head, except for 1 large supraocular scale over each eye (Fig. 28) **6**

5B. Nine large scales on top of head, 4 in front of eyes (Fig. 28); found throughout eastern two-thirds of Kansas; range extends westward in state along Arkansas and Cimarron river drainages ....*Sistrurus catenatus*

6A. Dorsal body pattern of dark crossbands, sometimes chevron-shaped (Fig. 29); tail normally a uniform dark color; found in two populations in eastern Kansas, one north and east of Nesoho River and one south and west of Verdigris River ....*Crotalus horridus*

6B. Dorsal body pattern of dark, oval blotches; tail darkly banded; found in western half of Kansas ....*Crotalus viridis*

7A. Rostral scale projecting forward or up, sharp edged, and with sharp, medial, longitudinal keel (Fig. 30) **8**

7B. Rostral scale rounded, not projecting forward or up, not sharp edged, nor with keel (Fig. 30) **9**

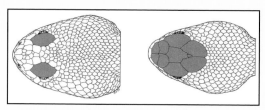

FIG. 28. Dorsal view of the heads of two rattlesnakes. **Left:** The head of a rattlesnake of the genus *Crotalus* showing the presence of two large supraocular scales (shaded) separated by much smaller scales. **Right:** The head of a Massasauga (*Sistrurus catenatus*) showing the presence of nine large scales (shaded) on top of the head.

FIG. 29. Dorsal view of a section of the body of an adult Timber Rattlesnake (*Crotalus horridus*), showing the pattern of dark chevron-shaped crossbands.

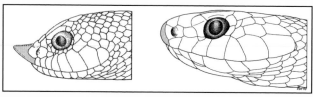

FIG. 30. Lateral views of the heads of two snakes. **Left:** The head of a snake of the genus *Heterodon* showing the forward projecting, up-turned rostral scale (shaded). **Right:** The rostral scale (shaded) is rounded, typical of other genera of snakes in Kansas.

8A. Rostral scale projecting forward; venter dusky, darker than underside of tail (Fig. 31); dorsal scale rows number 25 at midbody; found throughout Kansas except northwestern part of state ....*Heterodon platirhinos*

8B. Rostral scale projecting upward; venter and underside of tail mostly black (Fig. 31); dorsal scale rows number 23 or less at midbody; found throughout western three-quarters of Kansas......................................*Heterodon nasicus*

9A. No light neck ring, or, if present, the dorsal surface not uniformly dark.......**10**

9B. Entire dorsal surface of body and tail uniformly dark except for conspicuous light ring around neck; venter light colored with small, scattered black spots; found throughout eastern three-quarters of Kansas; range extends west along Smoky Hill River drainage..*Diadophis punctatus*

10A. All or most subcaudal scales divided (Fig. 32) **11**

10B. All or most subcaudal scales undivided (Fig. 32); found in southwestern Kansas south of Arkansas River drainage with an isolated colony in Logan County to north..*Rhinocheilus lecontei*

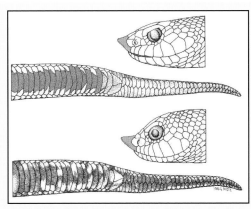

Fig. 31. Lateral view of the heads and ventral view of the tail and belly of two snakes of the genus *Heterodon*. **Upper:** An Eastern Hognose Snake (*Heterodon platirhinos*) showing the forward projecting, slightly up-turned rostral scale (shaded) and the light-colored underside of the tail in contrast to the darker belly. **Lower:** A Western Hognose Snake (*Heterodon nasicus*) showing the sharply up-turned rostral scale (shaded) and the dark-colored underside of the tail, similar to the belly.

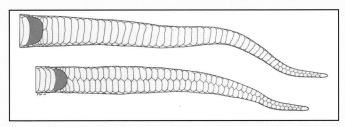

Fig. 32. Ventral view of the tails of two snakes. **Upper:** Most subcaudal scales posterior to the anal scale (shaded) are not divided, typical of the Longnose Snake (*Rhinocheilus lecontei*). **Lower:** All subcaudal scales posterior to the anal scale (shaded) are divided.

11A. Dorsal scale rows uniformly smooth (not keeled) (Fig. 33) **12**
11B. Dorsal scale rows either all strongly keeled (Fig. 33) or only weakly and faintly keeled on several middorsal rows **22**

12A. Anal scale entire (not divided) (Fig. 34) **13**
12B. Anal scale divided (Fig. 34)..**16**

13A. Venter with dark markings; dorsal scale rows number 27 or fewer at midbody **14**
13B. Venter uniform light color (no dark markings); dorsal scale rows number 29 or more at midbody; found throughout southwestern Kansas in Arkansas and Cimarron river drainages, from Hamilton County south to Morton County thence eastward to Sumner County; isolated colonies in Chase, Cheyenne and Gove counties ....*Arizona elegans*

14A. Dorsum blotched or banded..**15**
14B. Dorsum black with pattern consisting of small (smaller than a scale) light spots, either irregularly and profusely scattered or formed as faint, irregular outline of dark blotches; dorsal scale rows number 19–21 at midbody; found throughout Kansas...*Lampropeltis holbrooki*

15A. Dorsum consisting of pattern of 45–80 medium dark blotches that are black-edged; dorsal scale rows number 25–27 at midbody; found throughout south-eastern half of Kansas, from Washington County in northeast thence southwestward to Kearny and Seward counties in southwestern part of state; absent from northwestern Kansas ....*Lampropeltis calligaster*
15B. Dorsum consisting of pattern of 18–30 medium-light (red or orange-red in life) wide bands, each bordered by a pair of narrow black bands that, in turn, enclose narrow light (yellowish in life) bands; dorsal scale rows number 19–23 at midbody; found throughout Kansas ....*Lampropeltis triangulum*

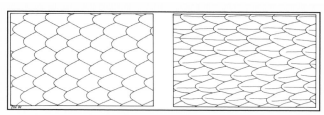

Fig. 33. Dorsal view of sections of the bodies of two snakes. **Left:** Smooth dorsal scales. **Right:** Keeled dorsal scales.

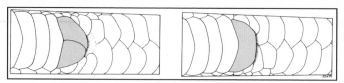

Fig. 34. Ventral view of two snakes at the juncture of the body and tail showing the anal plate (shaded). **Left:** Divided anal plate. **Right:** Single anal plate.

16A. Dorsum of body uniform color or banded; dorsal scale rows numbering 19 or fewer at midbody; pupil of eye not vertically slit **17**

16B. Dorsal pattern consisting of many small dark blotches; dorsal scale rows number 20 or more at midbody; pupil of eye slit vertically when exposed to strong light; restricted to Barber, Comanche, and Clark counties in southwest-central Kansas along Oklahoma border ....*Hypsiglena jani*

17A. Ventral scales generally number 160 or more; dorsal scale rows number 17 at midbody; adults attain 750–1750 mm in total length **18**

17B. Ventral scales generally number fewer than 160; dorsal scale rows number 15 or fewer at midbody; adults do not exceed 380 mm in total length **19**

18A. Dorsal scale rows number 15 about 1–2 inches in front of anus; preocular scale bordered below by second and third upper labial scales (Fig. 35); dorsum of body always uniformly colored; found statewide in every county...................
....*Coluber constrictor*

18B. Dorsal scale rows number 13 about 1–2 inches in front of anus; preocular scale bordered below by third and fourth upper labial scales (Fig. 35); dorsum may be uniformly light colored or anterior third to two-thirds may be black; found in southwestern half of Kansas, south of a line from Cheyenne County in northwest east to Smith County and southeast to Cherokee County.....................
....*Masticophis flagellum*

19A. Dorsum light colored, not sharply differentiated from light-colored venter; dorsal scale rows number 15 at midbody **20**

19B. Dorsum dark, sharply differentiated from light colored (coral) venter at third dorsal scale row (counting from venter up); dorsal scale rows number 13 at midbody; found in eastern third of Kansas, east of a line from Riley County in north to Cowley County in south.......................................*Carphophis vermis*

20A. Dorsum of head dark in contrast to light colored dorsum of body; dorsum of body uniform (no bands); 6–7 supralabial scales present **21**

20B. Dorsum of head same light color as dorsum of body; dorsum of body may or may not be darkly banded; 7 supralabial scales present; found in southern Kansas along the Oklahoma border, from Cherokee County in the southeast corner west to Clark County; an isolated colony to north in Russell County....
....*Sonora semiannulata*

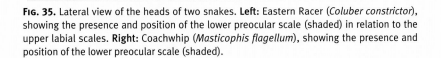

FIG. 35. Lateral view of the heads of two snakes. **Left:** Eastern Racer (*Coluber constrictor*), showing the presence and position of the lower preocular scale (shaded) in relation to the upper labial scales. **Right:** Coachwhip (*Masticophis flagellum*), showing the presence and position of the lower preocular scale (shaded).

21A. Six supralabial scales present on upper lip (Fig. 36); dark color on dorsum of head indented anteriorly and merges gradually into color of dorsum of body (Fig. 36); found in eastern third of Kansas, but absent from northern border counties ....*Tantilla gracilis*

21B. Seven supralabial scales present on upper lip (Fig. 36); dark color on dorsum of head sharply differentiated from light-colored dorsum of body, and terminates in a V-shape with apex on neck (Fig. 36); found in western three-quarters of Kansas, west of a line from Riley County In north to Cowley County in south ....*Tantilla nigriceps*

22A. Anal scale entire (not divided) (Fig. 37) **23**
22B. Anal scale divided (Fig. 37)..**28**

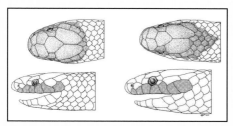

FIG. 36. Dorsal (upper) and lateral (lower) views of the heads of two snakes of the genus *Tantilla*. **Left:** A Flathead Snake (*Tantilla gracilis*) showing the shape of the dark head pattern (upper) and the presence of six upper labial scales (shaded, lower). **Right:** A Plains Blackhead Snake (*Tantilla nigriceps*) showing the shape of the dark head pattern (upper) and the presence of seven upper labial scales (shaded, lower).

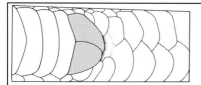 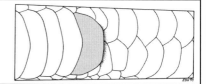

FIG. 37. Ventral view of two snakes at the juncture of the body and tail showing the anal plate (shaded). **Left:** Divided anal plate. **Right:** Single anal plate.

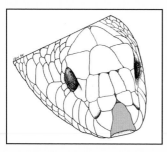

FIG. 38. Frontal view of the head of a Gopher Snake (*Pituophis catenifer*), showing the rostral scale (shaded) nearly twice as high as wide.

FIG. 39. Dorsolateral view of the head of a Great Plains Rat Snake (*Pantherophis emoryi*), showing the dorsum of head with spear-shaped mark, apex of which is between eyes, arms of which extend posteriorly.

23A. Dorsal scale rows number 24 or fewer at midbody; ventral scales number 190 or less; rostral scale nearly or fully as wide as high **24**

23B. Dorsal scale rows number 25 or more at midbody; ventral scales number 200 or more; rostral scale twice as high as wide (Fig. 38); statewide; found in every county in Kansas ....*Pituophis catenifer*

24A. Spots on venter, if present or in 2 rows, placed laterally along edges of belly; supralabial scales number 7 or more...**25**

24B. Two rows of black spots down middle of belly; supralabial scales number 6 or less; found throughout most of Kansas; absent from northwestern and southwestern parts of state ....*Tropidoclonion lineatum*

25A. Lateral light stripes placed on third and fourth dorsal scale rows (counting from venter up) on anterior portion of body **26**

25B. Lateral light stripes, sometimes indistinct, placed on second and/or third dorsal scale rows (never on fourth scale row) on anterior portion of body .**27**

26A. Light upper lips darkly barred; supralabial scales usually number 7 on upper lips; area between middorsal light stripe and lateral light stripes with dark spots; generally found in western three-quarters of Kansas, but ranges into northeastern part of state along major river valleys ....*Thamnophis radix*

26B. Upper lips uniform light color, not barred; supralabial scales usually number 8 on upper lips; area between middorsal light stripe and lateral light stripes uniformly dark; generally found across Kansas along major river valleys, but absent from northern border counties as well as northwestern and southwestern parts of state. ....*Thamnophis proximus*

27A. Middorsal light stripe with zigzag edges; lateral light stripes not placed on second dorsal scale row anteriorly, restricted to third dorsal scale row; dorsal scale rows number 21 at midbody; light upper lips distinctly barred; a large, conspicuous, crescent-shaped light mark behind angle of jaw; found in Kansas along the Oklahoma border from Sumner County in east to Morton County in west. ....*Thamnophis marcianus*

27B. Middorsal light stripe with straight edges; lateral light stripes placed on second and third dorsal scale rows anteriorly; dorsal scale rows number 19 at midbody; upper lips slate colored, feebly barred; no crescent-shaped light mark behind angle of jaw; found in eastern two-thirds of Kansas; ranges into northwestern and southwestern parts of state long major river valleys; absent from west-central Kansas..*Thamnophis sirtalis*

28A. Ventral scales number 190 or fewer **30**

28B. Ventral scales number 200 or more **29**

29A. Dorsum of body conspicuously patterned with 25–45 dark quadrangular blotches; dorsum of head with spear-shaped mark, apex of which is between eyes, arms of which extend posteriorly to unite with first dorsal body blotch (Fig. 39); dorsal scales very weakly keeled (sometimes not at all); found throughout most of Kansas; scarce in western quarter of state and absent from northeastern part..*Pantherophis emoryi*

29B. Dorsum of body uniform dark color (except in young adults which may have a subdued pattern of 28–38 dark body blotches); head uniformly dark; restricted to eastern half of Kansas, east of a line from Republic County in the north down through eastern Russell County to Barber County in the south ....*Scotophis obsoletus*

30A. Dorsal scale rows number 18 or fewer at midbody **34**
30B. Dorsal scale rows number 19 or more at midbody......................................**31**

31A. Dorsal scale rows number 21 or more at midbody; dorsum of body nowhere striped...**32**
31B. Dorsal scale rows number 19 at midbody; light lateral stripes placed on dorsal scale rows 1–3 and running entire length of body; found in eastern Kansas south of Kansas River valley and east of a line from Barton County in north to Barber County in south; records in northeastern Kansas may be introduced.. ....*Regina grahamii*

32A. Venter with dark markings; dorsal markings present and usually visible, always when held under water (Fig. 40) **33**
32B. Light-colored venter without dark markings or, if present, dark markings on only anterior edges of ventral scales; dorsal body pattern uniformly dark or with obscure bands or blotches (Fig. 40); found in southeastern half of Kansas, southeast of a line from Doniphan County in northeast to Seward County in southwest ....*Nerodia erythrogaster*

33A. Dorsal scale rows number 21–23 at midbody; large, complete dorsal bands on anterior portion of body behind neck; venter profusely and irregularly patterned with markings; found throughout eastern three-quarters of Kansas, but ranges west along major river valleys ....*Nerodia sipedon*
33B. Dorsal scale rows number 25–29 at midbody; dorsum of body behind neck with small alternating dorsal and lateral spots connected by oblique bars; venter light with small, dark, half-moon-shaped markings; found in southeastern half of Kansas, southeast of a line from northern Leavenworth County in northeast through Riley and Barton counties to Meade County in southwest......... ....*Nerodia rhombifer*

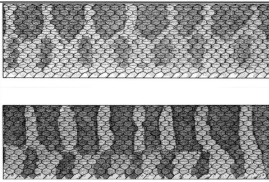

FIG. 40. Lateral view of sections of the bodies of two Water Snakes of the genus *Nerodia.* **Upper:** Body pattern the Plainbelly Water Snake (*Nerodia erythrogaster*), showing obscure bands or blotches. **Lower:** Body pattern the Northern Water Snake (*Nerodia sipedon*), showing distinct bands and blotches.

34A. One or more preocular scales present (Fig. 41); prefrontal scale not in contact with eye (Fig. 41) **35**

34B. Preocular scale absent; loreal scale present (Fig. 41); prefrontal scale in contact with eye (Fig. 41) **37**

35A. Dorsal scale rows number 17 at midbody; supralabial scales number 7.....**36**

35B. Dorsal scale rows number 15 at midbody; supralabial scales number 5–6; restricted to two tiers of counties in eastern Kansas that border Missouri, from Atchison in north to Cherokee in south..................*Storeria occipitomaculata*

36A. Dorsal and ventral surfaces of head, body, and tail uniformly colored (green in life), without pattern; restricted to southeastern Kansas, southeast of a line from Wyandotte County in northeast to eastern Sumner County in southwest ....*Opheodrys aestivus*

36B. Dorsum of body with light middorsal stripe, almost always bordered by small black dots; sides of venter stippled with very small dark dots; found throughout eastern three-quarters of Kansas ....*Storeria dekayi*

37A. Supralabial scales number 5 (Fig. 42); 1 postocular scale present (Fig. 42); usually 1 internasal scale present (Fig. 42); restricted to two colonies in extreme southeastern Kansas, one in Cherokee and Crawford counties and one to west in Chautauqua County ....*Virginia striatula*

37B. Supralabial scales number 6 (Fig. 42); 2 postocular scales present (Fig. 42); 2 internasal scales present (Fig. 42); restricted to northeastern Kansas in an area bounded in south by Linn County, to north by Atchison County, and to west by Shawnee County ....*Virginia valeriae*

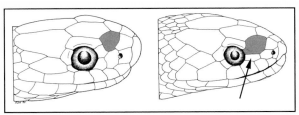

FIG. 41. Dorsolateral view of the heads of two snakes. **Left:** The head of a snake showing the prefrontal scale (shaded) not in contact with the eye because of the presence of preocular scales. **Right:** The head of a snake showing the prefrontal scale (shaded) in contact with the eye and the presence of a loreal scale (arrow), typical of snakes of the genus *Virginia*.

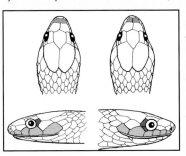

FIG. 42. Dorsal (upper) and lateral (lower) views of two snakes of the genus *Virginia*. **Left:** The Rough Earth Snake (*Virginia striatula*) showing the single internasal scale (shaded, upper), the presence of one postocular scale (darkly shaded, lower) and five upper labial scales (shaded, lower). **Right:** The Smooth Earth Snake (*Virginia valeriae*) showing the two internasal scales (shaded, upper), the presence of two postocular scales (darkly shaded, lower) and six upper labial scales (shaded, lower).

C. CLASS CHELONIA

ORDER CRYPTODIRA

·· **KEY TO SPECIES OF** TURTLES ··

1A. Carapace soft and leathery at least at edges; not covered by large plates; snout flexible, narrow, and pointed **2**
1B. Carapace hard and covered by large plates; snout not flexible, narrow, and pointed, instead blunt and hard **3**

2A. Anterior border of carapace with tubercles (Fig. 43); wall separating crescentic nostrils from each side has projections (ridges) into each nasal opening (Fig. 44); found statewide, but scarce along Nebraska border .....**Apalone spinifera**
2B. Anterior border of carapace smooth; wall separating round nostrils from each other smooth and even without projecting walls (ridges) (Fig. 44); generally found in eastern two-thirds of state in Arkansas and Kansas river drainages; single far west record from Kearney County in Arkansas River drainage ....**Apalone mutica**

3A. Bridge between carapace and plastron at least twice as wide as long (Fig. 45); tail length more than half length of carapace **4**

Fig. 43. View of the anterior edge of the leathery shell of a Spiny Softshell (*Apalone spinifera*), showing the bumpy tubercles or "spines," typical of the species.

Fig. 44. View of the nostrils of softshells (genus *Apalone*). **Left:** Smooth and even wall separating nostrils from each other, typical of the Smooth Softshell (*Apalone mutica*). **Right:** Projections extending from the wall separating the nostrils, typical of the Spiny Softshell (*Apalone spinifera*).

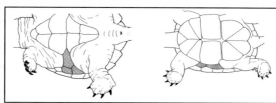

Fig. 45. View of the plastrons of two turtles. **Left:** A Common Snapping Turtle (*Chelydra serpentina*), showing the bridge (shaded) as wider than long. **Right:** A Yellow Mud Turtle (*Kinosternon flavescens*), showing the plates of the bridge (shaded) at least as long as wide.

3B. Bridge between carapace and plastron about as long as wide or longer (Fig. 45); tail length considerably less than half length of carapace **5**

4A. Head covered with smooth, symmetrical plates; scales on underside of tail irregularly arranged; tubercles on sides of neck branched; a row of 3–5 supramarginal plates above marginals on each side (Fig. 46); upper jaw strongly hooked; carapace with 3 prominent longitudinal ridges (Fig. 46); restricted to southeastern Kansas in Neosho, Verdigris, and Walnut river drainages............
....*Macrochelys temminckii*

4B. Head covered with uniformly rugose skin; scales on underside of tail arranged (in part) in 2 rows; tubercles on sides of neck unbranched; no extra row of plates above marginals on each side; upper jaw not strongly hooked; carapace without prominent ridges; found statewide, but least abundant in far west
....*Chelydra serpentina*

5A. Plastron (exclusive of the bridge) composed of 11 plates arranged as 5 pairs plus 1 anterior-most plate (Fig. 47) **6**
5B. Plastron (exclusive of the bridge) composed of 12 plates arranged in 6 pairs (Fig. 47) **7**

6A. Posterior plates on plastron squarish (Fig. 48); single indistinct transverse hinge on plastron; most anterior pair of plates on plastron widely separated from third pair (Fig. 48); tail without horny, clawlike tip; restricted to eastern Kansas south of Kansas River ....**Sternotherus odoratus**
6B. Posterior plates on plastron triangular (Fig. 48); 2 distinct transverse hinges on plastron; most anterior pair of plates on plastron reaching (or nearly so) third pair at midline (Fig. 48); tail with horny, clawlike tip (Fig. 48); generally found in western two-thirds of state, but with isolated colony in Cherokee and Labette counties; questionable record from Pottawatomie County in northeastern part of Kansas ....**Kinosternon flavescens**

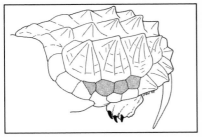

FIG. 46. Dorsolateral view of the carapace of an Alligator Snapping Turtle (*Macrochelys temminckii*), showing the placement of the three ridges on the shell and the location of the supramarginal plates (shaded).

FIG. 47. View of the plastron of two species of turtles. **Left:** Plastron (exclusive of the bridge) composed of 11 plates arranged as 5 pairs plus 1 anterior-most plate, typical of the Family Kinosternidae. **Right:** Plastron (exclusive of the bridge) composed of 12 plates arranged in 6 pairs.

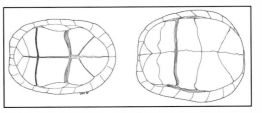

TURTLE KEYS

247

7A. Anterior portion of plastron hinged to posterior portion and is movable; hinge is deeply grooved, transverse seam **8**

7B. Entire plastron solid and immovable (no hinge); all plastral seams of shallow depth and uniform size...**9**

8A. Usually 3 claws on each rear foot (Fig. 49); plastron uniformly colored, without markings; median plates on carapace slightly keeled; restricted to southeastern Kansas, from Leavenworth County in north to Sumner County in west ....*Terrapene carolina*

8B. Usually 4 claws on each rear foot (Fig. 49); both carapace and plastron with distinct radiating light (yellow) lines; median plates on carapace not keeled; found statewide in every county ....*Terrapene ornata*

9A. Upper jaw without prominent notch bordered by projections; posterior edge of carapace notched and irregular (not smoothly rounded) **10**

9B. Upper jaw with a prominent notch anteriorly (beneath and between the nostrils) bordered on each side by projections; posterior edge of carapace smoothly rounded; found throughout Kansas...*Chrysemys picta*

10A. Crushing surface of upper jaw with ridge or tuberculate row extending parallel to its margin; nuchal plate at least twice as long as broad.....................**11**

10B. Crushing surface of upper jaw smooth or undulating, but without ridge; nuchal plate less than twice as long as broad..**12**

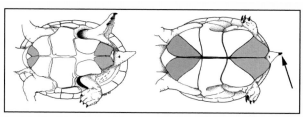

Fɪɢ. 48. Ventral view of the plastrons of turtles of the family Kinosternidae. **Left:** The plastron of a Common Musk Turtle (*Sternotherus odoratus*), showing the squarish shape of the posterior plates (shaded) and the most anterior pair of plates (shaded) being widely separated from the third pair. **Right:** The plastron of a Yellow Mud Turtle (*Kinosternon flavescens*), showing the triangular shape of the posterior pair of plates (shaded), the most anterior pair of plates (shaded) being in contact with the third pair at the midline, and the presence of a horny, clawlike tip on the tail (arrow).

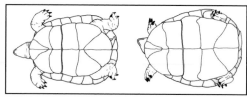

Fɪɢ. 49. View of the plastron of two species of turtles of the genus *Terrapene*. **Left:** The plastron of an Eastern Box Turtle (*Terrapene carolina*), showing the presence of three claws on each hind foot. **Right:** The plastron of an Ornate Box Turtle (*Terrapene ornata*), showing the presence of four claws on each hind foot.

11A. Plastron either unmarked and uniform light color or with a few irregular smudges or dark markings (not paired); broad light stripe on head (which originates at eye) not greatly widened posteriorly; underside of chin flat (Fig. 50); found in Marais des Cygnes, Little Osage, Marmaton, Spring, Neosho, Verdigris, Walnut, Caney, and southern Arkansas river drainages in southeastern Kansas ....*Pseudemys concinna*

11B. Plastron with large paired dark spots or blotches (except in melanistic individuals); broad light (orange-red in life) head stripe (which originates at eye) greatly widened posteriorly (Fig. 50); underside of chin rounded (Fig. 50); found throughout much of Kansas (in all major river drainages); absent from northwestern portion of state and along northern and western borders ....*Trachemys scripta*

12A. Vertebral keel of carapace low, without prominent ridge; small (smaller than diameter of orbit) light spot (sometimes triangular) behind eye with distinct, vertical light line between spot and eye (Fig. 51); restricted to Marais des Cygnes and Marmaton river drainages in east-central Kansas ....*Graptemys geographica*

12B. Vertebral keel of carapace well-developed, with a prominent ridge which may or may not be tipped with black; a transverse bar, oblong spot, or crescent behind the eye, usually with no (or indistinct) lines between them and the eye (Fig. 51); found in all major river drainages in eastern two-thirds of Kansas... ....*Graptemys ouachitensis* and *Graptemys pseudogeographica*

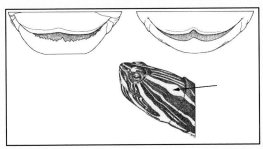

FIG. 50. Frontal view of the lower jaws of two species of turtles of the family Emydidae and lateral view of the head of a Slider (*Trachemys scripta*), showing the broad head stripe (arrow) greatly widened posteriorly. **Upper left:** Underside of chin flat, typical of the Eastern River Cooter (*Pseudemys concinna*). **Upper right:** Underside of chin rounded, typical of the Slider (*Trachemys scripta*).

FIG. 51. Dorsolateral view of the head patterns of turtles of the genus *Graptemys*. **Left:** Common Map Turtle (*Graptemys geographica*). **Middle** and **Right:** False Map Turtle (*Graptemys pseudogeographica*).

GLOSSARY

The line illustrations in this glossary were rendered by Errol D. Hooper, Jr.

Allopatric—When the ranges of two taxa or populations do not contact each other.

Anal scale—The last scale in the belly scale series of snakes. Usually it is larger than the other belly scales and is free along its rear margin. This scale is important in snake classification. It may be either single or divided in half along its midline (see Fig. 34 in the Technical Key). Lizards may have one or more anal scales; all are usually larger than surrounding scales.

Arboreal—Lives and climbs in trees.

Carapace—The upper shell of a turtle (Fig. 52).

Carnivorous—Feeding exclusively on animals but not plants.

Cloaca—An opening beneath the anal scale through which the waste and reproductive ducts discharge their contents. For the purpose of this book, the cloacal opening is synonymous with anus.

Constriction—A technique used by some species of nonvenomous snakes to capture and kill their prey. Constriction involves a snake biting and holding prey with its mouth and wrapping several coils very tightly about the prey's body. When the prey exhales, the coils are tightened further until suffocation occurs.

Diurnal—Active during daytime hours.

Dorsolateral fold—A fold of skin on each side of the back or dorsum of a frog or toad that extends from the eye down the back, often all the way to the to the groin (Fig. 53; see also Fig. 14 in the Technical Key).

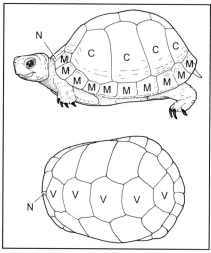

Fig. 52. Generalized drawing of the carapace of a turtle (*Terrapene carolina*), showing two aspects of the upper shell. **Upper:** Lateral view of turtle. **Lower:** Dorsal view of carapace. Scutes or plates labeled on the shell are: C = costal, M = marginal, N = nuchal, and V = vertebral.

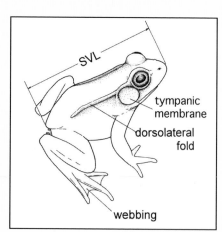

Fig. 53. Generalized drawing of an adult frog, showing various external features and the technique for determining the size of a frog or toad using the snout-vent length (SVL).

Dorsal scale rows—The rows of body scales on the top or dorsum and sides of a snake or lizard. These scales are particularly important in snake classification. They often must be counted circumferentially to make an identification (Fig. 54).

Fossorial—The habit of living beneath the ground; burrowing or digging beneath ground cover or substrate.

Gular fold—A fold of skin running transversely across the throat in lizards (Fig. 55).

Keeled scales—A small, raised ridge or keel found on the rows of scales on the upper part of the body of some snakes (see Fig. 31 in the Technical Key; see also *Smooth scales*).

Larva (plural—*larvae*)—An aquatic, preadult form of the salamander which possesses gills. This form occurs after the salamander hatches from its gelatinous egg, but before it metamorphoses into a more terrestrial adult (Fig. 56).

Metamorphosis—A transition by amphibians from aquatic larvae or tadpoles to primarily terrestrial adults (except salamanders of the genus *Necturus*). This change is accompanied by loss of external gills in salamanders and numerous body adjustments internally by both larvae and tadpoles.

Neoteny—A condition in some salamanders in which the aquatic larvae do not metamorphose into terrestrial adults but do attain sexual maturity as larvae and sometimes grow as large as terrestial adults. Such a salamander is neotenic (see Fig. 56).

Nocturnal—Active during the hours of night.

Omnivorous—Feeding on both plants and animals.

Phylogeny—A branching diagram depicting evolutionary history.

Plastron—The lower shell of a turtle (Fig. 57).

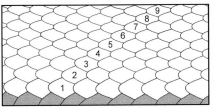

FIG. 54. Generalized drawing of a lateral section of the body of a snake, showing the technique used to count dorsal scale rows, starting with one at the bottom and counting diagonally upward, across the back, and down the other side of the body. Ventral scales of the belly are shaded.

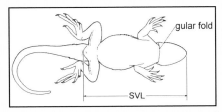

FIG. 55. Generalized drawing of the underside of an adult lizard, showing the location of the gular fold of a lizard, and the technique for making a measurement of the snout-vent length (SVL) of a lizard.

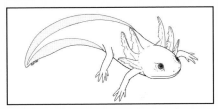

FIG. 56. Generalized drawing of the larva of a Barred Tiger Salamander (*Ambystoma mavortium*), showing the large gills on either side of the head.

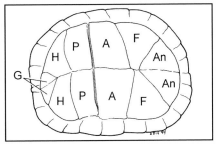

FIG. 57. Generalized drawing of the plastron of a turtle (*Terrapene carolina*). Scutes or plates labeled on the lower shell are: A = abdominal, An = anal, F = femoral, G = gular, H = humeral, and P = pectoral.

Plates—The large scales which compose the upper and lower shells of hard-shelled turtles (see Figs. 52 and 57). Softshells (genus *Apalone*) are the only turtles lacking plates.

Reticulate—A netlike pattern.

Rugose—Wrinkled.

Smooth scales—Scales on the upper part of the body of a snake which lack keels and feel smooth (see Fig. 31 in the Technical Key; see also *Keeled scales*).

Spermatophore—A gelatinous cone of jelly with a cap consisting of sperm. Spermatophores are secreted from the cloaca of male salamanders during breeding. They are picked up by the female with her cloacal lips to fertilize her eggs internally.

SVL—Snout to vent length—A length measurement from the tip of the nose to the cloacal or anal opening (Fig. 58; see also Figs. 53 and 55). Used as a measurement of body size for salamanders and lizards.

Tadpoles—An aquatic, preadult form of frogs and toads which, unlike salamander larvae, does not possess any external gills. This form occurs after the frog or toad hatches from its gelatinous egg, but before it metamorphoses into a more terrestrial adult (Fig. 59).

TL—Total length—A length measurement from the tip of the nose to the tip of the tail. Used to establish maximum size of salamanders, lizards, and snakes.

Tympanic membrane—The external membrane of frogs and toads that receives sound waves and transmits them to the middle ear (see Fig. 53).

Vent—The cloacal or anal opening of amphibians, reptiles, and turtles.

Webbing—Skin between the toes of frogs and turtles (see Fig. 53).

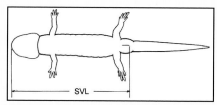

FIG. 58. Generalized drawing of the underside of an adult salamander, showing the technique used for determining the body size of a salamander using the snout-vent length (SVL).

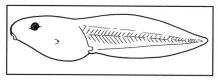

FIG. 59. Generalized drawing of a lateral view of a tadpole. Note the absence of gills.

METRIC CONVERSION CHART

The metric system is used in this publication because of its worldwide application, and in keeping with the system of measurements used in scientific investigation. Below is a chart for converting metric to U. S. units:

To convert	Multiply	To convert	Multiply
millimeters to inches	x 0.039	hectares to acres	x 2.472
centimeters to inches	x 0.394	hectares to square miles	x 0.004
meters to feet	x 3.281	grams to ounces	x 0.035
meters to yards	x 1.094	kilograms to pounds	x 2.205
kilometers to miles	x 0.621		

BIBLIOGRAPHY

The following list of books, papers, and other information sources is divided into four sections. The first part lists pertinent web sites that deal with Kansas herpetology. The second part is devoted to taxonomic and phylogenetic references, which may or may not directly mention Kansas amphibians, reptiles, and turtles. Readers may use these references to further their knowledge about the modern classification of amphibians, reptiles, and turtles. The third part is a short list of references to fossil herpetofauna. The fourth part is a bibliography of Kansas amphibians, reptiles, and turtles, an extensive list of over 1,700 books, papers, reports, and theses that in some way deals with or mentions (even if only briefly) amphibians, turtles, and reptiles in Kansas. This compilation is reasonably complete, with the exception of extremely old works (pre-1861). Kansas was admitted to the Union in 1861, and references prior to that year are less useful because the state's boundaries were not defined. Anonymous citations, abstracts, and the annual herpetofaunal counts reported by the Kansas Herpetological Society since 1989 are included in this bibliography, but are not cited in the references section of the species accounts. The annual updates begun by Joseph T. Collins in 1975 are all included in the bibliography, but are cited only when pertinent comments about a particular species are made therein. With few exceptions, this bibliography includes books and papers in press or appearing in print prior to 30 June 2009.

WEB SITES FOR KANSAS AMPHIBIANS, REPTILES, AND TURTLES

Brunson, K., J. T. Collins, and T. W. Taggart. 2004–2009 et seq. Kansas Anuran Monitoring Program. Electronic Database accessible at http://www.cnah.org/kamp/. Kansas Department of Wildlife and Parks, Pratt, Kansas (with calls by Coleman and Collins, 1998).

Collins, J. T., and T. W. Taggart. 1996–2009 et seq. Kansas Herpetological Society Web Site. Electronically accessible at http://www.cnah.org/khs/. Kansas Herpetological Society, Lawrence, Kansas.

Collins, J. T., and T. W. Taggart. 1998–2009 et seq. The Center for North American Herpetology Web Site.

Electronically accessible at http://www.cnah.org/. The Center for North American Herpetology, Lawrence, Kansas.

Taggart, T. W., J. T. Collins, and C. J. Schmidt. 1999–2009 et seq. Kansas Herpetofaunal Atlas: An On-line Reference. Electronic Database accessible at http://webcat.fhsu.edu/ksfauna/herps. Sternberg Museum of Natural History, Fort Hays State University, Hays, Kansas (with calls by Coleman and Collins, 1998).

TAXONOMIC AND PHYLOGENETIC REFERENCES TO CHORDATES

Ahlberg, P. E., and A. R. Milner. 1994. The origin and early diversification of tetrapods. Nature 368: 507–514.

Collins, J. T. 1990. Standard Common and Current Scientific Names for North American Amphibians and Reptiles. 3rd ed. Herpetological Circular 19, Society for the Study of Amphibians and Reptiles.

Collins, J. T. 1991. Viewpoint: a new taxonomic arrangement for some North American amphibians and reptiles. Herpetol. Rev. 22: 42–43.

Collins, J. T. 1992. The evolutionary species concept: a reply to Van Devender et al., and Montanucci. Herpetol. Rev. 23: 43–46.

Collins, J. T. 1997. Standard Common and Current Scientific Names for North American Amphibians and Reptiles. 4th ed. Herpetological Circular 25, Society for the Study of Amphibians and Reptiles.

Collins, J. T. 2006. A re-classification of snakes native to Canada and the United States. J. Kansas Herpetol. 19: 18–20.

Collins, J. T., R. Conant, J. E. Huheey, J. L. Knight, E. M. Rundquist, and H. M. Smith. 1982. Standard Common and Current Scientific Names for North American Amphibians and Reptiles. 2nd ed. Herpetological Circular 12, Society for the Study of Amphibians and Reptiles.

Collins, J. T., J. E. Huheey, J. L. Knight, and H. M. Smith. 1978. Standard Common and Current Scientific Names for North American Amphibians and Reptiles. 1st ed. Herpetological Circular 7, Society for the Study of Amphibians and Reptiles.

Collins, J. T., and T. W. Taggart. 2002. Standard Common and Current Scientific Names for North American

Amphibians, Turtles, Reptiles, and Crocodilians. 5th ed. The Center for North American Herpetology, Lawrence, Kansas.

COLLINS, J. T., AND T. W. TAGGART. 2009. Standard Common and Current Scientific Names for North American Amphibians, Turtles, Reptiles, and Crocodilians. 6th ed. The Center for North American Herpetology, Lawrence, Kansas.

COLLINS, J. T., AND T. W. TAGGART. 2006. Recent significant name changes for the amphibians of the United States. J. Kansas Herpetol. 18: 7.

COLLINS, J. T., AND T. W. TAGGART. 2008a. An alternative classification of the New World rat snakes (genus *Pantherophis* [Reptilia: Squamata: Colubridae]). J. Kansas Herpetol. 26: 16–18.

COLLINS, J. T., AND T. W. TAGGART. 2008b. A proposal to retain *Masticophis* as the generic name for the coachwhip and whipsnakes. J. Kansas Herpetol. 27: 12.

FROST, D. R. 1985. Amphibian Species of the World. Allen Press, Inc. and The Association of Systematics Collections, Lawrence, Kansas.

FROST, D. R., AND R. ETHERIDGE. 1989. A phylogenetic analysis and taxonomy of iguanian lizards (Reptilia: Squamata). Univ. Kansas Mus. Nat. Hist. Misc. Pub. 81: 1–65.

FROST, D. R., R. ETHERIDGE, D. JANIES, AND T. A. TITUS. 2001. Total evidence, sequence alignment, evolution of polychrotid lizards, and a reclassification of the iguania (Squamata: Iguania). American Mus. Novitates 3,343: 1–38.

FROST, D. R., T. GRANT, J. FAIVOVICH, R. H. BAIN, A. HAAS, C. F. B. HADDAD, R. O. DE SÁ, A. CHANNING, M. WILKINSON, S. C. DONNELLAN, C. J. RAXWORTHY, J. A. CAMPBELL, B. L. BLOTTO, P. MOLER, R. C. DREWES, R. A. NUSSBAUM, J. D. LYNCH, D. M. GREEN, AND W. C. WHEELER. 2006. The amphibian tree of life. Bull. American Mus. Nat. Hist. 297: 1–370.

FROST, D. R., AND D. M. HILLIS. 1990. Species in concept and practice: herpetological applications. Herpetologica 46: 87–104.

GAUTHIER, J. A., D. C. CANNATELLA, K. DE QUEIROZ, A. KLUGE, AND T. ROWE. 1989. Tetrapod phylogeny. Pp. 337–353 *In* B. Fernholm, K. Bremer, and H. Jorwall (Eds.), The Hierarchy of Life. Elsevier Science Publishers, Biomedical Division, Amsterdam, The Netherlands.

HEDGES, S. B., AND L. POLING. 1999. A molecular phylogeny of reptiles. Science 283: 998–1,001.

IWABE, N., Y. HARA, Y. KUMAZAWA, K. SHIBAMOTO, Y. SAITO, T. MIYATA, AND K. KATOH. 2005. Sister group relationship of turtles to the bird-crocodilian clade revealed by nuclear DNA-coded proteins. Mol. Biol. Evol. 22: 810–813.

JIANG, Z. J., T. A. CASTOE, C. C. AUSTIN, F. T. BURBRINK, M. D. HERRON, J. A. MCGUIRE, C. L. PARKINSON, AND D. D. POLLOCK. 2007. Comparative mitochondrial genomics of snakes: extraordinary substitution rate dynamics and functionality of the duplicate control region. BMC Evol. Biol. 7: 123.

KING, F. W., AND R. L. BURKE. 1989. Crocodilian, Tuatara, and Turtle Species of the World: A Taxonomic and Geographic Reference. The Association of Systematics Collections, Washington, D. C.

KIRSCH, J. W. A., AND G. C. MAYER. 1998. The Platypus is not a rodent: DNA hybridization, amniote phylogeny and the palimpsest theory. Philosoph. Trans. Royal Soc. London Series B 353: 1,221.

LAURIN, M., AND R. R. REISZ. 1995. A re-evaluation of early amniote phylogeny. Zool. J. Linnaean Soc. 113: 105–225.

POTTS, G. D., AND J. T. COLLINS. 1991. A Checklist of the Vertebrate Animals of Kansas. 1st ed. Special Publication Number 18, The University of Kansas Museum of Natural History, Lawrence.

POTTS, G. D., J. T. COLLINS, AND K. SHAW. 1999. A Checklist of the Vertebrate Animals of Kansas. 2nd ed. Bulletin of the Kansas Biological Survey, Lawrence.

POTTS, G. D., AND J. T. COLLINS. 2005. A Checklist of the Vertebrate Animals of Kansas. 3rd ed. Sternberg Museum of Natural History, Fort Hays State University, Hays.

REST, J. S., J. C. AST, C. C. AUSTIN, P. J. WADDELL, E. A. TIBBETTS, J. M. HAY, AND D. P. MINDELL. 2003. Molecular systematics of primary reptilian lineages and the Tuatara mitochondrial genome. Mol. Phylogenet. Evol. 29: 289–297.

UTIGER U., B. SCHÄTTI, AND N. HELFENBERGER. 2005. The Oriental colubrine genus *Coelognathus* Fitzinger, 1843 and classification of Old and New World racers and rat snakes (Reptilia, Squamata, Colubridae, Colubrinae). Russian J. Herpetol. 12: 39–60.

WILEY, E. O. 1978. The Evolutionary Species Concept reconsidered. Syst. Zool. 27: 17–26.

ZAHER, H., F. G. GRAZZIOTIN, J. E. CADLE, R. W. MURPHY, J. CESAR DE MOURA-LEITA, AND S. L. MONATTO. 2009. Molecular phylogeny of advanced snakes (Serpentes, Caenophidia) with an emphasis on South American Xenodontines: a revised classification and descriptions of new taxa. Papéis Avulsos Zool. 49: 115–153.

ZARDOYA, R., AND A. MEYER. 1998. Complete mitochondrial genome suggests diapsid affinities of turtles. Proc. Natl. Acad. Sci. USA 95: 14,226–14,231.

REFERENCES TO FOSSIL AMPHIBIANS, REPTILES, AND TURTLES

HAY, O. P. 1908. Fossil Turtles of North America. Publication 75, Carnegie Institution of Washington. 2006 Facsimile Reprint by The Center for North American Herpetology, Lawrence, Kansas.

HOLMAN, J. A. 1995. Pleistocene Amphibians and Reptiles in North America. Oxford University Press, New York.

HOLMAN, J. A. 2000. Fossil Snakes of North America. Indiana University Press, Bloomington.

HOLMAN, J. A. 2004. Fossil Frogs and Toads of North America. Indiana University Press, Bloomington.

HOLMAN, J. A. 2006. Fossil Salamanders of North America. Indiana University Press, Bloomington.

BIBLIOGRAPHY OF KANSAS AMPHIBIANS, REPTILES, AND TURTLES

ABBOTT, S. S. 2008. Cowley County herpetofaunal count. J. Kansas Herpetol. 27: 9.

ABERT, J. W. 1848 (dated 1847). Notes of Lieutenant J. W. Abert. Appendix Number 6. Pp. 386–414 In W. H. Emory. Notes of a Military Reconnaissance from Fort Leavenworth, in Missouri to San Diego, in California including Parts of Arkansas, Del Norte and Gila Rivers. Senate Executive Document 7, 30th Congress First Session. Washington, D. C.

ADALSTEINSSON, S. A., W. B. BRANCH, S. TRAPE, L. J. VITT, AND S. B. HEDGES. 2009. Molecular phylogeny, classification, and biogeography of snakes of the family Leptotyphlopidae (Reptilia, Squamata). 2009. Zootaxa 2,244: 1–50.

ALDRIDGE, R. D. 1993. Male reproductive anatomy and seasonal occurrence of mating and combat behavior of the rattlesnake Crotalus v. viridis. J. Herpetol. 27: 481–484.

ALDRIDGE, R. D., AND D. E. METTER. 1973. The reproductive cycle of the Western Worm Snake, Carphophis vermis, in Missouri. Copeia 1973: 472–477.

ALTIG, R., AND R. LOHOEFENER. 1983. Rana areolata. Cat. American Amph. Reptl. 324.1–324.4

ANDERSEN, K. W., AND E. D. FLEHARTY. 1967. Mammalian distribution within biotic communities of northeastern Jewell County, Kansas. Fort Hays Studies (n. s.) Sci. Series 6: 1–45.

ANDERSON, J. D. 1967a. Ambystoma texanum. Cat. America Amph. Reptl. 37.1–37.2.

ANDERSON, J. D. 1967b. Ambystoma maculatum. Cat. America. Amph. Reptl. 51.1–51.4.

ANDERSON, L. 1995. Butler County turtle rangers. Kansas Herpetol. Soc. News. 101: 18.

ANDERSON, L., M. SHAW, J. BLODIG, AND T. WALKER. 1995. Report to the Kansas Department of Wildlife and Parks: herps encountered during R-Emap project, summer 1994. Kansas Herpetol. Soc. News. 99: 10–17.

ANDERSON, L. R., AND J. A. ARRUDA. 1995. Land use and the biodiversity of frogs in selected breeding sites in southeast Kansas. Abstract. 22nd Annual Meeting of the Kansas Herpetological Society, Lawrence.

ANDERSON, L. R., AND J. A. ARRUDA. 2006. Land use and anuran biodiversity in southeast Kansas, USA. Amphib. Reptl. Conserv. 4(1): 46–59.

ANDERSON, M. L. 1983. Notes on thermal relations in Acris crepitans. Bull. Oklahoma Herpetol. Soc. 8: 48–53.

ANDERSON, P. 1942. Amphibians and reptiles of Jackson County, Missouri. Bull. Chicago Acad. Sci. 6: 203–222.

ANDERSON, P. 1945. New herpetological records for Missouri. Bull. Chicago Acad. Sci. 7: 271–275.

ANDERSON, P. 1950. The Greater Five-lined Skink, Eumeces laticeps (Schneider), in Kansas. Herpetologica 6: 53.

ANDERSON, P. 1965. The Reptiles of Missouri. University of Missouri Press, Columbia.

ANDERSON, R. E. 1996. Biochemical and Hemipenial Systematics of Coluber constrictor (Reptilia: Serpentes). Unpublished Master's thesis, Southeastern Louisiana University, Hammond.

ANDERSON, S. 1984. Areography of North American fishes, amphibians, and reptiles. American Mus. Novitates 2,802: 1–16.

ANDERSON, S., AND B. C. NELSON. 1958. Additional records of mammals in Kansas. Trans. Kansas Acad. Sci. 61: 302–312.

ANONYMOUS. 1957. Glimpses of Kansas wildlife: Kansas reptiles.Pp. 28–30 In Kansas Forestry, Fish and Game Commission, Pratt.

ANONYMOUS. 1961. Antivenin (Crotalidae) Polyvalent. Wyeth Laboratories, New York.

ANONYMOUS. 1967. Once upon a time. Int. Turtle Tortoise Soc. J. 1(4): 24–29.

ANONYMOUS. 1973. St. Louis Zoo Album (Reptile Section). Metropolitan Zoological Park and Museum District, St. Louis.

ANONYMOUS. 1975. A preliminary survey of the plants, aquatic invertebrates, and vertebrates of the Big Basin and St. Jacob's Well, Clark County, Kansas. Report (18 pp.), Kansas Park and Resources Authority. Topeka.

ANONYMOUS. 1977. A Plan for Kansas Wildlife, 1977–1982. Kansas Fish and Game Commission, Pratt.

ANONYMOUS. 1978. Your Guide to Endangered and Threatened Wildlife and Plants in the Lake Region Counties (Kansas). Ottawa University Print Center, Ottawa, Kansas.

ANONYMOUS. 1979. Herps observed or collected during the first three months of 1979. Kansas Herpetol. Soc. News. 30: 6–7.

ANONYMOUS. 1980. Collins photos in new Audubon book. Kansas Herpetol. Soc. News. 35: 4–6.

ANONYMOUS. 1986. These Bite and Sting: Be Careful. Kansas Department of Health and Environment, Topeka.

ANONYMOUS. 1987. Cheyenne Bottoms: An Environmental Assessment. Executive summary. Kansas Biological Survey and Kansas Geological Survey, Lawrence.

ANONYMOUS. 1998. KDWP herp sting so far nets nine on Kansas and Federal charges. Kansas Herpetol. Soc. News. 112: 5–6.

ANONYMOUS. 1998. Wildlife officer undercover. Kansas Wildlife and Parks 55(4): 6–9.

ANONYMOUS. 1999. Cherokee County fall 1999 herp count. Kansas Herpetol. Soc. News. 117: 6.

ANONYMOUS. 1999. Gerlanc studies Western Chorus Frogs on Konza Prairie. Kansas Maps and Gaps 3(1): 9.

ANONYMOUS. 1999. KHS spring field trip. Kansas Herpetol. Soc. News. 116: 3.

ANONYMOUS. 2000. KHS spring field trip sets record for attendance. Kansas Herpetol. Soc. News. 120: 4–5.

ARMSTRONG, H., J. DELANEY, C. FORGE, N. MEERPOHL, AND M. SIMON. 1995. Determination of recolonization by reptiles and amphibians of the Benedictine Bottoms Missouri River floodplain. Abstract. 127th Annual Meeting of the Kansas Academy of Science.

ARRUDA, J. A. 1996. Relationships between land use and amphibian biodiversity in southeast Kansas. Abstract. 128th Annual Meeting of the Kansas Academy of Science.

ASHTON, R. E. 1974a. It's raining salamanders. Zoo (Topeka), 10(1): 2 pp.

ASHTON, R. E. 1974b. Unusual flora and fauna of Kansas: The Central Newt (*Notophthalmus viridescens*) (Rafinesque). Kansas Assoc. Biol. Teachers News. 15: 2 pp.

ASHTON, R. E. 1977. The Central Newt, *Notophthalmus viridescens louisianensis* (Wolterstorff) in Kansas. Trans. Kansas Acad. Sci. 79: 15–19.

ASHTON, R. E., AND J. T. COLLINS. 1976. (Kansas account) Pp. 25–27 *In* R. E. Ashton (Chairman), S. R. Edwards, and G. R. Pisani (Eds.), Endangered and Threatened Amphibians and Reptiles in the United States. Herpetological Circular 5, Society for the Study of Amphibians and Reptiles.

ASHTON, R. E., J. T. COLLINS, AND G. W. FERGUSON. 1974a. Reptiles and Amphibians: Members Manual. Cooperative Extension Service, Environmental Education, Kansas State University, Unit 2B: 1–11.

ASHTON, R. E., J. T. COLLINS, AND G. W. FERGUSON. 1974b. Reptiles and Amphibians: Leaders Guide. Cooperative Extension Service, Environmental Education, Kansas State University, Unit 2B: 1–19.

ATHERTON, R. W. 1974. A gradient analysis of spermatogenesis in the toad *Bufo woodhousei* Girard (1854). Herpetologica 30: 240–244.

AUFFENBERG, W. 1955. A reconsideration of the Racer, *Coluber constrictor*, in eastern United States. Tulane Stud. Zool. 2: 89–155.

AUSTIN, J. D., AND K. R. ZAMUDIO. 2008. Incongruence in the pattern and timing of intra-specific diversification in Bronze Frogs and Bullfrogs (Ranidae). Mol. Phylogenet. Evol. 48: 1,041–1,053.

AXTELL, C. B. 1976. Comparisons of morphology, lactate dehydrogenase, and distribution of *Rana blairi* and *Rana utricularia* in Illinois and Missouri. Trans. Illinois Acad. Sci. 69: 37–48.

AXTELL, R. W. 1954. The Systematic Relationships of Certain Lizards in Two Species Groups of the Genus *Holbrookia*. Unpublished Master's thesis, University of Texas, Austin.

AXTELL, R. W. 1956. A solution to the long neglected *Holbrookia lacerta* problem, and the description of two new subspecies of *Holbrookia*. Bull. Chicago Acad. Sci. 10: 163–183.

AXTELL, R. W. 1958. A Monographic Revision of the Iguanid Genus *Holbrookia*. Unpublished Doctoral dissertation, University of Texas, Austin.

AXTELL, R. W. 1983. Range portrayal and reality: *Heterodon platyrhinos* distribution on the High Plains of Texas and Oklahoma. J. Herpetol. 17: 191–193.

AXTELL, R. W. 1993. Are Copperheads challenging both the Biological and the Evolutionary Species Concepts at the same time? Abstract. Annual Meeting of the Society for the Study of Amphibians and Reptiles, Bloomington, Indiana.

BAGDONAS, K. R., AND D. PETTUS. 1976. Genetic compatibility in wood frogs (Amphibia, Anura, Ranidae). J. Herpetol. 10: 105–112.

BAILEY, R. M. 1943. Four species new to the Iowa herpetofauna, with notes on their natural histories. Iowa Acad. Sci. 50: 347–352.

BAILEY, V., M. R. TERMAN, AND R. WALL. 1989. Noteworthy longevity in *Crotalus viridis viridis* (Rafinesque). Trans. Kansas Acad. Sci. 92: 116–117.

BAKER, R. E. 1983. The Western Cottonmouth in Osage County, Oklahoma. Bull. Oklahoma Herpetol. Soc. 8(2): 56–57.

BAKER, R. J., G. A. MENGDEN, AND J. J. BULL. 1972. Karyotypic studies of thirty-eight species of North American snakes. Copeia 1972: 257–265.

BALL, R. L. 1990. Captive propagation of the Kansas Glossy Snake, *Arizona e. elegans*. Pp. 6–12 *In* A. W. Zulich (Ed.), Proceedings of the 14th International Herpetological Symposium, Dallas/ Fort Worth, Texas.

BALL, R. L. 1992a. High Plains serpents: results of a long-term study in Texas County, Oklahoma and Morton County, Kansas. Kansas Herpetol Soc. News. 88: 16–17.

BALL, R. L. 1992b. Geographic distribution: *Lampropeltis getula*. Herpetol. Rev. 23: 27.

BALLINGER, R. E. 1974. Reproduction of the Texas Horned Lizard, *Phrynosoma cornutum*. Herpetologica 30: 321–327.

BALLINGER, R. E., D. L. DROGE, AND S. M. JONES. 1981. Reproduction in a Nebraska sandhills population of the Northern Prairie Lizard *Sceloporus undulatus garmani*. American Mid. Nat. 106: 157–164.

BALLINGER, R. E., AND J. D. LYNCH. 1983. How to Know the Amphibians and Reptiles. Wm. C. Brown, Dubuque, Iowa.

BAMMES, B. E. 1995. The Sharon Springs rattlesnake round-up: a report. Kansas Herpetol. Soc. News. 100: 31–32.

BARDEN, A. 1942. Activity of the lizard, *Cnemidophorus sexlineatus*. Ecology 23: 336–342.

BARTELS, B. C. 2007. Geographic distribution: *Lampropeltis calligaster*. J. Kansas Herpetol. 24: 15.

BARTEN, S. L. 1980. The consumption of turtle eggs by a Western Hognose Snake, *Heterodon nasicus*: a field observation. Bull. Chicago Herpetol. Soc. 15: 97–98.

BARTLETT, R. 1995. The captivating North American rat snakes, Reptiles. 3(6): 56–75.

BARTLETT, R. 1995. Venomous snakes: controversial captives. Reptiles 3(4): 24–47.

BARTLETT, R. 1995. Nature's bluffers: the American hognose snakes. Reptiles 3(5): 40–46.

BARTLETT, R. D., AND P. P. BARTLETT. 2006a. Guide and Reference to the Amphibians of Eastern and Central North America (North of Mexico). University Press of Florida, Gainesville.

BARTLETT, R. D., AND P. P. BARTLETT. 2006b. Guide and Reference to the Crocodilians, Turtles, and Lizards of Eastern and Central North America (North of Mexico). University Press of Florida, Gainesville.

BARTLETT, R. D., AND P. P. BARTLETT. 2006c. Guide and Reference to the Snakes of Eastern and Central North America (North of Mexico). University Press of Florida, Gainesville.

BARTLETT, R. D., AND A. TENNANT. 2000. Snakes of North America: Western Region. Gulf Publishing Company, Houston, Texas.

BARTLETT, W. E. 1920. Tragic death feint of a snake. Nature 5: 106, 503.

BARTON, A. J. 1948. Snake litters. Herpetologica 4: 198.

BAXTER, G. T., AND M. D. STONE. 1980. Amphibians and Reptiles of Wyoming. 1st ed. Bulletin 16, Wyoming Game and Fish Department, Cheyenne.

BAXTER, G. T., AND M. D. STONE. 1985. Amphibians and Reptiles of Wyoming. 2nd ed. Bulletin 16, Wyoming Game and Fish Department, Cheyenne.

BEARD, J. B. 1986. Salamanders of Schermerhorn Park Cave, Kansas. Kansas Herpetol. Soc. News. 66: 7–8.

BEATSON, R. R. 1976. Environmental and genetical correlates of disruptive coloration in the water snake, *Natrix s. sipedon*. Evolution 30: 241–252.

BEAVER, T., AND P. INGRAM. 2008. Geographic distribution: *Terrapene ornata*. J. Kansas Herpetol. 26: 6.

BECHTEL, H. B. 1978. Color and pattern in snakes (Reptilia, Serpentes). J. Herpetol. 12: 521–532.

BEHLER, J. L., AND F. W. KING. 1979. The Audubon Society Field Guide to North American Reptiles and Amphibians. A. A. Knopf, New York.

BELFIT, S. C., AND V. F. BELFIT. 1985. Notes on the ecology of a population of *Eumeces obsoletus* (Scincidae) in New Mexico. Southwest. Nat. 30: 612–614.

BELL, E. L. 1996. Descriptions of neotypes for *Sceloporus undulatus undulatus*, the Southern Fence Lizard, and *Sceloporus undulatus hyacinthinus*, the Northern Fence Lizard, and a lectotype for *Sceloporus undulatus garmani*, the Northern Prairie Lizard. Bull. Maryland Herpetol. Soc. 32: 81–103.

BELLOWS-BLAKELY, S. 2000. Observations on the snakes of the Red Hills, Kansas. Abstract. Missouri Herpetol. Assn. News. 13: 3.

BELLOWS-BLAKELY, S. 2001. Observations on the snakes of the Red Hills, Kansas. Abstract. Kansas Herpetol. Soc. News. 123: 3.

BELTZ, E. 1993. Distribution and status of the Eastern Massasauga, *Sistrurus catenatus catenatus* (Rafinesque, 1818), in the United States and Canada. Pp. 26–31 *In* B. Johnson and V. Menzies (Eds.), International Symposium and Workshop on the Conservation of the Eastern Massasauga *Sistrurus catenatus catenatus*. Metro Toronto Zoo, West Hill, Ontario, Canada.

BENDER, D. J. 2006. Graduate research on the Massasauga in Kansas. J. Kansas Herpetol. 18: 8–9.

BENDER, D. J., AND W. J. STARK. 2007. Diet of the Western Massasauga (*Sistrurus catenatus tergeminus*) in a grassland and wetland in Kansas. Abstract. Trans. Kansas Acad. Sci. 110: 294–295.

BENNETT, R. 2004. Life history. *Crotalus viridis*. Prairie Rattlesnake. Behavior. J. Kansas Herpetol. 12: 18.

BENTON, M. J. 1980. Geographic variation and validity of subspecies names for the Eastern Garter Snake, *Thamnophis sirtalis*. Bull. Chicago Herpetol. Soc. 15: 57–69.

BERGER, T. J. 1985. Spatial distribution and habitat utilization by tadpoles. Abstract. 65th Annual Meeting of the American Society of Ichthyologists and Herpetologists, Knoxville, Tennessee.

BERGLUND, L. A. 1967. Snakes on the Cheyenne Bottoms Area, Barton County, Kansas. Privately printed (1 p.; mimeograph).

BIRD, W., P. PEAK, AND J. T. COLLINS, 2005. *Lampropeltis calligaster*. New record length for the entire range. J. Kansas Herpetol. 15: 12.

BISHOP, S. C. 1943. Handbook of Salamanders. Comstock Publishing Company, Ithaca, New York.

BISHOP, S. C. 1944. A new neotenic plethodont salamander, with notes on related species. Copeia 1944: 1–5.

BLACK, J. H. 1981. Oklahoma rattlesnake hunts: 1981. Bull. Oklahoma Herpetol. Soc. 6: 39–43.

BLACK, J. H., AND A. N. BRAGG. 1968. New additions to the herpetofauna of Montana. Herpetologica 24: 247.

BLACK, J. H., AND J. T. COLLINS. 1977. An albino *Ambystoma tigrinum mavortium* from Kansas. Herpetol. Rev. 8: 5.

BLACK, J. H., AND R. LARDIE. 1976. A record-sized Barred Tiger Salamander in Oklahoma. Bull. Maryland Herpetol. Soc. 12: 54–55.

BLAIR, A. 1939. Records of the salamander *Typhlotriton*. Copeia 1939: 108–109.

BLAIR, M. 1989. Gallery. Kansas Wildlife and Parks 46(2): 38–39.

BLAIR, M. 2002. 2001 Photo Issue. Kansas Wildlife and Parks 59(1): 1–44 (pp. 10, 26, 30).

BLAIR, W. F. (Ed.). 1972. Evolution in the Genus *Bufo*. University of Texas Press, Austin.

BLAIR, W. F. 1958. Mating call in the speciation of anuran amphibians. American Nat. 92: 27–51.

BLAIR, W. F. 1976. Some aspects of the biology of the Ornate Box Turtle, *Terrapene ornata*. Southwest. Nat. 21: 89–104.

BLAIR, W. F., AND A. P. BLAIR. 1941. Food habits of the Collared Lizard in northeastern Oklahoma. American Mid. Nat. 26: 230–232.

BLAIR, W. F., A. P. BLAIR, P. BRODKORB, F. R. CAGLE, AND G. A. MOORE. 1968. Vertebrates of the United States. 2nd ed. McGraw Hill, New York.

BLANCHARD, F. N. 1920. A synopsis of the king snake genus *Lampropeltis* Fitzinger. Univ. Michigan Mus. Zool. Occas. Pap. 87: 1–8.

BLANCHARD, F. N. 1921. A revision of the kingsnakes: Genus *Lampropeltis*. Bull. U. S. Natl. Mus. 114: 1–260.

BLANCHARD, F. N. 1925. A key to the snakes of the United States, Canada, and lower California. Pap. Michigan Acad. Sci. Arts Letters 4: 1–65.

BLANCHARD, F. N. 1938. Snakes of the genus *Tantilla* in the United States. Zool. Ser. Field Mus. Nat. Hist. 20: 369–376.

BLANCHARD, F. N. 1942. The Ring-necked Snakes, genus *Diadophis*. Bull. Chicago Acad. Sci. 7: 1–144.

BLANEY, R. M. 1973. *Lampropeltis*. Cat. American Amph. Reptl. 150.1–150.2.

BLANEY, R. M. 1977. Systematics of the Common Kingsnake, *Lampropeltis getulus* (Linnaeus). Tulane Stud. Zool. Bot. 19: 47–103.

BLANEY, R. M. 1979. *Lampropeltis calligaster*. Cat. American Amph. Reptl. 229.1–229.2.

BLEM, C. R. 1981. *Heterodon platyrhinos*. Cat. American Amph. Reptl. 282.1–282.2.

BLYTHE, C. 1979. Poisonous Snakes of America. What You Need to Know about Them. Branch-Smith, Fort Worth, Texas.

BOCK, C. E., AND H. M. SMITH. 1982. Biogeography of North American amphibians: a numerical analysis. Trans. Kansas Acad. Sci. 85: 177–186.

BOGART, J. P. 1980. Evolutionary implications of polyploidy in amphibians and reptiles. Pp. 341–378 In W. H. Lewis (Ed.), Polyploidy. Plenum Press, New York.

BOLES, R. J. 1971. Alabaster. Kansas School Nat. 17(3): 1–15.

BOLES, R. J. 1973. Some questions about wildlife. Kansas School Nat. 19(3): 1–16.

BOLES, R. J. 1981. D. O. R. Kansas School Nat. 28(2): 1–16.

BONATI, R. 1995. Cave-dwellers of the Ozarks. Reptile & Amphibian Magazine. March–April: 50–58.

BONETT, R. M., AND P. T. CHIPPINDALE. 2004. Speciation, phylogeography and evolution of life history and morphology in plethodontid salamanders of the Eurycea multiplicata complex. Mol. Ecol. 13: 1,189–1,203.

BONIN, F., B. DEVAUX, AND A. DUPRÉ. 2006. Turtles of the World.The Johns Hopkins University Press, Baltimore, Maryland.

BOULENGER, G. A. 1920. A monograph of the American frogs of the genus Rana. Proc. American Acad. Arts Sci. 55: 413–480.

BOUNDY, J. 1995. Maximum lengths of North American snakes. Bull. Chicago Herpetol. Soc. 30: 109–122.

BOVEE, E. C. 1981. New epizoic suctorea (Protozoa) of the Smooth Softshell Turtle, Trionyx muticus, in northeastern Kansas. Trans. Kansas Acad. Sci. 84: 98–104.

BOYD, R. L. 1988. Baker University natural areas. Trans. Kansas Acad. Sci. 91: 52–54.

BOYD, R. 1991. Baker University wetlands. Pp. 106–125 In E. D. Kettle and D. O. Whittemore (Eds.), Ecology and Hydrology of Kansas Ecological Reserves and the Baker Wetlands. Multidisciplinary Guidebook 4, Kansas Academy of Science, Lawrence.

BRAGG, A. N. (ED.) 1950. Researches on the Amphibia of Oklahoma. University of Oklahoma Press, Norman.

BRAGG, A. N. 1936. The ecological distribution of some North American Anura. American Nat. 70: 459–466.

BRAGG, A. N. 1937. Observations on Bufo cognatus with special reference to breeding habits and eggs. American Mid. Nat. 18: 273–284.

BRAGG, A. N. 1940a. Observations on the ecology and natural history of Anura. I. Habits, habitat, and breeding of Bufo cognatus Say. American Nat. 74: 322–349.

BRAGG, A. N. 1940b. Observations on the ecology and natural history of Anura. II. Habits, habitat, and breeding of Bufo woodhousei woodhousei (Girard) in Oklahoma. American Mid. Nat. 24: 306–321.

BRAGG, A. N. 1940c. Observations on the ecology and natural history of Anura. III. The ecological distribution of Anura of Cleveland County, Oklahoma, including notes on the habits of several species. American Mid. Nat. 24: 322–335.

BRAGG, A. N. 1941. New records of frogs and toads for Oklahoma. Copeia 1941: 51.

BRAGG, A. N. 1943a. Observations on the ecology and natural history of Anura. XV. The hylids and microhylids in Oklahoma. Great Basin Nat. 4: 62–80.

BRAGG, A. N. 1943b. Observations on the ecology and natural history of Anura. XVI. Life-history of Pseudacris clarkii (Baird) in Oklahoma. Wasmann Coll. 5: 129–140.

BRAGG, A. N. 1944. The spadefoot toads in Oklahoma with a summary of our knowledge of the group. American Nat. 78: 517–533.

BRAGG, A. N. 1946. Some new county records of Salientia and a correction in the list from Oklahoma. Proc. Oklahoma Acad. Sci. 26: 16–17.

BRAGG, A. N. 1948a. Observations on Hyla versicolor in Oklahoma. Proc. Oklahoma Acad. Sci. 28: 31–35.

BRAGG, A. N. 1948b. Observations on the life history of Pseudacris triseriata (Wied) in Oklahoma. Wasmann Coll. 7: 149–168.

BRAGG, A. N. 1949. Observations on the Narrow-mouthed Salamander. Proc. Oklahoma Acad. Sci. 30: 21–24.

BRAGG, A. N. 1952. Amphibians of McCurtain County, Oklahoma. Wasmann J. Biol. 10: 241–250.

BRAGG, A. N. 1953. A study of Rana areolata in Oklahoma. Wasmann J. Biol. 11: 273–318.

BRAGG, A. N. 1954. Bufo terrestris charlesmithi, a new subspecies from Oklahoma. Wasmann J. Biol. 12: 245–254.

BRAGG, A. N. 1955. Taxonomic and physiological factors in the embryonic development of certain toads. Copeia 1955: 62.

BRAGG, A. N. 1965. Gnomes of the Night. University of Pennsylvania Press, Philadelphia.

BRAGG, A. N., AND C. C. SMITH. 1942. Observations on the ecology and natural history of Anura. IX. Notes on breeding behavior in Oklahoma. Great Basin Nat. 3: 33–50.

BRAGG, A. N., AND C. C. SMITH. 1943. Observations on the ecology and natural history of Anura. IV. The ecological distribution of toads in Oklahoma. Ecology 24: 285–309.

BRAGG, A. N., AND O. SANDERS. 1951. A new subspecies of the Bufo woodhousei group of toads (Salientia: Bufonidae). Wasmann J. Biol. 9: 363–378.

BRANDON, R. A. 1965. Typhlotriton, T. spelaeus, and T. nereus. Cat. American Amph. Reptl. 20.1–20.2.

BRANDON, R. A. 1966. A reevaluation of the status of the salamander, Typhlotriton nereus Bishop. Copeia 1966: 555–561.

BRANDON, R. A. 1970. Typhlotriton and T. spelaeus. Cat. American Amph. Reptl. 84.1–84.2.

BRANDON, R. A. 1971. North American troglobitic salamanders: some aspects of modification in cave habitats, with special reference to Gyrinophilus palleucus. Bull. Nat. Speleol. Soc. 33: 1–21.

BRANSON, B. A. 1966. A partial biological survey of the Spring River drainage in Kansas, Oklahoma and Missouri. Part I, collecting sites, basic limnological data, and mollusks. Trans. Kansas Acad. Sci. 69: 242–293.

Branson, B. A. 1992. Narrowmouth toads. Reptile & Amphibian Magazine. July–August: 64–68.

Branson, B. A. 1994. Snapping turtles. Reptile & Amphibian Magazine. May–June: 44–53.

Branson, B. A. 1996. Red-spotted newts. Reptile & Amphibian Magazine. July–August: 52–59.

Branson, B. A., J. Triplett, and R. Hartmann. 1969. A partial biological survey of the Spring River drainage in Kansas, Oklahoma and Missouri. Part II: the fishes. Trans. Kansas Acad. Sci. 72: 429–472.

Branson, E. B. 1904. Snakes of Kansas. Univ. Kansas Sci. Bull. 2: 353–430.

Brattstrom, B. H. 1967. A succession of Pliocene and Pleistocene snake faunas from the High Plains of the United States. Copeia 1967: 188–202.

Brennan, L. A. 1934. A check list of the amphibians and reptiles of Ellis County, Kansas. Trans. Kansas Acad. Sci. 37: 189–191.

Brennan, L. A. 1935. Notes on the Distribution of Amphibia and Reptilia of Ellis County, Kansas. Unpublished Master's thesis, Fort Hays State University, Hays, Kansas.

Brennan, L. A. 1937. A study of the habitat of reptiles and amphibians of Ellis County, Kansas. Trans. Kansas Acad. Sci. 40: 341–347.

Brest van Kempen, C. P. 2006. Rigor Vitae: Life Unyielding - The Art of Carel Pieter Brest van Kempen. Eagle Mountain Publishing, LC, Eagle Mountain, Utah.

Breukelman, J. 1940. The snake, *Haldea striatula*, in Kansas. Herpetologica 2: 56.

Breukelman, J. 1963. Kansas natural history in 1863. Kansas School Nat. 9(2): 1–16.

Breukelman, J., and R. F. Clarke. 1951. A revised list of amphibia and reptiles of Chase and Lyon Counties, Kansas. Trans. Kansas Acad. Sci. 54: 542–545.

Breukelman, J., and A. Downs. 1936. A list of amphibia and reptiles of Chase and Lyon Counties, Kansas. Trans. Kansas Acad. Sci. 39: 267–268.

Breukelman, J., T. A. Eddy, and E. L. Hartman. 1961. The F. B., and Rena G. Ross Natural History Reservation. Kansas School Nat. 7(4): 1–16.

Breukelman, J., and H. M. Smith. 1946. Selected records of reptiles and amphibians from Kansas. Univ. Kansas Publ. Mus. Nat. Hist. 1: 101–112.

Brewer, J. F. 1897. Prairie Rattlesnakes: their poison and its treatment. Kansas Medical J. 9(5): 57–60.

Brons, H. A. 1882. Notes on the habits of some western snakes. American Nat. 16: 564–567.

Brooks, G. R. 1975. *Scincella lateralis*. Cat. American Amph. Reptl. 169.1–169.4.

Broome, R. C. 1973. North American rat snakes of the genus *Elaphe*. Kentucky Herpetol. 4(10): 9–20.

Brown, A. E. 1901. A review of the genera and species of American snakes, north of Mexico. Proc. Acad. Nat. Sci. Philadelphia 53: 10–110.

Brown, B. C. 1950. An Annotated Check List of the Reptiles and Amphibians of Texas. Baylor University Studies, Baylor, Texas.

Brown, H. A. 1967. Embryonic temperature adaptations and genetic compatibility in two allopatric populations of the spadefoot toads (genus *Scaphiopus*). Los Angeles Co. Mus. Nat. Hist. Contrib. Sci. 286: 1–15.

Brown, J. H. 1973. Toxicology and Pharmacology of Venoms from Poisonous Snakes. C. C. Thomas, Springfield, Illinois.

Brown, K. L., V. G. Demarco, R. W. Drenner, G. W. Ferguson, and L. J. Smith. 1980. Factors influencing the sizes of prey ingested by the Northern Prairie Swift (*Sceloporus undulatus*). Abstract. 60th Annual Meeting of the American Society of Ichthyologists and Herpetologists, Fort Worth, Texas.

Brown, L. 1942. Propagation of the Spotted Channel Catfish (*Ictalurus lacustris punctatus*). Trans. Kansas Acad. Sci. 45: 311–314.

Brown, L. E. 1992. *Rana blairi*. Cat. American Amphib. Reptl. 536.1–536.6.

Brown, W. L. 1958. General adaptation and evolution. Syst. Zool. 7: 157–168.

Brown, W. S. 1987. Hidden life of the Timber Rattler. Natl. Geographic, July: 128–138.

Brown, W. S. 1993. Biology, Status, and Management of the Timber Rattlesnake (*Crotalus horridus*): A Guide for Conservation. Herpetological Circular 22, Society for the Study of Amphibians and Reptiles.

Brumwell, M. J. ca. 1933. Distributional records of the reptilia and amphibians of Kansas. Unpublished manuscript.

Brumwell, M. J. 1940. Notes on the courtship of the turtle, *Terrapene ornata*. Trans. Kansas Acad. Sci. 43: 391–392.

Brumwell, M. J. 1942. Establishment of *Anolis carolinensis* in Kansas. Copeia 1942: 54.

Brumwell, M. J. 1951. An ecological survey of the Fort Leavenworth Military Reservation. American Mid. Nat. 45: 187–231.

Brunson, K. 1979. Prairie rivers. Kansas Fish and Game 36(2): 7–22.

Brunson, K. 1986. Meet our state reptile. Kansas Wildlife 43(4): 5.

Brunson, K. 1989a. More on the Kansas endangered and threatened species list. Kansas Herpetol. Soc. News. 75: 17–19.

Brunson, K. 1989b. The rubber snake award. Kansas Herpetol. Soc. News. 75: 19.

Brunson, K. 2000a. Heartland hoppers. Kansas Wildlife and Parks 57(2): 2–7.

Brunson, K. 2000b. What's buggin' the frogs? Kansas Wildlife and Parks 57(4): 41.

Brush, S. W., and G. W. Ferguson. 1986. Predation on lark sparrow eggs by a Massasauga Rattlesnake. Southwest. Nat. 31: 260–261.

Bryson, R. W., Jr., J. Pastorini, F. T. Burbrink, and M. R. J. Forstner. 2007. A phylogeny of the *Lampropeltis mexicana* complex (Serpentes: Colubridae) based on mitochondrial DNA sequences suggests evidence for species-level polyphyly within *Lampropeltis*. Mol. Phylogenet. Evol. 43: 674–684.

Bugbee, R. E. 1945. A note on the mortality of snakes on highways in western Kansas. Trans. Kansas Acad. Sci. 47: 373–374.

Buhlmann, K., T. Tuberville, and J. W. Gibbons. 2008. Turtles of the Southeast. University of Georgia Press, Athens.

BUIKEMA, A. L., AND K. B. ARMITAGE. 1969. The effect of temperature on the metabolism of the Prairie Ring-neck Snake, *Diadophis punctatus arnyi* Kennicott. Herpetologica 25: 194–206.

BURBRINK, F. T. 2001. Systematics of the Eastern Rat Snake complex (*Elaphe obsoleta*). Herpetol. Monogr. 15: 1–53.

BURBRINK, F. T. 2002. Phylogeographic analysis of the Corn Snake (*Elaphe guttata*) complex as inferred from maximum likelihood and Bayesian analyses. Mol. Phylogenet. Evol. 25: 465–476.

BURBRINK, F. T., F. FONTANELLA, R. A. PYRON, T. J. GUIHER, AND C. JIMENEZ. 2008. Phylogeography across a continent: the evolutionary and demographic history of the North American Racer (Serpentes: Colubridae: *Coluber constrictor*). Mol. Phylogenet. Evol. 47: 274–288.

BURBRINK, F. T., R. LAWSON, AND J. B. SLOWINSKI. 2000. Mitochondrial DNA phylogeography of the polytypic North American Rat Snake (*Elaphe obsoleta*): a critique of the subspecies concept. Evolution 54: 2,107–2,118.

BURGER, W. L. 1952. A neglected subspecies of the turtle *Pseudemys scripta*. J. Tennessee Acad. Sci. 27: 75–79.

BURGER, W. L., P. W. SMITH, AND H. M. SMITH. 1949. Notable records of reptiles and amphibians in Oklahoma, Arkansas, and Texas. J. Tennessee Acad. Sci. 24: 130–134.

BURKE, R. L., AND G. DEICHSEL. 2008. Lacertid lizards introduced into North America: history and future. Pp. 347–353 *In* J. C. Mitchell, R. E. Jung Brown, and B. Bartholomew (Eds.), Urban Herpetology. Herpetological Conservation 3, Society for the Study of Amphibians and Reptiles.

BURKE, R. L., S. R. GOLDBERG, C. R. BURSEY, S. L. PERKINS, AND P. T., ANDREADIS. 2007. Depauperate parasite faunas in introduced populations of *Podarcis* (Squamata: Lacertidae) lizards in North America. J. Herpetol. 41: 755–757.

BURKETT, R. D. 1966. Natural history of Cottonmouth Moccasin, *Agkistrodon piscivorus* (Reptilia). Univ. Kansas Publ. Mus. Nat. Hist. 17: 435–491.

BURKETT, R. D. 1969. An Ecological Study of the Cricket Frog, *Acris crepitans*, in Northeastern Kansas. Unpublished Doctoral dissertation, University of Kansas, Lawrence.

BURKETT, R. D. 1984. An ecological study of the Cricket Frog (*Acris crepitans*). Pp. 89–103 *In* R. A. Seigel, L. E. Hunt, J. L. Knight, L. Malaret and N. L. Zuschlag (Eds.). Vertebrate Ecology and Systematics: A Tribute to Henry S. Fitch. Special Publication Number 10, The University of Kansas Museum of Natural History, Lawrence.

BURKHART, J. T. 1980. KHS field trip to Comanche and Kiowa counties. Kansas Herpetol. Soc. News. 37: 9–10.

BURKHART, J. T. 1984. Status of the Western Green Toad (*Bufo debilis insidior*) in Kansas. Contract 72, Final Report (24 pp.), Kansas Fish and Game Commission, Pratt.

BURR, A. 2003. Coffey County herp count 1. J. Kansas Herpetol. 7: 7.

BURR, A. 2003. Coffey County herp count 2. J. Kansas Herpetol. 7: 7.

BURR, A., AND C. BURR. 2003. Geographic distribution: *Rana areolata*. J. Kansas Herpetol. 6: 8.

BURRAGE, B. L. 1962. A new Kansas locality record for *Eumeces anthracinus pluvialis* Cope. Herpetologica 18: 210–211.

BURT, C. E. 1927a. On the type locality of the horned lizard (*Phrynosoma brevirostre* Girard). Copeia 163: 53–54.

BURT, C. E. 1927b. An annotated list of the amphibians and reptiles of Riley County, Kansas. Univ. Michigan Mus. Zool. Occas. Pap. 189: 1–9.

BURT, C. E. 1928a. The lizards of Kansas. Trans. Acad. Sci. St. Louis, 26: 1–81.

BURT, C. E. 1928b. Insect food of Kansas lizards with notes on feeding habits. J. Kansas Entomol. Soc. 1: 50–68.

BURT, C. E. 1928c. The synonymy, variation and distribution of the Collared Lizard, *Crotaphytus collaris* (Say). Univ. Michigan Mus. Zool. Occas. Pap. 196: 1–19.

BURT, C. E. 1928d. A key to the species of lizards definitely reported from Kansas. Privately printed.

BURT, C. E. 1928e. A new amphibian record from Kansas, *Hyla phaeocrypta* (Cope). Science 67(1,747): 630–631.

BURT, C. E. 1929a. The synonymy, variation, and distribution of the Sonoran Skink, *Eumeces obsoletus* (Baird and Girard). Univ. Michigan Mus. Zool. Occas. Pap. 201: 1–12.

BURT, C. E. 1929b. The sexual dimorphism of the Collared Lizard, *Crotaphytus collaris*. Pap. Michigan Acad. Sci. Arts Letters 10: 417–421.

BURT, C. E. 1931a. A report on some amphibians and reptiles from Kansas, Nebraska, and Oklahoma. Proc. Biol. Soc. Washington 44(6): 11–16.

BURT, C. E. 1931b. A study of the teiid lizards of the genus *Cnemidophorus*, with special reference to their phylogenetic relationships. Bull. U. S. Natl. Mus. 154: 1–286.

BURT, C. E. 1932. Records of amphibians from the eastern and central United States (1931). American Mid. Nat. 13: 75–85.

BURT, C. E. 1933. Some distributional and ecological records of Kansas reptiles. Trans. Kansas Acad. Sci. 36: 186–208.

BURT, C. E. 1935. Further records of the ecology and distribution of amphibians and reptiles in the middle west. American Mid. Nat. 16: 311–366.

BURT, C. E. 1936a. A key to the lizards of the United States and Canada. Trans. Kansas Acad. Sci. 38: 255–305.

BURT, C. E. 1936b. Contributions to Texan herpetology. IV. Sand snakes of the genus *Tantilla*. Trans. American Microscop. Soc. 55: 239–242.

BURT, C. E. 1937. The fauna: amphibians and reptiles of "Rock City." Trans. Kansas Acad. Sci. 40: 195.

BURT, C. E. 1949. Baby garter snake victim of garden spider. Herpetologica 5: 127.

BURT, C. E., AND M. D. BURT. 1929a. Field notes and locality records on a collection of amphibians and reptiles, chiefly from the western half of the United

States. I. Amphibians J. Washington Acad. Sci. 19: 428–434.

Burt, C. E., and M. D. Burt. 1929b. Field notes and locality records on a collection of amphibians and reptiles, chiefly from the western half of the United States. II. Reptiles. J. Washington Acad. Sci. 19: 448–460.

Burt, C. E., and M. D. Burt. 1929c. A collection of amphibians and reptiles from the Mississippi Valley, with field observations. American Mus. Novitates 381: 1–14.

Burt, C. E., and W. L. Hoyle. 1934. Additional records of the reptiles of the central prairie region of the United States. Trans. Kansas Acad. Sci. 37: 193–216.

Burton, P. R., and W. H. Vensel. 1966. Ultrastructural studies of normal and alloxan-treated islet cells of the pancreas of the lizard, *Eumeces fasciatus*. J. Morphol. 118: 91–118.

Bury, R. B., H. W. Campbell, and N. J. Scott, Jr. 1980. Role and importance of nongame wildlife. Pp. 197–207 *In* Transactions of the 45th North American Wildlife and Natural Resources Conference, Washington, D. C.

Bury, R. B., C. K. Dodd, Jr., and G. M. Fellers. 1980. Conservation of the Amphibia of the United States: A Review. Resource Publication 134. U. S. Fish and Wildlife Service, Washington, D. C.

Bury, R. B., and J. A. Whelan. 1984. Ecology and Management of the Bullfrog. Resource Publication 155. U. S. Fish and Wildlife Service, Washington, D. C.

Busby, W. H. 1988. The Kansas Natural Heritage Program: taking stock of Kansas' natural heritage. Kansas Herpetol. Soc. News. 71: 9–12.

Busby, W. H. 1990. An inventory of three prairie animals in eastern Kansas. Report 45 (34 pp.), Kansas Biological Survey, Lawrence.

Busby, W. H. 1996. Calling behavior and habitat associations of the Crawfish Frog in Kansas. Abstract. 23rd Annual Meeting of the Kansas Herpetological Society, Lawrence.

Busby, W. H. 1997. Mid-winter herp activity in Kansas. Kansas Herpetol. Soc. News. 108: 19.

Busby, W. H., and W. R. Brecheisen. 1997. Chorusing phenology and habitat associations of the Crawfish Frog, *Rana areolata* (Anura: Ranidae), in Kansas. Southwest. Nat. 42: 210–217.

Busby, W. H., J. T. Collins, and J. R. Parmelee. 1996. The Reptiles and Amphibians of Fort Riley and Vicinity. Kansas Biological Survey, Lawrence.

Busby, W. H., J. T. Collins, and G. Suleiman. 2005. The Snakes, Lizards, Turtles, and Amphibians of Fort Riley and Vicinity. 2nd (revised) ed. Kansas Biological Survey, Lawrence.

Busby, W. H., and J. R. Parmelee. 1996. Historical changes in a herpetofaunal assemblage in the Flint Hills of Kansas. American Mid. Nat. 135: 81–91.

Busby, W. H., J. R. Parmelee, C. M. Dwyer, E. D. Hooper, and K. J. Irwin. 1994. A survey of the herpetofauna on the Fort Riley Military Reservation, Kansas. Report 58 (79 pp.), Kansas Biological Survey, Lawrence.

Busby, W. H., and G. R. Pisani. 2007. Welda spring field trip. J. Kansas Herpetol. 22: 12.

Cagle, F. R. 1943. Two new subspecies of *Graptemys pseudogeographica*. Univ. Michigan Mus. Zool. Occas. Pap. 546: 1–17.

Cagle, F. R. 1950. The life history of the Slider turtle, *Pseudemys scripta troostii* (Holbrook). Ecol. Monogr. 20: 31–54.

Cagle, F. R. 1954. Two new species of the genus *Graptemys*. Tulane Stud. Zool. 1: 167–186.

Caldwell, J. P. 1975. Systematic status of leopard frogs (*Rana pipiens* complex) in Kansas. Abstract. 107th Annual Meeting of the Kansas Academy of Science.

Caldwell, J. P. 1977. Crawfish Frogs snore again in southeast Kansas. Kansas Herpetol. Soc. News. 17: 6–7.

Caldwell, J. P. 1978. Tail coloration as a defense mechanism in Cricket Frog tadpoles. Abstract. Trans. Kansas Acad. Sci. 81: 137.

Caldwell, J. P. 1982. Disruptive selection: a tail color polymorphism in *Acris* tadpoles in response to differential predation. Canadian J. Zool. 60: 2,818–2,827.

Caldwell, J. P. 1986. Selection of egg deposition sites: a seasonal shift in the Southern Leopard Frog, *Rana sphenocephala*. Copeia 1986: 249–253.

Caldwell, J. P., and J. T. Collins. 1977. New records of fishes, amphibians and reptiles in Kansas. Tech. Publ. State Biol. Surv. Kansas 4: 63–78.

Caldwell, J. P., and J. T. Collins. 1981. Turtles in Kansas. AMS Publishing, Lawrence, Kansas.

Caldwell, J. P., and G. Glass. 1976. Vertebrates of the Woodson County State Fishing Lake and Game Management Area. Pp. 62–76 *In* Preliminary Inventory of the Biota of Woodson County State Fishing Lake and Game Management Area. Report 56, State Biological Survey of Kansas, Lawrence.

Campbell, J. A., and E. D. Brodie, Jr. (Eds.). 1992. Biology of the Pitvipers. Selva, Tyler, Texas.

Campbell, J. A., and W. W. Lamar. 2004. The Venomous Reptiles of the Western Hemisphere. 2 Vols. Comstock Publishing Associates, Cornell University Press, Ithaca, New York.

Campbell, S. 2008. Eastern Collared Lizard (*Crotaphytus collaris*). Cross Timbers 11(3): 1–4.

Capranica, R. R., L. S. Frishkoff, and E. Nevo. 1973. Encoding of geographic dialects in the auditory system of the Cricket Frog. Science 182(4,118): 1,272–1,275.

Capron, M. 1975a. A trip through the Kansas Flint Hills. Kansas Herpetol. Soc. News. 8: 4–5.

Capron, M. 1975b. Observations on the Alligator Snapping Turtle in Kansas. Kansas Herpetol. Soc. News. 9: 11–13.

Capron, M. 1975c. Abuse of captive reptiles in Kansas. Kansas Herpetol. Soc. News. 10: 3–4.

Capron, M. 1976. A trip to the Ozarks. Kansas Herpetol. Soc. News. 15: 7–8.

Capron, M. 1977. An albino Eastern Yellow-bellied Racer from Kansas. Kansas Herpetol. Soc. News. 22: 7.

Capron, M. 1978a. Notes on the possible occurrence of Broad-banded Copperheads in Kansas. Kansas Herpetol. Soc. News. 26: 4–6.

CAPRON, M. 1978b. Four county collecting raid: a south central Kansas herping saga. Kansas Herpetol. Soc. News. 26: 9–12.

CAPRON, M. 1983. On snakes, snake hunting, and snake bites. Kansas Herpetol. Soc. News. 53: 19–21.

CAPRON, M. 1984. The Cottonmouth in Kansas. Kansas Herpetol. Soc. News. 58: 9–11.

CAPRON, M. 1985a. (Letter to Ed.). Kansas Wildlife 42(6): center section.

CAPRON, M. 1985b. A Western Diamondback Rattlesnake released in Sumner County, Kansas. Kansas Herpetol. Soc. News. 59: 5–6.

CAPRON, M. 1985c. Thunder snakes, blow vipers, and others. Kansas Herpetol. Soc. News. 60: 9–10.

CAPRON, M. 1985d. The quest for Kansas snappers. Kansas Herpetol. Soc. News. 61: 16–18.

CAPRON, M. 1985e. A Copperhead from Sumner County, Kansas. Kansas Herpetol. Soc. News. 62: 11–13.

CAPRON, M. 1986a. A radio telemetry study of an adult Alligator Snapping Turtle in Kansas. Report (14 pp.), Kansas Fish and Game Commision, Pratt.

CAPRON, M. 1986b. In praise of the tiger. Kansas Herpetol. Soc. News. 63: 12–13.

CAPRON, M. 1986c. Winter activity noted in southern Kansas herps. Kansas Herpetol. Soc. News. 64: 15–16.

CAPRON, M. 1986d. If its December it must be time for frogs. Kansas Herpetol. Soc. News. 66: 17.

CAPRON, M. 1987a. A study to determine the current presence and numbers of the Alligator Snapping Turtle at selected localities in southeastern Kansas. Report (14 pp.), Kansas Department of Wildlife and Parks, Pratt.

CAPRON, M. 1987b. Selected observations on south-central Kansas turtles. Kansas Herpetol. Soc. News. 67: 13–15.

CAPRON, M. 1988. Observations on box turtles, genus Terrapene, in captivity. Kansas Herpetol. Soc. News. 72: 17–19.

CAPRON, M. 1989. Threatened and endangered: a critique of the Kansas list. Kansas Herpetol. Soc. News. 76: 14–15.

CAPRON, M. 1991. Unusual foraging behavior in water snakes (Nerodia) around drying pools in south-central Kansas. Kansas Herpetol. Soc. News. 84: 14–15.

CAPRON, M. 1999. A case of predation by Bald Eagles on Spiny Softshells in Kansas. Kansas Herpetol. Soc. News. 117: 15.

CAPRON, M., AND J. PERRY 1976. A July weekend in Great Bend. Kansas Herpetol. Soc. News. 14: 1–2.

CAPRON, M., K. J. IRWIN, AND J. TOLLEFSON. 1982. KHS 1982 field trip for the fall. Kansas Herpetol. Soc. News. 50: 3–5.

CARAS, R. 1974. Venomous Animals of the World. Prentice-Hall, Englewood Cliffs, New Jersey.

CARL, G. 1978. Notes on worm-eating in the Prairie Ringneck Snake, Diadophis punctatus arnyi. Bull. Maryland Herpetol. Soc. 14: 95–97.

CARPENTER, C. C., AND C. C. VAUGHN. 1992. Determination of the distribution and abundance of the Texas Horned Lizard (Phrynosoma cornutum) in Oklahoma. Report (19 pp.), Oklahoma Department of Wildlife Conservation, Norman.

CARPENTER, J. R. 1940. The grassland biome. Ecological Monogr. 10: 617–684.

CARR, A. F. 1938. Notes on the Pseudemys scripta complex. Herpetologica 1: 131–135.

CARR, A. 1952. Handbook of Turtles of the United States, Canada, and Baja California. Comstock Publishing Associates, Cornell University Press, Ithaca, New York.

CARY, D. L., R. L. CLAWSON, AND D. GRIMES. 1981. An observation of snake predation on a bat. Trans. Kansas Acad. Sci. 84: 223–224.

CASE, D. 1984. Things that go cr-r-e-e-k in the night. Kansas Wildlife 41(4): 17–21.

CASLEY, S. W., AND G. SIEVERT. 2006. Graptemys ouachitensis. Geographic distribution. Herpetol. Rev. 37: 491.

CAVITT, J. F.2000a. Fire and a tallgrass prairie reptile community: effects on relative abundance and seasonal activity. J. Herpetol. 34: 12–20.

CAVITT, J. F. 2000b. Tallgrass prairie snake assemblage. Food habits. Herpetol. Rev. 31: 47–48.

CHANTELL, C. J. 1966. Late Cenozoic hylids from the Great Plains. Herpetologica 22: 259–264.

CHOATE, J. R. 1967. Wildlife in the Wakarusa Watershed of Northeastern Kansas. State Biological Survey of Kansas, Lawrence.

CHRAPLIWY, P. S. 1956. Taxonomy and Distribution of the Spadefoot Toads of North America (Salientia: Pelobatidae). Unpublished Master's thesis, University of Kansas, Lawrence.

CHRISTMAN, S. P. 1982. Storeria dekayi. Cat. American Amph. Reptl. 306.1–306.4.

CINK, C. L. 1991. Snake predation on nestling Eastern Phoebes followed by turtle predation on snake. Kansas Ornith. Soc. Bull. 42(3): 29.

CINK, C. L. 1995a. Changes in composition of a snake community with rehydration of a wetland tallgrass prairie in northeast Kansas. Abstract. 127th Annual Meeting of the Kansas Academy of Science.

CINK, C. L. 1995b. Seasonal activity of snakes on a Kansas floodplain tallgrass prairie. Pp. 83–86 In Proceedings of the 14th North American Prairie Conference, Kansas State University, Manhattan.

CLARK, B., E. I. SMITH, AND D. D. SMITH. 1983. Life history notes: Lampropeltis calligaster. Herpetol. Rev. 14: 120.

CLARK, D. R. 1966a. A funnel trap for small snakes. Trans. Kansas Acad. Sci. 69: 91–95.

CLARK, D. R. 1966b. Notes on sexual dimorphism in tail-length in American snakes. Trans. Kansas Acad. Sci. 69: 226–232.

CLARK, D. R. 1967. Experiments into selection of soil type, soil moisture level, and temperature by five species of small snakes. Trans. Kansas Acad. Sci. 70: 490–496.

CLARK, D. R. 1968. A proposal of specific status for the Western Worm Snake, Carphophis amoenus vermis (Kennicott). Herpetologica 24: 104–112.

CLARK, D. R. 1970a. Age-specific "reproductive effort" in the Worm Snake Carphophis vermis (Kennicott). Trans. Kansas Acad. Sci. 73: 20–24.

CLARK, D. R. 1970b. Ecological study of the Worm Snake *Carphophis vermis* (Kennicott). Univ. Kansas Publ. Mus. Nat. Hist. 19: 85–194.

CLARK, D. R. 1971. Branding as a marking technique for amphibians and reptiles. Copeia 1971: 148–151.

CLARK, D. R. 1976. Ecological observations on a Texas population of Six-lined Racerunners, *Cnemidophorus sexlineatus* (Reptilia, Lacertilia, Teiidae). J. Herpetol. 10: 133–138.

CLARK, D. R., JR., C. M. BUNCK, AND R. J. HALL. 1997. Female reproductive dynamics in a Maryland population of Ringneck Snakes (*Diadophis punctatus*). J. Herpetol. 31: 476–483.

CLARK, D. R., AND G. L. CALLISON. 1967. Vertebral and scute anomalies in a Racer, *Coluber constrictor.* Copeia 1967: 862–864.

CLARK, D. R., AND R. J. HALL. 1970. Function of the blue tail-coloration of the Five-lined Skink (*Eumeces fasciatus*). Herpetologica 26: 271–274.

CLARK, D. R., AND C. S. LIEB. 1973. Notes on reproduction in the Night Snake (*Hypsiglena torquata*). Southwest Nat. 18: 248–252.

CLARK, H. 1953. Eggs, egg-laying and incubation of the snake *Elaphe emoryi* (Baird and Girard). Copeia 1953: 90–92.

CLARKE, R. F. 1950. Notes on the Ornate Box Turtle. Herpetologica 6: 54.

CLARKE, R. F. 1953. Additional turtle records for Kansas. Trans. Kansas Acad. Sci. 56: 438–439.

CLARKE, R. F. 1954. Eggs and egg-laying of *Lampropeltis c. calligaster* (Harlan). Herpetologica 10: 15–16.

CLARKE, R. F. 1955. Observations on *Eumeces s. septentrionalis* in Kansas. Herpetologica 11: 161–164.

CLARKE, R. F. 1956a. Turtles in Kansas. Kansas School Nat. 2(4): 1–15.

CLARKE, R. F. 1956b. A case of possible overwintering of *Terrapene o. ornata* in a well. Herpetologica 12: 131.

CLARKE, R. F. 1956c. Distributional notes on some amphibians and reptiles of Kansas. Trans. Kansas Acad. Sci. 59: 213–219.

CLARKE, R. F. 1958. An ecological study of reptiles and amphibians in Osage County, Kansas. Emporia State Res. Stud. 7(1): 1–52.

CLARKE, R. F. 1959. Poisonous snakes of Kansas. Kansas School Nat. 5(3): 1–16.

CLARKE, R. F. 1965a. Lizards in Kansas. Kansas School Nat. 11(4): 1–16.

CLARKE, R. F. 1965b. An ethological study of the iguanid lizard genera *Callisaurus*, *Cophosaurus*, and *Holbrookia.* Emporia State Res. Stud. 13(4): 1–66.

CLARKE, R. F. 1970. Salamanders in Kansas and vicinity. Kansas School Nat. 16(4): 1–16.

CLARKE, R. F. 1977. A note from the president. Kansas Herpetol. Soc. News. 22: 1.

CLARKE, R. F. 1979. KHS September field meeting at Hays, Kansas, and Scott County State Lake. Kansas Herpetol. Soc. News. 33: 2–3.

CLARKE, R. F. 1980a. Herptiles and fishes of the western Arkansas River in Kansas. Report (55 pp.) to U. S. Army Corps of Engineers, Albuquerque, New Mexico.

CLARKE, R. F. 1980b. Snakes in Kansas. Kansas School Nat. 26(3): 1–15.

CLARKE, R. F. 1981. A record of the Alligator Snapping Turtle, *Macroclemys temminckii* (Testudines: Chelydridae), in Kansas. Trans. Kansas Acad. Sci. 84: 59–60.

CLARKE, R. F. 1984. Frogs and toads in Kansas. Kansas School Nat. 30(3): 1–15.

CLARKE, R. F. 1986a. Subspeciation of *Sceloporus undulatus* in the central United States. Abstract. 118th Annual Meeting of the Kansas Academy of Science.

CLARKE, R. F. 1986b. The invaders. Kansas School Naturalist 33(2): 1–16.

CLARKE, R. F., AND R. BOLES. 1980. I didn't know that! Amphibians and reptiles. Kansas School Nat. 27(1): 1–15.

CLARKE, R. F., J. BREUKELMAN, AND T. F., ANDREWS. 1958. An annotated check list of the vertebrates of Lyon County, Kansas. Trans. Kansas Acad. Sci. 61: 165–194.

CLARKE, R. F., AND J. W. CLARKE. 1984. New county records for Kansas fishes and amphibians. Trans. Kansas Acad. Science. 87: 71–72.

CLAY, W. M. 1936. The Taxonomy and Phylogenetic Relationships of the Water Snakes *Natrix erythrogaster* and *N. sipedon.* Unpublished Doctoral dissertation, University of Michigan, Ann Arbor.

CLAY, W. M. 1938. A synopsis of the North American water snakes of the genus *Natrix.* Copeia 1938: 173–182.

CLEVELAND, E. D. 1986. County record for Graham's Crayfish Snake (*Regina grahamii*). Trans. Kansas Acad. Sci. 89: 9.

CLIBURN, J. W. 1960. The Phylogeny and Zoogeography of North American *Natrix.* Unpublished Doctoral dissertation, University of Alabama, Tuscaloosa.

CLOUTMAN, D. G. 1982. The geology, folklore-history, morphometry, and biota of St. Jacob's Well, Clark County, Kansas. Final Report (21 pp.), Kansas Fish and Game Commission, Pratt.

COBB, V. A. 1990. Reproductive notes on the eggs and offspring of *Tantilla gracilis* (Serpentes: Colubridae), with evidence of communal nesting. Southwest. Nat. 35: 222–224.

COBB, V. A. 2004. Diet and prey size of the Flathead Snake, *Tantilla gracilis.* Copeia 2004: 397–402.

COCHRAN, D. M. 1961. Living Amphibians of the World. Doubleday, Garden City, New York.

COCHRAN, D. M., AND C. J. GOIN. 1970. The New Field Book of Reptiles and Amphibians. Putnam's Sons, New York.

COLE, C. J. 1966a. Femoral glands of the lizard, *Crotaphytus collaris.* J. Morphol. 118: 119–135.

COLE, C. J. 1966b. The chromosomes of *Acris crepitans blanchardi* Harper (Anura: Hylidae). Copeia 1966: 578–580.

COLE, C. J., AND L. M. HARDY. 1981. Systematics of North American colubrid snakes related to *Tantilla planiceps* (Blainville). Bull. American Mus. Nat. Hist. 171: 199–284.

COLEMAN, K. 1987. Annual KHS Field Trip held at Atchison State Lake. Kansas Herpetol. Soc. News. 68: 5–6.

COLEMAN, K. 2000. Geographic distribution. *Pseudacris crucifer.* Herpetol. Rev. 31: 50.

COLEMAN, K. 2002. Geographic distribution: *Pseudacris triseriata*. J. Kansas Herpetol. 4: 14.

COLEMAN, K. 2004. Life history. *Hyla chrysoscelis*. New Kansas maximum size. J. Kansas Herpetol. 10: 11.

COLEMAN, K. 2006. Central Kansas herpetofaunal count. J. Kansas Herpetol. 19: 6.

COLEMAN, K., AND J. T. COLLINS. 1998. The Calls of Kansas Frogs and Toads. A cassette. Kansas Department of Wildlife and Parks, Pratt.

COLLINS, H. H. 1959. Complete Field Guide to American Wildlife. Harper and Brothers, New York.

COLLINS, J. T. 1974a. A range extension and addition to the herpetofauna of Kansas. Trans. Kansas Acad. Sci. 76: 88–90.

COLLINS, J. T. 1974b. Amphibians and Reptiles in Kansas. Public Education Series No. 1, The University of Kansas Museum of Natural History, Lawrence.

COLLINS, J. T. 1975. Observations on reproduction in the Southern Coal Skink (*Eumeces anthracinus pluvialis* Cope). Trans Kansas Acad. Sci. 77: 126–127.

COLLINS, J. T. 1976. Kansas snakes: an educational experience. Kansas Fish and Game 33(4): 16–19.

COLLINS, J. T. 1977. Kansas frogs and toads. Kansas Fish and Game 34(3): 12–16.

COLLINS, J. T. 1978a. Rediscovery of the Western Cottonmouth (*Agkistrodon piscivorus leucostoma*) in southeastern Kansas. Trans. Kansas Acad. Sci. 80: 71–74.

COLLINS, J. T. 1978b. The Osage Copperhead. Univ. Kansas Mus. Nat. Hist. Assoc. News. 7(2): 71–74.

COLLINS, J. T. 1979. New records of fishes, amphibians, and reptiles in Kansas for 1978. Tech. Publ. State Biol. Surv. Kansas 8: 56–66.

COLLINS, J. T. 1980a. New records of fishes, amphibians, and reptiles in Kansas for 1979. Tech. Pub. State Biol. Surv. Kansas 9: 1–11.

COLLINS, J. T. 1980b. Petition to the Kansas Fish and Game Commission. Kansas Herpetol. Soc. News. 37: 12–14.

COLLINS, J. T. 1980c. Kansas turtles. Kansas Fish and Game 37(3): 11–14.

COLLINS, J. T. 1981. New records of fishes, amphibians, and reptiles in Kansas for 1980. Tech. Publ. State Biol. Surv. Kansas 10: 7–19.

COLLINS, J. T. 1982a. Amphibians and Reptiles in Kansas. 2nd ed. Public Education Series No. 8, The University of Kansas Museum of Natural History, Lawrence.

COLLINS, J. T. 1982b. Report to the Kansas Fish and Game Commission on the status of three amphibians in southeastern Kansas. Contract 46, Final Report (57 pp.), Kansas Fish and Game, Pratt.

COLLINS, J. T. 1983. New records of fishes, amphibians, and reptiles in Kansas for 1982. Tech. Publ. State Biol. Surv. Kansas 13: 9–21.

COLLINS, J. T. 1984a. New records of fishes, amphibians, and reptiles in Kansas for 1983. Kansas Herpetol. Soc. News. 56: 15–26.

COLLINS, J. T. 1984b. New records of amphibians and reptiles in Kansas for 1984. Kansas Herpetol. Soc. News. 58: 14–20.

COLLINS, J. T. 1984c. The lizards of Oz. Kansas Wildlife 41(2): 12–16.

COLLINS, J. T. 1985. Natural Kansas. University Press of Kansas, Lawrence.

COLLINS, J. T. 1986a. New records of amphibians and reptiles in Kansas for 1985. Kansas Herpetol. Soc. News. 63: 4–11.

COLLINS, J. T. 1986b. New records of amphibians and reptiles in Kansas for 1986. Kansas Herpetol. Soc. News. 66: 9–16.

COLLINS, J. T. 1987a. A numerical summary of herpetological county records and size maxima records for Kansas, 1975 to 1986. Kansas Herpetol. Soc. News. 67: 16–18.

COLLINS, J. T. 1987b. A report to the Kansas Fish and Game Commission summarizing the known voucher specimens for selected species of amphibians and reptiles in Kansas. Final Report (19 pp. + addendum), Kansas Fish and Game Commission, Pratt.

COLLINS, J. T. 1988. New records of amphibians and reptiles in Kansas for 1987. Kansas Herpetol. Soc. News. 71: 13–19.

COLLINS, J. T. 1989a. New records of amphibians and reptiles in Kansas for 1988. Kansas Herpetol. Soc. News. 75: 15–18.

COLLINS, J. T. 1989b. First Kansas herp counts held in 1989. Kansas Herpetol. Soc. News. 77: 11–14.

COLLINS, J. T. 1989c. New records of amphibians and reptiles in Kansas for 1989. Kansas Herpetol. Soc. News. 78: 16–21.

COLLINS, J. T. 1990a. Results of second Kansas herp count held during April–May 1990. Kansas Herpetol. Soc. News. 81: 10–12.

COLLINS, J. T. 1990b. Maximum size records for Kansas amphibians and reptiles. Kansas Herpetol. Soc. News. 81: 13–17.

COLLINS, J. T. 1991a. New records of amphibians and reptiles in Kansas for 1990. Kansas Herpetol. Soc. News. 83: 7–13.

COLLINS, J. T. 1991b. Results of third Kansas herp count held during April–May 1991. Kansas Herpetol. Soc. News. 85: 9–13.

COLLINS, J. T. 1992a. New records of amphibians and reptiles in Kansas for 1991. Kansas Herpetol. Soc. News. 87: 12–17.

COLLINS, J. T. 1992b. Results of fourth Kansas herp count held during April–May 1992. Kansas Herpetol. Soc. News. 89: 10–13.

COLLINS, J. T. 1993a. New records of amphibians and reptiles in Kansas for 1992. Kansas Herpetol. Soc. News. 92: 16–20.

COLLINS, J. T. 1993b. Foreword. Explication of Taylor's Latin name usage. Pp. 9–10 *In* E. H. Taylor, The Lizards of Kansas. Special Publication No. 2, Kansas Herpetological Society, Lawrence.

COLLINS, J. T. 1994a. New records of amphibians and reptiles in Kansas for 1993. Kansas Herpetol. Soc. News. 97: 15–19.

COLLINS, J. T. 1994b. Specific bibliographies for fifteen kinds of amphibians and reptiles that may occur in the Neosho River drainage. Final Report (134 pp.), Kansas Department of Wildlife and Parks, Pratt.

COLLINS, J. T. 1994c. A list of specimens of amphibians and reptiles from Geary and Riley counties in collections outside of Kansas. Report (68 pp.), Kansas Biological Survey, Lawrence.

COLLINS, J. T. 1995a. Geographic distribution: *Ophisaurus attenuatus*. Herpetol. Rev. 26: 109.

COLLINS, J. T. 1995b. New records of amphibians and reptiles in Kansas for 1994. Kansas Herpetol. Soc. News. 100: 24–47.

COLLINS, J. T. 1996a. New records of amphibians and reptiles in Kansas for 1995. Kansas Herpetol. Soc. News. 103: 13–15.

COLLINS, J. T. 1996b. Retail prices of Kansas amphibians and reptiles. Report (4 pp.), Kansas Department of Wildlife and Parks, Pratt.

COLLINS, J. T. 1997a. New records of amphibians and reptiles in Kansas for 1996. Kansas Herpetol. Soc. News. 107: 14–16.

COLLINS, J. T. 1997b. Creature feature: the Ringneck Snake. Fall Issue, Friends Hidden Valley News., Lawrence, Kansas.

COLLINS, J. T. 1997c. A report on the KHS fall field trip to the Marais des Cygnes wildlife refuges. Kansas Herpetol. Soc. News. 110: 2–3.

COLLINS, J. T. 1998a. New records of amphibians and reptiles in Kansas for 1997. Kansas Herpetol. Soc. News. 111: 12–14.

COLLINS, J. T. 1998b. *In Memoriam*. Alan H. Kamb. Kansas Herpetol. Soc. News. 114: 5.

COLLINS, J. T. 1998c. Results of the KHS silver anniversary fall field trip. Kansas Herpetol. Soc. News. 114: 6–7.

COLLINS, J. T. 1999a. New records of amphibians and reptiles in Kansas for 1998. Kansas Herpetol. Soc. News. 116: 14–15.

COLLINS, J. T. 1999b. Fear and loathing on the golf course: coping with snakes. Golf Course Management. July: 96, 98, 100, 102 and 104.

COLLINS, J. T. 2000a. A checklist of the amphibians, turtles, and reptiles of the Flint Hills Tallgrass Prairie Preserve in Butler and Greenwood counties, Kansas. Report (14 pp.), Kansas Chapter of The Nature Conservancy, Topeka.

COLLINS, J. T. 2000b. New records of amphibians and reptiles in Kansas for 1999. Kansas Herpetol. Soc. News. 119: 7–9.

COLLINS, J. T. 2001a. New records of amphibians and reptiles in Kansas for 2000. Kansas Herpetol. Soc. News. 124: 6–8.

COLLINS, J. T. 2001b. Frank Cross (1925–2001) *In Memoriam*. Kansas Herpetol. Soc. News. 125: 2.

COLLINS, J. T. 2002. New records of amphibians and reptiles in Kansas for 2001. J. Kansas Herpetol. 1: 10–11.

COLLINS, J. T. 2003a. Douglas County herp count. J. Kansas Herpetol. 7: 8.

COLLINS, J. T. 2003b. New records of amphibians and reptiles in Kansas for 2002. J. Kansas Herpetol. 5: 13–16.

COLLINS, J. T. 2004a. New records of amphibians, turtles, and reptiles in Kansas for 2003. J. Kansas Herpetol. 9: 8–11.

COLLINS, J. T. 2004b. Marais des Cygnes herp count. J. Kansas Herpetol. 11: 11.

COLLINS, J. T. 2004c. Geographic distribution. *Storeria dekayi*. Kansas. J. Kansas Herpetol. 12: 17.

COLLINS, J. T. 2005a. New records of amphibians, turtles, and reptiles in Kansas for 2004. J. Kansas Herpetol. 13: 13–16.

COLLINS, J. T. 2005b. Marais des Cygnes WMA herp count. J. Kansas Herpetol. 15: 9.

COLLINS, J. T. 2006a. New records of amphibians, turtles, and reptiles in Kansas for 2005. J. Kansas Herpetol. 17: 14–16.

COLLINS, J. T. 2006b. KHS spring field trip herp count. J. Kansas Herpetol. 19: 7–8.

COLLINS, J. T. 2007a. New records of amphibians, turtles, and reptiles in Kansas for 2006. J. Kansas Herpetol. 21: 11–13.

COLLINS, J. T. 2007b. A checklist of the amphibians, turtles, and reptiles of the Flint Hills Tallgrass Prairie Preserve in Butler and Greenwood counties, Kansas. J. Kansas Herpetol. 22: 9–10.

COLLINS, J. T. 2008a. Leap and learn year. Kansas Traveler, Winter issue, Page 1.

COLLINS, J. T. 2008b. New records of amphibians, turtles, and reptiles in Kansas for 2007. J. Kansas Herpetol. 25: 9–11.

COLLINS, J. T. 2008c. *In Memoriam*. Robert F. Clarke (1919–2008). J. Kansas Herpetol. 26: 5.

COLLINS, J. T. 2009a. New records of amphibians, turtles, and reptiles in Kansas for 2008. J. Kansas Herpetol. 29: 8–10.

COLLINS, J. T. (ED.). 2001c. Cheyenne Bottoms Amphibians, Turtles, and Reptiles. Kansas Department of Wildlife and Parks, Pratt.

COLLINS, J. T. (ED.). 2008d. Ringneck Snake revelations. J. Kansas Herpetol. 26: 7–8.

COLLINS, J. T. (ED.). 2008e. Racer re-arrangement revealed. J. Kansas Herpetol. 26: 8–9.

COLLINS, J. T., AND J. P. CALDWELL. 1976. New records of fishes, amphibians, and reptiles (in Kansas for 1975). Tech. Publ. State Biol. Surv. Kansas 1: 78–97.

COLLINS, J. T., AND J. P. CALDWELL. 1977. A bibliography of the amphibians and reptiles of Kansas (1854–1976). Reports State Biol. Surv. Kansas 12: 1–56.

COLLINS, J. T., AND J. P. CALDWELL. 1978. New records of fishes, amphibians, and reptiles in Kansas for 1977. Tech. Publ. State Biol. Surv. Kansas 6: 70–88.

COLLINS, J. T., AND S. L. COLLINS. 1991. Reptiles and Amphibians of the Cimarron National Grasslands, Morton County, Kansas. U. S. Forest Service, Elkhart, Kansas.

COLLINS, J. T., AND S. L. COLLINS. 1993a. Amphibians and Reptiles in Kansas. 3rd ed. University Press of Kansas, Lawrence.

COLLINS, J. T., AND S. L. COLLINS. 1993b. Reptiles and Amphibians of Cheyenne Bottoms. Hearth Publishing, Hillsboro, Kansas.

COLLINS, J. T., AND S. L. COLLINS. 1996. A Guide to Great Snakes of Kansas. Western Resources, Topeka, Kansas.

COLLINS, J. T., AND S. L. COLLINS. 2001. A Guide to Great Snakes of Kansas. 2nd ed. Western Resources, Topeka, Kansas.

COLLINS, J. T., AND S. L. COLLINS. 2006a. Amphibians, Turtles and Reptiles of Cheyenne Bottoms. 2nd ed. Sternberg Museum of Natural History, Fort Hays State University, Hays, Kansas.

COLLINS, J. T., AND S. L. COLLINS. 2006b. A Pocket Guide to Kansas Snakes. 1st ed. Great Plains Nature Center, Wichita, Kansas.

COLLINS, J. T., AND S. L. COLLINS. 2006c. Ninth annual running of the lizards. J. Kansas Herpetol. 20: 8.

COLLINS, J. T., AND S. L. COLLINS. 2006d. Washburn University herpetology class field trip. J. Kansas Herpetol. 20: 9.

COLLINS, J. T., AND S. L. COLLINS. 2009a. A Pocket Guide to Kansas Snakes. 2nd ed. Great Plains Nature Center, Wichita, Kansas.

COLLINS, J. T., AND S. L. COLLINS. 2009b. Geographic distribution: *Regina grahamii*. J. Kansas Herpetol. 30: 11.

COLLINS, J. T., S. L. COLLINS, AND B. GRESS. 1994. Kansas Wetlands: A Wildlife Treasury. University Press of Kansas, Lawrence.

COLLINS, J. T., S. L. COLLINS, B. GRESS, AND G. WIENS. 1991. Kansas Wildlife. University Press of Kansas, Lawrence.

COLLINS, J. T., S. L. COLLINS, J. HORAK, D. MULHERN, W. BUSBY, C. C. FREEMAN, AND G. WALLACE. 1995. An Illustrated Guide to Endangered or Threatened Species in Kansas. University Press of Kansas, Lawrence.

COLLINS, J. T., P. GRAY, H. GUARISCO, K. J. IRWIN, AND L. MILLER. 1981. The Kansas Herpetological Society Presents Endangered and Threatened Amphibians and Reptiles in Kansas. Kansas Herpetological Society, Lawrence.

COLLINS, J. T., AND D. M. HILLIS. 1985. Final report to the Kansas Fish and Game Commission on the Gray Treefrogs of Kansas. Contract 76, Final Report (12 pp.), Kansas Fish and Game, Pratt.

COLLINS, J. T., AND J. L. KNIGHT. 1980. *Crotalus horridus*. Cat. American Amph. Reptl. 253.1–253.2.

COLLINS, J. T., AND E. M. RUNDQUIST. 1993. Results of the fifth Kansas herp count held during April–June 1993. Kansas Herpetol. Soc. News. 94: 7–11.

COLLINS, J. T., AND T. W. TAGGART. 2006. Recent significant name changes for the amphibians of the United States. J. Kansas Herpetol. 18: 7.

COLLINS, J. T., T. W. TAGGART, C. SCHMIDT, AND S. L. COLLINS. 2002. Geographic distribution: *Phrynosoma cornutum*. J. Kansas Herpetol. 2: 10–11.

COLLINS, S. L. 2002. Geographic distribution: *Thamnophis radix*. J. Kansas Herpetol. 4: 14.

COLLINS, S. L. 2004. Geographic distribution. *Pituophis catenifer*. J. Kansas Herpetol. 11: 14.

COLLINS, S. L. 2006a. Barber County herpetofaunal count. J. Kansas Herpetol. 19: 6.

COLLINS, S. L. 2006b. East central Kansas herpetofaunal count. J. Kansas Herpetol. 19: 7.

COLLINS, S. L. 2007c. Ellis County herpetofaunal count. J. Kansas Herpetol. 23: 8.

COLLINS, S. L. 2007d. Geographic distribution. *Trachemys scripta*. J. Kansas Herpetol. 24: 15.

COLLINS, S. L. 2009b. Geographic distribution. *Phrynosoma cornutum*. J. Kansas Herpetol. 29: 7.

COLLINS, S. L., AND J. T. COLLINS. 1999. Geographic distribution: *Pseudacris crucifer*. Herpetol. Rev. 30: 107.

COLLINS, S. L., AND J. T. COLLINS. 2002. Size maxima: *Arizona elegans*. J. Kansas Herpetol. 4: 14.

COLLINS, S. L., AND J. T. COLLINS. 2004. Life history notes. *Storeria dekayi*. Reproduction. J. Kansas Herpetol. 12: 18.

COLLINS, S. L., J. T. COLLINS, AND E. KESSLER. 2002 Geographic distribution: *Lampropeltis calligaster*. J. Kansas Herpetol. 3: 13.

COLLINS, S. L., J. T. COLLINS, AND T. W. TAGGART. 2006. Longton area herp count. J. Kansas Herpetol. 19: 6–7.

COLLINS, S. L., AND T. W. TAGGART. 2002. KHS field trips: past and future. J. Kansas Herpetol. 2: 12–13.

COLLINS, S. L., T. W. TAGGART, AND J. T. COLLINS. 2007. Linn County herpetofaunal count. J. Kansas Herpetol. 23: 9.

COLT, M. D. 1862. Went to Kansas; Being a Thrilling Account of an Ill-fated Expedition. L. Ingalls and Company, Watertown.

CONANT, R. 1958. Peterson Field Guide to Reptiles and Amphibians. 1st ed. Houghton Mifflin Company, Boston, Massachusetts.

CONANT, R. 1960. The Queen Snake, *Natrix septemvittata*, in the Interior Highlands of Arkansas and Missouri, with comments upon similar disjunct distributions. Proc. Acad. Nat. Sci. Philadelphia 112: 25–40.

CONANT, R. 1963. Evidence for the specific status of the water snake *Natrix fasciata*. American Mus. Novitates 2,122: 1–38.

CONANT, R. 1975. Peterson Field Guide to Reptiles and Amphibians of Eastern and Central North America. 2nd ed. Houghton Mifflin Company, Boston, Massachusetts.

CONANT, R. 1978a. Distributional patterns of North American snakes: some examples of the effects of Pleistocene glaciation and subsequent climatic changes. Bull. Maryland Herpetol. Soc. 14: 241–259.

CONANT, R. 1978b. Semiaquatic reptiles and amphibians of the Chihuahuan Desert and their relationships to drainage patterns of the region. Pp. 455–491 *In* R. H. Wauer and D. H. Riskind (Eds.), Transactions of the Symposium on the Biological Resources of the Chihuahuan Desert Region, United States and Mexico. Proceedings and Transactions Series, National Park Service, No. 3, Sul Ross State University, Alpine, Texas.

CONANT, R. 1982. Herpetology in Ohio — Fifty Years Ago. Special Publication of the Toledo Herpetological Society, Ohio.

CONANT, R. 1992. Comments on the survival status of members of the *Agkistrodon complex*. Pp. 29–33 *In* P. D. Strimple and J. L. Strimple (Eds.), Contributions in Herpetology. Greater Cincinnati Herpetological Society, Ohio.

CONANT, R. 1995. Howard K. Gloyd (1902–1978). Reptile & Amphibian Magazine. May–June: 38–44.

CONANT, R., AND J. F. BERRY. 1978. Turtles of the family Kinosternidae in the southwestern United States and adjacent Mexico: identification and distribution. American Mus. Novitates 2,642: 1–18.

CONANT, R., AND J. T. COLLINS. 1991. Peterson Field Guide to Reptiles and Amphibians of Eastern and Central North America. 3rd ed. Houghton Mifflin Company, Boston, Massachusetts.

CONANT, R., AND J. T. COLLINS. 1998. Peterson Field Guide to Reptiles and Amphibians of Eastern and Central North America. 3rd ed., expanded. Houghton Mifflin Company, Boston, Massachusetts.

CONANT, R., AND C. J. GOIN. 1948. A new subspecies of soft-shelled turtle from the central United States, with comments on the application of the name *Amyda*. Univ. Michigan Mus. Zool. Occ. Pap. 510: 1–19.

CONANT, R., R. C. STEBBINS, AND J. T. COLLINS. 1992. Peterson First Guide to Reptiles and Amphibians. Houghton Mifflin Company, Boston, Massachusetts.

COOK, F. A. 1942. Alligator and Lizards of Mississippi. Survey Bulletin of the Mississippi Game and Fish Commission, Jackson.

COOK, F. A. 1954. Snakes of Mississippi. Survey Bulletin of the Mississippi Game and Fish Commission, Jackson.

COOK, F. A. 1957. Salamanders of Mississippi. Survey Bulletin of the Mississippi Game and Fish Commission, Jackson.

COOPER, J. G. 1849. Report upon the reptiles collected on the survey. Pp. 292–306 *In* Reports of Explorations and Surveys, to Ascertain the Most Practicable and Economical Route for a Railroad from the Mississippi River to the Pacific Ocean, Vol. X, Washington, D. C.

COOPER, W. E., JR., AND G. W. FERGUSON. 1972a. Steroids and color change during gravidity in the lizard *Crotaphytus collaris*. Gen. Comp. Endocrinol. 18: 69–72.

COOPER, W. E., JR., AND G. W. FERGUSON. 1972b. Relative effectiveness of progesterone and testosterone as inductors of orange spotting in female Collared Lizards. Herpetologica 28: 64–65.

COOPER, W. E., JR., AND G. W. FERGUSON. 1973. Estrogenic priming of color change induced by progesterone in the Collared Lizard, *Crotaphytus collaris*. Herpetologica 29: 107–110.

COPE, E. D. 1859. Catalogue of the venomous serpents in the Museum of the Academy of Natural Sciences of Philadelphia, with notes on the families, genera, and species. Proc. Acad. Nat. Sci. Philadelphia 11: 332–347.

COPE, E. D. 1860. Supplement to "A catalogue of the venomous serpents in the Museum of the Academy," etc. Proc. Acad. Nat. Sci. Philadelphia 12: 72–74.

COPE, E. D. 1866. On the reptilia and batrachia of the Sonoran province of the neartic region. Proc. Acad. Nat. Sci. Philadelphia 18: 300–314.

COPE, E. D. 1867. A review of the species of the Amblystomidae. Proc. Acad. Nat. Sci. Philadelphia 19: 166–211.

COPE, E. D. 1889. The Batrachia of North America. Bulletin No. 34, U. S. National Museum, Washington, D. C.

COPE, E. D. 1900. The crocodilians, lizards and snakes of North America. Pp. 153–1,270 *In* Report of the U. S. National Museum for the Year Ending June 30, 1898. Washington, D. C.

CORN, P. S. 1980. Comment on the occurrence of *Pseudacris clarki* in Montana. Bull. Chicago Herpetol. Soc. 15: 77–78.

CORTI, C., AND P. LO CASCIO. 2002. The Lizards of Italy and Adjacent Areas. Andreas S. Brahm, Frankfürt am Main, Germany.

COSTELLO, D. F. 1969. The Prairie World. Thomas Crowell Co., New York.

COUES, E., AND H. C. YARROW. 1878. Notes on the herpetology of Dakota and Montana. Bull. U. S. Geol. Geog. Surv. 4: 259–291.

COX, M. 1990. Frog tales from the dark side. Kansas Wildlife and Parks 47(4): 2–7.

COX, T. M. 1982. Milk Snakes and related species in the U. S. J. North. Ohio Assoc. Herpetol. 8: 43–49.

CRAGIN, F. W. 1881. A preliminary catalog of Kansas reptiles and batrachians. Trans. Kansas Acad. Sci. 7: 112–120.

CRAGIN, F. W. 1884. The faunal relations of Kansas. Bull. Washburn College Lab. Nat. Hist. 1: 103–105.

CRAGIN, F. W. 1885a. Recent additions to the list of Kansas reptiles and batrachians, with further notes on species previously reported. Bull. Washburn College Lab. Nat. Hist. 1: 100–103.

CRAGIN, F. W. 1885b. Editorial notes. Bull. Washburn College Lab. Nat. Hist. 1: 111–112.

CRAGIN, F. W. 1885c. Miscellaneous notes. Bull. Washburn College Lab. Nat. Hist. 1: 147–148.

CRAGIN, F. W. 1885d. Second contribution to the herpetology of Kansas. Trans. Kansas Acad. Sci. 9: 136–140.

CRAGIN, F. W. 1885e. On the Washburn Biological Survey of Kansas. Kansas City Rev. Sci. 8: 576–580.

CRAGIN, F. W. 1886a. Miscellaneous notes. Bull. Washburn College Lab. Nat. Hist. 1: 188.

CRAGIN, F. W. 1886b. Miscellaneous notes. Bull. Washburn College Lab. Nat. Hist. 1: 212.

CRAGIN, F. W. 1886c. Note on a new variety of a Sonoran serpent from Kansas. Trans. Kansas Acad. Sci. 10: 85–86.

CRAGIN, F. W. 1894. Herpetological notes from Kansas and Texas. Pp. 37–39 *In* Colorado College Studies, 5th Annual Publication, Colorado Springs.

CRAMPTON, L. 1983. Herpetological collecting in Sumner County, Kansas. Kansas Herpetol. Soc. News. 54: 8–9.

CRANSTON, T. 1994. Common Kingsnakes (*Lampropeltis getula*). The Vivarium 5(6): 30–33, 46–47, 54.

CROSS, F. B. 1971. Environmental Inventory and Assessment of the Grand (Neosho) River Basin, Kansas, Missouri, Oklahoma, Arkansas. State Biological Survey and Institute of Social Environmental Studies, Lawrence.

CROW, H. E. 1913. Some trematodes of Kansas snakes. Univ. Kansas Sci. Bull. 7: 125–136.

DANFORTH, G. W. 1956. What You Should Know about Snakes. Citadel Press, New York.

DANIEL, J. K. 2004. Cherokee County herp count. J. Kansas Herpetol. 11: 10.

DAVENPORT, J. W. 1943. Field Book of the Snakes of Bexar County, Texas and Vicinity. Witte Memorial Museum, San Antonio, Texas.

DAVIS, L. C. 1973. The Herpetofauna of Peccary Cave, Newton County, Arkansas. Unpublished Masters thesis, University of Arkansas, Fayetteville.

DAVIS, N., AND T. W. TAGGART. 2004. Geographic distribution. *Pseudacris streckeri*. J. Kansas Herpetol. 10: 10.

DAVIS, N., T. W. TAGGART, AND C. J. SCHMIDT. 2004. Geographic distribution. *Pseudacris streckeri*. J. Kansas Herpetol. 10: 10.

DAWSON, M. E. 1974. New turtle exhibit at Topeka Zoo. Kansas Herpetol. Soc. News. 3: 4–5.

DEICHSEL, G., AND L. L. MILLER. 2000. Change of specific status for the Green Lacerta, an alien lizard introduced in Topeka. Kansas Herpetol. Soc. News. 119: 10–11.

DEMARCO, V. G., R. W. DRENNER, AND G. W. FERGUSON. 1985. Maximum prey size of an insectivorous lizard, *Sceloporus undulatus garmani*. Copeia 1985: 1,077–1,080.

DERICKSON, W. K. 1976a. Ecological and physiological aspects of reproductive strategies in two lizards. Ecology 57: 445–458.

DERICKSON, W. K. 1976b. Lipid storage and utilization in reptiles. American Zool. 16: 711–723.

DESSAUER, H. C., AND E. NEVO. 1969. Geographic variation of blood and liver proteins in Cricket Frogs. Biochem. Genetics 3: 171–188.

DESSAUER, H. C., AND F. H. POUGH. 1975. Geographic variation of blood proteins and the systematics of kingsnakes (*Lampropeltis getulus*). Comp. Biochem. Physiol. 50B: 9–12.

DEVITT, T. J., B. I. CROTHER, A. MEIER, F. BURBRINK, AND J. BOUNDY. 2007. *Pituophis catenifer*. Maximum length. Herpetol. Rev. 38: 209–210.

DEWIT, C. A. 1982a. Resistance of the Prairie Vole (*Microtus ochrogaster*) and the Woodrat (*Neotoma floridana*) in Kansas, to venom of the Osage Copperhead (*Agkistrodon contortrix phaeogaster*). Toxicon 20: 709–714.

DEWIT, C. A. 1982b. Yield of venom from the Osage Copperhead, *Agkistrodon contortrix phaeogaster*. Toxicon 20: 525–527.

DI CANDIA, M. R., AND E. J. ROUTMAN. 2007. Cytonuclear discordance across a leopard frog contact zone. Mol. Phylogenet. Evol. 45: 564–575.

DICE, L. R. 1923. Notes on the communities of vertebrates of Riley County, Kansas, with especial reference to amphibians, reptiles, and mammals. Ecology 4: 40–53.

DICKERSON, M. C. 1913. The Frog Book. Doubleday, Page and Company, New York.

DIENER, R. A. 1956. New records of snakes in southwestern Kansas. Southwest. Nat. 1: 27–29.

DIENER, R. A. 1957a. A Western Hognose Snake eats a Collared Lizard. Herpetologica 13: 122.

DIENER, R. A. 1957b. An ecological study of the Plain-bellied Water Snake. Herpetologica 13: 203–211.

DILLENBECK, T. 1986. Snake killings. Kansas Herpetol. Soc. News. 64: 15.

DILLENBECK, T. 1988. Winter sightings. Kansas Herpetol. Soc. News. 71: 12.

DITMARS, R. L. 1905a. The batrachians of the vicinity of New York City. American Mus. J. 5(4): 161–206.

DITMARS, R. L. 1905b. The reptiles of the vicinity of New York City. American Mus. J. 5(3): 93–140.

DITMARS, R. L. 1907. Reptiles of North America. Doubleday and Company, New York.

DITMARS, R. L. 1915. The Reptile Book. Doubleday, Page and Co., New York.

DITMARS, R. L. 1930. The poisonous serpents of the New World. Bull. New York Zool. Soc. 33(3): 79–132.

DITMARS, R. L. 1934. A review of the box turtles. Zoologica 17(1): 1–44.

DITMARS, R. L. 1939. A Field Book of North American Snakes. Doubleday, Doran and Company, New York.

DIXON, J. R. 1959. Geographic variation and distribution of the long-tailed group of the Glossy Snake, *Arizona elegans* Kennicott. Southwest. Nat. 4: 20–29.

DIXON, J. R., AND R. R. FLEET. 1976. *Arizona* and *Arizona elegans*. Cat. American Amph. Reptl. 179.1–179.4.

DIXON, J. R., AND J. E. WERLER. 2005. Texas Snakes: A Field Guide. University of Texas Press, Austin.

DLOOGATCH, M. 1978. Eggs and hatchlings of the Worm Snake, *Carphophis vermis* (Kennicott). Bull. Chicago Herpetol. Soc. 13: 99–100.

DO AMARAL, A. 1927. Notes on Nearctic poisonous snakes and treatment of their bites. Bull. Antivenin Inst. America 1: 61–76.

DO AMARAL, A. 1929. Studies of Nearctic ophidia: V. On *Crotalus confluentus* Say, 1823, and its allied forms. Bull. Antivenin Inst. America 2: 86–97.

DOBIE, J. L. 1968. A new turtle species of the genus *Macroclemys* (Chelydridae) from the Florida Pliocene. Tulane Stud. Zool. Bot. 15: 59–63.

DODD, C. K., JR. 2001. North American Box Turtles: A Natural History. University of Oklahoma Press, Norman.

DOLAC, J., AND M. SIMON. 1991. Survey of reptiles and amphibians of Fort Leavenworth, Kansas. Abstract. 123rd Annual Meeting of the Kansas Academy of Science.

DOLMAN, H. 1929. Studies of Kansas Water Snakes. Unpublished Master's thesis, University of Kansas, Lawrence.

DOUGLAS, M. E. M. R. DOUGLAS, G. W. SCHUETT, L. W. PORRAS, AND A. T. HOLYCROSS. 2002. Phylogeography of the Western Rattlesnake (*Crotalus viridis*) complex, with emphasis on the Colorado Plateau. Pp. 11–50 *In* Schuett, G. W., M. Höggren, M. E. Douglas, and H. W. Greene (Eds.), Biology of the Vipers. Eagle Mountain Publishing, Eagle Mountain, Utah.

DOWLING, H. G. 1951. A Taxonomic Study of the American Representatives of the Genus *Elaphe* Fitzinger, with Particular Attention to the Forms Occurring in Mexico and Central America. Unpublished Doctoral dissertation, University of Michigan, Ann Arbor.

DOWLING, H. G. 1952. A taxonomic study of the rat snakes, genus *Elaphe* Fitzinger. IV. A check list of the American forms. Univ. Michigan Mus. Zool. Occas. Pap. 541: 1–12.

DOWLING, H. G. 1956. Geographic relations of Ozarkian amphibians and reptiles. Southwest. Nat. 1: 174–189.

DRANOFF, R. W. 1981. The family Trionychidae, soft-shelled turtles, with special interest in the species indigenous to the United States. HERP (Bull. New York Herpetol. Soc.) 16(2): 2–14.

DUELLMAN, W. E. 1970. The Hylid Frogs of Middle America. 2 Vols. Monograph Number 1, Museum of Natural History, The University of Kansas, Lawrence.

DUELLMAN, W. E. 1977. Liste der Rezenten Amphibien und Reptilien: Hylidae, Centrolenidae, Pseudidae. Das Tierreich 95. Friedlander and Son, Berlin, Germany.

DUELLMAN, W. E., AND B. BERG. 1962. Type specimens of amphibians and reptiles in the Museum of Natral History, University of Kansas. Univ. Kansas Publ. Mus. Nat. Hist. 15: 183–204.

DUELLMAN, W. E., AND L. TRUEB. 1985. Biology of Amphibians. McGraw Hill, New York.

DUELLMAN, W. E., AND R. G. ZWEIFEL. 1962. A synopsis of the lizards of the *sexlineatus* group (Genus *Cnemidophorus*). Bull. American Mus. Nat. Hist. 123: 155–210.

DUGAN, E. 2006. Observations on native and alien *Podarcis*. J. Kansas Herpetol. 19: 10.

DUNDEE, H. A. 1965. *Eurycea multiplicata*. Cat. American Amph. Reptl. 21.1–21.2.

DUNDEE, H. A. 1971. *Cryptobranchus, C. alleganiensis*. Cat. American Amph. Reptl. 101.1–101.4.

DUNDEE, H. A. 1988. *Ambystoma tigrinum* locality records—be wary. Herpetol. Rev. 19: 53.

DUNDEE, H. A., AND W. L. BURGER. 1948. A denning aggregation of the Western Cottonmouth. Nat. Hist. Misc. 21: 1–2.

DUNDEE, H. A., AND M. C. MILLER. 1968. Aggregative behavior and habitat conditioning by the Prairie Ring-neck Snake, *Diadophis punctatus arnyi*. Tulane Stud. Zool. Bot. 15: 41–58.

DUNLAP, D. G., AND K. C. KRUSE. 1976. Frogs of the *Rana pipiens* complex in the northern and central Plains states. Southwest. Nat. 20: 559–571.

DUNN, E. R. 1918. The collection of amphibia of the Museum of Comparative Zoology. Bull. Mus. Comp. Zool. 62: 445–471.

DUNN, E. R. 1926. The Salamanders of the Family Plethodontidae. Smith College, 50th Anniversary Publication, Northhampton, Massachusetts.

DUNN, E. R. 1932. The status of *Tropidoclonion lineatum*. Proc. Biol. Soc. Washington, 45: 195–198.

DUNN, E. R. 1938. Notes on frogs of the genus *Acris*. Proc. Acad. Nat. Sci. Philadelphia 90: 153–154.

DUNN, E. R., AND G. C. WOOD. 1939. Notes on eastern snakes of the genus *Coluber*. Notulae Naturae 5: 1–4.

DUVALL, D., R. HERSKOWITZ, AND J. TRUPIANO-DUVALL. 1980. Responses of Five-lined Skinks (*Eumeces fasciatus*) and Ground Skinks (*Scincella lateralis*) to conspecific and interspecific chemical cues. J. Herpetol. 14: 121–127.

DYCHE, L. L. 1908. The poison-glands of a rattlesnake during the period of hibernation. Trans. Kansas Acad. Sci. 22: 312–313.

DYCHE, L. L. 1914. Enemies of fish. Pp. 145–158 *In* Ponds, Pond Fish, and Pond Fish Culture. State Department Fish and Game Bulletin No. 1, Kansas State Printing Office, Topeka.

DYRKACZ, S. 1981. Recent instances of albinism in North American Amphibians and Reptiles. Herpetological Circular 11, Society for the Study of Amphibians and Reptiles.

EARLE, A. M. 1958. Albinism in the Prairie Ring-necked Snake. Herpetologica 13: 272.

EASTER, J. 1996. Notes on the Ground Snake in California and in captivity. Herpetology 25(3): 1–3 + 4 (part).

EATON, T. H., AND R. M. IMAGAWA. 1948. Early development of *Pseudacris clarkii*. Copeia 1948: 263–266.

EDDS, D. R. 1991. Conservation status of the Common Map Turtle in Kansas. Report (45 pp.), Kansas Department of Wildlife and Parks, Pratt.

EDDS, D. R. 1992a. Population status and incidence of anatomical abnormalities in semi-aquatic turtles of the Walnut and lower Arkansas river basins. Contract 279, Report (58 pp.), Kansas Department of Wildlife and Parks, Pratt.

EDDS, D. R. 1992b. Arkansas River turtle community ecology. Abstract. 124th Annual Meeting of the Kansas Academy of Science.

EDDS, D. R. 1992c. Observations of the 1992 Sharon Springs rattlesnake roundup. Kansas Herpetol. Soc. News. 90: 11.

EDDS, D. R. 1993a. Presentation to the Kansas Wildlife and Parks Commission. Kansas Herpetol. Soc. News. 91: 13–14.

EDDS, D. 1993b. Rattlesnake commercialization update. Kansas Herpetol. Soc. News. 93: 13.

EDDS, D., AND P. SHIPMAN. 1991. Investigation of aquatic turtle communities in Kansas. Abstract. 123rd Annual Meeting of the Kansas Academy of Science.

EDDS, D., W. VOORHEES, J. SCHNELL, AND L. SHIPMAN. 1990. Common Map Turtle rediscovered in Kansas. Kansas Herpetol. Soc. News. 82: 12.

EDDY, S., AND A. C. HODSON. 1961. Taxonomic Keys to the Common Animals of the North Central States. Burgess Publishing Company, Minneapolis, Minnesota.

EDDY, T., AND K. BABCOCK. 1973. A bit of Kansas history. Kansas School Nat. 20(1): 3–15.

EDGREN, R. A. 1952a. Geographic variability in the Common Hognose Snake, *Heterodon platyrhinos* Latreille. Syst. Zool. 1: 184.

EDGREN, R. A. 1952b. A synopsis of the snakes of the genus *Heterodon* with the diagnosis of a new race of *Heterodon nasicus* Baird and Girard. Nat. Hist. Misc. 112: 1–4.

EDGREN, R. A. 1957. Melanism in hog-nosed snakes. Herpetologica 13: 131–135.

EDGREN, R. A. 1959. Hormonal control of red head coloration in the Five-lined Skink, *Eumeces fasciatus* Linnaeus. Herpetologica 15: 155–157.

EDGREN, R. A. 1961. A simplified method for analysis of clines: Geographic variation in the hog-nose snake *Heterodon platyrhinos* Latreille. Copeia 1961: 125–132.

EDGREN, R. A. 2001. *Heterodon* to 1950, an historical assessment. Bull. Chicago Herpetol. Soc. 36: 121–123.

EDGREN, R. A. 2004. *Heterodon kennerlyi* revisited. J. Kansas Herpetol. 11: 16–17.

ELICK, G. E., AND J. A. SEALANDER. 1972. Comparative water loss in relation to habitat selection in small colubrid snakes. American Mid. Nat. 88: 429–439.

ELLIS, M. M., AND J. HENDERSON. 1913. The amphibia and reptilia of Colorado. Part I. Univ. Colorado Stud. 10: 39–129.

ELLIS, M. R. 2001. Reproduction in the Common Garter Snake in Shawnee County, Kansas. Kansas Herpetol. Soc. News. 125: 12.

ELLIS, M. R. 2002. Fall 2002 KHS field trip to Washington County. J. Kansas Herpetol. 2: 4–5.

ELLIS, M. R., AND K. ELLIS. 2004. Wakarusa herp count. J. Kansas Herpetol. 11: 12.

ELLNER, L. R., AND W. H. KARASOV. 1993. Latitudinal variation in the thermal biology of Ornate Box Turtles. Copeia 1993: 447–455.

ERNST, C. H. 1971. *Chrysemys picta.* Cat. American Amph. Reptl. 106.1–106.4.

ERNST, C. H. 1992. Venomous Reptiles of North America. Smithsonian Institution Press, Washington, D. C.

ERNST, C. H. 2002. *Storeria occipitomaculata.* Cat. American Amphib. Reptl. 759.1–759.8.

ERNST, C. H. 2008. Systematics, taxonomy, and geographic distribution of the snapping turtles, family Chelydridae. Pp. 1–13 *In* A. G. Steyermark, M. S. Finkler, and R. J. Brooks (Eds.), Biology of the Snapping Turtle (*Chelydra serpentina*). The Johns Hopkins University Press, Baltimore.

ERNST, C. H., AND R. W. BARBOUR. 1972. Turtles of the United States. University of Kentucky Press, Lexington.

ERNST, C. H., AND E. M. ERNST. 2003. Snakes of the United States and Canada. Smithsonian Institution Press, Washington, D. C.

ERNST, C. H., J. W. GIBBONS, AND M. E. DORCAS. 2002. *Regina.* Cat. American Amphib. Reptl. 756.1–756.4.

ERNST, C. H., AND J. E. LOVICH. 2009. Turtles of the United States and Canada. 2nd ed. The Johns Hopkins University Press, Baltimore, Maryland.

ERNST, C. H., J. E. LOVICH, AND R. W. BARBOUR. 1994. Turtles of the United States and Canada. Smithsonian Institution Press, Washington, D. C.

ERNST, C. H., J. M. ORR, AND T. R. CREQUE. 2003. *Carphophis.* Cat. American Amphib. Reptl. 773.1–2.

ERNST, C. H., J. M. ORR, AND T. R. CREQUE. 2003. *Carphophis vermis.* Cat. American Amphib. Reptl. 775.1–4.

ESHELMAN, R., AND M. HAGER. 1984. Two Irvingtonian (Medial Pleistocene) vertebrate faunas from north-central Kansas. Pp. 384–404 *In* H. H. Genoways and M. R. Dawson (Eds.), Contributions in Quaternary Vertebrate Paleontology: A Volume in Memorial to John E. Guilday. Special Publication Number 8, Carnegie Museum of Natural History, Pittsburgh, Pennsylvania.

ETHERIDGE, R. 1960. Additional notes on the lizards of the Cragin Quarry fauna. Pap. Michigan Acad. Sci. Arts Letters 45: 113–117.

EVANS, L. T. 1959. A motion picture study of maternal behavior of the lizard *Eumeces obsoletus* Baird and Girard. Copeia 1959: 103–110.

EVANS, P. D., AND H. K. GLOYD. 1948. The subspecies of the Massasauga, *Sistrurus catenatus*, in Missouri. Bull. Chicago Acad, Sci 8: 225–232.

EWERT, M. A. 1979. The embryo and its egg: development and natural history. Pp. 333–413 *In* Harless, M., and H. Morlock (Eds.), Turtles: Perspectives and Research. John Wiley & Sons, New York.

FALLS, O. 1933. An Analysis of the Habitat Distribution of the Vertebrate Fauna of a Streambank Associated in Western Kansas. Unpublished Master's thesis, Fort Hays State University, Hays, Kansas.

FARLEY, G. H. 1999. Geographic distribution: *Storeria dekayi.* Herpetol. Rev. 30: 114.

FASSBENDER, R., AND D. J. WATERMOLEN. 2002. Bird predation on the Mudpuppy (*Necturus maculosus maculosus*). Bull. Chicago Herpetol. Soc. 37: 137–138.

FEDER, J. H. 1979. Natural hybridization and genetic divergence between the toads *Bufo boreas* and *Bufo punctatus.* Evolution 33: 1,089–1,097.

FERGUSON, G. W. 1976. Color change and reproductive cycling in female Collared Lizards (*Crotaphytus collaris*). Copeia 1976: 491–494.

FERGUSON, G. W., AND C. H. BOHLEN. 1978. Demographic analysis: a tool for the study of natural selection of behavioral traits. Pp. 227–243 *In* N. Greenberg and P. S. Maclean (Eds.), Behavior and Neurology of Lizards: An Interdisciplinary Colloquium. Publication No. (ADM) 77-491. U. S. Department of Health, Education, and Welfare, Washington, D. C.

FERGUSON, G. W., C. H. BOHLEN, AND H. P. WOOLLEY. 1980. *Sceloporus undulatus:* comparative life history and regulation of a Kansas population. Ecology 61: 313–322.

FERGUSON, G. W., AND T. BROCKMAN. 1980. Geographic differences in growth rate of *Sceloporus* lizards (Sauria: Iguanidae). Copeia 1980: 259–264.

FERGUSON, G. W., K. L. BROWN, AND V. G. DEMARCO. 1982. Selective basis for the evolution of variable egg and hatchling size in some iguanid lizards. Herpetologica 38: 178–188.

FERGUSON, G. W., J. L. HUGHES, AND K. L. BROWN. 1983. Food availability and territorial establishment of juvenile *Sceloporus undulatus.* Pp. 134–148 *In* R. B. Huey, E. R. Pianka, and T. W. Schoener (Eds.), Lizard Ecology. Harvard University Press, Cambridge, Massachusetts.

FERGUSON, G. W., AND H. L. SNELL. 1986. Endogenous control of seasonal change of egg, hatchling, and clutch size of the lizard *Sceloporus undulatus garmani.* Herpetologica 42: 185–191.

FERRIER, W. 1997. Natural history and captive care of the American Bullfrog. Reptile & Amphibian Magazine. November–December: 38–43.

FICHTER, L. S. 1969. Geographical distribution and osteological variation in fossil and recent specimens of two species of *Kinosternon* (Testudines). J. Herpetol. 3: 113–119.

FIRSHEIN, I. L. 1951. The range of *Cryptobranchus bishopi* and remarks on distribution of the genus *Cryptobranchus.* American Mid. Nat. 45: 455–459.

FISH, S. A. 1965. Analysis of Lizard Sera for Use as a Taxonomic Tool. Unpublished Master's thesis, Emporia State University, Emporia, Kansas.

FITCH, A. V. 1966. Sensory cues in the feeding of the Ornate Box Turtle. Trans. Kansas Acad. Sci. 68: 422–532.

FITCH, H. S. 1951. A simplified type of funnel trap for reptiles. Herpetologica 7: 77–80.

FITCH, H. S. 1954. Life history and ecology of the Five-lined Skink, *Eumeces fasciatus*. Univ. Kansas Publ. Mus. Nat. Hist. 8: 1–156.

FITCH, H. S. 1955. Habits and adaptations of the Great Plains Skink (*Eumeces obsoletus*). Ecol. Monogr. 25: 59–83.

FITCH, H. S. 1956a. Early sexual maturity and longevity under natural conditions in the Great Plains Narrow-mouthed Frog. Herpetologica 12: 281–282.

FITCH, H. S. 1956b. A ten-year-old skink? Herpetologica 12: 328.

FITCH, H. S. 1956c. An ecological study of the Collared Lizard (*Crotaphytus collaris*). Univ. Kansas Publ. Mus. Nat. Hist. 8: 213–274.

FITCH, H. S. 1956d. A field study of the Kansas ant-eating frog, *Gastrophryne olivacea*. Univ. Kansas Publ. Mus. Nat. Hist. 8: 275–306.

FITCH, H. S. 1956e. Temperature responses in free-living amphibians and reptiles of northeastern Kansas. Univ. Kansas Publ. Mus. Nat. Hist. 8: 417–476.

FITCH, H. S. 1958a. Natural history of the Six-lined Racerunner (*Cnemidophorus sexlineatus*). Univ. Kansas Publ. Mus. Nat. Hist. 11: 11–62.

FITCH, H. S. 1958b. Home. ranges, territories, and seasonal movements of vertebrates of the Natural History Reservation. Univ. Kansas Publ. Mus. Nat. Hist. 11: 63–326.

FITCH, H. S. 1959. A patternless phase of the Copperhead. Herpetologica 15: 21–24.

FITCH, H. S. 1960a. Criteria for determining sex and bre-eding maturity in snakes. Herpetologica 16: 49–51.

FITCH, H. S. 1960b. Autecology of the Copperhead. Univ. Kansas Publ. Mus. Nat. Hist. 13: 85–288.

FITCH, H. S. 1963a. Natural History of the Racer, *Coluber constrictor*. Univ. Kansas Publ. Mus. Nat. Hist. 15: 351–468.

FITCH, H. S. 1963b. Spiders of the University of Kansas Natural History Reservation and Rockefeller Experimental Tract. Univ. Kansas Mus. Nat. Hist. Misc. Publ. 33: 1–202.

FITCH, H. S. 1963c. Natural history of the Black Rat Snake (*Elaphe o. obsoleta*) in Kansas. Copeia 1963: 649–658.

FITCH, H. S. 1965a. An ecological study of the garter snake, *Thamnophis sirtalis*. Univ. Kansas Publ. Mus. Nat. Hist. 15: 493–564.

FITCH, H. S. 1965b. The University of Kansas Natural History Reservation in 1965. Univ. Kansas Mus. Nat. Hist. Misc. Publ. 42: 1–60.

FITCH, H. S. 1967. Ecological studies of lizards on the University of Kansas Natural History Reservation. Pp. 30–44 *In* W. W. Milstead (Ed.), Lizard Ecology: A Symposium. University of Missouri Press, Columbia.

FITCH, H. S. 1970. Reproductive cycles in lizards and snakes. Univ. Kansas Mus. Nat. Hist. Misc. Publ. 52: 1–247.

FITCH, H. S. 1975. A demographic study of the Ringneck Snake (*Diadophis punctatus*) in Kansas. Univ. Kansas Mus. Nat. Hist. Misc. Publ. 62: 1–53.

FITCH, H. S. 1978. Sexual size differences in the genus *Sceloporus*. Univ. Kansas Sci. Bull. 51: 441–461.

FITCH, H. S. 1979. A field study of the Prairie Kingsnake (*Lampropeltis calligaster*). Trans. Kansas Acad. Sci. 81: 353–363.

FITCH, H. S. 1980a. Reproductive strategies of reptiles. Pp. 25–31 *In* J. B. Murphy and J. T. Collins (Eds.), Reproductive Biology and Diseases of Captive Reptiles. Contributions to Herpetology, Number 1, Society for the Study of Amphibians and Reptiles. Meseraull Printing, Inc., Lawrence, Kansas.

FITCH, H. S. 1980b. *Thamnophis sirtalis*. Cat. American Amph. Reptl. 270. 1–270.4.

FITCH, H. S. 1981. Sexual size differences in reptiles. Univ. Kansas Mus. Nat. Hist. Misc. Publ. 70: 1–72.

FITCH, H. S. 1982. Resources of a snake community in prairie-woodland habitat of northeastern Kansas. Pp. 83–97 *In* N. J. Scott, Jr. (Ed.), Herpetological Communities. Wildlife Research Report Number 13, U. S. Fish and Wildlife Service, Washington, D. C.

FITCH, H. S. 1985a. Observations on rattle size and demography of Prairie Rattlesnakes (*Crotalus viridis*) and Timber Rattlesnakes (*Crotalus horridus*) in Kansas. Univ. Kansas Mus. Nat. Hist. Occas. Pap. 118: 1–11.

FITCH, H. S. 1985b. Variation in clutch and litter size in New World reptiles. Univ. Kansas Mus. Nat. Hist. Misc. Pub. 76: 1–76.

FITCH, H. S. 1987. Collecting and life-history techniques. Pp. 143–164 *In* Seigel, R. A., J. T. Collins, and S. S. Novak (Eds.), Snakes: Ecology and Evolutionary Biology. Blackburn Press, Caldwell, New Jersey.

FITCH, H. S. 1989. A field study of the Slender Glass Lizard, *Ophisaurus attenuatus*, in northeastern Kansas. Univ. Kansas Mus. Nat. Hist. Occas. Pap. 125: 1–50.

FITCH, H. S. 1991. Reptiles and amphibians of the Kansas ecological reserves. Pp. 71–74 *In* E. D. Kettle and D. O. Whittemore (Eds.), Ecology and Hydrology of Kansas Ecological Reserves and the Baker Wetlands. Multidisciplinary Guidebook 4, Kansas Academy of Science, Lawrence.

FITCH, H. S. 1992. Methods of sampling snake populations and their relative success. Herpetol. Rev. 23: 17–19.

FITCH, H. S. 1993a. Distribution and abundance of snakes in Kansas. Abstract No. 62. Annual Meeting of the Southwestern Association of Naturalists, Springfield, Missouri.

FITCH, H. S. 1993b. Relative abundance of snakes in Kansas. Trans. Kansas Acad. Sci. 96: 213–224.

FITCH, H. S. 1995a. The Sharon Springs rattlesnake roundup May 12, 13, 14, 1995. Report (12 pp.), Kansas Department of Wildlife and Parks, Pratt.

FITCH, H. S. 1995b. A Report on a sample of Prairie Rattlesnakes (*Crotalus viridis*) taken at the Wallace County rattlesnake roundup. Abstract. 22nd Annual Meeting of the Kansas Herpetological Society, Lawrence.

FITCH, H. S. 1995c. 47-year changes in a snake community. Abstract. 75th Annual Meeting of the American

Society of Ichthyologists and Herpetologists, Edmonton, Alberta, Canada.

FITCH, H. S. 1999. A Kansas Snake Community: Composition and Changes over 50 Years. Krieger Publishing Company, Malabar, Florida.

FITCH, H. S. 2000a. Female Copperhead aggregations in northeastern Kansas. Abstract. Missouri Herpetol. Assn. News. 13: 5.

FITCH, H. S. 2000b. Population structure and biomass of some common snakes in central North America. Univ. Kansas Nat. Hist. Mus. Sci. Pap. 17: 1–7.

FITCH, H. S. 2001a. Female Copperhead aggregations in northeastern Kansas. Abstract. Kansas Herpetol. Soc. News. 123: 4.

FITCH, H. S. 2001b. Further study of the garter snake, Thamnophis sirtalis, in northeastern Kansas. Univ. Kansas Nat. Hist. Mus. Sci. Pap. 19: 1–6.

FITCH, H. S. 2003. Reproduction in rattlesnakes of the Sharon Springs, Kansas, roundup. J. Kansas Herpetol. 8: 23–24.

FITCH, H. S. 2004a. The effect of female size on number of eggs or young in snakes. J. Kansas Herpetol. 9: 11–12.

FITCH, H. S. 2004b. Food surplus and body size in local populations of snakes. J. Kansas Herpetol. 10: 14–16.

FITCH, H. S. 2004c. Observations on Ringneck Snakes (Diadophis punctatus). J. Kansas Herpetol. 12: 19.

FITCH, H. S. 2004d. Observations on Osage Copperheads in northeastern Kansas. J. Kansas Herpetol. 12: 20.

FITCH, H. S. 2005. Observations on wandering of juvenile snakes in northeastern Kansas. J. Kansas Herpetol. 13: 11–12.

FITCH, H. S. 2006a. Collapse of a fauna: reptiles and turtles of the University of Kansas Natural History Reservation. J. Kansas Herpetol. 17: 10–13.

FITCH, H. S. 2006b. Gopher Snakes, Bullsnakes and Pine Snakes. J. Kansas Herpetol. 17: 16–17.

FITCH, H. S., AND R. O. BARE. 1978. A field study of the Red-tailed Hawk in eastern Kansas. Trans. Kansas Acad. Sci. 81: 1–13.

FITCH, H. S., W. S. BROWN, AND W. S. PARKER. 1981. Coluber mormon, a species distinct from C. constrictor. Trans. Kansas Acad. Sci. 84: 196–203.

FITCH, H. S., AND J. T. COLLINS. 1985. Intergradation of Osage and Broad-banded Copperheads in Kansas. Trans. Kansas Acad. Sci. 88: 135–137.

FITCH, H. S., AND A. F. ECHELLE. 2006. Abundance and biomass of twelve species of snakes native to northeastern Kansas. Herpetol. Rev. 37: 161–165.

FITCH, H. S., AND A. V. FITCH. 1967. Preliminary experiments on physical tolerances of the eggs of lizards and snakes. Ecology 48: 160–165.

FITCH, H. S., AND V. R. FITCH. 1966. Spiders from Meade County, Kansas. Trans. Kansas Acad. Sci. 69(1): 11–22.

FITCH, H. S., AND R. R. FLEET. 1970. Natural history of the Milk Snake (Lampropeltis triangulum) in northeastern Kansas. Herpetologica 26: 387–396.

FITCH, H. S., AND H. W. GREENE 1965. Breeding cycle in the Ground Skink, Lygosoma laterale. Univ. Kansas Publ. Mus. Nat. Hist. 15: 565–575.

FITCH, H. S., AND E. R. HALL 1978. A 20-year record of succession on reseeded fields of tallgrass prairie on the Rockefeller Experimental Tract. Univ. Kansas Mus. Nat. Hist. Spec. Publ. 4: 1–15.

FITCH, H. S., AND T. P. MASLIN. 1961. Occurrence of the garter snake, Thamnophis sirtalis, in the Great Plains and Rocky Mountains. Univ. Kansas Publ. Mus. Nat. Hist. 13: 289–308.

FITCH, H. S., AND R. L. PACKARD. 1955. The Coyote on a natural area in northeastern Kansas. Trans. Kansas Acad. Sci. 58: 211–221.

FITCH, H. S., AND G. R. PISANI. 2002. Longtime recapture of a Timber Rattlesnake (Crotalus horridus) in Kansas. J. Kansas Herpetol. 3: 15–16.

FITCH, H. S., AND G. R. PISANI. 2005. Disappearance of radio-monitored Timber Rattlesnakes. J. Kansas Herpetol. 14: 14–15.

FITCH, H. S., AND G. R. PISANI. 2006. The Timber Rattlesnake in northeastern Kansas. J. Kansas Herpetol. 19: 11–15.

FITCH, H. S., G. R. PISANI, HARRY W. GREENE, ALICE F. ECHELLE, AND MICHAEL ZERWEKH. 2004. A field study of the Timber Rattlesnake in Leavenworth County, Kansas. J. Kansas Herpetol. 11: 18–24.

FITCH, H. S., AND M. V. PLUMMER. 1975. A preliminary ecological study of the soft-shelled turtle Trionyx muticus in the Kansas River. Israel J. Zool. 24: 28–42.

FITCH, H. S., S. SHARP, AND K. SHARP. 2003. Snakes of the University of Kansas biotic succession area. J. Kansas Herpetol. 8: 20–21.

FITCH, H. S., AND H. W. SHIRER. 1971. A radiotelemetric study of spatial relationships in some common snakes. Copeia 1971: 118–128.

FITCH, H. S., AND W. W. TANNER. 1951. Remarks concerning the systematics of the Collared Lizard (Crotaphytus collaris), with a description of a new subspecies. Trans. Kansas Acad. Sci. 54: 548–559.

FITCH, H. S., AND P. L. VON ACHEN. 1977. Spatial relationships and seasonality in the skinks Eumeces fasciatus and Scincella laterale in northeastern Kansas. Herpetologica 33: 303–313.

FITZGERALD, E. C. 1994. Habitat Suitability Index Models for Three Threatened Snake Species in an Urban County. Unpublished Master's thesis, University of Missouri, Columbia.

FITZGERALD, E. C., AND C. NILON. 1994. Classification of habitats for endangered and threatened species in Wyandotte County, Kansas. Unpublished Final Report (98 pp.), Kansas Department of Wildlife and Parks, Pratt.

FLEET, R. R., AND J. R. DIXON. 1971. Geographic variation with the long-tailed group of the Glossy Snake, Arizona elegans Kennicot. Herpetologica 27: 295–302.

FLEET, R. R., AND R. J. HALL. 1969. A skink of record size. Trans. Kansas Acad. Sci. 72: 403.

FLEHARTY, E. D. 1995. Wild Animals and Settlers on the Great Plains. University of Oklahoma Press, Norman.

FLEHARTY, E. D., AND G. K. HULETT. 1988. Fort Hays State University natural areas. Trans. Kansas Acad. Sci. 91: 41–43.

FLEHARTY, E. D., AND D. R. ITTNER. 1967. Additional locality records for some Kansas herptiles. Southwest. Nat. 12: 199–200.

Fleharty, E. D., and J. D. Johnson. 1975. Distributional records of herptiles from the Chautauqua Hills of southeastern Kansas. Trans. Kansas Acad. Sci. 77: 65–67.

Flowers, T. L. 2005. Life history notes: *Rana catesbeiana*. Diet. J. Kansas Herpetol. 14: 10.

Fogell, D. D. 2008a. Marshall County herpetofaunal count. J. Kansas Herpetol. 27: 9.

Fogell, D. D. 2008b. Riley County herpetofaunal count. J. Kansas Herpetol. 27: 9.

Fogell, D. D. 2008c. Wabaunsee County herpetofaunal count. J. Kansas Herpetol. 27: 10.

Fogell, D. D. 2009. Geographic distribution: *Lampropeltis triangulum*. J. Kansas Herpetol. 30: 11.

Fontanella, Frank M., Chris R. Feldman, Mark E. Siddall, and Frank T. Burbrink. 2008. Phylogeography of *Diadophis punctatus*: extensive lineage diversity and repeated patterns of historical demography in a trans-continental snake. Mol. Phylogenet. Evol. 46: 1,049–1,070.

Force, E. R. 1935. A local study of the opisthoglyph snake *Tantilla gracilis* Baird and Girard. Pap. Michigan Acad. Sci. Arts Letters 20: 645–659.

Ford, N. B. 1976. Pheromone Trailing Behavior in Three Species of Garter Snakes (*Thamnophis*). Unpublished Master's thesis, University of Oklahoma, Norman.

Ford, N. B., and C. W. Schofield. 1984. Species specificity of sex pheromone trails in the Plains Garter Snake, *Thamnophis radix*. Herpetologica 40: 51–55.

Forge, C., A. Rogers, and M. P. Simon. 1996. Anuran amphibian colonization of the Benedictine Bottoms. Abstract. 128th Annual Meeting of the Kansas Academy of Science.

Forney, E. A. 1926. The Fauna of an Artificial Pond. Unpublished Master's thesis, University of Kansas, Lawrence.

Foster, S., and R. A. Caras. 1994. Peterson Field Guide to Venomous Animals and Poisonous Plants. Houghton Mifflin Company, Boston, Massachusetts.

Fowler, H. W., and E. R. Dunn. 1917. Notes on salamanders. Proc. Acad. Nat. Sci. Philadelphia, 69: 7–28.

Frank, N., and E. Ramus. 1994. State, Federal, and C.I.T.E.S. Regulations for Herpetologists. NG Publishing, Pottsville, Pennsylvania.

Fraser, J. 1983a. First KHS field trip of 1983 a success. Kansas Herpetol. Soc. News. 52: 1–2.

Fraser, J. 1983b. A trip to the "Trans-Pecos." Kansas Herpetol. Soc. News. 54: 18–23.

Fraser, J. C. 1987. The egg and eye. Kansas Herpetol. Soc. News. 70: 10–11.

Frederickson, L. (Ed.). 1987. (Untitled article on the Crawfish Frog in Douglas County, Kansas). Jayhawk Audubon Society News. (June).

Freeman, C. C., and K. Kindscher. 1992. Endangered and threatened species in the southeast Kansas highway corridor: supporting documentation for Tasks 4 and 5 (in) Environmental Segments 2, 3, and 5. Kansas Biol. Surv. Report 51: 1–23.

Freiburg, R. E. 1951. An ecological study of the Narrow-mouthed Toad (*Microhyla*) in northeastern Kansas. Trans. Kansas Acad. Sci. 54: 374–386.

Fritz, U., and P. Havas. 2007. Checklist of chelonians of the world. Vert. Zool. 57: 149–368.

Frost, D. R. 1983. *Sonora semiannulata*. Cat. American Amph. Reptl. 333.1–333.4.

Fuerst, G. S., and C. C. Austin. 2004. Population genetic structure of the Prairie Skink (*Eumeces septentrionalis*): nested clade analysis of Post Pleistocene populations. J. Herpetol. 38: 257–268.

Fuller, C. C. 1966. Observations of the Behavior of the Lizard *Sceloporus undulatus* in Captivity. Unpublished Special Education thesis, Emporia State University, Emporia, Kansas.

Funk, R. S. 1975. The leopard frogs of Missouri. St. Louis Herpetol. Soc. News. 2(3): 2–6.

Fusilier, L. C. 1993. Habitat partitioning among three sympatric species of map turtles, genus *Graptemys* (Testudines, Emydidae). Abstract. Annual Meeting of the Southwestern Association of Naturalists, Springfield, Missouri.

Fuselier, L., and D. R. Edds. 1992. Niche comparisons of map turtles in Kansas. Abstract. 124th Annual Meeting of the Kansas Academy of Science.

Fuselier, L., and D. Edds. 1994. Habitat partitioning among three species of map turtles, genus *Graptemys*. J. Herpetol. 28: 154–158.

Gablehouse, D. W., Jr., R. L. Hager, and H. E. Klaassen. 1982. Producing Fish and Wildlife from Kansas Ponds. Kansas Fish and Game, Pratt.

Galbreath, E. C. 1948. Pliocene and Pleistocene records of fossil turtles from western Kansas and Oklahoma. Univ. Kansas Publ. Mus. Nat. Hist. 1: 281–284.

Galligan, J. H., and W. A. Dunson. 1979. Biology and status of Timber Rattlesnake (*Crotalus horridus*) populations in Pennsylvania. Biol. Conserv. 15: 13–58.

Gamble, T., P. B. Berendzen, H. B. Shaffer, D. E. Starkey, and A. Simons. 2008. Species limits and phylogeography of North American cricket frogs (*Acris*: Hylidae). Mol. Phylogenet. Evol. 48: 112–125.

Gans, C. 1975. Reptiles of the World. Bantam Books, New York.

Garman, H. 1892. A synopsis of the reptiles and amphibians of Illinois. Illinois Lab. Nat. Hist. Bull. 3: 215–403.

Garman, S. 1883. The reptiles and batrachians of North America. Mem. Mus. Comp. Zool. Harvard Univ. 8(3): *xxxi* + 1–185.

Garman, S. 1884. The North American reptiles and batrachians. Bull. Essex Inst. 16: 1–46.

Garrett, J. M., and D. G. Barker. 1987. A Field Guide to Reptiles and Amphibians of Texas. Texas Monthly Press, Austin.

Garrigues, N. W. 1962. Placement of internal organs in snakes in relation to ventral scalation. Trans. Kansas Acad. Sci. 65: 297–300.

Gatten, R. E. 1974. Effect of nutritional status on the preferred body temperature of the turtles *Pseudemys scripta* and *Terrapene ornata*. Copeia 1974: 912–917.

Gehlbach, F. R. 1956. Annotated records of southwestern amphibians and reptiles. Trans. Kansas Acad. Sci. 59: 364–372.

GEHLBACH, F. R. 1967. *Ambystoma tigrinum*. Cat. American Amph. Reptl. 52.1–52.4.

GEHLBACH, F. R. 1974. Evolutionary relationships of southwestern Ringneck Snakes (*Diadophis punctatus*). Herpetologica 30: 140–148.

GERLANC, N. M. 1999. Effects of Breeding Pool Permanence on Developmental Rate of Western Chorus Frogs, *Pseudacris triseriata*, in Tallgrass Prairie. Unpublished Master's thesis, Kansas State University, Manhattan.

GERLANC, N. M. 2000. The Kansas-Gap Project: implications for Kansas herpetology. Abstract. Missouri Herpetol. Assn. News. 13: 5–6.

GERLANC, N. M. 2001. The Kansas-GAP Project: implications for Kansas Herpetology. Abstract. Kansas Herpetol. Soc. News. 123: 5.

GERLANC, N. M., AND G. A. KAUFMAN. 1997. Bison wallows as ephemeral breeding pools for Western Chorus Frogs. Abstract. Annual Meeting of the Society for the Study of Amphibians and Reptiles, The Herpetologists' League, and the American Association of Ichthyologists and Herpetologists, Seattle, Washington.

GIBBONS, J. W. 1987. Why do turtles live so long? BioScience 37: 262–269.

GIBBONS, J. W. 1990. Life History and Ecology of the Slider Turtle. Smithsonian Institution Press, Washington, D. C.

GIBBONS, J. W., AND M. E. DORCAS. 2004. North American Water Snakes. University of Oklahoma Press, Norman.

GIBBONS, J. W., R. R. HAYNES, AND J. L. THOMAS. 1990. Poisonous Plants and Venomous Animals of Alabama and Adjoining States. University of Alabama Press, Tuscaloosa.

GIER, H. T. 1967. Vertebrates of the Flint Hills. Trans. Kansas Acad. Sci. 70: 51–59.

GILLINGHAM, J. C. 1979. Reproductive behavior of the rat snakes of eastern North America, genus *Elaphe*. Copeia 1979: 319–331.

GISH, C. D. 1962. The Herpetofauna of Ellis County, Kansas. Unpublished Master's thesis, Fort Hays State University, Hays, Kansas.

GISSER, K. 2000. Tracking the elusive Italian Wall Lizard. Colorado Herpetol. Soc. News. 27(10): 1–3.

GLENN, A., AND P. STINSON. 1988. World's oldest Prairie Rattler lives in Hillsboro. Kansas Too! 5(3): 2.

GLENN, J. L., AND R. C. STRAIGHT. 1982. The rattlesnakes and their venom yield and lethal toxicity. Pp. 3–119 *In* A. T. Tu (Ed.), Rattlesnake Venoms: Their Actions and Treatment. Marcel Dekker, New York.

GLOYD, H. K. 1928. The amphibians and reptiles of Franklin County, Kansas. Trans. Kansas Acad. Sci. 31: 115–141.

GLOYD, H. K. 1929. Two additions to the herpetological fauna of Riley County, Kansas. Science 66(1,776): 44.

GLOYD, H. K. 1932. The herpetological fauna of the Pigeon Lake Region, Miami County, Kansas. Pap. Michigan Acad. Sci. Arts Letters 15: 389–409.

GLOYD, H. K. 1934. Studies on the breeding habits and young of the Copperhead, *Agkistrodon mokasen*

Beauvois. Pap. Michigan Acad. Sci. Arts Letters 19: 587–604.

GLOYD, H. K. 1935. Some aberrant color patterns in snakes. Pap. Michigan Acad. Sci. Arts Letters 20: 661–668.

GLOYD, H. K. 1938. Snake poisoning in the United States: a review of present knowledge. Bull. Jackson Park Branch Chicago Med. Soc. 15(7): 3–11 and 15(8): 13–17.

GLOYD, H. K. 1940. The Rattlesnakes, Genera *Sistrurus* and *Crotalus*. A Study in Zoogeography and Evolution. Special Publication Number 4. The Chicago Academy of Sciences, Chicago, Illinois.

GLOYD, H. K. 1947. Notes on the courtship and mating behavior of certain snakes. Nat. Hist. Misc. 12: 1–4.

GLOYD, H. K. 1955. A review of the Massasaugas, *Sistrurus catenatus*, of the south-western United States (Serpentes: Crotalidae). Bull. Chicago Acad. Sci. 10: 83–98.

GLOYD, H. K. 1958. Aberrations in the color patterns of some crotalid snakes. Bull. Chicago Acad. Sci. 10: 185–195.

GLOYD, H. K. 1969. Two additional subspecies of North American crotalid snakes, genus *Agkistrodon*. Proc. Biol. Soc. Washington 82: 219–232.

GLOYD, H. K., AND R. CONANT. 1938. The subspecies of the Copperhead, *Agkistrodon mokasen* Beauvois. Bull. Chicago Acad. Sci. 5: 163–166.

GLOYD, H. K., AND R. CONANT. 1943. A synopsis of the American forms of *Agkistrodon* (Copperheads and Moccasins). Bull. Chicago Acad. Sci. 7: 147–170.

GLOYD, H. K., AND R. CONANT. 1990. Snakes of the *Agkistrodon* complex: A Monographic Review. Contributions to Herpetology, Number 6, Society for the Study of Amphibians and Reptiles.

GODWIN, G. J., AND S. M. ROBLE. 1983. Mating success in male treefrogs, *Hyla chrysoscelis* (Anura: Hylidae). Herpetologica 39: 141–146.

GOIN, C. J., AND M. G. NETTING. 1940. A new gopher frog from the Gulf Coast, with comments upon the *Rana areolata* group. Ann. Carnegie Mus. 28: 137–168.

GOLDBERG, S. R., C. R. BURSEY, AND J. E. PLATZ. 2000. Helminths of the Plains Leopard Frog, *Rana blairi* (Ranidae). Southwestern Nat. 45: 362–366.

GOMEZ, N. 2008a. Geographic distribution: *Gastrophryne olivacea*. J. Kansas Herpetol. 26: 6.

GOMEZ, N. 2008b. Geographic distribution: *Thamnophis proximus*. J. Kansas Herpetol. 26: 6.

GORMAN, W. L. 1985. Patterns of color polymorphism in the Cricket Frog, *Acris crepitans*, in Kansas. Abstract. 65th Annual Meeting of the American Society of Ichthyologists and Herpetologists, Knoxville, Tennessee.

GORMAN, W. L. 1986. Patterns of color polymorphism in the Cricket Frog, *Acris crepitans*, in Kansas. Copeia 1986: 995–999.

GORMAN, W. L., AND M. S. GAINES. 1987. Patterns of genetic variation in the Cricket Frog, *Acris crepitans*, in Kansas. Copeia 1987: 352–360.

GRAHAM, T. E., AND A. A. GRAHAM. 1997. Ecology of the Eastern Spiny Softshell, *Apalone spinifera spinifera*, in the Lamoille River, Vermont. Chelonian Conserv. Biol. 2: 363–369.

GRAVENSTEIN, T., AND A. GRAVENSTEIN. 2003. Geographic distribution: *Eumeces septentrionalis*. J. Kansas Herpetol. 6: 8.

GRANT, C. 1927. The Blue-tail Skink of Kansas. Copeia 164: 67–69.

GRANT, C. 1937. Herpetological notes from central Kansas. American Mid. Nat. 18: 370–372.

GRAY, L. J. 1988. Ottawa University natural area. Trans. Kansas Acad. Sci. 91: 55.

GRAY, L. J., AND M. E. DOUGLAS. 1989. Predation by terrestrial vertebrates on stranded fish and crayfish in a tallgrass prairie stream. Abstract. 121st Annual Meeting of the Kansas Academy of Science.

GRAY, P. 1978. Geographical distribution: *Pseudacris streckeri streckeri*. Herpetol. Rev. 9: 21–22.

GRAY, P. 1979. Low attendance slows KHS. Kansas Herpetol. Soc. News. 32: 1.

GRAY, P. 1980. An albino *Thamnophis radix haydeni* (Western Plains Garter Snake). Herpetol. Rev. 11: 112.

GRAY, P. 1982. Distribution and status of Strecker's Chorus Frog (*Pseudacris streckeri streckeri*) in south-central Kansas. Contract 48, Final Report (23 pp.), Kansas Fish and Game, Pratt.

GRAY, P., AND E. STEGALL. 1979. A field trip to the Red Hills. Kansas Herpetol. Soc. News. 29: 6–8.

GRAY, P., AND E. STEGALL. 1986. Distribution and status of Strecker's Chorus Frog (*Pseudacris streckeri streckeri*) in Kansas. Trans. Kansas Acad. Sci. 89: 81–85.

GREENE, H. W. 1984. Taxonomic status of the Western Racer, *Coluber constrictor mormon*. J. Herpetol. 18: 210–211.

GREENE, H. W., AND G. V. OLIVER. 1965. Notes on the natural history of the Western Massasauga. Herpetologica 21: 225–228.

GREER, A. E. 1974. The generic relationships of the scincid lizard genus *Leiolopisma* and its relatives. Australian J. Zool. Suppl. Ser. 31: 1–67.

GRESS, B. 2003. Faces of the Great Plains. University Press of Kansas, Lawrence.

GRESS, B., AND G. POTTS. 1993. Watching Kansas Wildlife. Kansas Department of Wildlife and Parks, Pratt.

GROBMAN, A. B. 1941. A contribution to the knowledge of variation in *Opheodrys vernalis* (Harlan) with the description of a new subspecies. Univ. Michigan Mus. Zool. Misc. Publ. 50: 1–38.

GROBMAN, A. B. 1984. Scutellation variation in *Opheodrys aestivus*. Bull. Florida St. Mus. 29: 153–170.

GROBMAN, A. B. 1992. Metamerism in the snake *Opheodrys vernalis*, with a description of a new subspecies. J. Herpetol. 26: 175–186.

GROW, D. 1976a. Large garter snake caught at zoo. Kansas Herpetol. Soc. News. 12: 8.

GROW, D. 1976b. The KHS goes to Chetopa. Kansas Herpetol. Soc. News. 13: 2–3.

GROW, D. 1976c. A record size Bullfrog. Kansas Herpetol. Soc. News. 16: 4.

GROW, D. 1977. Clark County visited by the Society. Kansas Herpetol. Soc. News. 19: 1–2.

GUARISCO, H. 1979a. Historical note. Kansas Herpetol. Soc. News. 32: 6–7.

GUARISCO, H. 1979b. Kansas herpetology during 1979. Kansas Herpetol. Soc. News. 34: 9–10.

GUARISCO, H. 1980. Human fatalities caused by venomous animals in Kansas from 1959 to 1978. Kansas Herpetol. Soc. News. 40: 9.

GUARISCO, H. 1981a. The Black Rat Snake (*Elaphe obsoleta obsoleta*). Kansas Herpetol. Soc. News. 41: 6–8.

GUARISCO, H. 1981b. Amphibians and reptiles in Kansas, 2. The Red-spotted Toad (*Bufo punctatus*). Kansas Herpetol. Soc. News. 42: 13–14.

GUARISCO, H. 1981c. Fall field trip of the KHS at Wilson Lake very enjoyable. Kansas Herpetol. Soc. News. 45: 2–3.

GUARISCO, H. 1982a. KHS field trip to Atchison County a great success. Kansas Herpetol. Soc. News. 48: 4.

GUARISCO, H. 1982b. An incidence of albinism in the Eastern Yellowbelly Racer (*Coluber constrictor flaviventris*). Kansas Herpetol. Soc. News. 49: 11.

GUARISCO, H. 1983a. Discovery of the remains of a Boa Constrictor in Johnson County, Kansas. Kansas Herpetol. Soc. News. 53: 21–23.

GUARISCO, H. 1983b. Repair of the plastron of an Ornate Box Turtle using a rapid polymerizing polyester resin. Animal Keeper's Forum 10: 115–116.

GUARISCO, H. 1985. Opportunistic scavenging by the Bullfrog, *Rana catesbeiana* (Amphibia, Anura, Ranidae). Trans. Kansas Acad. Sci. 88: 38–39.

GUARISCO, H. 2001. Ode to an ophidion autumn. Kansas Herpetol. Soc. News. 123: 19.

GUARISCO, H. (ED.). 1980a. Low attendance at July KHS field trip. Kansas Herpetol. Soc. News. 38: 4–5.

GUARISCO, H. (ED.). 1980b. September meeting at Atchison County Lake: September 26–28. Kansas Herpetol. Soc. News. 38: 1.

GUARISCO, H. (ED.). 1980c. Record size Great Plains Toad collected in Sumner County. Kansas Herpetol. Soc. News. 38: 5.

GUARISCO, H. (ED.). 1980d. 1980 KHS annual meeting at the University of Kansas a big success. Kansas Herpetol. Soc. News. 40: 2–4.

GUARISCO, H., P. GRAY, AND J. T. COLLINS. 1982. Focus of 1982 KHS field trips. Kansas Herpetol. Soc. News. 47: 5–6.

GUBANYI, J. E. 1992. An observation on the stomach contents of a Texas Longnose Snake (*Rhinocheilus lecontei tessellatus*). Kansas Herpetol. Soc. News. 89: 17.

GUBANYI, J. E. 1996. Green Lacerta rediscovered in Topeka, Kansas. Kansas Herpetol. Soc. News. 106: 15.

GUBANYI, J. E. 2000a. A breeding colony of Western Green Lacerta (*Lacerta bilineata*) confirmed in southwestern Topeka (Kansas). Trans. Kansas Acad. Sci. 103(3–4): 191–192.

GUBANYI, J. E. 2000b. Update on *Lacerta* in Topeka, Kansas. Kansas Herpetol. Soc. News. 118: 13–14.

GUBANYI, J. E. 2001. Notes on reproduction of the Western Green Lacerta (*Lacerta bilineata*) and the Italian Wall Lizard (*Podarcis sicula*) in Kansas. Kansas Herpetol. Soc. News. 126: 15.

GUBANYI, J. E. 2002a. Notes on the Italian Wall Lizard (*Podarcis sicula*) when maintained in captivity with native Kansas lizards. J. Kansas Herpetol. 3: 14.

BIBLIOGRAPHY

275

GUBANYI, J. E. 2002b. Osage County herp count I. J. Kansas Herpetol. 4: 15.

GUBANYI, J. E. 2002c. Size maxima: *Lacerta bilineata*. J. Kansas Herpetol. 4: 14.

GUBANYI, J. E. 2003a. Osage County herp count. J. Kansas Herpetol. 7: 8.

GUBANYI, J. E. 2003b. Shawnee County herp count. J. Kansas Herpetol. 7: 9.

GUBANYI, J. E. 2003c. Western Ribbon Snake reproduction. J. Kansas Herpetol. 7: 12.

GUBANYI, J. E. 2003d. Additional notes on reproduction in the Italian Wall Lizard (*Podarcis sicula*). J. Kansas Herpetol. 8: 22.

GUBANYI, J. E. 2003e. *Lacerta bilineata:* new maximum size. J. Kansas Herpetol. 8: 22.

GUBANYI, J. E. 2004a. Osage County herp count. J. Kansas Herpetol. 11: 11.

GUBANYI, J. E. 2004b. Wilson County herp count. J. Kansas Herpetol. 11: 12.

GUBANYI, J. E. 2005. Egg-laying and hatching in a captive Slider (*Trachemys scripta*) from Kansas. J. Kansas Herpetol. 13: 12.

GUBANYI, J. E. 2006. Shawnee County herp count. J. Kansas Herpetol. 19: 8.

GUBANYI, J. E. 2007a. Shawnee County herpetofaunal count. J. Kansas Herpetol. 23: 9.

GUBANYI, J. E. 2007b. Wabaunsee County herpetofaunal count. J. Kansas Herpetol. 23: 10.

GUBANYI, J. E. 2008a. Shawnee County herpetofaunal count 1. J. Kansas Herpetol. 27: 9.

GUBANYI, J. E. 2008b. Shawnee County herpetofaunal count 2. J. Kansas Herpetol. 27: 10.

GUBANYI, J. E., AND KEITH COLEMAN. 2002. Size maxima: *Thamnophis proximus*. J. Kansas Herpetol. 4: 14.

GUBANYI, M. A. 2004c. Geographic distribution. *Chelydra serpentina*. J. Kansas Herpetol. 12: 17.

GUIHER, T. J., AND F. T. BURBRINK. 2008. Demographic and phylogeographic histories of two venomous North American snakes of the genus *Agkistrodon*. Mol. Phylogenet. Evol. 48: 543–553.

GUILLETTE, L. J., JR., AND W. P. SULLIVAN. 1983. Life history notes: *Coluber constrictor flaviventris*. Herpetol. Rev. 14: 19.

GUNTHORP, H. 1923. Results of Feeding Thyroid Glands of Various Types of Vertebrates to Tadpoles. Unpublished Doctoral dissertation, University of Kansas, Lawrence.

HAHN, D. E. 1979. *Leptotyphlops dulcis*. Cat. American Amph. Reptl. 231.1–231.2.

HAHN, D. E. 1980. Liste der Rezenten Amphibian and Reptilien. Anomalepididae, Leptotyphlopidae, Typhlopidae. Das Tierreich. Walter de Gruyter, New York.

HALL, E. R. 1953. A western extension of known geographic range for the Timber Rattlesnake in southern Kansas. Trans. Kansas Acad. Sci. 56: 89.

HALL, H. H., AND H. M. SMITH. 1947. Selected records of reptiles and amphibians from southeastern Kansas. Trans. Kansas Acad. Sci. 49: 447–454.

HALL, R. J. 1968. A simplified live-trap for reptiles. Trans. Kansas Acad. Sci. 70: 402–404.

HALL, R. J. 1969. Ecological observations on Graham's Water Snake (*Regina grahami* Baird and Girard). American Mid. Nat. 81: 156–163.

HALL, R. J. 1971. Ecology of a population of the Great Plains Skink (*Eumeces obsoletus*). Univ. Kansas Sci. Bull. 48: 357–388.

HALL, R. J. 1972. Food habits of the Great Plains Skink (*Eumeces obsoletus*). American Mid. Nat. 87: 258–263.

HALL, R. J. 1976. *Eumeces obsoletus*. Cat. American Amph. Reptl. 186.1–186.3.

HALL, R. J., AND H. S. FITCH. 1972. Further observations on the demography of the Great Plains Skink (*Eumeces obsoletus*). Trans. Kansas Acad. Sci. 74: 93–98.

HALLOWELL, E. 1857a. Notice of a collection of reptiles from Kansas and Nebraska presented to the Academy of Natural Sciences, by Doctor Hammond, U. S. A. Proc. Acad. Nat. Sci. Philadelphia 8: 238–253.

HALLOWELL, E. 1857b. Note on the collection of reptiles from the neighborhood of San Antonio, Texas, recently presented to the Academy of Natural Sciences by Dr. A. Heerman. Proc. Acad. Nat. Sci. Philadelphia 8: 306–310.

HALLOWELL, E. 1859. Report upon the reptiles collected on the survey. Pp. 1–27 In Reports of Explorations and Surveys, to Ascertain the Most Practicable and Economical Route for a Railroad from the Mississippi River to the Pacific Ocean. Vol. X, Part IV, No. 1, Washington, D. C.

HALPIN, Z. T. 1983. Naturally occurring encounters between Black-tailed Prairie Dogs (*Cynomys ludovicianus*) and snakes. American Mid. Nat. 109: 50–54.

HALTOM, W. L. 1931. Alabama reptiles. Alabama Mus. Nat. Hist. Pap. 11: 1–145.

HAMILTON, B. 2007. Rarity as an ecological paradigm. J. Kansas Herpetol. 22: 7–8.

HAMMERSON, G. A. 1982. Amphibians and Reptiles in Colorado. Colorado Division of Wildlife, Denver.

HAMMERSON, G. A. 1999. Amphibians and Reptiles in Colorado. 2nd ed. University Press of Colorado & Colorado Division of Wildlife, Niwot.

HAMMERSON, G. A., AND L. J. LIVO. 1999. Conservation status of the Northern Cricket Frog (*Acris crepitans*) in Colorado and adjacent areas at the northwestern extent of the range. Herpetol. Rev. 30: 78–80.

HARDING, K. A., AND K. R. G. WELCH. 1980. Venomous Snakes of the World: A Checklist. Pergamon Press, New York.

HARDY, D. F. 1959. Chorus structure in the Striped Chorus Frog, *Pseudacris nigrita*. Herpetologica 15: 14–16.

HARDY, D. F. 1962. Ecology and behavior of the Six-lined Racerunner, *Cnemidophorus sexlineatus*. Univ. Kansas Sci. Bull. 43: 3–73.

HARDY, L. M., AND C. J. COLE. 1968. Morphological variation in a population of the snake, *Tantilla gracilis* Baird and Girard. Univ. Kansas Publ. Mus. Nat. Hist. 17: 613–629.

HARE, A. 2006. Exotic lizard discovered in Kansas. J. Kansas Herpetol. 19: 9.

HARKAVY, R. 1994. Mudpuppies. Reptile & Amphibian Magazine. May–June: 59–63.

HARKSEN, J. C. 1963. A Bibliography and Catalogue of the Reptiles and Birds of the Kansas Cretaceous with Descriptions of New Species. Unpublished Master's thesis, Fort Hays State University, Hays, Kansas.

HARLESS, M., AND H. MORLOCK. 1979. Turtles: Perspectives and Research. John Wiley & Sons, New York.

HARPER, F. 1939. A southern subspecies of the Spring Peeper (Hyla crucifer). Notulae Naturae 27: 1–4.

HARRIS, H. S., AND R. S. SIMMONS. 1978. A preliminary account of the rattlesnakes with descriptions of four new subspecies. Bull. Maryland Herpetol. Soc. 14: 105–211.

HARRISON, R. 1977. A plan for Kansas wildlife. Kansas Fish and Game 34(6): 5–24.

HARTMAN, F. A. 1906. Food habits of Kansas lizards and batrachians. Trans. Kansas Acad. Sci. 20: 225–229.

HARTMANN, B. 1984. Permit programs (commercial—for amphibians and reptiles). Pp. 8–9 In Project Synopsis. 1983–1984. Fisheries Investigation and Development Section. Fisheries Division, Kansas Fish and Game Commission, Pratt.

HARTMANN, R. F. 1986. For sale: frogs, turtles and snakes. Kansas Wildlife 44(1): 10–13.

HARTWEG, N. 1938. Kinosternon flavescens stejnegeri, a new turtle from northern Mexico. Univ. Michigan Mus. Zool. Occas. Pap. 371: 1–5.

HAWTHORN, K. 1971. Kingsnakes and Milk Snakes: identification and care in captivity. HERP (Bull. New York Herpetol. Soc.) 8(1–2): 13–21.

HAWTHORN, K. 1972. Rat snakes: Genus Elaphe. HERP (Bull. New York Herpetol. Soc.) 9(1–2): 11–16.

HAY, D. W. 1972. Snake hunt. Kansas Alumni 71(3): 16–18.

HAY, O. P. 1892. The batrachians and reptiles of the state of Indiana. Pp. 409–624 In 17th Annual Report of the Indiana Department of Geology and Natural Resources, Indianapolis.

HAYNES, C. M., AND S. D. AIRD. 1981. The distribution and habitat requirements of the Wood Frog (Ranidae: Rana sylvatica Le Conte) in Colorado. Special Report Number 50, Colorado Division of Wildlife, Denver.

HAYWARD, S., AND M. HAYWARD. 1989. Walks and Rambles on the Cimarron National Grassland. Tri-State News, Elkhart, Kansas.

HECHT, M. K. 1958. A synopsis of the mudpuppies of eastern North America. Proc. Staten Island Inst. Arts Letters 21: 1–38.

HECHT, M. K., AND B. L. MATALAS. 1946. A review of middle North American toads of the genus Microhyla. American Mus. Novitates 1,315: 1–21.

HEGER, N. A., AND J. SHERRIN. 1991. Life history notes: Sceloporus undulatus. Rafting. Herpetol. Rev. 22: 59–60.

HEINRICH, M. L. 1985. Life history notes: Pseudacris triseriata triseriata. Herpetol. Rev. 16: 24.

HEINRICH, M. L., AND D. W. KAUFMAN. 1985. Herpetofauna of the Konza Prairie Research Natural Area, Kansas. Prairie Nat. 17: 101–112.

HENDERSON, F. R., AND W. F., ANDELT. 1982. Leaders guide to Endangered and Threatened wildlife in Kansas. Kansas St. Univ. Coop. Ext. Serv. 4-H466: 1–8.

HENDERSON, F. R., AND C. LEE. 1992. Snakes. Urban wildlife damage control. Kansas St. Univ. Coop. Ext. Serv. L-864: 1–5.

HENDERSON, F. R., AND M. SCHWILLING. 1982. Endangered and threatened wildlife of Kansas. A manual for Kansas 4-H'ers. Kansas St. Univ. Coop. Ext. Serv. 4-H465: 1–8.

HENDERSON, R. W. 1970. Feeding behavior, digestion, and water requirements of Diadophis punctatus arnyi Kennicott. Herpetologica 26: 520–526.

HENDERSON, R. W. 1974. Resource partitioning among snakes of the University of Kansas Natural History Reservation: a preliminary analysis. Contrib. Biol. Geol. Milwaukee Pub. Mus. 1: 1–11.

HENSLEY, M. 1959. Albinism in North American amphibians and reptiles. Publ. Mus. Michigan St. Univ. Biol. Series 1: 133–159.

HERIN, K. C. 1987. South Lawrence Trafficway Wetlands Functional Assessment. Kansas Department of Transportation, Topeka.

HIBBARD, C. W. 1934. Notes on some cave bats of Kansas. Trans. Kansas Acad. Sci. 37: 235–238.

HIBBARD, C. W. 1937. Hypsiglena ochrorhynchus in Kansas and additional notes on Leptotyphlops dulcis. Copeia 1937: 74.

HIBBARD, C. W. 1963. The presence of Macroclemys and Chelydra in the Rexroad fauna from the Upper Pliocene of Kansas. Copeia 1963: 708–709.

HIBBARD, C. W. 1964. A brooding colony of the blind snake, Leptotyphlops dulcis dissecta Cope. Copeia 1964: 222.

HIBBARD, C. W., AND A. B. LEONARD. 1936. The occurrence of Bufo punctatus in Kansas. Copeia 1936: 114.

HIGHTON, R. 1962. Revision of North American salamanders of the genus Plethodon. Bull. Florida St. Mus. 6: 235–367.

HILL, J. E. 1931. An addition to the herpetological fauna of Kansas. Science 74(1,926): 547–548.

HILLIS, D. M. 1981. Premating isolating mechanisms among three species of the Rana pipiens complex in Texas and southern Oklahoma. Copeia 1981: 312–319.

HILLIS, D. M. 1985. Evolutionary Genetics and Systematics of New World Frogs of the Genus Rana: An Analysis of Ribosomal DNA, Allozymes, and Morphology. Unpublished Doctoral dissertation, University of Kansas, Lawrence.

HILLIS, D. M., J. T. COLLINS, AND J. P. BOGART. 1987. Distribution of diploid and tetraploid species of Gray Treefrogs (Hyla chrysoscelis and Hyla versicolor) in Kansas. American Mid. Nat. 117: 214–217.

HILLIS, D. M., J. S. FROST, AND D. A. WRIGHT. 1983. Phylogeny and biogeography of the Rana pipiens complex: a biochemical evaluation. Syst. Zool. 32: 132–143.

HILLIS, D. M., AND T. P. WILCOX. 2005. Phylogeny of the New World true frogs (Rana). Mol. Phylogenet. Evol. 34: 299–314.

HISAW, F. L., AND H. K. GLOYD. 1926. The Bull Snake as a natural enemy of injurious rodents. J. Mammal. 7: 200–205.

HLAVACHICK, B. D. 1978. Rare, threatened and endangered. Kansas Fish and Game 35(1): 18–24.

HOBERT, J. P. 1997. The Massasauga (*Sistrurus catenatus*) in Colorado. Unpublished Master's thesis, University of Northern Colorado, Greeley.

HOLMAN, J. A. 1967. The age of the turtle. Int. Turtle Tortoise Soc. J. 1(4): 15–21, 45.

HOLMAN, J. A. 1971. *Ophisaurus attenuatus*. Cat. American Amph. Reptl. 111.1–111.3.

HOLMAN, J. A. 1972. Herpetofauna of the Kanopolis local fauna (Pleistocene: Yarmouth) of Kansas. Michigan Academic. 5: 87–98.

HOLMAN, J. A. 1974. A late Pleistocene herpetofauna of southwestern Missouri. J. Herpetol. 8: 343–346.

HOLMAN, J. A. 1976a. Owl predation on *Ambystoma tigrinum*. Herpetol. Rev., 7: 114.

HOLMAN, J. A. 1976b. Snakes and stratigraphy. Michigan Academic. 8: 387–396.

HOLMAN, J. A. 1976c. Snakes of the Split Rock Formation (Middle Miocene), central Wyoming. Herpetologica 32: 419–426.

HOLMAN, J. A. 1977. America's northernmost Pleistocene herpetofauna (Java, northcentral South Dakota). Copeia 1977: 191–193.

HOLMAN, J. A. 1979. Herpetofauna of the Nash Local fauna (Pleistocene: Aftonian) of Kansas. Copeia 1979: 747–749.

HOLMAN, J. A. 1980. Paleoclimatic implications of Pleistocene herpetofaunas of eastern and central North America. Trans. Nebraska Acad. Sci. 8: 131–140.

HOLMAN, J. A. 1984. Herpetofaunas of the Duck Creek and Williams Local Faunas (Pleistocene: Illinoian) of Kansas. Pp. 20–38 In H. H. Genoways and M. R. Dawson (Eds.).Contributions in Quaternary Vertebrate Paleontology: A Volume in Memorial to John E. Guilday. Special Publication Number 8, Carnegie Museum of Natural History, Pittsburgh, Pennsylvania.

HOLMAN, J. A. 1986. Butler Spring herpetofauna of Kansas (Pleistocene: Illinoian) and its climatic significance. J. Herpetol. 20: 568–569.

HOLMAN, J. A., AND R. L. RICHARDS. 1981. Late Pleistocene occurrence in southern Indiana of the Smooth Green Snake, *Opheodrys vernalis*. J. Herpetol. 15: 123–125.

HOLYCROSS, A. 1995. Serpents of the sandhills. Nebraskaland 73(6): 29–35.

HOLYCROSS, A. T., T. G. ANTON, M. E. DOUGLAS, AND D. R. FROST. 2008. The type localities of *Sistrurus catenatus* and *Crotalus viridis* (Serpentes: Viperidae), with the unraveling of a most unfortunate tangle of names Copeia 2008: 421–424.

HOLYCROSS, A. T., AND S. P. MACKESSY. 2002. Variation in the diet of *Sistrurus catenatus* (Massasauga), with emphasis on *Sistrurus catenatus edwardsii* (Desert Massasauga). J. Herpetol. 36: 454–464.

HOOPER, E. D., JR. 2001. Female brooding in a Northern Prairie Skink from Kansas. Kansas Herpetol. Soc. News. 124: 15.

HOOPER, E. D., JR., AND J. F. WHIPPLE. 1986. Smallmouth Salamander coloration. Kansas Herpetol. Soc. News. 65: 18.

HOOVER, F. S. 1931. The Myology of *Eumeces obsoletus*. Unpublished Master's thesis, University of Kansas, Lawrence.

HORR, D. 1949. Some lower vertebrates. Trans. Kansas Acad. Sci. 52: 143–144.

HOUSEAL, T. W., J. W. BICKHAM, AND M. D. SPRINGER. 1982. Geographic variation in the Yellow Mud Turtle, *Kinosternon flavescens*. Copeia 1982: 567–580.

HOUSEHOLDER, V. H. 1917. The Lizards and Turtles of Kansas with Notes on Their Distribution and Habitat. Unpublished Master's thesis, University of Kansas, Lawrence.

HOWES, B. J., B. LINDSAY, AND S. C. LOUGHEED. 2006. Range-wide phylogeography of a temperate lizard, the Five-lined Skink (*Eumeces fasciatus*). Mol. Phylogenet. Evol. 40: 183–194.

HOYLE, W. L. 1936. Notes on faunal collection in Kansas. Trans. Kansas Acad. Sci. 39: 283–293.

HOYLE, W. L. 1937. Drought in Cowley County, Kansas, 1931–1936. Trans. Kansas Acad. Sci. 40: 211–214.

HOYT, D. L. 1960. Mating behavior and eggs of the Plains Spadefoot. Herpetologica 16: 199–201.

HUBBS, B. 2009. Common Kingsnakes, a Natural History of *Lampropeltis getula*. Tricolor Books, Tempe, Arizona.

HUBBS, B., AND B. O'CONNOR. 2009. A Guide to the Rattlesnakes of the United States. Tricolor Books, Tempe, Arizona.

HUDSON, G. E. 1942. The amphibians and reptiles of Nebraska. Nebraska Conserv. Bull. 24: 1–146.

HUDSON, R., AND G. CARL. 1983. Life history notes: *Coluber constrictor flaviventris*. Herpetol. Rev. 14: 19.

HUGHES, J. L., L. C. FITZPATRICK, AND G. W. FERGUSON. 1980. Energy metabolism in the Northern Prairie Swift, *Sceloporus undulatus garmani*, from Kansas. Abstract. 60th Annual Meeting of the American Society of Ichthyologists and Herpetologists, Fort Worth, Texas.

HUNTER, M. L., JR., J. ALBRIGHT, AND J. ARBUCKLE (EDS.). 1992. The amphibians and reptiles of Maine. Maine Agric. Exp. Sta. Bull. 838: 1–188.

HURTER, J. 1897. A contribution to the herpetology of Missouri. Trans. St. Louis Acad. Sci. 7(19): 499–503.

HURTER, J. 1903. Second contribution to the herpetology of Missouri. Trans. St. Louis Acad. Sci. 13(3): 77–86.

HURTER, J. 1911. Herpetology of Missouri. Trans. St. Louis Acad. Sci. 20(5): 59–274.

HURTER, J., AND J. K. STRECKER. 1909. The amphibians and reptiles of Arkansas. Trans. St. Louis Acad. Sci. 18(2): 11–27.

HUTCHISON, J. H. 2008. History of fossil Chelydridae. Pp. 14–30 In A. G. Steyermark, M. S. Finkler, and R. J. Brooks (Eds.), Biology of the Snapping Turtle (*Chelydra serpentina*). The Johns Hopkins University Press, Baltimore.

HUTCHISON, V. H. 1956. Notes on the plethodontid salamanders, *Eurycea lucifuga* (Rafinesque) and *Eurycea longicauda longicauda* (Green). Natl. Speleol. Soc. Occas. Pap. 3: 1–24.

HUTCHISON, V. H. 1958. The distribution and ecology of the Cave Salamander, *Eurycea lucifuga*. Ecol. Monogr. 28: 1–20.

HUTCHISON, V. H. 1966. *Eurycea lucifuga*. Cat. American Amph. Reptl. 24.1–24.2.

HUTCHISON, V. H., AND J. L. LARIMER. 1960. Reflectivity of the integuments of some lizards from different habitats. Ecology 41: 199–209.

HVASS, H. 1964. Reptiles and Amphibians of the World. Methuen/Two Continents, New York.

INGRAM, W., AND W. W. TANNER. 1971. A taxonomic study of *Crotaphytus collaris* between the Rio Grande and Colorado rivers. Brigham Young Univ. Sci. Bull. 13(2): 1–29.

IRELAND, P. H. 1970. Rediscovery of the Grey-bellied Salamander, *Eurycea multiplicata griseogaster* Moore and Hughes, in southeastern Kansas. Southwest. Nat. 14: 366.

IRELAND, P. H. 1979. *Eurycea longicauda.* Cat. American Amph. Reptl. 221.1–221.4.

IRELAND, P. H., AND R. ALTIG. 1983. Key to the gilled salamander larvae and larviform adults of Arkansas, Kansas, Missouri, and Oklahoma. Southwest. Nat. 28: 271–274.

IRSCHICK, D., AND H. B. SHAFFER. 1997. The polytypic species revisited: morphological differentiation among Tiger Salamanders (*Ambystoma tigrinum*) (Amphibia: Caudata). Herpetologica 53: 30–49.

IRWIN, K. J. 1977. KHS Ottawa County meeting profitable. Kansas Herpetol. Soc. News. 20: 1–2.

IRWIN, K. J. 1979. Two aberrant crotalid snakes from Kansas. Herpetol. Rev. 10: 85.

IRWIN, K. J. 1980. Hitchhike herping in December. Kansas Herpetol. Soc. News. 36: 1–14.

IRWIN, K. J. 1982a. Geographic distribution: *Leptotyphlops dulcis dissectus.* Herpetol. Rev. 13: 82.

IRWIN, K. J. 1982b. Life history notes: *Eumeces anthracinus pluvialis.* Herpetol. Rev. 13: 125–126.

IRWIN, K. J. 1983. Meade County KHS field trip successful. Kansas Herpetol. Soc. News. 53: 4–5.

IRWIN, K. J. 1985a. Distribution, abundance, and habitat preference of the Alligator Snapping Turtle in southeastern Kansas. Contract 50, Final Report (43 pp.), Kansas Fish and Game, Pratt.

IRWIN, K. J. 1985b. Field trip time again. Kansas Herpetol. Soc. News. 59: 1–2.

IRWIN, K. J. 1992a. Geographic distribution: *Macroclemys temminckii.* Herpetol. Rev. 23: 25.

IRWIN, K. J. 1992b. Geographic distribution: *Thamnophis sirtalis annectens.* Herpetol. Rev. 23: 28.

IRWIN, K. J. 1995a. Geographic distribution: *Notophthalmus viridescens.* Herpetol. Rev. 26: 104.

IRWIN, K. J. 1995b. Geographic distribution: *Eumeces laticeps.* Herpetol. Rev. 26: 108.

IRWIN, K. J., AND J. T. COLLINS. 1987. Amphibians and Reptiles of Cheyenne Bottoms. Pp. 401–432 *In* Cheyenne Bottoms: An Environmental Assessment. Kansas Biological Survey and Kansas Geological Survey, Pratt.

IRWIN, K. J., AND J. T. COLLINS. 1994. A survey for threatened and endangered herpetofauna in the lower Marais des Cygnes River valley. Final Report (5 pp.), Kansas Department of Wildlife and Parks, Pratt.

IRWIN, K. J., AND J. T. COLLINS. 2000. Geographic distribution. *Thamnophis radix* (Plains Garter Snake). Herpetol. Rev. 31: 58.

IRWIN, K. J., AND J. T. COLLINS. 2005. A survey for selected species of herpetofauna in the lower Marais des Cygnes river valley, Linn and Miami counties, Kansas. J. Kansas Herpetol. 14: 12–13.

IRWIN, K. J., L. MILLER, AND T. W. TAGGART. 1992. Geographic distribution: *Elaphe obsoleta lindheimerii.* Herpetol. Rev. 23: 27.

IVERSON, J. B. 1976. Notes on Nebraska reptiles. Trans. Kansas Acad. Sci. 78: 51–62.

IVERSON, J. B. 1977. *Kinosternon subrubrum.* Cat. American Amph. Reptl. 193.1–193.4.

IVERSON, J. B. 1978. Further notes on Nebraska reptiles. Trans. Kansas Acad. Sci. 80: 55–59.

IVERSON, J. B. 1979. A taxonomic reappraisal of the Yellow Mud Turtle, *Kinosternon flavescens* (Testudines: Kinosternidae). Copeia 1979: 212–225.

IVERSON, J. B. 1982. Biomass in turtle populations: a neglected subject. Oecologia 55: 69–76.

IVERSON, J. B. 1986. A Checklist with Distribution Maps of Turtles of the World. Privately printed, Richmond, Indiana.

IVERSON, J. B. 1992. A Revised Checklist with Distribution Maps of Turtles of the World. Privately printed, Richmond, Indiana.

IVERSON, J. B. 2001. Reproduction of the River Cooter, *Pseudemys concinna*, in Arkansas and across its range. Southwestern Nat. 46: 364–370.

JACKSON, J. A. 1976. Relative climbing tendencies of Gray (*Elaphe obsoleta spiloides*) and Black Rat Snakes (*E. o. obsoleta*). Herpetologica 32: 359–361.

JADIN, R. C., AND J. L. COLEMAN. 2007. New county records of the Mediterranean Gecko (*Hemidactylus turcicus*) in northeastern Texas, with comments on range expansion. Applied Herpetol. 4: 90–94.

JAGELS, J. L. 2003. Geographic distribution: *Notophthalmus viridescens.* J. Kansas Herpetol. 8: 19.

JAMESON, E. W. 1947. The food of the Western Cricket Frog. Copeia 1947: 212.

JAN, G., AND F. SORDELLI. 1865. Iconographie Generale des Ophidiens. Vol. 1. Livrais 10. Milan, Italy, and Paris, France.

JAN, G., AND F. SORDELLI. 1867. Iconographie Generale des Ophidiens. Vol. 2. Livrais 21. Milan, Italy, and Paris, France.

JANTZEN, P. G. 1960. The ecology of a boggy march in Stafford County, Kansas. Emporia State Res. Stud. 9(2): 1–47.

JANTZEN, P. G. 1993. Prairie Wanderings: The Land and Creatures of the Grasslands. Hearth Publishing, Hillsboro, Kansas.

JEWELL, M. E. 1927. Aquatic biology of the prairie. Ecology 8: 289–298.

JINKS, J. L., AND J. C. JOHNSON. 1970. Trematodes of *Rana catesbeiana* from three strip-mine lakes in southeast Kansas. Trans. Kansas Acad. Sci. 73: 519–520.

JOHNSON, C. F. 1961. Cryptic Speciation in the *Hyla versicolor* Complex. Unpublished Doctoral dissertation, University of Texas, Austin.

JOHNSON, J. D., AND T. LADUC. 1994. Geographic distribution. *Ophisaurus attenuatus.* Herpetol. Rev. 25: 33.

JOHNSON, T. R. 1974. Rare and endangered herpetofauna of Kansas. St. Louis Herpetol. Soc. News. 1(10): 4–5.

JOHNSON, T. R. 1986. In search of the "horny toad." Missouri Conservationist (September issue). Unpaginated.

JOHNSON, T. R. 1987. The Amphibians and Reptiles of Missouri. Missouri Department of Conservation, Jefferson City.

JOHNSON, T. R. 2000. The Amphibians and Reptiles of Missouri. 2nd ed. Missouri Department of Conservation, Jefferson City.

JONES, J. M. 1976. Variations of venom proteins in *Agkistrodon* snakes from North America. Copeia 1976: 558–562.

JONES, J. M., AND P. M. BURCHFIELD. 1971. Relationship of specimen size to venom extracted from the Copperhead, *Agkistrodon contortrix*. Copeia 1971: 162–163.

JONES, R. E., T. SWAIN, L. J. GUILLETTE, JR., AND K. T. FITZGERALD. 1982. The comparative anatomy of lizard ovaries, with emphasis on the number of germinal beds. J. Herpetol. 16: 240–252.

JONES, S. M., AND R. E. BALLINGER. 1987. Comparative life histories of *Holbrookia maculata* and *Sceloporus undulatus* in western Nebraska. Ecology 68: 1,828–1,838.

JONES-BURDICK, W. H. 1963. Guide to the Snakes of Colorado. University of Colorado Museum, Boulder.

JORDAN, D. S. 1929. Manual of the Vertebrate Animals of the Northeastern United States. 13th ed., revised. World Book, New York.

KAMB, A. 1978a. Food consumption in the Red Milk Snake *Lampropeltis triangulum syspila*. Kansas Herpetol. Soc. News. 25: 5–13.

KAMB, A. H. 1978b. Unusual feeding behavior of the Red Milk Snake, *Lampropeltis triangulum syspila* (Lacepede). Trans. Kansas Acad. Sci. 81: 273.

KARNS, D., R. E. ASHTON, AND T. SWEARINGEN. 1974. Illustrated Guide to Amphibians and Reptiles in Kansas: An Identification Manual. Public Education Series, Number 2. The University of Kansas Museum of Natural History and State Biological Survey, Lawrence.

KAUFFELD, C. F. 1942. Rattlesnakes of the United States. Don't tread on me. Fauna (September). 4 pp.

KAUFFELD, C. F. 1969.Snakes. The Keeper and the Kept. Doubleday, Garden City, New York.

KAZMAIER, R. T., AND R. J. ROBEL. 2001. Scute anomalies of Ornate Box Turtles in Kansas. Trans. Kansas Acad. Sci. 104: 178–182.

KEARNEY, M., AND L. M. SIEVERT. 2000. Digestive parameters of Great Plains Skinks (*Eumeces obsoletus*) in Kansas. Abstract. Missouri Herpetol. Assn. News. 13: 8.

KEARNEY, M., AND L. M. SIEVERT. 2001. Digestive parameters of Great Plains Skinks (*Eumeces obsoletus*) in Kansas. Abstract. Kansas Herpetol. Soc. News. 123: 6–7.

KELLOGG, R. 1932. Mexican tailless amphibians in the United States National Museum. Bull. U. S. Natl. Mus. 160: 1–224.

KENNICOTT, R. 1859. Notes on *Coluber calligaster* of Say, and a description of new species of serpents in the collection of the North Western University of Evanston, Illinois. Proc. Acad. Nat. Sci. Philadelphia 1859: 98–100.

KERN, A., L. RICE, AND M. WARNER. 1978. The turtles of Sumner County, Kansas. Kansas Herpetol. Soc. News. 27: 10–11.

KESSLER, E. M. 1998. Retreat-Site Selection by *Diadophis punctatus* and *Carphophis vermis* in Eastern Kansas. Unpublished Master's thesis, Emporia State University, Emporia, Kansas.

KING, F. W., AND R. L. BURKE. 1989. Crocodilian, Tuatara, and Turtle Species of the World. A Taxonomic and Geographic Reference. The Association of Systematics Collections, Washington, D. C.

KINGMAN, R. H. 1932. A comparative study of the skull in the genus *Eumeces* of the family Scincidae. Univ. Kansas Sci. Bull. 20: 273–295.

KIRK, J. D. 2000. Reintroduction of the Pickerel Frog in Cherokee County, Kansas. Abstract. Missouri Herpetol. Assn. News. 13: 8.

KIRK, J. D. 2001a. Reintroduction of the Pickerel Frog (*Rana palustris*) to Cherokee County, Kansas. Unpublished Master's thesis, Friends University, Wichita, Kansas.

KIRK, J. D. 2001b. Reintroduction of the Pickerel Frog in Cherokee County, Kansas. Abstract. Kansas Herpetol. Soc. News. 123: 7.

KIRN, A. J., W. L. BURGER, AND H. M. SMITH. 1949. The subspecies of *Tantilla gracilis*. American Mid. Nat. 42: 238–251.

KIRKPATRICK, D. 1993. Map turtles of the United States. Reptile & Amphibian Magazine November–December: 6–17.

KLAUBER, L. M. 1930. New and renamed subspecies of *Crotalus confluentus* Say, with remarks on related species. Trans. San Diego Soc. Nat. Hist. 6: 95–144.

KLAUBER, L. M. 1935. A new subspecies of *Crotalus confluentus*, the Prairie Rattlesnake. Trans. San Diego Soc. Nat. Hist. 8: 75–90.

KLAUBER, L. M. 1940. The worm snakes of the genus *Leptotyphlops* in the United States and northern Mexico. Trans. San Diego Soc. Nat. Hist. 9: 87–162.

KLAUBER, L. M. 1941. The Longnosed Snakes of the genus *Rhinocheilus*. Trans. San Diego Soc. Nat. Hist. 9: 289–332.

KLAUBER, L. M. 1946. The Glossy Snake, *Arizona*, with descriptions of new subspecies. Trans. San Diego Soc. Nat. Hist. 10: 311–398.

KLAUBER, L. M. 1947. Classification and ranges of the Gopher Snakes of the genus *Pituophis* in the western United States. Bull. Zool. Soc. San Diego 22: 1–83.

KLAUBER, L. M. 1972. Rattlesnakes. Their Habits, Life Histories, and Influence on Mankind. 2 Vols. 2nd ed. University of California Press, Berkeley and Los Angeles.

KLEMENS, M. W. 1993. Amphibians and reptiles of Connecticut and adjacent regions. St. Geol. Nat. Hist. Surv. Connecticut Bull. 112: 1–318.

KLUGE, A. G. 1966. A new pelobatine frog from the lower Miocene of South Dakota with a discussion of the evolution of the *Scaphiopus-Spea* complex. Los Angeles Co. Mus. Contrib. Sci. 113: 1–26.

KNIGHT, A., L. D. DENSMORE III, AND E. D. RAEL. 1992. Molecular systematics of the *Agkistrodon* complex.

Pp. 49–69 *In* J. A. Campbell and E. D. Brodie, Jr. (Eds.), Biology of the Pitvipers. Selva, Tyler, Texas.

KNIGHT, H. 1935. A Key to the Amphibians and Reptiles of Kansas. Unpublished Master's thesis, Emporia State University, Emporia, Kansas.

KNIGHT, J. L. 1992. *In memoriam.* George Toland. Kansas Herpetol. Soc. News. 88: 12.

KNIGHT, J. L., AND J. T. COLLINS. 1977. The amphibians and reptiles of Cheyenne County, Kansas. Report Number 15 (18 pp) of the State Biological Survey of Kansas, Lawrence.

KNIGHT, J. L., E. D. FLEHARTY, AND J. D. JOHNSON. 1973. Noteworthy records of distribution and habits of some Kansas herptiles. Trans. Kansas Acad. Sci. 75: 273–275.

KORB, R. M. 2001. Wisconsin Frogs. Northeastern Wisconsin Audubon, Green Bay.

KRAUS, F. 2009. Alien Reptiles and Amphibians: A Scientific Compendium and Analysis. Springer-Verlag, Heidelberg, Germany.

KRAUS, F., AND R. A. NUSSBAUM. 1989. The status of the Mexican Salamander, *Ambystoma schmidti* Taylor. J. Herpetol. 23: 78–79.

KRAUS, F., AND J. W. PETRANKA. 1989. A new sibling species of *Ambystoma* from the Ohio River drainage. Copeia 1989: 94–110.

KRETZER, J. E., AND J. F. CULLY, JR. 2001. Effects of Black-tailed Prairie Dogs on reptiles and amphibians in Kansas shortgrass prairie. Southwest. Nat. 46: 171–177.

KRUPA, J. J. 1990. *Bufo cognatus.* Cat. American Amphib. Reptl. 457.1–457.8.

KRYSKO, K. L., AND W. S. JUDD. 2006. Morphological systematics of kingsnakes, *Lampropeltis getula* complex (Serpentes: Colubridae), in the eastern United States. Zootaxa 1,193: 1–39.

KUMBERG, M. 1996. The skinny on salamanders. Kansas Wildlife and Parks 53(5): 20–23.

LAFORCE, R. W. 2004. Life history. *Nerodia sipedon.* Diet. J. Kansas Herpetol. 10: 11.

LAMB, T., C. LYDEARD, R. B. WALKER, AND J. W. GIBBONS. 1994. Molecular systematics of map turtles (*Graptemys*): a comparison of mitochondrial restriction site versus sequence data. Syst. Biol. 43: 543–559.

LAMSON, G. H. 1935. The reptiles of Connecticut. Connecticut Geol. Nat. Hist. Surv. Bull. 54: 1–35.

LANE, H. H. 1945. A survey of the fossil vertebrates of Kansas, Part II. Amphibia. Trans. Kansas Acad. Sci. 48: 286–316.

LANE, H. H. 1946. A survey of the fossil vertebrates of Kansas, Part III. The reptiles. Trans. Kansas Acad. Sci. 49: 289–332.

LANGLEY, W. M., H. W. LIPPS, AND J. F. THEIS. 1989. Responses of Kansas motorists to snake models on a rural highway. Trans. Kansas Acad. Sci. 92: 43–48.

LARDIE, R. L. 1975a. Three new herpetological records for Kansas. Kansas Herpetol. Soc. News. 5: 6.

LARDIE, R. L. 1975b. Geographic distribution: *Thamnophis sirtalis annectens.* Herpetol. Rev. 6: 116.

LARDIE, R. L. 1976. Distributional notes on the Checkered Garter Snake and other *Thamnophis* in north-central Oklahoma. Bull. Oklahoma Herpetol. Soc. 4: 56–57.

LARDIE, R. L. 1979. Eggs and young of the plains Yellow Mud Turtle. Bull. Oklahoma Herpetol. Soc. 4(2–3): 24–32.

LARDIE, R. L. 1981. Some new and important county records for Oklahoma. Bull. Oklahoma Herpetol. Soc. 6(2): 44–45.

LARDIE, R. L. 1990. Kansas threatened species and protection of the Gypsum Hills habitat. Kansas Herpetol. Soc. News. 80: 14–15.

LARDIE, R. L. 1999. The subspecific status and western distribution of the Eastern Rat Snake, *Elaphe obsoleta*, in Oklahoma. Kansas Herpetol. Soc. News. 115: 16–17.

LARDIE, R. L. 2001. The subspecific status of the Common Garter Snake, *Thamnophis sirtalis*, in western Oklahoma. Kansas Herpetol. Soc. News. 123: 16–18.

LARDIE, R. L., AND J. H. BLACK. 1981. The amphibians and reptiles of the Cimarron Gypsum Hills region in northwestern Oklahoma. Bull. Oklahoma Herpetol. Soc. 5(4): 76–125.

LAYHER, W. G. 2002. Recovery plan for four salamander species of Cherokee County, Kansas. Report (18 pp.), Kansas Department of Wildlife and Parks, Pratt.

LAYHER, W. G., K. L. BRUNSON, J. SCHAEFER, M. D. SCHWILLING, AND R. D. WOOD. 1986. Summary of Nongame Task Force Actions Relative to Developing Three Species Lists: Species in Need of Conservation, Threatened, and Endangered. Kansas Fish and Game Commission, Pratt.

LEACHÉ, A. D., AND T. W. REEDER. 2002. Molecular systematics of the Eastern Fence Lizard (*Sceloporus undulatus*): a comparison of parsimony, likelihood, and bayesian approaches. Syst. Biol. 51: 44–68.

LEE, W. F. 1967. Preferential Body Temperature of the Collared Lizard, *Crotaphytus collaris collaris.* Unpublished Master's thesis, Emporia State University, Emporia, Kansas.

LEGLER, J. M. 1955. Observations on the sexual behavior of captive turtles. Lloydia 18: 95–99.

LEGLER, J. M. 1956. A social relation between snapping and painted turtles. Trans. Kansas Acad. Sci. 59: 461–462.

LEGLER, J. M. 1960a. Natural history of the Ornate Box Turtle, *Terrapene ornata ornata* Agassiz. Univ. Kansas Publ. Mus. Nat. Hist. 11: 527–669.

LEGLER, J. M. 1960b. Distributional records of amphibians and reptiles in Kansas. Trans. Kansas Acad. Sci. 63: 40–43.

LEGLER, J. M., AND H. S. FITCH. 1957. Observations on hibernation and nests of the Collared Lizard, *Crotaphytus collaris.* Copeia 1957: 305–307.

LEMMON, E. M., A. R. LEMMON, J. T. COLLINS, J. A. LEE-YAW, AND D. C. CANNATELLA. 2007. Phylogeny-based delimitation of species boundaries and contact zones in the trilling chorus frogs (*Pseudacris*). Mol. Phylogenet. Evol. 44: 1,068–1,082.

LEMMON, E. M., A. R. LEMMON, J. T. COLLINS, AND D. C. CANNATELLA. 2008. A new North American chorus frog species (*Pseudacris*: Hylidae: Amphibia) from the south-central United States. Zootaxa 1,675: 1–30.

LEMOS-ESPINAL, J. A., G. R. SMITH, AND R. E. BALLINGER. 1996. Covariation of egg size, clutch size, and offspring survivorship in the genus *Sceloporus*. Bull. Maryland Herpetol. Soc. 32: 58–66.

LEVELL, J. P. 1995. A Field Guide to Reptiles and the Law. Serpent's Tale, Excelsior, Minnesota.

LEVELL, J. P. 1997. A Field Guide to Reptiles and the Law. 2nd Revised ed. Serpent's Tale, Lanesboro, Minnesota.

LEVITON, A. E. 1971. Reptiles and Amphibians of North America. Doubleday, New York.

Li, A. 1978a. Comparative reproduction studies of two colubrid snakes (*Thamnophis sirtalis parietalis, Lampropeltis triangulum syspila*). Advanced Biology Report (15 pp.), Lawrence (Kansas) High School.

Li, A. 1978b. Comparative reproduction studies of two colubrid snakes. Kansas Herpetol. Soc. News. 24: 5–8.

LILLYWHITE, H. B. 1985a. Behavioral control of arterial pressure in snakes. Physiol. Zool. 58: 159–165.

LILLYWHITE, H. B. 1985b. Postural edema and blood pooling in snakes. Physiol. Zool. 58: 759–766.

LILLYWHITE, H. B. 1985c. Trailing movements and sexual behavior in *Coluber constrictor*. J. Herpetol. 19: 306–308.

LILLYWHITE, H. B. 2008. Dictionary of Herpetology. Krieger Publishing Company, Malabar, Florida.

LILLYWHITE, H. B., AND L. H. SMITH. 1981. Haemodynamic responses to haemorrhage in the snake, *Elaphe obsoleta obsoleta*. J. Exp. Biol. 94: 275–283.

LINDEMAN, P. V. 1997. A comparative spotting-scope study of the distribution and relative abundance of River Cooters (*Pseudemys concinna*) in western Kentucky and southern Mississippi. Chelonian Conserv. Biol. 2: 378–383.

LINSDALE, J. M. 1925. Land Vertebrates of a Limited Area in Eastern Kansas. Unpublished Master's thesis, University of Kansas, Lawrence.

LINSDALE, J. M. 1927. Amphibians and reptiles of Doniphan County, Kansas. Copeia 164: 75–81.

LINZEY, D. W., AND M. J. CLIFFORD. 1981. Snakes of Virginia. University Press of Virginia, Charlottesville.

LIPPS, H. W., AND J. F. THEIS. 1987. Wildlife killed on the highway: Are some people deliberately hitting them? Abstract. 119th Annual Meeting of the Kansas Academy of Science.

LIST, J. C. 1966. Comparative osteology of the snake families Typhlopidae and Leptotyphlopidae. Illinois Biol. Monogr. 36: 1–112.

LIVEZEY, R. L., AND A. H. WRIGHT. 1947. A synoptic key to the salientian eggs of the United States. American Mid. Nat. 37: 178–222.

LOEWEN, S. L. 1941. A polydactylous lizard. Copeia 1941: 48–49.

LOKKE, J. 1981. Featured herp: *Agkistrodon contortrix phaeogaster*. Nebraska Herpetol. Soc. News. 3(1): 4–5.

LOKKE, J. 1982. Featured herp: *Elaphe obsoleta obsoleta*. Nebraska Herpetol. Soc. News. 3(2): 4.

LOKKE, J. 1983a. Featured herp: *Crotalus horridus horridus*. Nebraska Herpetol. Soc. News. 3(5): 4–5.

LOKKE, J. 1983b. Featured herp: The western rat snakes. Nebraska Herpetol. Soc. News. 4(3): 5–7.

LOKKE, J. 1984. The Speckled King. Nebraska Herpetol. News 4(6): 13–16.

LOKKE, J. 1989. Massasauga Rattlesnake *Sistrurus catenatus*: a vanishing Nebraskan. Privately Printed (6 pp.), Omaha, Nebraska.

LOKKE, J. 1992. Some thoughts on the status of the Western Fox Snake (*Elaphe vulpina*) in southeast Nebraska, southwest Iowa, northwest Missouri, and northeast Kansas. Kansas Herpetol. Soc. News. 89: 14–16.

LOKKE, J. 2003. Geographic distribution: *Carphophis vermis*. J. Kansas Herpetol. 5: 10.

LOKKE, J., AND J. LOKKE. 2003. Cowley County herp count 2. J. Kansas Herpetol. 7: 8.

LOOMIS, R. B. 1945. *Microhyla olivacea* (Hallowell) in Nebraska. Herpetologica 2: 211–212.

LOOMIS, R. B. 1956. The chigger mites of Kansas (*Acarina*, Trombiculidae). Univ. Kansas Sci. Bull. 37: 1,195–1,443.

LONG, D. R. 1992. Blind snakes. Reptile & Amphibian Magazine. November–December: 14–19.

LONG, D. R. 1993. Yellow Mud Turtles. Reptile & Amphibian Magazine. March–April: 22–26.

LORAINE, R. 1980. Snakes vs. cars. Iowa Herpetol. Soc. News. 3(2): 7.

LORAINE, R. K. 1983. Report to the Kansas Fish and Game Commission on the status of two amphibians in southeastern Kansas. Contract 76, Final Report (56 pp.), Kansas Fish and Game, Pratt.

LORAINE, R. K. 1984. Life history notes: *Hyla crucifer crucifer*. Herpetol. Rev. 15: 16–17.

LOVICH, J. E. 1993. *Macroclemys, M. temminckii*. Cat. American Amph. Reptl. 562.1–562.4.

LOW, B. 2008. Geographic distribution: *Gastrophyne olivacea*. J. Kansas Herpetol. 26: 6.

LOW, J. 2005. Synthetic netting nabs serpents. J. Kansas Herpetol. 13: 9.

LOWE, C. H. 1966. The Prairie Lined Race Runner. J. Arizona Acad. Sci. 4: 44–45.

LUETH, F. X. 1941. Manual of Illinois Snakes. Illinois Department of Conservation, Springfield.

LUND, D. 1974. Amphibians and reptiles of the Nebraska mid-state division. Section V. *In* Identification and Evaluation of Present Zoologic Resources Nebraska Mid-state Division Pick-Sloan Missouri Basin Program and Associated Areas. Final Report (54 pp.), United States Department of the Interior, Bureau of Reclamation, Grand Island, Nebraska.

LYNCH, J. D. 1964. The toad *Bufo americanus charlesmithi* in Indiana with remarks on the range of the subspecies. J. Ohio Herpetol. Soc. 4: 103–104.

LYNCH, J. D. 1978. The distribution of Leopard Frogs (*Rana blairi* and *Rana pipiens*) (Amphibia, Anura, Ranidae) in Nebraska. J. Herpetol. 12: 157–162.

LYNCH, J. D. 1985. Annotated checklist of the amphibians and reptiles of Nebraska. Trans. Nebraska Acad. Sci. 13: 33–57.

MAHER, B. W., AND G. SIEVERT. 2006. *Agkistrodon contortrix*. Geographic distribution. Herpetol. Rev. 37: 496.

MAHER, M. J. 1967. Response to thyroxine as a function of environmental temperature in the toad, *Bufo woodhousei*, and the frog, *Rana pipiens*. Copeia 1967: 361–365.

MALARET, L. 1979. Notas sobre la ecologia de la Lagartija de Collar (*Crotaphytus collaris*) en Nuevo León, México. Unpublished manuscript (9 pp.).

MALARET, L. 1980. Reproductive and fatbody cycles in a Kansas population of the Collared Lizard, *Crotaphytus collaris* (Say). Abstract. Trans. Kansas Acad. Sci. 83: 142–143.

MALARET, L. 1985. Geographic and Temporal Variation in the Life History of *Crotaphytus collaris* (Sauria, Iguanidae) in Kansas and Mexico. Unpublished Doctoral dissertation, University of Kansas, Lawrence.

MALDONADO-KOERDELL, M. 1950. Notas anfibiologicas, IV. Urodelos de la colección de anfibios y reptiles de la Universidad de Ottawa, Kansas, E. E. U. U. Rev. Soc. México Hist. Nat. 11: 217–222.

MALDONADO-KOERDELL, M., AND I. L. FIRSCHEIN. 1947. Notes on the ranges of some North American salamanders. Copeia 1947: 140.

MANES, R. 1986. 100-year snapper. Kansas Wildlife 43(4): 23.

MANN, A. M. 2007. A Taxonomic Investigation of the Black Rat Snake, *Elaphe o. obsoleta* (Say) [Reptilia, Squamata, Colubridae], in West Virginia Using Morphometric Analyses. Unpublished Master's thesis, Marshall University, Huntington, West Virginia.

MANSFIELD, R. R. 1932. A Comparative Study of the Helminthes of the Anura from Five Given Localities. Unpublished Master's thesis, University of Kansas, Lawrence.

MANSUETI, R. 1941. A descriptive catalogue of the amphibians and reptiles found in and around Baltimore City, Maryland, within a radius of twenty miles. Proc. Nat. Hist. Soc. Maryland 7: 1–53.

MARA, W. P. 1994. Garter and Ribbon Snakes. TFH Publications, Inc., Neptune City, New Jersey.

MARA, W. P. 1995. Water Snakes of North America. TFH Publications, Inc., Neptune City, New Jersey.

MARKEL, R. 1972. Checklist and distribution of the genus *Lampropeltis*. Herpetology (Southwest. Herpetol. Soc.), 6(4): 5–7.

MARKEL, R. 1979. The kingsnakes: an annotated checklist. Bull. Chicago Herpetol. Soc. 14: 101–116.

MARKEL, R. G. 1990. Kingsnakes and Milk Snakes. TFH Publications, Inc., Neptune City, New Jersey.

MARKEL, R. G., AND R. D. BARTLETT. 1995. Kingsnakes and Milk Snakes. Barron's, Hauppauge, New York.

MARR, J. C. 1944. Notes on amphibians and reptiles from the central United States. American Mid. Nat. 32: 478–490.

MARSHALL, S. 2007. Reno County herpetofaunal count. J. Kansas Herpetol. 23: 9.

MARSHALL, S. 2007. Russell County herpetofaunal count. J. Kansas Herpetol. 23: 9.

MARTIN, E. P. 1956. A population study of the Prairie Vole (*Microtus ochrogaster*) in northeastern Kansas. Univ. Kansas Publ. Mus. Nat. Hist. 8: 361–416.

MARTIN, E. P. 1960. Distribution of native mammals among the communities of the mixed prairie. Fort Hays Studies (new series) Sci. Series 1: *iv* + 1–26.

MARTIN, R. (ED.). 1994. Life in Kansas. Pp. 16–17 *In* Explore (Winter). University of Kansas, Lawrence.

MARTIN, W. H. 1992. Phenology of the Timber Rattlesnake (*Crotalus horridus*) in an unglaciated section of the Appalachian Mountains. Pp. 259–277 *In* J. A. Campbell and E. D. Brodie, Jr. (Eds.), Biology of the Pitvipers. Selva, Tyler, Texas.

MARTIN, W. H. 2002. Life history constraints on the Timber Rattlesnake (*Crotalus horridus*) at its climatic limits. Pp. 285–306 *In* G. W. Schuett, M. Höggren, M. E. Douglas, and H. W. Greene (Eds.), Biology of the Vipers. Eagle Mountain Publishing, LC, Eagle Mountain, Utah.

MARTOF, B. S. 1970. *Rana sylvatica*. Cat. American Amph. Reptl. 86.1–86.4.

MARTOF, B. S., AND R. L. HUMPHRIES. 1959. Geographic variation in the Wood Frog, *Rana sylvatica*. American Mid. Nat. 61: 350–389.

MARTOF, B. S., W. M. PALMER, J. R. BAILEY, AND J. R. HARRISON III. 1980. Amphibians and Reptiles of the Carolinas and Virginia. University of North Carolina Press, Chapel Hill.

MASLIN, T. P. 1953. The status of the whipsnake *Masticophis flagellum* (Shaw) in Colorado. Herpetologica 9: 193–200.

MASLIN, T. P. 1959. An annotated checklist of the amphibians and reptiles of Colorado. Univ. Colorado Stud. 6: 1–98.

MASLOWSKI, K. H. 1964. Did you ever see a blue Bullfrog? National Wildlife 2(2): 8–9.

MASTA, S. E., B. K. SULLIVAN, T. LAMB, AND E. J. ROUTMAN. 2002. Molecular systematics, hybridization, and phylogeography of the *Bufo americanus* complex in eastern North America. Mol. Phylogenet. Evol. 24: 302–314.

MATLACK, R. S., AND R. L. REHMEIER. 2002. Status of the Western Diamondback Rattlesnake (*Crotalus atrox*) in Kansas. Southwest. Nat. 47: 312–314.

MATTHEWS, D. C. 1935. A Survey of the Parasites of Turtles. Unpublished Master's thesis, University of Kansas, Lawrence.

MAURER, E. F. 1995. *Chrysemys picta belli*. Feeding behavior. Herpetol. Rev. 26: 34.

MAYERS, Z. 2008. Geographic distribution: *Plestiodon obsoletus*. J. Kansas Herpetol. 26: 6.

McALLISTER, C. T. 1983. Aquatic behavior of Collared Lizards, *Crotaphytus c. collaris*, from Arkansas. Herpetol. Rev. 14: 11.

McALLISTER, C. T. 1985. Food habits and feeding behavior of *Crotaphytus collaris collaris* (Iguanidae) from Arkansas and Missouri. Southwest. Nat. 30: 597–600.

McCAULEY, R. H. 1945. The Reptiles of Maryland and the District of Columbia. Privately printed. Hagerstown, Maryland.

McCONKEY, E. H. 1954. A systematic study of the North American lizards of the genus *Ophisaurus*. American Mid. Nat. 51: 133–171.

McCoy, C. J. 1961. Distribution of the subspecies of *Sceloporus undulatus* (Reptilia: Iguanidae) in Oklahoma. Southwest. Nat. 6: 79–85.

McCoy, C. J. 1975a. Notes of Oklahoma reptiles. Proc. Oklahoma Acad. Sci. 55: 53–54.

McCoy, C. J. 1975b. Cave-associated snakes, *Elaphe guttata*, in Oklahoma. Bull. Natl. Speleol. Soc. 37: 41.

McCoy, C. J. 1975c. Timber Rattlesnake. Coop. Ext. Serv. Pennsylvania Forest Resources 21: 1–4.

McCoy, C. J., and R. C. Vogt. 1994. *Graptemys*. Cat. American Amphib. Reptl. 584.1–584.3.

McDowell, A. 1951. Bull Snake active in December. Herpetologica 7: 142.

McGuire, J. A. 1996. Phylogenetic systematics of crotaphytid lizards (Reptilia: Iguania: Crotaphytidae). Bull. Carnegie Mus. Nat. Hist. 32: *iv* + 1–143.

McKay, J. L. 2006. A Field Guide to the Amphibians and Reptiles of Bali. Krieger Publishing Company, Malabar, Florida.

McKenna, M. G., and R. W. Seabloom. 1979. Endangered, threatened, and peripheral wildlife of North Dakota. Inst. Ecol. Stud., Univ. North Dakota Res. Report 28: 1–62.

McLeod, D. 1994. Observations of growth after injury in the Slider Turtle, *Trachemys scripta elegans*. Herpetol. Rev. 25: 116–117.

McLeran, V. 1971. Of frogs and froggin'. Kansas Fish and Game 28(3): 3–4.

McLeran, V. 1972a. Copperhead. Kansas Fish and Game 29(2): 1–5.

McLeran, V. 1972b. Kansas rattlesnakes. Kansas Fish and Game 29(3): 1–4.

McLeran, V. 1972c. The Moccasin myth. Kansas Fish and Game 29(4): 1–4.

McLeran, V. 1973. Friendly constrictors. Kansas Fish and Game 30(2): 8–11.

McLeran, V. 1974a. The camouflaged Copperhead. Sports Afield (August): 56, 110–112.

McLeran, V. 1974b. The southwest. Kansas Fish and Game 31(5): 15–19.

McLeran, V. 1975a. Blue and gold impressions. Kansas Fish and Game 32(1): 15–19.

McLeran, V. (Ed.). 1975b. Kansas Fish and Game News (Record size softshell turtle). *In* Kansas Fish and Game 32(5): 7.

McNearney, R. 2004. Life history. *Diadophis punctatus*. Albino. J. Kansas Herpetol. 10: 11.

Meachen, W. R. 1962. Factors affecting secondary intergradation between two allopatric populations in the *Bufo woodhousei* complex. American Mid. Nat. 67: 282–304.

Mecham, J. S. 1954. Geographic variation in the Green Frog, *Rana clamitans* Latreille. Texas J. Sci. 6: 1–24.

Mecham, J. S. 1958. Some Pleistocene amphibians and reptiles from Friesenhahn Cave, Texas. Southwest. Nat. 3: 17–27.

Mecham, J. S. 1967. *Notophthalmus viridescens*. Cat. American Amph. Reptl. 53.1–53.4.

Mecham, J. S. 1980. *Eumeces multivirgatus*. Cat. American Amph. Reptl. 241.1–241.2.

Mecham, J. S., M. J. Littlejohn, R. S. Oldham, L. E. Brown, and J. R. Brown. 1973. A new species of leopard frog (*Rana pipiens* complex) from the plains of the central United States. Texas Tech Univ. Mus. Occas. Pap. 18: 1–11.

Medden, R. V. 1929. Tales of the rattlesnake: from the works of early travelers in America. Bull. Antivenin Inst. America 3(3): 82–87, 3(4): 17–23, 4(2): 43–50, and 4(3): 71–75.

Medica, P. A. 1975. *Rhinocheilus*. Cat. American Amph. Reptl. 175.1–175.4.

Medica, P. A. 1980. Locality records of *Rhinocheilus lecontei* in the United States and Mexico. Herpetol. Rev. 11: 42.

Mehrtens, J. M. 1952. Notes on the eggs and young of *Pituophis melanoleucus sayi*. Herpetologica 8: 69–70.

Merkle, D. A., and S. I. Guttman. 1977. Geographic variation in the Cave Salamander *Eurycea lucifuga*. Herpetologica 33: 313–321.

Merkle, D. A., S. I. Guttman, and M. A. Nickerson. 1977. Genetic uniformity throughout the range of the Hellbender, *Cryptobranchus alleganiensis*. Copeia 1977: 549–553.

Merrell, D. J. 1965. The distribution of the dominant *burnsi* gene in the Leopard Frog, *Rana pipiens*. Evolution 19: 69–85.

Merriman, T. 1975. Large litter sizes in intergrade Copperheads. St. Louis Herpetol. Soc. News. 2(12): 3–5.

Meshaka, W. E., Jr. 2008. Seasonal activity, reproduction, and growth of the Ringneck Snake (*Diadophis punctatus*) in Pennsylvania. J. Kansas Herpetol. 28: 17–20.

Meshaka, W. E., Jr., S. D. Marshall, L. R. Raymond, and L. M. Hardy. 2009. Seasonal activity, reproductive cycles, and growth of the Bronze Frog (*Lithobates clamitans clamitans*) in northern Louisiana: the long and short of it. J. Kansas Herpetol. 29: 12–20.

Metcalf, A. L., and E. Metcalf. 1978. An experiment with homing in Ornate Box Turtles (*Terrapene ornata ornata* Agassiz). J. Herpetol. 12: 411–412.

Metcalf, A. L., and E. L. Metcalf. 1985. Longevity in some Ornate Box Turtles (*Terrapene ornata ornata*). J. Herpetol. 19: 157–158.

Metcalf, E., and A. L. Metcalf. 1970. Observation on Ornate Box Turtles (*Terrapene ornata ornata* Agassiz). Trans. Kansas Acad. Sci. 73: 96–117.

Metcalf, E., and A. L. Metcalf. 1979. Mortality in hibernating Ornate Box Turtles, *Terrapene ornata*. Herpetologica 35: 93–96.

Mielke, E. 2002. Geographic distribution: *Crotaphytus collaris*. J. Kansas Herpetol. 4: 14.

Miles, D. B., R. Noecker, W. M. Roosenburg, and M. M. White. 2002. Genetic relationships among populations of *Sceloporus undulatus* fail to support present subspecific designations. Herpetologica 58: 277–292.

Miller, C. 1997a. Grace Olive Wiley: cobra queen. Reptile & Amphibian Magazine. March–April: 26–30.

Miller, E. J., and B. Gress. 2005. A Pocket Guide to Kansas Threatened and Endangered Species. Great Plains Nature Center, Wichita, Kansas.

Miller, L. 1976a. Amphibians and reptiles that might be found in the Sumner County, Kansas area. Privately printed (1 p.; mimeographed).

Miller, L. 1976b. Pesticides kill fish and other animals. Kansas Herpetol. Soc. News. 12: 2–3.

Miller, L. 1976c. Pesticides kill many animals. Bull. Oklahoma Herpetol. Soc. 3: 36.

MILLER, L. 1976d. KHS visits Elk County. Kansas Herpetol. Soc. News. 15: 1–2.

MILLER, L. 1977a. Five days in February. Kansas Herpetol. Soc. News. 18: 10–11.

MILLER, L. 1977b. Elementary students find large turtle. Kansas Herpetol. Soc. News. 19: 4.

MILLER, L. 1977c. Who cares about snakes? Heartland 1(2): 10.

MILLER, L. 1978. The status of the Checkered Garter Snake in Kansas. Kansas Herpetol. Soc. News. 27: 6–8.

MILLER, L. 1979. Herps abundant during third annual Chikaskia River wildlife study. Bull. Oklahoma Herpetol. Soc. 4(3): 46–48.

MILLER, L. 1980a. KHS joins KABT on banks of Chikaskia for first spring meeting. Kansas Herpetol. Soc. News. 37: 4–7.

MILLER, L. 1980b. Observations of the Red-spotted Toad in Barber County, Kansas. Kansas Herpetol. Soc. News. 38: 7–9.

MILLER, L. 1980c. September KHS meeting enjoyable for those who attended. Kansas Herpetol. Soc. News. 39: 3.

MILLER, L. 1981a. KHS and KABT meet at Clark County State Lake for spring meeting. Kansas Herpetol. Soc. News. 43: 2–4.

MILLER, L. 1981b. Record length Bullsnake from Harper County, Kansas. Kansas Herpetol. Soc. News. 43: 4.

MILLER, L. 1981c. KHS members attend annual wildlife study in Sumner County, Kansas. Kansas Herpetol. Soc. News. 43: 8.

MILLER, L. 1981d. Fifth and sixth grade students learn about herpetology in Caldwell. Kansas Herpetol. Soc. News. 46: 3.

MILLER, L. 1982a. Report of the first spring KHS field trip to Sumner County, Kansas. Kansas Herpetol. Soc. News. 48: 2–3.

MILLER, L. 1982b. New herpetological records for Sumner County, Kansas. Kansas Herpetol. Soc. News. 48: 5.

MILLER, L. 1982c. 1982 fall field trip held in Butler County, Kansas. Kansas Herpetol. Soc. News. 50: 2–3.

MILLER, L. 1983a. Bourbon County field trip well attended and successful. Kansas Herpetol. Soc. News. 54: 6–7.

MILLER, L. 1983b. Kansas protest against rattlesnake roundups. Oklahoma Herpetol. Soc. News. 4: 21.

MILLER, L. 1983c. The status of the Red-spotted Toad in Barber County, Kansas. Contract, Final Report (15 pp.), Kansas Fish and Game, Pratt.

MILLER, L. 1985a. Investigation of the Green Frog Rana clamitans melanota in southeastern Kansas. Contract 2A, Final report (16 pp.), Kansas Fish and Game, Pratt.

MILLER, L. 1985b. KHS 1985 field trip to Kirwin Reservoir. Kansas Herpetol. Soc. News. 61: 11–12.

MILLER, L. 1986a. KHS 1986 spring field trip to Cheyenne Bottoms. Kansas Herpetol. Soc. News. 64: 4–5.

MILLER, L. 1986b. The status of the Black Rat Snake in Sumner County, Kansas. Kansas Herpetol. Soc. News. 64: 12.

MILLER, L. 1987a. An investigation of four rare snakes in south-central Kansas. Report (24 pp.), Kansas Wildlife and Parks Commission, Pratt.

MILLER, L. 1987b. Another Bullsnake story. Kansas Herpetol. Soc. News. 68: 17–18.

MILLER, L. 1988. Harper County KHS field trip well attended. Kansas Herpetol. Soc. News. 72: 5–6.

MILLER, L. 1991. Study of the Eastern Narrowmouth Toad in southeast Kansas. Report (12 pp.), Kansas Department of Wildlife and Parks, Pratt.

MILLER, L. L. 1996a. Many amphibian and reptile species identified during KHS 1996 fall field trip to Wabaunsee County. Kansas Herpetol. Soc. News. 106: 2–3.

MILLER, L. L. 1996b. Third graders conduct amphibian and reptile field study. Kansas Herpetol. Soc. News. 106: 15.

MILLER, L. L. 1997b. Topeka Collegiate School summer research class yields specimen of Green Lacerta. Kansas Herpetol. Soc. News. 109: 13.

MILLER, L. L. 2000a. A winter snake in Shawnee County. Kansas Herpetol. Soc. News. 119: 11.

MILLER, L. L. 2000b. Amphibians, turtles, and reptiles of the Smoky Hill Ranch and surrounding areas of western Kansas. Abstract. Missouri Herpetol. Assn. News. 13: 10.

MILLER, L. L. 2000c. February amphibian and turtle observations in Shawnee County, Kansas. Kansas Herpetol. Soc. News. 119: 11.

MILLER, L. L. 2001. Amphibians, turtles, and reptiles of the Smoky Hill Ranch and surrounding areas of western Kansas. Abstract. Kansas Herpetol. Soc. News. 123: 8.

MILLER, L. L. 2002a. Geographic distribution: Hyla chrysoscelis. J. Kansas Herpetol. 4: 14.

MILLER, L. L. 2002b. Osage County herp count II. J. Kansas Herpetol. 4: 15.

MILLER, L. L. 2002c. Shawnee County herp count. J. Kansas Herpetol. 4: 15.

MILLER, L. L. 2002d. Sumner County herp count. J. Kansas Herpetol. 4: 15.

MILLER, L. L. 2003a. Indian Creek herp count. J. Kansas Herpetol. 7: 9.

MILLER, L. L. 2003b. Sumner County herp count. J. Kansas Herpetol. 7: 10.

MILLER, L. L. 2004a. Life history. Podarcis sicula. Winter activity. J. Kansas Herpetol. 10: 11.

MILLER, L. L. 2004b. Notes on Strecker's Chorus Frog activity in Kansas. J. Kansas Herpetol. 10: 13.

MILLER, L. L. 2004c. Sumner County herp count. J. Kansas Herpetol. 11: 11–12.

MILLER, L. L. 2004d. Life history. Lampropeltis triangulum. Reproduction. J. Kansas Herpetol. 12: 17.

MILLER, L. L. 2005. Life history notes: Podarcis sicula. Winter activity. J. Kansas Herpetol. 14: 10.

MILLER, L. L. 2006a. Late fall Ringneck Snake activity. J. Kansas Herpetol. 17: 6.

MILLER, L. L. 2006b. Thirtieth annual herpetological survey of southern Sumner County, Kansas. J. Kansas Herpetol. 18: 9.

MILLER, L. L. 2006c. Sumner County herp count. J. Kansas Herpetol. 19: 8.

MILLER, L. L. 2007a. Herpetological collecting in Kansas: the law and the herpetologist. J. Kansas Herpetol. 21: 10.

MILLER, L. L. 2007b. Sumner County herpetofaunal count. J. Kansas Herpetol. 23: 9–10.

MILLER, L. L. 2008. 32nd Sumner County herpetofaunal count. J. Kansas Herpetol. 27: 10.

MILLER, L. L. 2009. Geographic distribution: *Hyla chrysoscelis.* J. Kansas Herpetol. 30: 11.

MILLER, L. L., AND J. T. COLLINS. 1993. History, distribution and habitat requirements for three species of threatened reptiles in eastern Kansas. Final Report (29 pp.), Kansas Department of Wildlife and Parks, Pratt.

MILLER, L. L., AND J. T. COLLINS. 1995. *In memoriam.* Gene David Trott. Kansas Herpetol. Soc. News. 100: 22.

MILLER, L. L., AND R. L. LARDIE. 1976. Geographic distribution: *Chrysemys picta belli.* Herpetol. Rev. 7: 178.

MILLER, L. L., AND S. L. MILLER. 2003. Wakarusa herp count. J. Kansas Herpetol. 7: 10.

MILLER, L. L., AND S. L. MILLER. 2006. Ellsworth County herpetofaunal count. J. Kansas Herpetol. 19: 7.

MILLER, L. L., AND M. WILLIAMS. 2002. Geographic distribution: *Sceloporus consobrinus.* J. Kansas Herpetol. 3: 13.

MILLER, M. 1987c. Poison snakes. Kansas Wildlife 44(3): 22–23.

MILSTEAD, W. W. 1969. Studies on the evolution of box turtles (genus *Terrapene*). Bull. Florida St. Mus. 14: 1–113.

MINCKLEY, W. L. 1959. An atypical *Eurycea lucifuga* from Kansas. Herpetologica 15: 240.

MINTON, S. A. 1957. Snake bite. Sci. American 196(1): 114–122.

MINTON, S. A. 1971. Snake Venoms and Envenomation. Marcel Dekker, New York.

MINTON, S. A. 1972. Amphibians and Reptiles of Indiana. Monograph Number 3, The Indiana Academy of Science, Indianapolis.

MINTON, S. A. 1983. *Sistrurus catenatus.* Cat. American Amph. Reptl. 332.1–332.2.

MINTON, S. A. 2001. Amphibians and Reptiles of Indiana. Revised 2nd ed. The Indiana Academy of Science, Indianapolis..

MINTON, S. A., AND D. BROWN. 1997. Distribution and variation of the lizard *Cnemidophorus sexlineatus* in Indiana. Bull. Chicago Herpetol. Soc. 32: 102–104.

MINX, P. 1996. Phylogenetic relationships among the box turtles, genus *Terrapene*. Herpetologica 52: 584–597.

MITCHELL, E. W. 1965. Behavior Patterns of the Great Plains Skink, *Eumeces obsoletus.* Unpublished Master's thesis, Emporia State University, Emporia, Kansas.

MITCHELL, J. C. 1979. The concept of phenology and its application to the study of amphibian and reptile histories. Herpetol. Rev. 10: 51–54.

MITCHELL, J. C. 1994. The Reptiles of Virginia. Smithsonian Institution Press, Washington, D. C.

MITTLEMAN, M. B. 1945a. An additional note on a supposed specimen of *Hyla phaeocrypta.* Herpetologica 3: 20–21.

MITTLEMAN, M. B. 1945b. The status of *Hyla phaeocrypta* with notes on its variation. Copeia 1945: 31–37.

MITTLEMAN, M. B. 1949. Geographic variation in Marcy's Garter Snake, *Thamnophis marcianus* (Baird and Girard). Bull. Chicago Acad. Sci. 8: 235–249.

MITTLEMAN, M. B. 1950. Cavern-dwelling salamanders of the Ozark Plateau. Bull. Nat. Speleol. Soc. 12: 1–4.

MOEHN, L. D. 1981a. Microhabitat preference in the Broadheaded Skink. Bull. Chicago Herpetol. Soc. 15: 49–53.

MOEHN, L. D. 1981b. Differential abundance of the Broadheaded Skink, *Eumeces laticeps,* and the Five-lined Skink, *Eumeces fasciatus,* in southern Illinois. Bull. Chicago Herpetol. Soc. 16: 6–11.

MOELLER, D., A. KOSSOY, W. HAMILTON, AND L. L. MILLER. 2000. Geographic distribution. *Leptotyphlops dulcis.* Herpetol. Rev. 31: 56.

MOLL, D. 1976. Environmental influence on growth rate in the Ouachita Map Turtle, *Graptemys pseudogeographica ouachitensis.* Herpetologica 32: 439–443.

MOLL, D., AND J. B. IVERSON. 2008. Geographic variation in life-history traits. Pp. 181–192 *In* A. G. Steyermark, M. S. Finkler, and R. J. Brooks (Eds.), Biology of the Snapping Turtle (*Chelydra serpentina*). The Johns Hopkins University Press, Baltimore, Maryland.

MOLL, D., AND E. O. MOLL. 2004. The ecology, exploitation, and conservation of river turtles. Oxford University Press, New York.

MOLL, E. O. 1979. Reproductive cycles and adaptations. Pp. 305–331 *In* M. Harless and H. Morlock (Eds.). Turtles: Perspectives and Research. John Wiley & Sons, New York.

MOLL, E. O., AND J. M. LEGLER. 1971. The life history of a Neotropical slider turtle, *Pseudemys scripta* (Schoepff), in Panama. Bull. Los Angeles Co. Mus. Nat. Hist. Sci. 11: 1–102.

MONELLO, R. J., C. R. FORGE, AND M. P. SIMON. 1995. An examination of the physiognomic and nutritional characteristics of prehibernating Blanchard's Cricket Frogs *Acris crepitans blanchardi.* Abstract. 127th Annual Meeting of the Kansas Academy of Science.

MONRO, D. F. 1947. Effect of a bite by *Sistrurus* on *Crotalus.* Herpetologica 4: 57.

MONRO, D. F. 1949a. Vertical position on the pupil in the Crotalidae. Herpetologica 5: 106–108.

MONRO, D. F. 1949b. Food of *Heterodon nasicus nasicus.* Herpetologica 5: 133.

MONRO, D. F. 1949c. Gain in size and weight of *Heterodon* eggs during incubation. Herpetologica 5: 133–134.

MONRO, D. F. 1949d. Hatching of a clutch of *Heterodon* eggs. Herpetologica 5: 134–136.

MONTANUCCI, R. R. 1974. Convergence, polymorphism or introgressive hybridization? An analysis of interaction between *Crotaphytus collaris* and *C. reticulatus* (Sauria: Iguanidae). Copeia 1974: 87–101.

MONTANUCCI, R. R., R. W. AXTELL, AND H. C. DESSAUER. 1975. Evolutionary divergence among collared lizards (*Crotaphytus*), with comments on the status of *Gambelia.* Herpetologica 31: 336–347.

MOORE, J. A. 1944. Geographic variation in *Rana pipiens* Schreber of eastern North America. Bull. American Mus. Nat. Hist. 82: 345–369.

MORAFKA, D. J. 1977. A Biogeographical Analysis of the Chihuahuan Desert through its Herpetofauna. Dr. W. Junk bv, Publisher, The Hague, Netherlands.

MORIARTY, E. C. 1996. A preliminary report on number, seasonal and daily activity, and growth rate of a population of Painted Turtles (*Chrysemys picta*) in northeastern Kansas: a three year study. Kansas Herpetol. Soc. News. 105: 5–15.

MORIARTY, E. C. 1997. A molecular comparison of *Pseudacris triseriata* and *Pseudacris maculata* from sympatric populations in Douglas County, Kansas. Abstract. Annual Meeting of the Society for the Study of Amphibians and Reptiles, The Herpetologists' League, and the American Society of Ichthyologists and Herpetologists, Seattle, Washington.

MORIARTY, E. C., AND D. C. CANNATELLA. 2004. Phylogenetic relationships of the North American chorus frogs (*Pseudacris:* Hylidae). Mol. Phylogenet. Evol. 30: 409–420.

MORIARTY, E. C., AND J. T. COLLINS. 1995a. First known occurrence of amphibian species in Kansas. Kansas Herpetol. Soc. News. 100: 28–30.

MORIARTY, E. C., AND J. T. COLLINS. 1995b. An estimate of numbers of Plains Leopard Frogs at a site in northeastern Kansas. Kansas Herpetol. Soc. News. 102: 14–15.

MORIARTY, J. J. 1997. Amphibian and reptile diversity and distribution in the United States. Minnesota Herpetol. Soc. News. 17(8): 4–5.

MOSHER, T. 1997. Another Western Diamondback Rattlesnake in Kansas. Kansas Herpetol. Soc. News. 108: 19.

MOSIER, D., AND J. T. COLLINS. 1998. Geographic distribution: *Storeria occipiomaculata.* Herpetol. Rev. 29: 116.

MOSS, R. E., AND K. BRUNSON. 1981. Kansas Stream and River Fishery Resource Evaluation. Kansas Fish and Game Commission, Pratt.

MOUNT, R. H. 1975. The Reptiles and Amphibians of Alabama. Auburn University, Agricultural Experiment Station, Auburn, Alabama.

MOZLEY, A. E. 1978. List of Kansas snakes in the museum of the Kansas State University. Trans. Kansas Acad. Sci. 6: 34–35.

MULCAHY, D. G. 2008. Phylogeography and species boundaries of the western North American Night Snake (*Hypsiglena torquata*): revisiting the subspecies concept. Mol. Phylogenet. Evol. 46: 1,095–1,115.

MULHERN, D., AND J. T. COLLINS. 2005. Cherokee County herp count. J. Kansas Herpetol. 15: 7.

MULHERN, D., AND J. T. COLLINS. 2005. Marais des Cygnes NWR herp count. J. Kansas Herpetol. 15: 9.

MULLEN, K. 1979. A new faunal record for Kansas? Kansas Herpetol. Soc. News. 29: 3.

MULVANEY, P. S. 1983. Blind snakes of the United States, their natural history with a discussion of climate and physiography as limiting factors to their range. Bull. Oklahoma Herpetol. Soc. 8: 2–45.

MURPHY, J. 1976. The natural history of the box turtle. Bull. Chicago Herpetol. Soc. 11: 2–45.

MURPHY, J. B., AND B. L. ARMSTRONG. 1978. Maintenance of Rattlesnakes in Captivity. Special Publication Number 3. The University of Kansas, Museum of Natural History, Lawrence.

MURPHY, J. B., AND J. T. COLLINS (EDS.). 1980. Reproductive Biology and Diseases of Captive Reptiles. Contribu-

tions to Herpetology, Number 1, Society for the Study of Amphibians and Reptiles. Meseraull Printing, Inc., Lawrence, Kansas.

MURPHY, J. C., AND S. L. BARTEN. 1979. Two abnormal patterns in the Copperhead, *Agkistrodon contortrix.* Bull. Chicago Herpetol. Soc. 14: 93.

MURPHY, T. D. 1964. Box turtle, *Terrapene carolina*, in stomach of Copperhead, *Agkistrodon contortrix.* Copeia 1964: 221.

MURRELL, M. A. 1993. Snakes alive! Kansas Wildlife and Parks 50(3): 18–21.

MURROW, D. G. 2004. Geographic distribution. *Thamnophis proximus.* Kansas. J. Kansas Herpetol. 11: 14.

MURROW, D. G. 2007a. Jefferson County herpetofaunal count. J. Kansas Herpetol. 23: 8.

MURROW, D. G. 2007b. Wyandotte County herpetofaunal count. J. Kansas Herpetol. 23: 10.

MURROW, D. G. 2008a. KHS spring 2008 field trip results. J. Kansas Herpetol. 26: 2–4.

MURROW, D. G. 2008b. Report on the KHS fall field trip to Smith County, Kansas. J. Kansas Herpetol. 28: 7–9.

MURROW, D. G. 2009a. KHS 2009 spring field trip. J. Kansas Herpetol. 29: 2–3.

MURROW, D. G. 2009b. Results of the KHS 2009 spring field trip. J. Kansas Herpetol. 30: 2–3.

NADEAU, M. R. 1979. Comparisons of Heart Rate and Duration of Submergence During Voluntary Diving in Three Species of Turtles. Unpublished Master's thesis, University of Kansas, Lawrence.

NAZARAN, S. L., AND L. J. GUILLETTE, JR. 1983. Ovarianoviductal interactions in the Collared Lizard *Crotaphytus collaris.* Abstract. Kansas Acad. Sci. 2: 3–24.

NEILL, W. T. 1951. The taxonomy of North American softshelled turtles, genus *Amyda.* Res. Div. Ross Allen's Reptl. Inst. 1: 7–24.

NELSON, C. E. 1972a. *Gastrophryne carolinensis.* Cat. American Amph. Reptl. 120.1–120.4.

NELSON, C. E. 1972b. *Gastrophryne olivacea.* Cat. American Amph. Reptl. 122.1–122.4.

NELSON, C. E. 1972c. Systematic studies of the North American microhylid genus *Gastrophryne.* J. Herpetol. 6: 111–137.

NELSON, C. E. 1973a. Mating calls of the microhylinae: descriptions and phylogenetic and ecological considerations. Herpetologica 29: 163–176.

NELSON, C. E. 1973b. *Gastrophryne.* Cat. American Amph. Reptl. 134.1–134.2.

NEVO, E. 1973a. Adaptive variation in size of cricket frogs. Ecology 54: 1,271–1,281.

NEVO, E. 1973b. Adaptive color polymorphism in cricket frogs. Evolution 27: 353–367.

NEVO, E., AND R. R. CAPRANICA. 1985. Evolutionary origin of ethological reproductive isolation in cricket frogs, *Acris.* Evol. Biol. 10: 147–214.

NICKERSON, M. A., AND R. KRAGER. 1973. Additional noteworthy records of Missouri amphibians and reptiles with a possible addition to the herpetofauna. Trans. Kansas Acad. Sci. 75: 276–277.

NICKERSON, M. A., AND C. E. MAYS. 1974. The Hellbenders: North American "Giant Salamanders." Milwaukee Public Museum Press, Milwaukee, Wisconsin.

NIEUWOLT, P. M. 1996. Movement, activity, and micro-habitat selection in the Western Box Turtle, *Terrapene ornata luteola*, in New Mexico. Herpetologica 52: 487–495.

NOBLE, G. K., AND E. R. MASON. 1933. Experiments on the brooding habits of the lizards *Eumeces* and *Ophisaurus*. American Mus. Novitates 619: 1–29.

NOECKER, R., W. ROOSENBURG, D. MILES, AND M. WHITE. 1995. Relationships within *Sceloporus undulatus* based on allozyme electrophoresis. Abstract. 75th Annual Meeting of the American Society of Ichthyologists and Herpetologists, Edmonton, Alberta, Canada.

NULTON, M. T., AND M. S. RUSH. 1988. New county records of amphibians and reptiles in Gray County, Kansas. Kansas Herpetol. Soc. News. 74: 10–12.

NUR, V., AND E. NEVO. 1969. Supernumerary chromosomes in the cricket frog *Acris crepitans*. Caryologia 22: 97–102.

O'ROKE, E. C. 1922. Frogs and frogging. Trans. Kansas Acad. Sci. 30: 448–451.

OELRICH, T. M. 1952. A new *Testudo* from the upper Pliocene of Kansas with additional notes on associated Rexroad mammals. Trans. Kansas Acad. Sci. 55: 300–311.

OELRICH, T. M. 1953. A new box turtle from the Pleistocene of southwestern Kansas. Copeia 1953: 53.

OELRICH, T. M. 1954. A horned toad, *Phrynosoma cornutum*, from the Upper Pliocene of Kansas. Copeia 1954: 262–263.

OLDFIELD, B., AND J. J. MORIARTY. 1994. Amphibians and Reptiles Native to Minnesota. University of Minnesota Press, Minneapolis.

OLSON, R. E. 1977. Evidence for the species status of Baird's Rat Snake. Texas J. Sci. 29: 79–84.

ORR, J. M. 2007. Microhabitat use by the Eastern Worm Snake, *Carphophis amoenus*. Herpetol. Bull. 97: 29–35.

ORTENBURGER, A. I. 1926a. The whipsnakes and racers: genera *Masticophis* and *Coluber*. Mem. Univ. Michigan Mus. 1: *xviii* + 1–247.

ORTENBURGER, A. I. 1926b. A report on the amphibians and reptiles of Oklahoma. Proc. Oklahoma Acad. Sci. 6: 89–100.

ORTENBURGER, A. I., AND B. FREEMAN. 1930. Notes on some reptiles and amphibians from western Oklahoma. Publ. Univ. Oklahoma Biol. Surv. 2(4): 175–188.

OWENS, A., M. TRAGER AND E. HORNE. 2002. Natural history notes: *Phrynosoma cornutum*. Herpetol. Rev. 33: 308–309.

PACE, A. E. 1974. Systematic and biological studies of the Leopard Frogs (*Rana pipiens* complex) of the United States. Univ. Michigan Mus. Zool. Misc. Publ. 148: 1–140.

PACKARD, G. C., M. J. PACKARD, AND W. H. N. GUTZKE. 1985. Influence of hydration of the environment on eggs and embryos of the terrestrial turtle *Terrapene ornata*. Physiol. Zool. 58: 564–575.

PALENSKE, N., AND D. SAUNDERS. 2001. Blood viscosity comparisons between endotherms and ectotherms at two temperatures. Abstract. Kansas Herpetol. Soc. News. 123: 8.

PALMER, W. M., AND F. F. SNELSON. 1966. Eastern Hognose Snake. Wildlife in North Carolina (January): 23.

PARKER, W. S., AND W. S. BROWN. 1973. Species composition and population changes in two complexes of snake hibernacula in northern Utah. Herpetologica 29: 319–326.

PARKS, L. H. 1969. An active Bullsnake in near-freezing temperature. Trans. Kansas Acad. Sci. 72: 266.

PARMELEE, J. R. 1994. Geographic distribution. *Trachemys scripta*. Herpetol. Rev. 25: 33.

PARMELEE, J. R., AND H. S. FITCH. 1995a. An experiment with artificial shelters for snakes. Abstract. 22nd Annual Meeting of the Kansas Herpetological Society, Lawrence.

PARMELEE, J. R., AND H. S. FITCH. 1995b. An experiment with artificial shelters for snakes: effects of material, age, and surface preparation. Herpetol. Nat. Hist. 3: 187–191.

PARRISH, H. M. 1980. Poisonous Snakebites in the United States. Vantage Press, New York.

PATTEN, T. J., J. D. FAWCETT, AND D. D. FOGELL. 2009. Natural history of the Western Massasauga (*Sistrurus catenatus tergeminus*) in Nebraska. J. Kansas Herpetol. 30: 13–20.

PATTON, L. G. 1978. Checkered Garter Snakes found. Bull. Oklahoma Herpetol. Soc. 3: 53.

PAULY, G. B., D. M. HILLIS, AND D. C. CANNATELLA. 2004. The history of a Nearctic colonization: molecular phylogenetics and biogeography of the Nearctic toads (*Bufo*). Evolution 58: 2,517–2,535.

PEABODY, F. E. 1958. A Kansas drouth recorded in growth zones of a Bullsnake. Copeia 1958: 91–94.

PEARCE, M. 1991. Of skinks and worm snakes. Wall Street Journal, 26 August.

PEREZ, R. 1985. Some say supplier of Slimy Salamanders shouldn't stop sales. Chicago Herpetol. Soc. News. (September).

PERKINS, C. B. 1940. A key to the snakes of the United States. Bull. Zool. Soc. San Diego 16: 1–63.

PERRY, J. 1974a. KHS members take trip to southwest Kansas. Kansas Herpetol. Soc. News. 3: 2–3.

PERRY, J. 1974b. An unusual frog in Kansas. Kansas Herpetol. Soc. News. 4: 2–3.

PERRY, J. 1975. A trip to southeastern Kansas. Kansas Herpetol. Soc. News. 7: 4.

PERRY, J. 1977a. The Kansas Herpetological Society: a brief history. Kansas Herpetol. Soc. News. 20: 4–6.

PERRY, J. 1977b. KHS members achieve goal: get Cottonmouth. Kansas Herpetol. Soc. News. 21: 3–4.

PERRY, J. 1978. KHS successful at Miami County State Lake. Kansas Herpetol. Soc. News. 27: 5.

PERRY, J. (ED.). 1977c. Kansas herps needed. Kansas Herpetol. Soc. News. 18: 2–3.

PERRY, J. (ED.). 1977d. KHS to hunt Cottonmouths. Kansas Herpetol. Soc. News. 20: 2.

PETERS, J. A. 1952. Catalogue of type specimens in the herpetological collections of the University of Michigan Museum of Zoology. Univ. Michigan Mus. Zool. Occas. Pap. 539: 1–55.

PETERS, J. A. 1953. A fossil snake of the genus *Heterodon* from the Pliocene of Kansas. J. Paleontol. 27: 328–331.

PETERSON, C. 1992. Maxwell Wildlife Refuge: a glimpse of Kansas past. Kansas Wildlife and Parks 49(1): 10–14.

PETERSON, C. L., R. F. WILKINSON, JR., M. S. TOPPING, AND D. E. METTER. 1983. Age and growth of the Ozark Hellbender (*Cryptobranchus alleganiensis bishopi*). Copeia 1983: 225–231.

PETRANKA, J. W. 1984. Breeding migrations, breeding season, clutch size, and oviposition of stream-breeding *Ambystoma texanum*. J. Herpetol. 18: 106–112.

PETRANKA, J. W. 1998. Salamanders of the United States and Canada. Smithsonian Institution Press, Washington, D. C.

PEYTON, M. M. 1989. Geographic distribution: *Storeria occipitomaculata*. Herpetol. Rev. 20: 13.

PIERCE, B. A., AND P. H. WHITEHURST. 1990. *Pseudacris clarkii*. Cat. American Amphib. Reptl. 458.1–458.3.

PILCH, J. A. 1982. Notes on a brood of Northern Water Snakes from Kansas. Kansas Herpetol. Soc. News. 47: 22.

PILCH, J. A. 2004. *Lampropeltis calligaster*. Winter activity. J. Kansas Herpetol. 9: 7.

PINDER, A. W., K. B. STOREY, AND G. R. ULTSCH. 1992. Estivation and hibernation. Pp. 250–274 In M. E. Feder and W. W. Burggren (Eds.). Environmental Physiology of the Amphibians. The University of Chicago Press, Chicago, Illinois.

PINNEY, R. 1981. The Snake Book. Doubleday, New York.

PISANI, G. R. 1974. Herpetology in the KU Division of Biological Sciences. Kansas Herpetol. Soc. News. 4: 3–4.

PISANI, G. R. 1976. Comments on the courtship and mating mechanics of *Thamnophis* (Reptilia, Serpentes, Colubridae). J. Herpetol. 10: 139–142.

PISANI, G. R. 2003. Natural history notes: *Lampropeltis calligaster calligaster*. Herpetol. Rev. 34: 150.

PISANI, G. R. 2004. Life history. *Chelydra serpentina*. Mating behavior. J. Kansas Herpetol. 11: 15.

PISANI, G. R. 2005. A new Kansas locality for *Virginia valeriae*. J. Kansas Herpetol. 16: 25.

PISANI, G. R. 2007. New Kansas maximum sizes for *Virginia valeriae* and *Carphophis vermis*. J. Kansas Herpetol. 22: 11.

PISANI, G. R. 2009a. Early activity of *Storeria dekayi* in Jefferson County, Kansas. J. Kansas Herpetol. 29: 10–11.

PISANI, G. R. 2009b. Use of an active ant nest as a hibernaculum by small snake species. Trans. Kansas Acad. Sci. 112: 113–118.

PISANI, G. R., J. T. COLLINS, AND S. R. EDWARDS. 1973. A re-evaluation of the subspecies of *Crotalus horridus*. Trans. Kansas Acad. Sci. 75: 255–263.

PISANI, G. R., AND H. S. FITCH. 1993. A survey of Oklahoma's rattlesnake roundups. Kansas Herpetol. Soc. News. 92: 7–15.

PISANI, G. R., AND H. S. FITCH. 2006. Rapid early growth in northeastern Kansas Timber Rattlesnakes. J. Kansas Herpetol. 20: 19–20.

PLATT, D. R. 1969. Natural history of the hognose snakes *Heterodon platyrhinos* and *Heterodon nasicus*. Univ. Kansas Publ. Mus. Nat. Hist. 18: 253–420.

PLATT, D. R. 1975. Reptiles of Sand Prairie Natural History Reservation, Harvey Co., Kansas. Privately printed (mimeographed).

PLATT, D. R. 1983. *Heterodon*. Cat. American Amph. Reptl. 315.1–315.2.

PLATT, D. R. 1984a. Population trends and habitat assessment of snakes and lizards in south central Kansas. Contract 80, Report (6 pp.), Kansas Fish and Game Commission, Pratt.

PLATT, D. R. 1984b. Growth of Bullsnakes (*Pituophis melanoleucus sayi*) on a sand prairie in south central Kansas. Pp. 41–55 In R. A. Seigel, L. E. Hunt, J. L. Knight, L. Malaret and N. L. Zuschlag (Eds.), Vertebrate Ecology and Systematics: A Tribute to Henry S. Fitch. Special Publication Number 10. The University of Kansas Museum of Natural History, Lawrence.

PLATT, D. R. 1985a. Changes in size of populations of snakes on sand prairies in south central Kansas. Abstract. 117th Annual Meeting of the Kansas Academy of Science.

PLATT, D. R. 1985b. Population trends and habitat assessment of snakes and lizards in south central Kansas. Contract 80, Final Report (37 pp.), Kansas Fish and Game, Pratt.

PLATT, D. R. 1986. Habitat use by snakes and lizards in Harvey County, Kansas. Abstract. 118th Annual Meeting of the Kansas Academy of Science.

PLATT, D. R. 1988. Bethel College natural area. Trans. Kansas Acad. Sci. 91: 48–49.

PLATT, D. R. 1989. Seasonal activity of snakes on a sand prairie. Pp. 251–254 In Proceedings of the 11th North American Prairie Conference. University of Nebraska Press, Lincoln.

PLATT, D. R. 1998. Monitoring population trends of snakes and lizards in Harvey County, Kansas. Final Report (42 pp.), Kansas Department of Wildlife and Parks, Pratt.

PLATT, D. R. 2000. Food habits of snakes on a sand prairie in south-central Kansas. Abstract. Missouri Herpetol. Assn. News. 13: 10.

PLATT, D. R. 2001. Food habits of snakes on a sand prairie in south-central Kansas. Abstract. Kansas Herpetol. Soc. News. 123: 8–9.

PLATT, D. R. 2003. Lizards and snakes (Order Squamata) of Harvey County, Kansas. J. Kansas Herpetol. 6: 13–20.

PLATT, D. R., J. T. COLLINS, AND R. E. ASHTON. 1974. Rare, endangered and extirpated species in Kansas. II. Amphibians and reptiles. Trans. Kansas Acad. Sci. 76: 185–192.

PLATT, D. R., AND C. H. ROUSELL. 1963. County records of snakes from southcentral Kansas. Trans. Kansas Acad.Sci. 66: 551.

PLATT, S. G. 1994. The biology, status, and captive propagation of the Alligator Snapping Turtle (*Macroclemys temminckii*). The Vivarium 6(2): 22–23, 42–45.

PLATZ, J. E. 1989. Speciation within the Chorus Frog *Pseudacris triseriata*: morphometric and mating call analyses of the boreal and western subspecies. Copeia 1989: 704–712.

PLATZ, J. E., AND A. LATHROP. 1993. Body size and age assessment among advertising male Chorus Frogs. J. Herpetol. 27: 109–111.

PLESHKEWYCH, A. 1964. Effects of Stimulation of the Motor Centers in the Cerebral Hemispheres of *Pseudemys scripta* and *Crotaphytus collaris*. Unpublished Master's thesis, Emporia State University, Emporia, Kansas.

PLUMMER, M. V. 1975. Population Ecology of the Softshell Turtle, *Trionyx muticus*. Unpublished Doctoral dissertation, University of Kansas, Lawrence.

PLUMMER, M. V. 1976. Some aspects of nesting success in the turtle, *Trionyx muticus*. Herpetologica 32: 353–359.

PLUMMER, M. V. 1977a. Observations on breeding migrations of *Ambystoma texanum*. Herpetol. Rev., 8: 79–80.

PLUMMER, M. V. 1977b. Predation by Black Rat Snakes in bank swallow colonies. Southwest. Nat. 22: 147–148.

PLUMMER, M. V. 1977c. Notes on the courtship and mating behavior of the softshell turtle, *Trionyx muticus* (Reptilia, Testudines, Trionychidae). J. Herpetol. 11: 90–92.

PLUMMER, M. V. 1977d. Activity, habitat and population structure in the turtle, *Trionyx muticus*. Copeia 1977: 431–440.

PLUMMER, M. V. 1977e. Reproduction and growth in the turtle, *Trionyx muticus*. Copeia 1977: 440–447.

PLUMMER, M. V. 1987. Geographic variation in body size of Green Snakes (*Opheodrys aestivus*). Copeia 1987: 483–485.

PLUMMER, M. V., AND D. B. FARRAR. 1981. Sexual dietary differences in a population of *Trionyx muticus*. J. Herpetol. 15: 175–179.

PLUMMER, M. V., C. E. SHADRIX, AND R. C. COX. 1994. Thermal limits of incubation in embryos of softshell turtles (*Apalone mutica*). Chelonian Conserv. Biol. 1: 141–144.

PLUMMER, M. V., AND H. W. SHIRER. 1975. Movement patterns in a river population of the softshell turtle, *Trionyx muticus*. Univ. Kansas Mus. Nat. Hist. Occas. Pap. 43: 1–26.

POPE, C. H. 1937. Snakes Alive and How they Live. Viking Press, New York.

POPE, C. H. 1939. Turtles of the United States and Canada. A. A. Knopf, New York.

POPE, C. H. 1944a. The Poisonous Snakes of the New World. New York Zoological Society, New York.

POPE, C. H. 1944b. Amphibians and Reptiles of the Chicago Area. Chicago Natural History Museum Press, Chicago, Illinois.

POPE, C. H. 1956. The Reptile World. A. A. Knopf, New York.

POPE, T. E. B. 1927. A two-headed Bull Snake. Yearbook Publ. Mus. Milwaukee (1925) 5: 161–167.

PORTEL, M. (ED.) 1977. Snakebite. Heartland 1(2): 3.

PORTER, K. R. 1972. Herpetology. W. B. Saunders Co., Philadelphia, Pennsylvania.

POUND, A. E. 1968. Gypsum caves in Kansas. Oklahoma Underground (Oklahoma Grotto Natl. Speleol. Soc.) 1(2): 56.

POWELL, R. 1980a. Kansas size record for Black Rat Snake. Kansas Herpetol. Soc. News. 35: 12.

POWELL, R. 1980b. Geographic Distribution: *Pseudacris triseriata triseriata*. Herpetol. Rev. 11: 38.

POWELL, R. 1990. *Elaphe vulpina*. Cat. American Amphib. Reptl. 470.1–470.3.

POWELL, R., J. T. COLLINS, AND L. D. FISH. 1992. *Virginia valeriae*. Cat. American Amphib. Reptl. 552.1–552.6

POWELL, R., J. T. COLLINS, AND L. D. FISH. 1994. *Virginia striatula*. Cat. American Amphib. Reptl. 599.1–599.6.

POWELL, R., AND H. GREGORY. 1978. Emergency snakebite: a quick guide to recognizing poisonous snakes in western Missouri and eastern Kansas with notes on the treatment of their bites. Privately printed (7 pp.), Kansas City, Missouri.

PRESTON, R. E. 1971. Pleistocene turtles from the Arkalon Local Fauna of southwestern Kansas. J. Herpetol. 5: 208–211.

PRESTON, W. B. 1982. The Amphibians and Reptiles of Manitoba. Manitoba Museum of Man and Nature, Winnipeg, Manitoba, Canada.

PRICE, A. H. 1990. *Phrynosoma cornutum*. Cat. American Amphib. Reptl. 469.1–469.7.

PRIOR, K. A., G. BLOUIN-DEMERS, AND P. J. WEATHERHEAD. 2001. Sampling biases in demographic analyses of Black Rat Snakes (*Elaphe obsoleta*). Herpetologica 57: 460–469.

PRITCHARD, P. C. H. 1967. Living Turtles of the World. T. F. H. Publications, Inc., Jersey City, New Jersey.

PRITCHARD, P. C. H. 1979a. Encyclopedia of Turtles. T. F. H. Publications, Inc., Neptune City, New Jersey.

PRITCHARD, P. C. H. 1979b. Taxonomy, evolution, and zoogeography. Pp. 1–42 In M. Harless and H. Morlock (Eds.), Turtles: Perspectives and Research. John Wiley & Sons, New York.

PRITCHARD, P. C. H. 1980. Record size turtles from Florida and South America. Chelonologica 1: 113–123.

PRITCHARD, P. C. H. 1989. The Alligator Snapping Turtle: Biology and Conservation. Milwaukee Public Museum, Milwaukee, Wisconsin.

PRITCHARD, P. C. H. 2006. The Alligator Snapping Turtle: Biology and Conservation. Revised ed. Krieger Publishing Company, Malabar, Florida.

PROPHET, C. W. (ED.). 1957. Along the roadside. Kansas School Nat. 4(1): 1–16.

PROPHET, C. W. 1959. Life in a stream. Kansas School Nat. 5(4): 1–16.

PYRON, R. A. 2007. A revised distribution record of *Masticophis* in Kansas. J. Kansas Herpetol. 24: 16.

PYRON, R. A. 2009. Systematics and Historical Biogeography of the Lampropeltinine Snakes (Serpentes: Colubridae). Unpublished Doctoral dissertation, City University of New York, College of Staten Island, New York.

PYRON, R. A., AND F. T. BURBRINK. 2009a. Lineage diversification in a widespread species: roles for niche divergence and conservatism in the Common Kingsnake, *Lampropeltis getula*. Mol. Ecol. 18: 3,443–3,457.

PYRON, R. A., AND F. T. BURBRINK. 2009b. Systematics of the Common Kingsnake (*Lampropeltis getula*; Serpentes: Colubridae) and the burden of heritage in taxonomy. Zootaxa 2,241: 22–37.

RAFFAELLI, J. 2007. Les Urodèles du Monde. Penclen ed., France.

RAINEY, D. G. 1953. Death of an Ornate Box Turtle parasitized by dipterous larvae. Herpetologica 9: 109–110.

RAINEY, D. G. 1956. Eastern Woodrat, *Neotoma floridana:* Life history and ecology. Univ. Kansas Publ. Mus. Nat. Hist. 8: 535–646.

RAKESTRAW, J. 1996. Spring herp counts: a Kansas tradition. Reptile & Amphibian Magazine. March–April: 75–80.

RALIN, D. B. 1968. Ecological and reproductive differentiation in the cryptic species of the *Hyla versicolor* complex (Hylidae). Southwest. Nat. 13: 283–300.

RALIN, D. B. 1977. Evolutionary aspects of mating call variation in a diploid-tetraploid species complex of treefrogs (Anura). Evolution 31: 721–736.

RAMSEY, L. W. 1953. The Lined Snake, *Tropidoclonion lineatum* (Hallowell). Herpetologica 9: 7–24.

REAGAN, D. P. 1974a. Threatened native amphibians of Arkansas. Pp. 93–99 *In* Arkansas Natural Area Plan. Arkansas Department of Planning, Little Rock.

REAGAN, D. P. 1974b. Threatened native reptiles of Arkansas. Pp. 101–105 *In* Arkansas natural area plan. Arkansas Department of Planning, Little Rock.

REBER, D. L., AND A. SMITH-REBER. 1994. Kansas Herpetological Society position paper regarding rattlesnake roundups. Kansas Herpetol. Soc. News. 96: 9–20.

REDDER, A. J., C. K. DODD, JR., D. KEINATH, D. MCDONALD, AND T. ISE. 2006. Ornate Box Turtle (*Terrapene ornata ornata*): a technical conservation assessment. Species Conservation Project (54 pp.), United States Department of Agriculture, Rocky Mountain Region.

REEVE, W. L. 1952. Taxonomy and distribution of the horned lizards genus *Phrynosoma*. Univ. Kansas Sci. Bull. 34: 817–960.

REGAN, G. T. 1972. Natural and Man-made Conditions Determining the Range of *Acris crepitans* in the Missouri River Basin. Unpublished Doctoral dissertation, University of Kansas, Lawrence.

REHMEIER, R. L., AND R. S. MATLACK. 2004. Life history. *Thamnophis sirtalis.* Diet. J. Kansas Herpetol. 11: 15.

REICHARD, K., T. DUNCAN, H. M. SMITH, AND D. CHISZAR. 1995a. Herpetological microbiogeography of Kansas I: Quantitative summary. Kansas Herpetol. Soc. News. 102: 6–10.

REICHARD, K., T. DUNCAN, H. M. SMITH, AND D. CHISZAR. 1995b. Herpetological microbiogeography of Kansas. Abstract. 22nd Annual Meeting of the Kansas Herpetological Society, Lawrence.

REICHLING, S. B. 1995. The taxonomic status of the Louisiana Pine Snake (*Pituophis melanoleucus ruthveni*) and its relevance to the Evolutionary Species Concept. J. Herpetol. 29: 186–189.

REICHMAN, O. J. 1987. Konza Prairie: A Tallgrass Natural History. University Press of Kansas, Lawrence.

REILLY, S. M. 1990. Biochemical systematics and evolution of the eastern North American newts, genus *Notophthalmus* (Caudata: Salamandridae). Herpetologica 46: 51–59.

REINERT, H. K. 1978. The Ecology and Morphological Variation of the Massasauga Rattlesnake, *Sistrurus catenatus.* Unpublished Master's thesis, Pennsylvania State University, Clarion.

REISERER, R. S. 2002. Stimulus control of caudal luring and other feeding responses: a program for research on visual perception in vipers. Pp. 361–383 *In* G. W. Schuett, M. Höggren, M. E. Douglas and H. W. Greene (Eds.), Biology of the Vipers. Eagle Mountain Publishing, LC, Eagle Mountain, Utah.

REISERER, R. S., AND D. L. REBER. 1992. Comments on the Wallace County rattlesnake roundup. Kansas Herpetol. Soc. News. 90: 12–15.

REYNOLDS, S. L., AND M. E. SEIDEL. 1982. *Sternotherus odoratus.* Cat. American Amph. Reptl. 287.1–287.4.

REYNOLDS, S. L., AND M. E. SEIDEL. 1983. Morphological homogeneity in the turtle *Sternotherus odoratus* (Kinosternidae) throughout its range. J. Herpetol. 17: 113–120.

RICHARDSON, C., AND C. PHILLIPS. 1997. Hazardous Animals and Plants of Fort Riley. Fort Riley Integrated Training Area Management Program, Fort Riley, Kansas.

RICHARDSON, M. 1972. The Fascination of Reptiles. Hill and Wang, New York.

RICKART, E. A. 1976. A new horned lizard (*Phrynosoma adinognathus*) from the early Pleistocene of Meade County, Kansas, with comments on the herpetofauna of the Borchers locality. Herpetologica 32: 64–67.

RIEDLE, J. D. 1994a. A survey of reptiles and amphibians at Montgomery County State Fishing Lake. Kansas Herpetol. Soc. News. 98: 11–13.

RIEDLE, J. D. 1994b. Distribution of the Timber Rattlesnake (*Crotalus horridus*) in Chautauqua, Elk, and Montgomery counties, Kansas. Kansas Herpetol. Soc. News. 95: 11–12.

RIEDLE, J. D. 1996. Some occurrences of the Western Diamondback Rattlesnake (*Crotalus atrox*) in Kansas. Kansas Herpetol. Soc. News. 105: 18–19.

RIEDLE, J. D. 1998a. Water snake feeding records. Kansas Herpetol. Soc. News. 111: 16.

RIEDLE, J. D. 1998b. Winter snake activity. Kansas Herpetol. Soc. News. 111: 16.

RIEDLE, J. D., AND K. GRIMM. 1995. The amphibians and reptiles of Reading Woods Natural Area. Abstract. 22nd Annual Meeting of the Kansas Herpetological Society, Lawrence.

RIEDLE, J. D., AND A. HYNEK. 2002. Amphibian and reptile inventory of the Kansas Army Ammunition Plant, Labette County, Kansas. J. Kansas Herpetol. 2: 18–20.

RIEDLE, J. DAREN, DAY B. LIGON, AND KERRY GRAVES. 2008. Distribution and management of Alligator Snapping Turtles, *Macrochelys temminckii*, in Kansas and Oklahoma. Trans. Kansas Acad. Sci. 111: 21–28.

RIEDLE, J. D., P. A. SHIPMAN, S. F. FOX, AND D. M. LESLIE, JR. 2005. Status and distribution of the Alligator Snapping Turtles, *Macrochelys temminckii*, in Oklahoma. Southwest. Nat. 50: 79–84.

RITKE, M. E., J. G. BABB, AND M. K. RITKE. 1991. Annual growths rates of adult Gray Treefrogs (*Hyla chrysoscelis*). J. Herpetol. 25: 382–385.

ROBISON, W. G., AND W. W. TANNER. 1962. A comparative study of the species of the genus *Crotaphytus* Holbrook (Iguanidae). Brigham Young Univ. Sci. Bull. 2(1): 1–31.

ROBLE, S. M. 1979. Dispersal movements and plant associations of juvenile Gray Treefrogs, *Hyla versicolor* LeConte. Trans. Kansas Acad. Sci. 82: 235–245.

ROBLE, S. M. 1985. Observations on satellite males in *Hyla chrysoscelis*, *Hyla picta*, and *Pseudacris triseriata*. J. Herpetol. 19: 432–436.

RODECK, H. G. 1950. Guide to the turtles of Colorado. Univ. Colorado Mus., Leaflet 7: 1–9.

RODRÍGUEZ-ROBLES, J. A., C. J. BELL, AND H. W. GREENE. 1999. Food habits of the Glossy Snake, *Arizona elegans*, with comparisons to the diet of sympatric Longnose Snakes, *Rhinocheilus lecontei*. J. Herpetol. 33: 87–92.

ROGERS, J. S. 1973a. Protein polymorphism, genic heterozygosity and divergence in the toads *Bufo cognatus* and *B. speciosus*. Copeia 1973: 322–330.

ROGERS, J. S. 1973b. Biochemical and morphological analysis of potential introgression between *Bufo cognatus* and *B. speciosus*. American Mid. Nat. 90: 127–142.

ROGERS, K. L. 1976. Herpetofauna of the Beck Ranch Local Fauna (Upper Pliocene: Blancan) of Texas. Publ. Mus. Michigan St. Univ. Paleo. Ser. 1: 163–200.

ROGERS, K. L. 1982. Herpetofaunas of the Courtland Canal and Hall Ash Local faunas (Pleistocene: Early Kansas) of Jewell Co., Kansas. J. Herpetol. 16: 174–177.

ROSE, R. 1978. Observations on natural history of the Ornate Box Turtle (*Terrapene o. ornata*). Abstract. Trans. Kansas Acad. Sci. 81: 171–172.

ROSEN, P. C. 1991. Comparative ecology and life history of the Racer (*Coluber constrictor*) in Michigan. Copeia 1991: 897–909.

ROSSI, J. W. 1992. Snakes of the United States and Canada: Keeping Them Healthy in Captivity. Vol. 1, Eastern Area. Krieger Publishing Company, Malabar, Florida.

ROSSI, J. W., AND R. ROSSI. 1995. Snakes of the United States and Canada: Keeping Them Healthy in Captivity. Vol. 2, Western Area. Krieger Publishing Company, Malabar, Florida.

ROSSMAN, D. A. 1963. The colubrid snake genus *Thamnophis:* a revision of the *sauritus* group. Bull. Florida St. Mus. 7: 99–178.

ROSSMAN, D. A. 1970. *Thamnophis proximus.* Cat. American Amph. Reptl. 98.1–98.3.

ROSSMAN, D. A., N. B. FORD, AND RICHARD A. SEIGEL. 1996. The Garter Snakes: Evolution and Ecology. University of Oklahoma Press, Norman.

ROTH, S. D. 1959. The Comparative Microclimate of Four Habitats in Chase County, Kansas. Unpublished Master's thesis, Emporia State University, Emporia, Kansas.

ROTH, S. D., AND J. T. COLLINS. 1979. Geographic distribution: *Bufo debilis insidior.* Herpetol. Rev., 10: 118.

ROYAL, S. M. 1982. Herpetofauna of a Sandsage Prairie Near Holcomb, Kansas. Unpublished Master's thesis, Fort Hays State University, Hays, Kansas.

RUDOLPH, D. C. 1978. Aspects of the larval ecology of five plethodontid salamanders of the western Ozarks. American Mid. Nat. 100: 141–159.

RUES, T. 1986. The Kaw River Nature and History Guide. Kansas State Historical Society, Topeka.

RUNDQUIST, E. M. 1975. First KHS field trip yields three county records. Kansas Herpetol. Soc. News. 7: 1–3.

RUNDQUIST, E. M. 1976. Field checklist (of) amphibians and reptiles of Kansas. Publication (2 pp.) of the Kansas Herpetological Society, Lawrence.

RUNDQUIST, E. M. 1977. Field checklist (of) amphibians and reptiles in Kansas. Revised. Publication (2 pp.) of the Kansas Herpetological Society, Lawrence.

RUNDQUIST, E. M. 1978. The Spring Peeper, *Hyla crucifer* Wied (Anura, Hylidae) in Kansas. Trans. Kansas Acad. Sci. 80: 155–158.

RUNDQUIST, E. M. 1979. The status of *Bufo debilis* and *Opheodrys vernalis* in Kansas. Trans. Kansas Acad. Sci. 82: 67–70.

RUNDQUIST, E. M. 1992. Results of the KHS 1992 fall field trip. Kansas Herpetol. Soc. News. 90: 4.

RUNDQUIST, E. M. 1993. A 20-year retrospective of the Kansas Herpetological Society: past, present, and future. Kansas Herpetol. Soc. News. 94: 12–16.

RUNDQUIST, E. M. 1994a. Results of the sixth annual KHS herp counts held 1 April–31 May 1994. Kansas Herpetol. Soc. News. 97: 5–14.

RUNDQUIST, E. M. 1994b. Reptiles and Amphibians: Management in Captivity. TFH Publications, Inc., Neptune City, New Jersey.

RUNDQUIST, E. M. 1995a. Additional KHS herp counts for 1995. Kansas Herpetol. Soc. News. 102: 11–13.

RUNDQUIST, E. M. 1995b. Results of the seventh annual KHS herp counts held 1 April–31 May 1995. Kansas Herpetol. Soc. News. 101: 11–17.

RUNDQUIST, E. M. 1996a. Notes on the natural history of some Kansas amphibians and reptiles: Parasites. Kansas Herpetol. Soc. News. 105: 16–17.

RUNDQUIST, E. M. 1996b. Results of the eighth annual KHS herp counts Held 1 April–31 May 1996. Kansas Herpetol. Soc. News. 104: 6–17.

RUNDQUIST, E. M. 1997a. Addendum to 1997 KHS herp counts. Kansas Herpetol. Soc. News. 109: 14–15.

RUNDQUIST, E. M. 1997b. Results of the ninth annual KHS herp counts held 1 April–31 May 1997. Kansas Herpetol. Soc. News. 108: 12–17.

RUNDQUIST, E. M. 1998a. Blind snake reproductive activity. Kansas Herpetol. Soc. News. 111: 16–17.

RUNDQUIST, E. M. 1998b. Racer reproduction and diet observation. Kansas Herpetol. Soc. News. 113: 15.

RUNDQUIST, E. M. 1998c. Results of the tenth annual KHS herp counts for 1998, held 1 April–31 May. Kansas Herpetol. Soc. News. 112: 11–18.

RUNDQUIST, E. M. 1998d. Winter snake activity. Kansas Herpetol. Soc. News. 111: 16.

RUNDQUIST, E. M. 1999. Kansas Herpetological Society herp counts: a 10-year summary and evaluation. Kansas Herpetol. Soc. News. 115: 8–15.

RUNDQUIST, E. M. 2000a. Garter snake hybridization. Kansas Herpetol. Soc. News. 118: 15.

RUNDQUIST, E. M. 2000b. Results of the eleventh and twelfth annual KHS herpetofaunal counts for 1999–2000, held 1 April–31 May. Kansas Herpetol. Soc. News. 122: 11–16.

RUNDQUIST, E. M. 2001. Results of the thirteenth annual KHS herp counts for 2001, held 1 April–30 June. Kansas Herpetol. Soc. News. 125: 13–16.

RUNDQUIST, E. M. 2002. Natural history of the Night Snake, *Hypsiglena torquata*, in Kansas. J. Kansas Herpetol. 4: 16–20.

RUNDQUIST, E. M. (ED.). 1994. Additions and corrections [to the results of the sixth annual KHS herp counts held 1 April–31 May 1994]. Kansas Herpetol. Soc. News. 98: 4.

RUNDQUIST, E. M., AND J. T. COLLINS. 1977. The amphibians of Cherokee County, Kansas. Reptl. St. Biol. Surv. Kansas 14: 1–12.

RUNDQUIST, E. M., AND J. T. COLLINS. 1993. Results of the fifth Kansas herp count held during April–June 1993. Kansas Herpetol. Soc. News. 94: 7–11.

RUNDQUIST, E. M., E. STEGALL, D. GROW, AND P. GRAY. 1978. New herpetological records from Kansas. Trans. Kansas Acad. Sci. 81: 73–77.

RUNDQUIST, E. M., AND J. TRIPLETT. 1993. Additional specimens of the Western Cottonmouth (*Agkistrodon piscivorus leucostoma*, Reptilia: Squamata) from Kansas. Trans. Kansas Acad. Sci. 96: 148–151.

RUSH, M. S. 1981. The Effectiveness of Seven Trapping Techniques for Amphibians, Reptiles, and Incidental Mammals in the Sandsage Prairie. Unpublished Master's thesis, Fort Hays State University, Hays, Kansas.

RUSH, M. S., AND E. D. FLEHARTY. 1980. Geographic distribution: *Gastrophryne carolinensis*. Herpetol. Rev. 11: 80.

RUSH, M. S., AND E. D. FLEHARTY. 1981. New county records of amphibians and reptiles in Kansas. Trans. Kansas Acad. Sci. 84: 204–208.

RUSH, M. S., S. M. ROYAL AND E. D. FLEHARTY. 1982. New county records and habitat preferences of amphibians and reptiles from the sandsage prairie in Finney County, Kansas. Trans. Kansas Acad. Sci. 85: 165–173.

RUSSELL, A. P., AND A. M. BAUER. 1993. The Amphibians and Reptiles of Alberta. University of Alberta Press, Edmonton, Canada.

RUSSELL, F. E. 1980. Snake Venom Poisoning. Lippincott, Philadelphia, Pennsylvania.

RUTHVEN, A. G. 1907. A collection of reptiles and amphibians from southern New Mexico and Arizona. Bull. American Mus. Nat. Hist. 23: 483–604.

RUTHVEN, A. G. 1908. Variations and genetic relationships of the garter snakes. Bull. U. S. Natl. Mus. 61: 1–201.

SAGE, R. D., AND R. K. SELANDER. 1979. Hybridization between species of the *Rana pipiens* complex in central Texas. Evolution 33: 1,069–1,088.

SALAZAR, B. M. 1968a. Field trips. Inter. Turtle Tortoise Soc. J. 2(4): 4–5.

SALAZAR, B. M. 1968b. Finding the turtles. Inter. Turtle Tortoise Soc. J. 2(5): 4–5.

SALTHE, S. N., AND E. NEVO. 1969. Geographic variation of lactate dehydrogenase in the cricket frog, *Acris crepitans*. Biochem. Genetics 3: 335–341.

SANDERS, O. 1987. Evolutionary hybridization and speciation in North American indigenous bufonids. Privately Printed (*viii* + 110 pp.), Dallas, Texas.

SANDERS, O., AND H. M. SMITH. 1951. Geographic variation in toads of the *debilis* group of *Bufo*. Field and Laboratory 19(4): 141–160.

SANDERS, R., B. WOLHUTER, AND J. T. COLLINS. 2005. Geographic distribution: *Sternotherus odoratus* from Kansas. J. Kansas Herpetol. 16: 22.

SATTLER, P. W. 1985. Introgressive hybridization between the spadefoot toads *Scaphiopus bombifrons* and *S. multiplicatus* (Salientia: Pelobatidae). Copeia 1985: 324–332.

SATTLER, P. W., AND J. S. RIES. 1995. Intraspecific genetic variation among four populations of the Texas Horned Lizard, *Phrynosoma cornutum*. J. Herpetol. 29: 137–141.

SAVAGE, J. 1877. On the bite of the rattlesnake. Trans. Kansas Acad. Sci. 6: 36–38.

SAVAGE, J. M. 1954. A revision of the toads of the *Bufo debilis* complex. Texas J. Sci. 6: 83–112.

SAVAGE, J. M. 1959. An illustrated key to the lizards, snakes and turtles of the west. Naturegraph Pocket Keys 2: 1–36.

SCHAAF, R. T., AND P. W. SMITH. 1970. Geographic variation in the Pickerel Frog. Herpetologica 26: 240–254.

SCHAAF, R. T., AND P. W. SMITH. 1971. *Rana palustris*. Cat. American Amph. Reptl. 117.1–117.3.

SCHAEFER, J. 1984. New life for nongame. Kansas Wildlife 41(2): 26–30.

SCHAEFER, J. 1985. The extinct: losses and lessons. Kansas Wildlife 42(2): 26–30.

SCHEFFER, T. H. 1911. Distribution, natural enemies and breeding habits of the Kansas Pocket Gopher. Trans. Kansas Acad. Sci. 23: 109–114.

SCHMAUS, J. W. 1959. Envenomation in Kansas. The poisonous varieties of animals in the state. J. Kansas Med. Soc. 60(7): 237–248.

SCHMID, W. D. 1965. High temperature tolerances of *Bufo hemiophrys* and *Bufo cognatus*. Ecology 46: 559–560.

SCHMIDT, C. J. 1998. Herpetological observations at Cheyenne Bottoms, Barton County, Kansas. Kansas Herpetol. Soc. News. 113: 15.

SCHMIDT, C. J. 2000. Observations on reptilian predation. Kansas Herpetol. Soc. News. 120: 18.

SCHMIDT, C. J. 2001. The amphibians, turtles, and reptiles of the Smoky Valley Ranch, Logan County, Kansas. Kansas Herpetol. Soc. News. 124: 9–11.

SCHMIDT, C. J. 2002a. A demographic analysis of the Prairie Rattlesnakes collected for the 2000 and 2001 Sharon Springs, Kansas, rattlesnake roundups. J. Kansas Herpetol. 1: 12–18.

SCHMIDT, C. J. 2002b. Organization and background of the 2000 and 2001 Sharon Springs, Kansas, rattlesnake roundups. J. Kansas Herpetol. 1: 9.

SCHMIDT, C. J. 2003a. Geographic distribution: *Nerodia erythrogaster*. J. Kansas Herpetol. 6: 8.

SCHMIDT, C. J. 2003b. Geographic distribution: *Regina grahamii*. J. Kansas Herpetol. 6: 8.

SCHMIDT, C. J. 2003c. Geographic distribution: *Ambystoma mavortium*. J. Kansas Herpetol. 8: 19.

SCHMIDT, C. J. 2003d. Geographic distribution: *Eumeces septentrionalis*. J. Kansas Herpetol. 8: 19.

SCHMIDT, C. J. 2004a. Geographic distribution: *Chelydra serpentina*. Kansas. J. Kansas Herpetol. 10: 10.

SCHMIDT, C. J. 2004b. Geographic distribution: *Chrysemys picta*. Kansas. J. Kansas Herpetol. 10: 10.

SCHMIDT, C. J. 2004c. Geographic distribution: *Trachemys scripta*. Kansas. J. Kansas Herpetol. 10: 10.

SCHMIDT, C. J. 2004d. Attempted predation on a hatch-ling Texas Horned Lizard (*Phrynosoma cornutum*) by an adult Plains Leopard Frog (*Rana blairi*). J. Kansas Herpetol. 10: 12.

SCHMIDT, C. J. 2004e. Geographic distribution: *Hyla chrysoscelis/Hyla versicolor*. Kansas. J. Kansas Herpetol. 11: 13.

SCHMIDT, C. J. 2004f. Geographic distribution: *Bufo cognatus*. Kansas. J. Kansas Herpetol. 11: 13.

SCHMIDT, C. J. 2004g. Geographic distribution: *Apalone spinifera*. Kansas. J. Kansas Herpetol. 11: 13.

SCHMIDT, C. J. 2004h. Geographic distribution: *Ophisau-rus attenuatus*. Kansas. J. Kansas Herpetol. 11: 13.

SCHMIDT, C. J. 2004i. Geographic distribution: *Pituophis catenifer*. Kansas. J. Kansas Herpetol. 11: 14.

SCHMIDT, C. J. 2004j. Geographic distribution: *Storeria dekayi*. Kansas. J. Kansas Herpetol. 11: 14.

SCHMIDT, C. J. 2004k. Geographic distribution: *Thamnophis sirtalis*. Kansas. J. Kansas Herpetol. 11: 14.

SCHMIDT, C. J. 2004l. Natural history and status of the exploited Prairie Rattlesnake (*Crotalus viridis*) in western Kansas and a herpetofaunal inventory of the Smoky Valley Ranch, Logan County, Kansas. Unpublished Master's thesis, Fort Hays State University, Hays, Kansas.

SCHMIDT, C. J. 2005. Geographic distribution: *Chelydra serpentina*. Kansas. J. Kansas Herpetol. 16: 22.

SCHMIDT, C. J. 2006. An addition to the herpetofauna of Cheyenne Bottoms, Barton County, Kansas. J. Kansas Herpetol. 20: 9.

SCHMIDT, C. J. 2007. First record of the Bullfrog from Osborne County, Kansas. J. Kansas Herpetol. 22: 11.

SCHMIDT, C. J. 2008a. Geographic distribution: *Spea bombifrons*. Kansas. J. Kansas Herpetol. 27: 7.

SCHMIDT, C. J. 2008b. Geographic distribution: *Anaxyrus cognatus*. Kansas. J. Kansas Herpetol. 27: 7.

SCHMIDT, C. J. 2009a. Geographic distribution: *Nerodia sipedon*. J. Kansas Herpetol. 30: 11.

SCHMIDT, C. J. 2009b. Geographic distribution: *Tropido-clonion lineatum*. J. Kansas Herpetol. 30: 11.

SCHMIDT, C. J., AND R. HAYES. 2004. Life history. *Gastro-trophryne olivacea*. New maximum length. J. Kansas Herpetol. 11: 15.

SCHMIDT, C. J., D. MURROW, AND T. W. TAGGART. 2005. Geographic distribution: *Rana areolata*. Kansas. J. Kansas Herpetol. 14: 11.

SCHMIDT, C. J., AND K. PHELPS. 2007. Geographic distribu-tion: *Thamnophis sirtalis*. J. Kansas Herpetol. 24: 15.

SCHMIDT, C. J., AND W. J. STARK. 2002. An assessment of the harvest of Prairie Rattlesnakes (*Crotalus viridis*) during the Sharon Springs rattlesnake roundups in 2000 and 2001, and an investigation of unexploited populations within the Smoky Valley Ranch, a Nature Conservancy property in Logan County, Kansas. Report (46 pp.), Kansas Department of Wildlife and Parks, Pratt.

SCHMIDT, C. J., AND T. W. TAGGART. 2002a. Geographic distribution: *Gastrophryne olivacea*. Kansas. J. Kansas Herpetol. 2: 10.

SCHMIDT, C. J., AND T. W. TAGGART. 2002b. Geographic distribution: *Trachemys scripta*. Kansas. J. Kansas Herpetol. 2: 11.

SCHMIDT, C. J., AND T. W. TAGGART. 2004a. Life history. *Kinosternon flavescens*. Abnormal characteristic. J. Kansas Herpetol. 9: 7.

SCHMIDT, C. J., AND T. W. TAGGART. 2004b. Life history. *Rana blairi*. New state maximum length. J. Kansas Herpetol. 12: 17.

SCHMIDT, C. J., AND T. W. TAGGART. 2005a. Life history: *Sceloporus consobrinus*. New state maximum length. J. Kansas Herpetol. 14: 10.

SCHMIDT, C. J., AND T. W. TAGGART. 2005b. Geographic distribution: *Nerodia erythrogaster*. Kansas. J. Kansas Herpetol. 14: 11.

SCHMIDT, D. 2006. County record Milk Snake from Kiowa County. J. Kansas Herpetol. 20: 9.

SCHMIDT, D., AND D. MURROW. 2006. Kansas Herpetological Society fall field trip results. J. Kansas Herpetol. 20: 2–3.

SCHMIDT, D., AND D. MURROW. 2007a. KHS spring field trip to Seward County. J. Kansas Herpetol. 21: 2–3.

SCHMIDT, D., AND D. MURROW. 2007b. KHS spring 2007 field trip results. J. Kansas Herpetol. 22: 2–3.

SCHMIDT, D., AND D. MURROW. 2007c. Report on the KHS fall field trip to Miami County, Kansas. J. Kansas Herpetol. 24: 11–14.

SCHMIDT, D., AND L. SCHMIDT. 2009. Geographic distribu-tion: *Trachemys scripta*. J. Kansas Herpetol. 30: 11.

SCHMIDT, K. P. 1922. A review of the North American genus of lizards *Holbrookia*. Bull. American Mus. Nat. Hist. 46: 709–725.

SCHMIDT, K. P. 1938a. Turtles of the Chicago area. Field Mus. Nat. Hist. Zool. Leaflet 14: 1–24.

SCHMIDT, K. P. 1938b. Herpetological evidence for the postglacial eastward extension of the steppe in North America. Ecology 19: 396–407.

SCHMIDT, K. P. 1953. A Check List of North American Amphibians and Reptiles. 6th ed. American Society of Ichthyologists and Herpetologists. Composed and Printed by The University of Chicago Press, Chicago, Illinois.

SCHMIDT, K. P., AND D. D. DAVIS. 1941. Field Book of Snakes of the United States and Canada. Putnam's Sons, New York.

SCHMIDT, K. P., AND R. F. INGER. 1957. Living Reptiles of the World. Hanover House, Garden City, New York.

SCHOENER, T. W. 1977. Competition and niche. Pp. 35–136 *In* C. Gans and D. W. Tinkle (Eds.), Biology of the Reptilia. Vol. 7. Academic Press, New York.

SCHOEWE, W. H. 1937. Rock City, a proposed national monument in Ottawa County, Kansas. National Monument Project of the Kansas Academy of Science. 15 pp.

SCHROEDER, E. E., AND T. S. BASKETT. 1965. Frogs and Toads of Missouri. Missouri Conservationist (11 pp.).

SCHUETT, G. W. 1992. Is long-term sperm storage an important component of the reproductive biology of temperate pitvipers? Pp. 169–184 *In* J. A. Campbell and E. D. Brodie, Jr. (Eds.), Biology of the Pitvipers. Selva, Tyler, Texas.

SCHULENBERG PTACEK, M. 1984a. Reproductive Ecology and Habitat Analysis of the Northern Spring Peeper (*Hyla crucifer crucifer*) in Cherokee County, Kansas. Unpublished Master's thesis, Emporia State University, Emporia, Kansas.

SCHULENBERG PTACEK, M. 1984b. Reproductive ecology and habitat analysis of the Northern Spring Peeper (*Hyla crucifer crucifer*) in Cherokee County, Kansas. Kansas Herpetol. Soc. News. 61: 12–14.

SCHULTZ, K. D. 1996. A Monograph of the Colubrid Snakes of the Genus *Elaphe* Fitzinger. Koeltz Scientific Books, Havlickuv Brod, Czech Republic.

SCHUMANN, M. 1923. A Vestigial Maturation Cycle in the Male Larvae of *Rana catesbeiana*. Unpublished Master's thesis, University of Kansas, Lawrence.

SCHWANER, T. D. 1978. KHS field trip to Grant County, Kansas, 12–14 May 1978. Kansas Herpetol. Soc. News. 25: 3–4.

SCHWARDT, H. H. 1938. Reptiles of Arkansas. Univ. Arkansas Agric. Exp. Sta. Bull. 357: 1–47.

SCHWARTING, N. 1984. KHS field trip—May 1984. Kansas Herpetol. Soc. News. 57: 3–4.

SCHWARTZ, F. J., AND B. L. DUTCHER. 1961. A record of the Mississippi Map Turtle, *Graptemys kohni*, in Maryland. Chesapeake Science 2(1–2): Unpaginated.

SCHWENKE, Z. J. 2008. Geographic distribution: *Lithobates catesbeianus*. Kansas. J. Kans as Herpetol. 27: 7.

SCHWILLING, M. 1981. Kansas nongame and endangered wildlife. Kansas School Nat. 27(3): 3–15.

SCHWILLING, M. 1985. Cheyenne Bottoms. Kansas School Nat. 32(3): 1–15.

SCHWILLING, M., AND C. NILON. 1988. Kansas nongame and endangered wildlife: Annual report (*i* + 35 pp.), Federal Aid Project Number FW-9-P-6.

SCHWILLING, M., AND J. M. SCHAEFER. 1983. Kansas nongame, threatened, and endangered species: Annual report (*i* + 28 pp.), Kansas Fish and Game Commission, Pratt.

SCOTT, A. F., AND J. KOONS. 1994. Natural history notes. *Macroclemys temminckii*. Herpetol. Rev. 25: 64.

SCOTT, C. 1996. Snake Lovers' Lifelist and Journal. University of Texas Press, Austin.

SCOTT, N. J., JR. (EDS.). 1982. Herpetological communities. Wildlife Service Report 13 (*iv* + 239 pp.), U. S. Fish and Wildlife Service, Washington, D. C.

SECOR, S. M. 1987. Courtship and mating behavior of the Speckled Kingsnake, *Lampropeltis getulus holbrooki*. Herpetologica 43: 15–28.

SEIDEL, M. E. 1994. Morphometric analysis and taxonomy of cooter and redbelly turtles in the North American genus *Pseudemys* (Emydidae). Chelonian Conserv. Biol. 1: 117–130.

SEIDEL, M. E., AND M. J. DRESLIK. 1996. *Pseudemys concinna*. Cat. American Amph. Reptl. 626.1–626.12.

SEIDEL, M. E., AND C. H. ERNST. 1996. *Pseudemys*. Cat. American Amph. Reptl. 625.1–625.7.

SEIDEL, M. E., AND C. H. ERNST. 2006. *Trachemys scripta*. Cat. American Amph. Reptl. 831.1–831.94.

SEIGEL, R. A., J. T. COLLINS AND S. S. NOVAK (EDS.). 1987. Snakes: Ecology and Evolutionary Biology. Blackburn Press, Caldwell, New Jersey.

SEIGEL, R. A., AND H. S. FITCH. 1985. Annual variation in reproduction in snakes in a fluctuating environment. J. Anim. Ecol. 54: 497–505.

SEIGEL, R. A., L. E. HUNT, J. L. KNIGHT, L. MALARET, AND N. L. ZUSCHLAG (EDS.). 1984. Vertebrate Ecology and Systematics: A Tribute to Henry S. Fitch. Special Publication Number 10. The University of Kansas Museum of Natural History, Lawrence.

SEVER, S. G. 1966. Sexual and Brooding Behavior of the Lizard, *Eumeces fasciatus*. Unpublished Master's thesis, Emporia State University, Emporia, Kansas.

SEXTON, O. J., N. SHANNON, AND S. SHANNON. 1976. Late season hatching success of *Elaphe o. obsoleta*. Herpetol. Rev. 7: 171.

SHAFFER, L. L., AND C. J. McCOY. 1991. Pennsylvania Amphibians and Reptiles. Pennsylvania Fish Commission, Harrisburg.

SHANTZ, H. L., F. W. ALBERTSON, W. A. ALBRECHT, R. DAUBENMIRE, H. S. FITCH, R. L. PIEMEISEL, AND W. G. WHALEY. 1959. Aspects of needed research on North American grasslands. Trans. Kansas Acad. Sci. 62: 175–183.

SHAW, C. E., AND S. CAMPBELL. 1974. Snakes of the American West. A. A. Knopf, New York.

SHEPHERD, J. L. 1973. Osteology of the Worm Snakes (*Carphophis*). Unpublished Doctoral dissertation, Ball State University, Muncie, Indiana.

SHERBROOKE, W. C. 1981. Horned Lizards. Unique Reptiles of Western North America. Southwest Parks and Monuments Association, Globe, Arizona.

SHINE, R., AND D. CREWS. 1988. Why male garter snakes have small heads: the evolution and endocrine control of sexual dimorphism. Evolution 42: 1,105–1,110.

SHIPMAN, P. A. 1993a. Behavioral ecology of the Alligator Snapping Turtle (*Macroclemys temminckii*) in southeast Kansas. Abstract No. 6. Annual Meeting of the Southwestern Association of Naturalists, Springfield, Missouri.

SHIPMAN, P. A. 1993b. Natural history of the Alligator Snapping Turtle (*Macroclemys temminckii*) in Kansas. Kansas Herpetol. Soc. News. 93: 14–17.

SHIPMAN, P. A., AND D. R. EDDS. 1992. Ecology of the Alligator Snapping Turtle in Kansas. Abstract. 124th Annual Meeting of the Kansas Academy of Science.

SHIPMAN, P. A., D. R. EDDS, AND D. BLEX. 1991. Report on the recapture of an Alligator Snapping Turtle (*Macroclemys temminckii*) in Kansas. Kansas Herpetol. Soc. News. 85: 8–9.

SHIPMAN, P. A., D. R. EDDS, AND D. BLEX. 1994. Natural history notes. *Macroclemys temminckii* and *Chelydra serpentina*. Herpetol. Rev. 25: 24–25.

SHIPMAN, P. A., D. R. EDDS, AND L. E. SHIPMAN. 1995. Distribution of the Alligator Snapping Turtle (*Macroclemys temminckii*) in Kansas. Trans. Kansas Acad. Sci. 98: 83–91.

SHIPMAN, P. A., D. R. EDDS, L. E. SHIPMAN, AND D. BLEX. 1993. Alligator Snapping Turtle (*Macroclemys temminckii*) habitat selection, movements, and natural history in southeast Kansas. Contract 279, Final Report (*ix* + 91 pp.), Kansas department of Wildlife and Parks, Pratt.

SHIPMAN, P. A., AND J. D. RIEDLE. 2008. Status and distribution of the Alligator Snapping Turtle (*Macrochelys temminckii*) in southeastern Missouri. Southeastern Nat. 7: 331–338.

SHIRER, H. W., AND H. S. FITCH. 1970. Comparison from radiotracking of movements and denning habits of the Raccoon, Striped Skunk, and Opossum in northeastern Kansas. J. Mammal. 51(3): 491–503.

SHOUP, J. M. 1992a. Confirmed Cottonmouth. Kansas Wildlife and Parks 49(1): 41.

SHOUP, J. M. 1992b. Blue shadow in the shallows. Kansas Wildlife and Parks 49(4): 2–7.

SHOUP, J. M. 1996a. Treefrogs, indeed! Kansas Wildlife and Parks Magazine 53(4): 34.

SHOUP, J. M. 1996b. Wise as serpents. Kansas Wildlife and Parks Magazine 53(4): 39.

SIEVERT, G., M. GILKERSON, V. TUTTLE, AND P. TUTTLE. 2006. *Arizona elegans.* Geographic distribution. Herpetol. Rev. 37: 496.

SIEVERT, L. M., AND J. K. BAILEY. 2000. Specific dynamic action in the toad, *Bufo woodhousii.* Copeia 2000: 1,076–1,078.

SIMBOTWE, P. M. 1978. Geographic Variation and Natural Selecton in *Eumeces obsoletus* (Reptilia: Sauria: Scincidae). Unpublished Master' thesis, University of Kansas, Lawrence.

SIMBOTWE, P. M. 1981. Natural selection in the lizard *Eumeces obsoletus* (Lacertilia: Scincidae). Amphibia-Reptilia 2: 143–151.

SIMMONS, J. E. 1983. Fan fare. Kansas Herpetol. Soc. News. 54: 16–18.

SIMMONS, J. E. 1989. Endangered and threatened in Kansas. Kansas Herpetol. Soc. News. 75: 4–5.

SIMMONS, J. E. 1997. Report on a clutch of rat snake eggs (*Elaphe obsoleta*) from Kansas. Kansas Herpetol. Soc. News. 108: 10–11.

SIMMONS, J. E. 1998. A technique to remove reptiles from glue boards. Kansas Herpetol. Soc. News. 114: 18.

SIMMONS, J. E. (ED.). 1984. Record turtle from Kansas. Kansas Herpetol. Soc. News. 58: 11–13.

SIMMONS, J. E. (ED.). 1985. Commercial herping in Kansas. Kansas Herpetol. Soc. News. 59: 4.

SIMMONS, J. E. (ED.). 1986a. The Kansas state reptile. Kansas Herpetol. Soc. News. 64: 7–10.

SIMMONS, J. E. (ED.). 1986b. KHS brings you news of the world—special giant turtle edition. Kansas Herpetol. Soc. News. 65: 8–12.

SIMMONS, J. E. (ED.). 1986c. Study shows most truckers aim at reptiles. Kansas Herpetol. Soc. News. 66: 6.

SIMMONS, J. E. (ED.). 1987a. To remember our state reptile. Kansas Herpetol. Soc. News. 67: 12.

SIMMONS, J. E. (ED.). 1987b. Gray Treefrogs in Kansas. Kansas Herpetol. Soc. News. 68: 7.

SIMMONS, J. E. (ED.). 1987c. September 1987 field trip report. Kansas Herpetol. Soc. News. 69: 6–8.

SIMMONS, J. E. (ED.). 1988. Is this any way to divide a highway? Kansas Herpetol. Soc. News. 73: 19–20.

SIMON, H. 1973. Snakes, the Facts and the Folklore. Viking Press, New York.

SIMON, M. P. 1988. Report on the status of selected amphibian species of special interest in northeastern Kansas. Final Report (11 pp.), Kansas Department of Wildlife and Parks, Pratt.

SIMON, M. P., AND J. H. DORLAC. 1990. The results of a faunistic survey of reptiles and amphibians of Fort Leavenworth, Kansas. Project Report (11 pp.), Kansas Department of Wildlife and Parks, Pratt.

SITES, J. W., JR., J. W. ARCHIE, C. J. COLE, AND O. FLORES VILLELA. 1992. A review of phylogenetic hypotheses for lizards of the genus *Sceloporus* (Phrynosomatidae): implications for ecological and evolutionary studies. Bull. American Mus. Nat. Hist. 213: 1–110.

SKIE, S., AND M. BICKFORD. 1978. KHS takes to the field in July at Winfield. Kansas Herpetol. Soc. News. 26: 3–4.

SMITH, A. G. 1940. Notes on the reproduction of the Northern Copperhead, *Agkistrodon mokasen cupreus* (Rafinesque), in Pennsylvania. Annals Carnegie Mus. 28: 77–82.

SMITH, A. G. 1949. The subspecies of the Plains Garter Snake, *Thamnophis radix.* Bull. Chicago Acad. Sci. 8: 285–300.

SMITH, A. K. 1973. Feeding Strategies and Population Dynamics of the Bullfrog (*Rana catesbeiana*). Unpublished Doctoral dissertation, University of Kansas, Lawrence.

SMITH, A. K. 1976. Incidence of tail coiling in a population of Ringneck Snakes (*Diadophis punctatus*). Trans. Kansas Acad. Sci. 77(4): 237–238.

SMITH, A. K. 1977. Attraction of Bullfrogs (Amphibia, Anura, Ranidae) to distress calls of immature frogs. J. Herpetol. 11: 234–235.

SMITH, B. E., AND N. T. STEPHENS. 2003. Conservation assessment of the Pale Milk Snake in the Black Hills National Forest, South Dakota and Wyoming. Report (17 pp.), U. S. Department of Agriculture Forest Service, Custer, South Dakota.

SMITH, D. D. 1980. The Copperhead. AALAS Kansas City Branch News. 3(2): 8–10.

SMITH, D. D. 1983. Life history notes: *Nerodia erythrogaster transversa.* Herpetol. Rev. 14: 120–121.

SMITH, D. D., D. J. PFLANZ, AND R. POWELL. 1993. Observations of autohemorrhaging in *Tropidophis haetianus, Rhinocheilus lecontei, Heterodon platyrhinos,* and *Nerodia erythrogaster.* Herpetol. Rev. 24: 130–131.

SMITH, D. D., R. POWELL, T. R. JOHNSON, AND H. L. GREGORY. 1983. Life history observations of Missouri amphibians and reptiles with recommendations for standardized data collection. Trans. Missouri Acad. Sci. 17: 37–58.

SMITH, H. M. 1931. Additions to the herpetological fauna of Riley County, Kansas. Copeia 1931: 143.

SMITH, H. M. 1932. A report upon amphibians hitherto unknown from Kansas. Trans. Kansas Acad. Sci. 35: 93–96.

SMITH, H. M. 1933a. On the proper name for the breviciptid frog *Gastrophryne texensis* (Girard). Copeia 1933: 217.

SMITH, H. M. 1933b. The Amphibians of Kansas. Unpublished Master's thesis, University of Kansas, Lawrence.

SMITH, H. M. 1934. The amphibians of Kansas. American Mid. Nat. 15: 377–528.

SMITH, H. M. 1938. Remarks on the status of the subspecies of *Sceloporus undulatus*, with descriptions of new species and subspecies of the *undulatus* group. Univ. Michigan Mus. Zool. Occas. Pap. 387: 1–17.

SMITH, H. M. 1942. Another case of species versus subspecies. American Mid. Nat. 28: 201–203.

SMITH, H. M. 1946a. The systematic status of *Eumeces pluvialis* Cope, and noteworthy records of other amphibians and reptiles from Kansas and Oklahoma. Univ. Kansas Publ. Mus. Nat. Hist. 1: 85–89.

SMITH, H. M. 1946b. The tadpoles of *Bufo cognatus* Say. Univ. Kansas Publ. Mus. Nat. Hist. 1: 93–96.

SMITH, H. M. 1946c. Handbook of Lizards: Lizards of the United States and Canada. Comstock Publishing Company, Ithaca, New York.

SMITH, H. M. 1946d. Hybridization between two species of garter snakes. Univ. Kansas Publ. Mus. Nat. Hist. 1: 97–100.

SMITH, H. M. 1947a. Kyphosis and other variations in softshelled turles. Univ. Kansas Publ. Mus. Nat. Hist. 1: 119–124.

SMITH, H. M. 1947b. *Microhyla carolinensis* in Kansas. Herpetologica 4: 13–14.

SMITH, H. M. 1950. Handbook of Amphibians and Reptiles of Kansas. 1st ed. Miscellaneous Publication Number 2. University of Kansas Museum of Natural History, Lawrence.

SMITH, H. M. 1956a. Handbook of Amphibians and Reptiles of Kansas. 2nd ed. Miscellaneous Publication Number 9. University of Kansas Museum of Natural History and State Biological Survey, Lawrence.

SMITH, H. M. 1978. Howard Kay Gloyd: biographical notes and the genesis of his "Rattlesnakes." Pp. *vii–xvi In* The Rattlesnakes, Genera *Sistrurus* and *Crotalus*: A study in Zoogeography and Evolution. 1940. Special Publication, Number 4, The Chicago Academy of Sciences, Chicago Illinois. Fascimile Reprints in Herpetology, Society for the Study of Amphibians and Reptiles.

SMITH, H. M. 2001. Searching for herps in Mexico in the 1930s: I. Bull. Chicago Herpetol. Soc. 36: 1–17.

SMITH, H. M., E. L. BELL, J. S. APPLEGARTH, AND D. CHISZAR. 1992a. Adaptive convergence in the lizard superspecies *Sceloporus undulatus*. Bull. Maryland Herpetol. Soc. 28: 123–149.

SMITH, H. M., AND D. CHISZAR. 1994. Variation in the Lined Snake (*Tropidoclonion lineatum*) in northern Texas. Bull. Maryland Herpetol. Soc. 30: 6–14.

SMITH, H. M., D. CHISZAR, C. M. ECKERMAN, AND H. D. WALLEY. 2003. The taxonomic status of the Mexican Hognose Snake *Heterodon kennerlyi* Kennicott (1860). J. Kansas Herpetol. 5: 17–20.

SMITH, H. M., D. CHISZAR, AND J. HUBBELL. 1992b. Another Colorado communal nest of the Blue Racer (*Coluber constrictor*). Cold Blooded News (Colorado Herpetol. Soc.) 19(11): 4.

SMITH, H. M., D. CHISZAR, J. A. LEMOS-ESPINAL. 1995. A new subspecies of the polytypic lizard species *Sceloporus undulatus* (Sauria: Iguanidae) from northern Mexico. Texas J. Sci. 47: 117–143.

SMITH, H. M., D. CHISZAR, J. A. LEMOS-ESPINAL, G. J. WATKINS-COLWELL, AND F. VAN BREUKELEN. 2002. Concurrent existence of relictual and contemporary subspecific intergrade populations. Bull. Chicago Herpetol. Soc. 37: 139–140.

SMITH, H. M., D. CHISZAR, AND R. B. SMITH. 1983. Comparison of regional taxonomic densities. Bull. Philadelphia Herpetol. Soc. 29(1981): 9–13.

SMITH, H. M., D. CHISZAR, J. R. STALEY II, AND K. TEPEDELEN. 1994. Populational relationships in the Corn Snake *Elaphe guttata* (Reptilia: Serpentes). Texas J. Sci. 46: 259–292.

SMITH, H. M., AND A. J. KOHLER. 1977. A survey of herpetological introductions in the United States and Canada. Trans. Kansas Acad. Sci. 80: 1–24.

SMITH, H. M., AND A. B. LEONARD. 1934. Distributional records of reptiles and amphibians in Oklahoma. American Mid. Nat. 15: 190–196.

SMITH, H. M., C. W. NIXON, AND S. A. MINTON. 1949. Observations on constancy of color and pattern in softshelled turtles. Trans. Kansas Acad. Sci. 52: 92–98.

SMITH, H. M., C. W. NIXON, AND P. E. SMITH. 1948. A partial description of the tadpole of *Rana areolata circulosa* and notes on the natural history of the race. American Mid. Nat. 39: 608–614.

SMITH, H. M., M. S. RAND, J. D. DREW, B. D. SMITH, D. CHISZAR, AND C. M. DWYER. 1991. Relictual intergrades between the Northern Prairie Lizard (*Sceloporus undulatus garmani*) and the Red-lipped Plateau Lizard (*S. u. erythrocheilus*) in Colorado. Northwest. Nat. 72: 1–11.

SMITH, H. M., AND O. SANDERS. 1952. Distributional data on Texas amphibians and reptiles. Texas J. Sci. 4: 204–219.

SMITH, H. M., AND J. A. SLATER. 1949. The southern races of *Eumeces septentrionalis* (Baird). Trans. Kansas Acad. Sci. 52: 438–448.

SMITH, J. L. 1961. An Ecological Study of the Vertebrates of a Streambank Community in Ellis County, Kansas. Unpublished Master's thesis, Fort Hays State University, Hays, Kansas.

SMITH, N. M., AND W. W. TANNER. 1974. A taxonomic study of the western collared lizards, *Crotaphytus collaris* and *Crotaphytus insularis*. Brigham Young Univ. Sci. Bull. 19(4): 1–29.

SMITH, P. W. 1951. A new frog and a new turtle from the western Illinois Sand Prairies. Bull. Chicago Acad. Sci. 9: 189–199.

SMITH, P. W. 1956b. The status, correct name, and geographic range of the Boreal Chorus Frog. Proc. Biol. Soc. Washington 69: 169–176.

SMITH, P. W. 1957. An analysis of post-Wisconsin biogeography of the prairie peninsula region based on distributional phenomena among terrestrial vertebrate populations. Ecology 38: 205–218.

SMITH, P. W., AND D. M. SMITH. 1952a. The relationships of the Chorus Frogs, *Pseudacris nigrita feriarum* and *Pseudacris n. triseriata*. American Mid. Nat. 48: 165–180.

SMITH, P. W., AND H. M. SMITH. 1952b. Geographic variation in the lizard *Eumeces anthracinus*. Univ. Kansas Sci. Bull. 34: 679–694.

SMITH, P. W., AND H. M. SMITH. 1962. The systematic and biogeographic status of two Illinois snakes. C. C. Adams Center Ecol. Stud. Occas. Pap. 5: 1–10.

SMITH, P. W., AND H. M. SMITH. 1963. The systematic status of the Lined Snake of Iowa. Proc. Biol. Soc. Washington 76: 297–304.

SMITS, A. W., AND H. B. LILLYWHITE. 1985. Maintenance of blood volume in snakes: transcapillary shifts of extravascular fluids during acute hemorrhage. J. Comp. Physiol. B. 155: 305–310.

SNELSON, F. F., AND W. M. PALMER. 1966a. The Eastern Coachwhip. Wildlife in North Carolina (February issue): 26.

SNELSON, F. F., AND W. M. PALMER. 1966b. The Copperhead. Wildlife in North Carolina (April issue): 20.

SNELSON, F. F., AND W. M. PALMER. 1967. The Rough Green Snake. Wildlife in North Carolina (May issue): 13.

SNIDER, A. T., AND J. K. BOWLER. 1992. Longevity of Reptiles and Amphibians in North American Collections. 2nd ed. Herpetological Circular 21, Society for the Study of Amphibians and Reptiles.

SOMMA, L. A., AND P. A. COCHRAN. 1989. Bibliography and subject index of the Prairie Skink, *Eumeces septentrionalis* (Baird) (Sauria: Scincidae). Great Basin Nat. 49: 525–534.

SPANBAUER, M. K. 1984. Snakes alive! Kansas Wildlife 41(4): 26–30.

SPANBAUER, M. K. 1988. Little Balkans. Kansas Wildlife and Parks 45(4): 6–10.

SPARKS, D. W., A. G. BURR, M. N. BASS, AND G. A. LIGGETT. 1999. New county distribution records of amphibians and reptiles from southwestern Kansas. Herpetol. Rev. 30: 120–121.

SPENCER, D. 1988. Emporia State University natural areas. Trans. Kansas Acad. Sci. 91: 37–40.

SPERRY, T. M. 1960. An ecological paradox. Trans. Kansas Acad. Sci. 63: 215–227.

SPERRY, T. M. 1963. The Natural History Research Reserve of the Kansas State College of Pittsburg. Trans. Kansas Acad. Sci. 66: 76–81.

SPRACKLAND, R. G. 1985. Thoughts of a midwestern naturalist. Kansas Herpetol. Soc. News. 59: 14–15.

SPRACKLAND, R. G. 1993. Husbandry and breeding of collared lizards. The Vivarium 4(4): 23–26.

STAINS, H. J. 1954. A western extension of the known geographic range of the glass lizard, *Ophisaurus attenuatus attenuatus* Baird, in south-central Kansas. Trans. Kansas Acad. Sci. 57(4): 482.

STAINS, H. J., AND J. OZMENT. 1962. A record of the Brown Skink (*Scincella laterale*) and Prairie Skink (*Eumeces septentrionalis*) from Barber County, Kansas. Trans. Kansas Acad. Sci. 65: 143.

STARKEY, D. E., H. B. SHAFFER, R. L. BURKE, M. R. J. FORSTNER, J. B. IVERSON, F. J. JANZEN, A. G. J. RHODIN, AND G. R. ULTSCH. 2003. Molecular systematics, phylogeography, and the effects of Pleistocene glaciation in the Painted Turtle (*Chrysemys picta*) complex. Evolution 57: 119–128.

STASZKO, R., AND J. G. WALLS. 1994. Rat Snakes: A Hobbyist's Guide to *Elaphe* and Kin. T.F.H. Publications, Inc., Neptune City, New Jersey.

STEBBINS, R. C. 1951. Amphibians of Western North America. University of California Press, Berkeley.

STEBBINS, R. C. 1954. Amphibians and Reptiles of Western North America. McGraw Hill, New York.

STEBBINS, R. C. 1966. Peterson Field Guide to Western Reptiles and Amphibians. 1st ed. Houghton Mifflin Company, Boston, Massachusetts.

STEBBINS, R. C. 1985. Peterson Field Guide to Western Reptiles and Amphibians. 2nd ed. Houghton Mifflin Company, Boston, Massachusetts.

STEBBINS, R. C. 2003. Peterson Field Guide to Western Reptiles and Amphibians. 3rd ed. Houghton Mifflin Company, Boston, Massachusetts.

STEGALL, E. 1977. First Strecker's Chorus Frog collected in Kansas. Kansas Herpetol. Soc. News. 21: 11–12.

STEINER, S. L., AND R. M. LEHTINEN. 2008. Occurrence of the amphibian pathogen *Batrachochytrium dendrobatidis* in Blanchard's Cricket Frog (*Acris crepitans blanchardi*) in the U. S. Midwest. Herpetol. Rev. 39: 193–196.

STEJNEGER, L. 1891. Notes on some North American snakes. Proc. U. S. Natl. Mus. 14(876): 501–505.

STEJNEGER, L. 1938. Restitution of the name *Ptychemys hoyi* Agassiz for a western river tortoise. Proc. Biol. Soc. Washington 51(40): 173–176.

STEJNEGER, L. 1944. Notes on the American soft-shelled turtles with special reference to *Amyda agassizii*. Bull. Mus. Comp. Zool. 94: 1–175.

STEJNEGER, L., AND T. BARBOUR. 1917. A Checklist of North American Amphibians and Reptiles. 1st ed. Harvard University Press, Cambridge, Massachusetts.

STEJNEGER, L., AND T. BARBOUR. 1923. A Checklist of North American Amphibians and Reptiles. 2nd ed. Harvard University Press, Cambridge, Massachusetts.

STEJNEGER, L., AND T. BARBOUR. 1933. A Checklist of North American Amphibians and Reptiles. 3rd ed. Harvard University Press, Cambridge, Massachusetts.

STEJNEGER, L., AND T. BARBOUR. 1939. A Checklist of North American Amphibians and Reptiles. 4th ed. Harvard University Press, Cambridge, Massachusetts.

STEJNEGER, L., AND T. BARBOUR. 1943. A Checklist of North American Amphibians and Reptiles. 5th ed. Bulletin of the Museum of Comparative Zoology Number 93. Harvard University, Cambridge, Massachusetts.

STEWART, M. M. 1983. *Rana clamitans*. Cat. American Amph. Reptl. 337.1–337.4.

STEWART, P. L. 1960. Lung-flukes of snakes, genera *Thamnophis* and *Coluber*, in Kansas. Univ. Kansas Sci. Bull. 41: 877–890.

STICKEL, L. F., W. H. STICKEL, AND F. C. SCHMID. 1980. Ecology of a Maryland population of Black Rat Snakes (*Elaphe o. obsoleta*). American Mid. Nat. 103: 1–14.

STICKEL, W. H. 1938. The snakes of the genus *Sonora* in the United States and lower California. Copeia 1938: 182–190.

STICKEL, W. H. 1943. The Mexican snakes of the genera *Sonora* and *Chionactis* with notes on the status of other colubrid genera. Proc. Biol. Soc. Washington 56: 109–128.

STILLE, W. T. 1954. Observations on the reproduction and distribution of the Green Snake, *Opheodrys vernalis* (Harlan). Nat. Hist. Misc. 127: 1–11.

STORER, T. I. 1925. A synopsis of the amphibia of California. Univ. California Publ. Zool. 27: 1–342.

STRECKER, J. K. 1915. Reptiles and amphibians of Texas. Baylor Bull. 18(4): 1–82.

STRIMPLE, P. 1986. The Cottonmouth or Water Moccasin *Agkistrodon piscivorus* (Lacepede). Part II. Greater Cincinnati Herpetol. Soc. News. 11(10): 8–14.

STRIMPLE, P. 1988. The Six-lined Racerunner, *Cnemidophorus sexlineatus sexlineatus* (Linnaeus) 1766. The Forked Tongue 13(8): 7–12.

STUART, J. N., AND C. W. PAINTER. 1996. Natural history notes on the Great Plains Narrowmouth Toad, *Gastrophryne olivacea*, in New Mexico. Bull. Chicago Herpetol. Soc. 31: 44–47.

STUART, S. N., M. HOFFMANN, J. S. CHANSON, N. A. COX, R. J. BERRIDGE, P. RAMANI, AND B. E. YOUNG (EDS.) 2008. Threatened Amphibians of the World. Lynx Edicions, Barcelona, Spain.

STULL, O. G. 1940. Variations and relationships in the snakes of the genus *Pituophis*. Bull. U. S. Natl. Mus. 175: 1–225.

SULEIMAN, G. 2003a. Fort Riley herpetofaunal count. J. Kansas Herpetol. 5: 11–12.

SULEIMAN, G. 2003b. Fort Riley herp count. J. Kansas Herpetol. 7: 9.

SULEIMAN, G. 2005. A summary of Fort Riley herpetofaunal counts from 2002–2005. J. Kansas Herpetol. 16: 23–24.

SULLIVAN, B. K., K. B. MALMOS, AND M. F. GIVEN. 1996. Systematics of the *Bufo woodhousii* complex (Anura: Bufonidae): advertisement call variation. Copeia: 274–280.

SWAIN, T. A., AND H. M. SMITH. 1978. Communal nesting in *Coluber constrictor* in Colorado (Reptilia: Serpentes). Herpetologica 34: 175–177.

SWEENEY, R. 1992. Garter Snakes: Their Natural History and Care in Captivity. Blandford, London.

SWEET, S. S., AND W. S. PARKER. 1990. *Pituophis melanoleucus*. Cat. American Amphib. Reptl. 474.1–474.8.

SWITAK, K. H. No date. The Care of Desert Reptiles. Dexter Press, West Nyack, New York.

TAGGART, T. W. 1991a. Geographic distribution: *Ophisaurus attenuatus attenuatus*. Herpetol. Rev. 22: 66.

TAGGART, T. W. 1991b. Geographic distribution: *Lampropeltis calligaster calligaster*. Herpetol. Rev. 22: 67.

TAGGART, T. W. 1992a. Observations on Kansas amphibians and reptiles. Kansas Herpetol. Soc. News. 88: 13–15.

TAGGART, T. W. 1992b. Results of the KHS annual field trip to Sheridan County State Lake. Kansas Herpetol. Soc. News. 90: 2–4.

TAGGART, T. W. 1992c. *Bufo debilis*. Geographic distribution. Herpetol. Rev. 23: 85.

TAGGART, T. W. 1992d. *Crotalus viridis*. Geographic distribution. Herpetol. Rev. 23: 91.

TAGGART, T. W. 1992e. *Eumeces obsoletus*. Geographic distribution. Herpetol. Rev. 23: 89.

TAGGART, T. W. 1992f. *Graptemys pseudogeographica*. Geographic distribution. Herpetol. Rev. 23: 88.

TAGGART, T. W. 1992g. *Heterodon platirhinos*. Geographic distribution. Herpetol. Rev. 23: 91.

TAGGART, T. W. 1992h. *Kinosternon flavescens*. Geographic distribution. Herpetol. Rev. 23: 88.

TAGGART, T. W. 1992i. *Lampropeltis getula*. Geographic distribution. Herpetol. Rev. 23: 91.

TAGGART, T. W. 1992j. *Nerodia sipedon*. Geographic distribution. Herpetol. Rev. 23: 91.

TAGGART, T. W. 1992k. *Thamnophis sirtalis*. Geographic distribution. Herpetol. Rev. 23: 92.

TAGGART, T. W. 1993. *Elaphe emoryi*. Geographic distribution. Herpetol. Rev. 24: 67.

TAGGART, T. W. 1994. The natural history and distribution of the Green Toad (*Bufo debilis*) in Kansas, with a report on an effort to reintroduce the species into the Cimarron National Grasslands. Final Report (12 pp.), Kansas Department of Wildlife and Parks, Pratt.

TAGGART, T. W. 1997. Status of *Bufo debilis* (Anura: Bufonidae) in Kansas. Kansas Herpetol. Soc. News. 109: 7–12.

TAGGART, T. W. 2000a. Biogeographic analysis of the reptiles (Squamata) in Ellis County, Kansas. Kansas Herpetol. Soc. News. 121: 7–16.

TAGGART, T. W. 2002a. Geographic distribution: *Arizona elegans*. J. Kansas Herpetol. 2: 10.

TAGGART, T. W. 2002b. Geographic distribution: *Lampropeltis getula*. J. Kansas Herpetol. 4: 14.

TAGGART, T. W. 2003a. Geographic distribution: *Necturus louisianensis*. J. Kansas Herpetol. 5: 10.

TAGGART, T. W. 2003b. Communal brumaculum. J. Kansas Herpetol. 5: 16.

TAGGART, T. W. 2003c. Logan County herp count. J. Kansas Herpetol. 7: 8.

TAGGART, T. W. 2003d. Dangerous Diamondbacks in Kansas. J. Kansas Herpetol. 8: 18.

TAGGART, T. W. 2003e. Geographic distribution: *Lampropeltis triangulum*. J. Kansas Herpetol. 8: 19.

TAGGART, T. W. 2004a. Geographic distribution. *Bufo woodhousii*. Kansas. J. Kansas Herpetol. 10: 10.

TAGGART, T. W. 2004b. Geographic distribution. *Kinosternon flavescens*. Kansas. J. Kansas Herpetol. 10: 10.

TAGGART, T. W. 2004c. Geographic distribution. *Podarcis sicula*. Kansas. J. Kansas Herpetol. 10: 10.

TAGGART, T. W. 2004d. Life history. *Necturus louisianensis*. New maximum length for entire range. J. Kansas Herpetol. 10: 11.

TAGGART, T. W. 2004e. Geographic distribution. *Scincella lateralis*. Kansas. J. Kansas Herpetol. 11: 13.

TAGGART, T. W. 2004f. Geographic distribution. *Ophisaurus attenuatus*. Kansas. J. Kansas Herpetol. 11: 13.

TAGGART, T. W. 2004g. Geographic distribution. *Pantherophis emoryi*. Kansas. J. Kansas Herpetol. 11: 14.

TAGGART, T. W. 2006a. Where have all the *Holbrookia* gone? J. Kansas Herpetol. 19: 10.

TAGGART, T. W. 2006b. Distribution and status of Kansas herpetofauna in need of information. Final Report (vii + 106 pp.), State Wildlife Grant T-7, Kansas Department of Wildlife and Parks, Pratt.

TAGGART, T. W. 2006c. Addendum report to biological inventory of the sandsage prairie near Holcomb, Kansas. Final Report (31 pp.), Kansas Department of Wildlife and Parks, Pratt.

TAGGART, T. W. 2007a. Grumpy turtle. Kansas Wildlife and Parks 64(2): 41.

TAGGART, T. W. 2007b. Brief herpetological history of Pigeon Lake. J. Kansas Herpetol. 22: 6.

TAGGART, T. W. 2007c. A biological inventory of the Sunflower Electric Site near Holcomb, Kansas. J. Kansas Herpetol. 23: 11–16.

TAGGART, T. W. 2009a. Geographic distribution. Holbrookia maculata. J. Kansas Herpetol. 29: 7.

TAGGART, T. W. 2009b. Geographic distribution: Lampropeltis getula. J. Kansas Herpetol. 30: 11.

TAGGART, T. W. (ED.). 2000b. Results of the KHS 2000 fall field trip. Kansas Herpetol. Soc. News. 122: 6–8.

TAGGART, T. W. (ED.). 2001a. The KHS 2001 spring field trip: a rainy rendezvous. Kansas Herpetol. Soc. News. 124: 12–15.

TAGGART, T. W. (ED.). 2001b. Results of the KHS spring field trip west. Kansas Herpetol. Soc. News. 125: 10–11.

TAGGART, T. W. (ED.). 2002c. Results of the spring 2002 KHS field trip. J. Kansas Herpetol. 3: 6–7.

TAGGART, T. W. (ED.). 2002d. Results of the KHS 2002 fall field Trip. J. Kansas Herpetol. 4: 11–13.

TAGGART, T. W. (ED.). 2003f. Kansas Herpetological Society 2003 spring field trip. J. Kansas Herpetol. 5: 3–4.

TAGGART, T. W. (ED.). 2003g. Results of the KHS 2003 fall field trip. J. Kansas Herpetol. 8: 14–15.

TAGGART, T. W. (ED.). 2004h. Results of the KHS 2004 fall feld trip. J. Kansas Herpetol. 12: 15–16.

TAGGART, T. W. (ED.). 2005. Results of the KHS 2005 fall field trip [to Crawford County]. J. Kansas Herpetol. 16: 19–21.

TAGGART, T. W. (ED.). 2006d. Results of the KHS spring field trip to Kiowa County. J. Kansas Herpetol. 18: 2–5.

TAGGART, T. W. (ED.). 2007d. The 10th annual running of the lizards. J. Kansas Herpetol. 24: 15.

TAGGART, T. W. (ED.). 2008. KHS 2008 spring field trip. J. Kansas Herpetol. 25: 2–3.

TAGGART, T. W., AND J. T. COLLINS. 2000. Geographic distribution. Eumeces laticeps. Herpetol. Rev. 31: 52.

TAGGART, T. W., AND J. T. COLLINS. 2008. Wicked serpent of the west. J. Kansas Herpetol. 27: 7.

TAGGART, T. W., J. T. COLLINS, AND C. J. SCHMIDT. 2006. Herpetological collections and collecting in Kansas. J. Kansas Herpetol. 17: 17–20.

TAGGART, T. W., J. T. COLLINS, AND C. J. SCHMIDT. 2007. Estimates of amphibian, reptile, and turtle mortality if Phostoxin is applied to 10,000 acres of prairie dog burrows in Logan County, Kansas. Report (5 pp.), Kansas Department of Wildlife and Parks, Pratt.

TAGGART, T. W., AND C. J. SCHMIDT. 2002a. Geographic distribution: Coluber constrictor. J. Kansas Herpetol. 2: 10.

TAGGART, T. W., AND C. J. SCHMIDT. 2002b. Geographic distribution: Leptotyphlops dulcis. J. Kansas Herpetol. 2: 10.

TAGGART, T. W., AND C. J. SCHMIDT. 2004. Life History. Crotalus viridis. New maximum size for entire range. J. Kansas Herpetol. 12: 18.

TAGGART, T. W., AND C. J. SCHMIDT. 2005a. Life history notes: Phrynosoma cornutum. New state maximum length. J. Kansas Herpetol. 14: 10.

TAGGART, T. W., AND C. J. SCHMIDT. 2005b. Geographic distribution: Rana areolata (Chautauqua County, Kansas). J. Kansas Herpetol. 14: 11.

TAGGART, T. W., AND C. J. SCHMIDT. 2005c. Geographic distribution: Rana areolata (Montgomery County, Kansas). J. Kansas Herpetol. 14: 11.

TAGGART, T. W., AND C. J. SCHMIDT. 2005d. Geographic distribution: Chelydra serpentina. J. Kansas Herpetol. 14: 11.

TAGGART, T. W., AND C. J. SCHMIDT. 2005e. Geographic distribution: Phrynosoma cornutum. J. Kansas Herpetol. 14: 11.

TAGGART, T. W., AND C. J. SCHMIDT. 2005f. Geographic distribution: Diadophis punctatus. J. Kansas Herpetol. 14: 11.

TAGGART, T. W., AND C. J. SCHMIDT. 2005g. Geographic distribution: Crotalus horridus. J. Kansas Herpetol. 14: 11.

TAGGART, T. W., C. SCHMIDT, AND J. T. COLLINS. 2002. A range extension of the Texas Longnose Snake in western Kansas. J. Kansas Herpetol. 1: 8.

TAGGART, T. W., C. J. SCHMIDT, AND R. S. HAYES. 2005a. Geographic distribution: Bufo woodhousii. J. Kansas Herpetol. 13: 10.

TAGGART, T. W., C. J. SCHMIDT, AND R. S. HAYES. 2005b. Geographic distribution: Rana catesbeiana. J. Kansas Herpetol. 13: 10.

TAGGART, T. W., C. J. SCHMIDT, AND R. S. HAYES. 2005c. Geographic distribution: Gastrophryne olivacea. J. Kansas Herpetol. 13: 10.

TAGGART, T. W., C. J. SCHMIDT, AND R. S. HAYES. 2005d. Geographic distribution: Chrysemys picta. J. Kansas Herpetol. 13: 10.

TAGGART, T. W., C. J. SCHMIDT, AND R. S. HAYES. 2005e. Geographic distribution: Apalone spinifera. J. Kansas Herpetol. 13: 10.

TAGGART, T. W., C. J. SCHMIDT, AND R. S. HAYES. 2005f. Geographic distribution: Sceloporus consobrinus. J. Kansas Herpetol. 13: 10.

TAGGART, T. W., C. J. SCHMIDT, AND R. S. HAYES. 2005g. Geographic distribution: Masticophis flagellum. J. Kansas Herpetol. 13: 10.

TAGGART, T. W., C. J. SCHMIDT, AND R. S. HAYES. 2005h. Geographic distribution: Tropidoclonion lineatum. J. Kansas Herpetol. 13: 10.

TANNER, S. 1992. Comparison of the arrival times and tenure at the chorus of two populations of Gray Treefrogs (Hyla chrysoscelis and Hyla versicolor). Missouri Herpetol. Assn. News. 5: 2.

TANNER, W. W. 1946. A taxonomic study of the genus Hypsiglena. Great Basin Nat. 5: 25–92.

TANNER, W. W. 1950. Notes on the habits of Microhyla carolinensis olivacea (Hallowell). Herpetologica 6: 47–48.

TANNER, W. W. 1978. Zoogeography of reptiles and amphibians in the Intermountain region. Pp. 43–53 In Intermountain Biogeography: A Symposium. Great Basin Nat. Mem. 2: 1–268.

TANNER, W. W. 1988. Status of *Thamnophis sirtalis* in Chihuahua, Mexico (Reptilia: Colubridae). Great Basin Nat. 48: 499–507.

TANNER, W. W. 1989. Status of *Spea stagnalis* Cope (1875), *Spea intermontanus* Cope (1889), and a systematic review of *Spea hammondii* Baird (1839) (Amphibia: Anura). Great Basin Nat. 49: 503–510.

TANNER, W. W., AND R. B. LOOMIS. 1957. A taxonomic and distributional study of the western subspecies of the Milk Snake, *Lampropeltis doliata*. Trans. Kansas Acad. Sci. 60: 12–42.

TAYLOR, D. R. 1965. Display Patterns of Various Sub-species of the Eastern Fence Lizard *Sceloporus undulatus*. Unpublished Master's thesis, Emporia State University, Emporia, Kansas.

TAYLOR, E. H. 1916. The Lizards of Kansas. Unpublished Master's thesis, University of Kansas, Lawrence.

TAYLOR, E. H. 1929a. A revised check-list of the snakes of Kansas. Univ. Kansas Sci. Bull. 19: 53–62.

TAYLOR, E. H. 1929b. List of reptiles and batrachians of Morton County, Kansas, reporting species new to the state fauna. Univ. Kansas Sci. Bull. 19: 63–65.

TAYLOR, E. H. 1933. Observations on the courtship of turtles. Univ. Kansas Sci. Bull. 21: 269–271.

TAYLOR, E. H. 1936. A taxonomic study of the cosmopolitan scincoid lizards of the genus *Eumeces*, with an account of the distribution and relationships of its species. Univ. Kansas Sci. Bull. 23: 1–643.

TAYLOR, E. H. 1937. Notes and comments on certain American and Mexican snakes of the genus *Tantilla*, with descriptions of new species. Trans. Kansas Acad. Sci. 39: 335–348.

TAYLOR, E. H. 1938a. A new anuran amphibian from the Pliocene of Kansas. Univ. Kansas Sci. Bull. 24: 407–419.

TAYLOR, E. H. 1938b. Notes on the herpetological fauna of the Mexican state of Sinaloa. Univ. Kansas Sci. Bull. 24: 505–537.

TAYLOR, E. H. 1939. On North American snakes of the genus *Leptotyphlops*. Copeia 1939: 1–7.

TAYLOR, E. H. 1940. Palatal sesamoid bones and palatal teeth in *Cnemidophorus*, with notes on these teeth in other saurian genera. Proc. Biol. Soc. Washington 53: 119–124.

TAYLOR, E. H. 1942. Extinct toads and frogs from the Upper Pliocene deposits of Meade County, Kansas. Univ. Kansas Sci. Bull. 28: 199–235.

TAYLOR, E. H. 1993. The lizards of Kansas. Kansas Herpetol. Soc. Spec. Publ. 2: 1–72.

TAYLOR, E. H., A. B. LEONARD, H. M. SMITH, AND G. R. PISANI. 1975. Edward H. Taylor: Recollections of an Herpetologist. Monograph 4, The University of Kansas Museum of Natural History, Lawrence.

TAYLOR, W. E. 1892. The ophidia of Nebraska. Ann. Report State Board of Agriculture, Nebraska. Pp. 310–357.

TAYLOR, W. E. 1895. The box tortoises of North America. Proc. U. S. Natl. Mus. 17(1,019): 573–588.

TELLER, J. J. 2007. Record Slider from Pottawatomie County, Kansas. J. Kansas Herpetol. 22: 10.

TENNANT, A., AND R. D. BARTLETT. 1999. Snakes of North America: Eastern and Central Regions. Gulf Publishing Company, Houston, Texas.

TERMAN, M. R. 1988. Terman Environmental Study Area. Trans. Kansas Acad. Sci. 91: 50–51.

THOMAS, R. B., I. M. NALL, AND W. J. HOUSE. 2008. Relative efficacy of three different baits for trapping pond-dwelling turtles in east-central Kansas. Herpetol. Rev. 39: 186–188.

THOMASSON, J. R. 1980. Ornate Box Turtle, *Terrapene ornata* (Testudinae), feeding on Pincushion Cactus, *Coryphantha vivipara* (Cataceae). Southwest. Nat. 25: 438.

THOMPSON, J. N., AND E. M. GRIGSBY. 1971. The Pickerel Frog in northeastern Oklahoma. Southwest. Nat. 16: 219–220.

TIEMEIER, O. W. 1962. Increasing size of fingerling Channel Catfish by supplemental feeding. Trans. Kansas Acad. Sci. 65: 144–153.

TIHEN, J. A. 1937. Additional distributional records of amphibians and reptiles in Kansas counties. Trans. Kansas Acad. Sci. 40: 401–409.

TIHEN, J. A. 1942. A colony of fossil neotenic *Ambystoma tigrinum*. Univ. Kansas Sci. Bull. 28: 189–198.

TIHEN, J. A. 1960. Notes on late Cenozoic hylid and leptodactylid frogs from Kansas, Oklahoma and Texas. Southwest. Nat. 5: 66–70.

TIHEN, J. A. 1962. A review of New World fossil bufonids. American Mid. Nat. 68: 1–50.

TIHEN, J. A., AND J. M. SPRAGUE. 1939. Amphibians, reptiles and mammals of Meade County State Park. Trans. Kansas Acad. Sci. 42: 499–512.

TINKLE, D. W. 1961. Geographic variation in reproduction, size, sex ratio and maturity of *Sternothaerus odoratus* (Testudinata: Chelydridae). Ecology 42: 68–76.

TINKLE, D. W. 1972. The role of environment in the evolution of life history differences within and between lizard species. Pp. 77–100 *In* R. T. Allen and R. C. James, Eds.), A Symposium on Ecosystematics. Occasional Papers of the University of Arkansas Museum, Number 4, Fayetteville.

TINKLE, D. W., AND A. E. DUNHAM. 1986. Comparative life histories of two syntopic sceloporine lizards. Copeia 1986: 1–18.

TINKLE, D. W., AND J. W. GIBBONS. 1977. The distribution and evolution of viviparity in reptiles. Univ. Michigan Mus. Zool. Misc. Publ. 154: 1–55.

TINKLE, D. W., AND N. F. HADLEY. 1975. Lizard reproductive efforts: caloric estimates and comments on its evolution. Ecology 56: 427–434.

TITUS, T. A. 1991. Use of road-killed amphibians in allozyme electrophoresis. Herpetol. Rev. 22: 14–16.

TOAL, K. R. 1992. Geographic distribution. *Storeria dekayi*. Herpetol. Rev. 23: 92.

TOEPFER, K. 1993a. A comprehensive assessment of aquatic habitat and known herpetofauna of Ellis County, Kansas. Report (48 pp.), Kansas Department of Wildlide and Parks, Pratt.

TOEPFER, K. 1993b. Geographic distribution. *Elaphe emoryi*. Herpetol. Rev. 24: 156.

TOLAND, G. F., AND G. L. DEHNE. 1960. Laboratory observations of the anatomy of the snake. American Biol. Teach. 22: 278–285.

TOLLEFSON, J. 2000. Preliminary observations on turtle assemblages in the Kansas River drainage. Abstract. Missouri Herpetol. Assn. News. 13: 12.

TOLLEFSON, J. 2001. Preliminary observations on turtle assemblages in the Kansas River drainage. Abstract. Kansas Herpetol. Soc. News. 123: 9–10.

TRAPIDO, H. 1944. The snakes of the genus *Storeria*. American Mid. Nat. 31: 1–84.

TRAUTH, J. B., R. L. JOHNSON, AND S. E. TRAUTH. 2007. Conservation implications of a morphometric comparison between the Illinois Chorus Frog (*Pseudacris streckeri illinoensis*) and Strecker's Chorus Frog (*P. s. streckeri*) (Anura: Hylidae) from Arkansas, Illinois, Missouri, Oklahoma, and Texas. Zootaxa 1,589: 23–32.

TRAUTH, S. E. 1978. Ovarian cycle of *Crotaphytus collaris* (Reptilia, Lacertilia, Iguanidae) from Arkansas with emphasis on *Corpora albicantia*, follicular atresia, and reproductive potential. J. Herpetol. 12: 461–470.

TRAUTH, S. E. 1992. A new subspecies of Six-lined Racerunner, *Cnemidophorus sexlineatus* (Sauria: Teiidae), from southern Texas. Texas J. Sci. 44: 437–443.

TRAUTH, S. E. 1994. Reproductive cycles in two Arkansas skinks in the genus *Eumeces* (Sauria: Scincidae). Proc. Arkansas Acad. Sci. 48: 210–218.

TRAUTH, S. E., R. L. COX, JR., W. E. MESHAKA, JR., B. P. BUTTERFIELD, AND A. HOLT. 1994. Female reproductive traits in selected Arkansas snakes. Proc. Arkansas Acad. Sci. 48: 196–209.

TRAUTH, S. E., AND C. T. MCALLISTER. 1996. *Cnemidophorus sexlineatus*. Cat. American Amph. Reptl. 628.1–628.12.

TRAUTH, S. E., H. W. ROBISON, AND M. V. PLUMMER. 2004. The Amphibians and Reptiles of Arkansas. University of Arkansas Press, Fayetteville.

TROTT, G. 1977. Chikaskia River wildlife study. Kansas Herpetol. Soc. News. 19: 2–3.

TROTT, G. 1983. Chikaskia River wildlife study. Kansas Herpetol. Soc. News. 52: 3–4.

TROWBRIDGE, A. H., AND M. S. TROWBRIDGE. 1937. Notes on the cleavage rate of *Scaphiopus bombifrons* Cope, with additional remarks on certain aspects of its life history. American Nat. 71: 460–480.

TRYON, B. W., AND J. B. MURPHY. 1982. Miscellaneous notes on the reproductive biology of reptiles. 5. Thirteen varieties of the genus *Lampropeltis*, species *mexicana*, *triangulum* and *zonata*. Trans. Kansas Acad. Sci. 85: 96–119.

TUBBS, A. A., AND G. W. FERGUSON. 1976. Effects of artificial crowding on behavior, growth and survival of juvenile spiny lizards. Copeia 1976: 820–823.

TUCKER, B. J. 1998. October Activity of an Italian Wall Lizard (*Podarcis sicula*) Community at 1880 S. W. Gage Boulevard, Topeka, Kansas. Unpublished Herpetology thesis, Washburn University, Topeka, Kansas.

TUCKER, H. 1912. A review of the dangerously poisonous snakes of the United States. Therapeutic Gazette 36: 313–323.

TUCKER, J. K. 1976. Erroneous usage of the name *Rana pipiens*. Copeia 1976: 412–413.

TURNER, F. B. 1952. The mouth parts of tadpoles of the spadefoot toad, *Scaphiopus hammondi*. Copeia 1952: 172–175.

TURNER, F. B. 1960. Postmetamorphic growth in anurans. American Mid. Nat. 64: 327–338.

TURNER, F. B. 1977. The dynamics of populations of squamates, crocodilians and rhynchocephalians. Pp. 157–264 *In* C. Gans and D. W. Tinkle (Eds.), Biology of the Reptilia. Vol. 7. Academic Press, New York.

TWEET, F., AND L. TWEET. 2006. A Crawfish Frog from Crawford County. J. Kansas Herpetol. 17: 6.

TWENTE, J. W., JR. 1952. Pliocene lizards from Kansas. Copeia 1952: 70–73.

TWENTE, J. W., JR. 1955. Aspects of a population study of cavern dwelling bats. J. Mammal. 36: 379–390.

TYLER, J. D. 1979. A case of swimming in *Terrapene carolina* (Testudines; Emydidae). Southwest. Nat. 24: 189–190.

U. S. ARMY CORPS OF ENGINEERS. 1972a. Environmental statement (for) El Dorado Lake, Walnut River, Kansas. Vol. 2 (Appendices), Appendix II, Section 3. Pp. 3–43 and 3–51.

U. S. ARMY CORPS OF ENGINEERS 1972b. Draft environmental statement (for) Big Hill Lake, Big Hill Creek, Kansas. Pp. 1–1 to 8–1 + 4 appendices.

U. S. NUCLEAR REGULATORY COMMISSION. 1975. Draft environmental statement related to construction of Wolf Creek Generating Station. Appendix B. Pp. B6–7 and B14–15.

UHLER, F. M., AND F. A. WARREN. 1929. Cheyenne Bottoms, Barton County, Kansas. Unpublished Report.

UHLER, F. M., AND F. A. WARREN. 2008. [A biological survey of] Cheyenne Bottoms, Barton County, Kansas. J. Kansas Herpetol. 26: 10–13.

VANCE, T. 1983. The Prairie-lined Racerunner, *Cnemidophorus sexlineatus viridis*, in Arkansas. Trans. Dallas Herpetol. Soc. 1: 1–2.

VAN DEVENDER, T. R., AND J. E. KING. 1975. Fossil Blanding's Turtles, *Emydoidea blandingii* (Holbrook), and the late Pleistocene vegetation of western Missouri. Herpetologica 31: 208–212.

VAN DOREN, M. D., AND C. J. SCHMIDT. 2000. A herpetological survey of the Fort Larned National Historic Site, Pawnee County, Kansas. Kansas Herpetol. Soc. News. 120: 8–11.

VETTER, H. 2004. Turtles of the World. Vol. 2. North America. A. C. S. Glaser, Germany.

VIAL, J. L. 1974. Demographic characteristics of a Kansas population of Copperheads (*Agkistrodon contortrix*). Abstract. Herpetol. Rev. 5: 79.

VIAL, J. L., T. J. BERGER, AND W. T. MCWILLIAMS, JR. 1977. Quantitative demography of Copperheads, *Agkistrodon contortrix* (Serpentes, Viperidae). Res. Pop. Ecol. 18: 223–234.

VIAL, J. L., AND L. SAYLOR. 1993. The status of amphibian populations: a compilation and analysis. Declining Amphibian Populations Task Force, IUCN/SSC, Working Document 1: 1–98.

VIETS, B. E. 1993. An annotated list of the herpetofauna of the F. B., and Rena G. Ross Natural History Reservation. Trans. Kansas Acad. Sci. 96: 103–113.

VITT, C. G. 2000. New records for aquatic turtles in Brown County, Kansas. Kansas Herpetol. Soc. News. 121: 17–18.

VITT, L. J. 1975. Observations on reproduction in five species of Arizona snakes. Herpetologica 31: 83–84.

VITT, L. J., AND W. E. COOPER, JR. 1986. Skink reproduction and sexual dimorphism: *Eumeces fasciatus* in the southeastern United States, with notes on *Eumeces inexpectatus*. J. Herpetol. 20: 65–76.

VOGT, R. C. 1993. Systematics of the False Map Turtles (*Graptemys pseudogeographica* complex: Reptilia, Testudines, Emydidae). Ann. Carnegie Mus. Nat. Hist. 62: 1–46.

VOGT, R. C. 1995. *Graptemys pseudogeographica*. Cat. American Amph. Reptl. 604.1–604.6.

VOLKMANN, A. 2003. Cowley County herp count 1. J. Kansas Herpetol. 7: 7.

VOLKMANN, A. 2004. Cowley County herp count. J. Kansas Herpetol. 11: 10.

VOLKMANN, A. 2006. Cowley County herpetofaunal count. J. Kansas Herpetol. 19: 6.

VOLKMANN, A. 2007. Cowley County herpetofaunal count. J. Kansas Herpetol. 23: 8.

VOLPE, E. P. 1955. Intensity of reproductive isolation between sympatric and allopatric populations of *Bufo americanus* and *Bufo fowleri*. American Nat. 89: 303–317.

VOLPE, E. P. 1957. The early development of *Rana capito sevosa*. Tulane Stud. Zool. 5: 207–225.

VON ACHEN, P. 1987. Population status and habitat preference of the threatened Northern Crawfish frog in Baker University Wetlands. Final Report (14 pp.), Kansas Fish and Game Commission, Pratt.

VON ACHEN, P. H., AND J. L. RAKESTRAW. 1984. The role of chemoreception in the prey selection of neonate reptiles. Pp. 163–172 *In* R. A. Seigel, L. E. Hunt, J. L. Knight, L. Malaret, and N. L. Zuschlag (Eds.). Vertebrate Ecology and Systematics: A Tribute to Henry S. Fitch. Special Publication Number 10. The University of Kansas Museum of Natural History, Lawrence.

VON FRESE, V. 1973. To catch a skink. Zoo (Publ. Topeka Zoo) 9(1): 4 pp.

VOORHEES, W., J. SCHNELL, AND D. EDDS. 1991. Bait preferences of semi-aquatic turtles in southeast Kansas. Kansas Herpetol. Soc. News. 85: 13-15.

WAGNER, B. 1984. Information for exhibit E. Environmental Report (15 pp.), Bowersock Dam, Lawrence, Kansas.

WALKER, M. L., J. A. DORR, AND G. R. PISANI. 2008. Observation of aberrant growth in a Timber Rattlesnake. Trans. Kansas Acad. Sci. 111: 156–158.

WALLEY, H. D., AND C. M. ECKERMAN. 1999. *Heterodon nasicus*. Cat. American Amph. Reptl. 698: 1–10.

WALLEY, H. D., R. B. KING, J. M. RAY, AND J. ROBINSON. 2005. What should be done about erosion mesh netting and its destruction of herpetofauna? J. Kansas Herpetol. 16: 26–28.

WALLEY, H. D., T. L. WUSTERBARTH, AND K. M. STANFORD. 2003. *Thamnophis radix*. Cat. American Amph. Reptl. 779.1–13.

WARD, J. P. 1984. Relationships of chrysemyd turtles of North America (Testudines: Emydidae). Spec. Publ. Texas Tech. Univ. 21: 1–50.

WARNER, M., AND R. WENCEL. 1978. Chikaskia River study held near Caldwell. Kansas Herpetol. Soc. News. 25: 15–16.

WASHBURNE, J. 2003a. Geographic distribution: *Sistrurus catenatus*. J. Kansas Herpetol. 6: 8.

WASHBURNE, J., AND M. WASHBURNE. 2003. Geographic distribution: *Bufo woodhousii*. J. Kansas Herpetol. 6: 8.

WASHBURNE, M. 2003b. Geographic distribution: *Eumeces septentrionalis*. J. Kansas Herpetol. 6: 8.

WASHBURNE, M. 2004. Ellsworth County herp count. J. Kansas Herpetol. 11: 10.

WASSERMAN, A. O. 1968. *Scaphiopus holbrookii*. Cat. American Amph. Reptl. 70.1–70.4.

WASSERSUG, R. J. 1976. The identification of leopard frog tadpoles. Copeia 1976: 413–414.

WASSERSUG, R. J., AND E. A. SEIBERT. 1975. Behavioral responses of amphibian larvae to variation in dissolved oxygen. Copeia 1975: 86–103.

WATKINS, L. C., AND L. L. HINESLEY. 1970. Notes on the distribution and abundance of the Sonoran Skink, *Eumeces obsoletus*, in western Missouri. Trans. Kansas Acad. Sci. 73: 118–119.

WEBB, R. G. 1962. North American Recent soft-shelled turtles (Family Trionychidae). Univ. Kansas Publ. Mus. Nat. Hist. 13: 429–611.

WEBB, R. G. 1970. Reptiles of Oklahoma. Stovall Museum, University of Oklahoma Press, Norman.

WEBB, R. G. 1973a. *Trionyx muticus*. Cat. American Amph. Reptl. 139.1–139.2.

WEBB, R. G. 1973b. *Trionyx spiniferus*. Cat. American Amph. Reptl. 140.1–140.4.

WEBSTER, C. 1986. Substrate preference and activity in the turtle, *Kinosternon flavescens flavescens*. J. Herpetol. 20: 477–482.

WELCH, D. 2005. Geographic distribution: *Eumeces laticeps*. J. Kansas Herpetol. 14: 11.

WELCH, D., AND C. J. SCHMIDT. 2006. A world record Crawfish Frog. J. Kansas Herpetol. 17: 7.

WELCH, K. R. G. 1994a. Snakes of the World: A Checklist. 1. Venomous Snakes. KCM Books, Somerset, England.

WELCH, K. R. G. 1994b. Turtles, Tortoises, and Terrapins: A Checklist. KCM Books, Somerset, England.

WELLS, W. 1997. Collared lizards of the genus *Crotaphytus*. Reptiles Magazine 5(4): 48–68.

WERTH, R. J. 1969. Ecology of Four Sympatric Lizards. Unpublished Doctoral dissertation, University of Colorado, Boulder.

WERTH, R. J. 1972. Lizard ecology: evidence of competition. Trans. Kansas Acad. Sci. 75: 283–300.

WERTH, R. J. 1974. Implications of aquatic behavior in the Collared Lizard, *Crotaphytus c. collaris*. HISS News-Journal 1: 192.

WESTERMAN, L. 2006. A remarkable record of prey ingestion by a Common Kingsnake. J. Kansas Herpetol. 19: 9.

WHEELER, G. C., AND J. WHEELER. 1966. The Amphibians and Reptiles of North Dakota. University of North Dakota Press, Grand Forks.

WHIPPLE, J. F., AND J. T. COLLINS. 1988. First complete clutch record for the Central Plains Milk Snake (*Lampro-*

peltis triangulum gentilis) in Kansas. Trans. Kansas Acad. Sci. 91: 187–188.

WHIPPLE, J. F., AND J. T. COLLINS. 1990a. First Kansas record of reproduction in the Broadhead Skink (*Eumeces laticeps*). Trans. Kansas Acad. Sci. 93: 138–139.

WHIPPLE, J. F., AND J. T. COLLINS. 1990b. A unique pattern variant of the Bullfrog (*Rana catesbeiana*). Trans. Kansas Acad. Sci. 93: 140–141.

WHITNEY, C. 2009. Geographic distribution. *Lampropeltis getula*. J. Kansas Herpetol. 29: 7.

WHITNEY, C., AND J. T. COLLINS. 2005. Franklin County herp count. J. Kansas Herpetol. 15: 8.

WHITNEY, C., C. J. SCHMIDT, AND T. W. TAGGART. 2004. Life history. *Kinosternon flavescens*. New state maximum length. J. Kansas Herpetol. 12: 17.

WIENS, J. J. 1989. Ontogeny of the skeleton of *Spea bombifrons* (Anura: Pelobatidae). J. Morphol. 202: 29–51.

WIENS, J. J., AND T. A. TITUS. 1991. A phylogenetic analysis of *Spea* (Anura: Pelobatidae). Herpetologica 47: 21–28.

WILEY, E. O., AND R. L. MAYDEN. 1985. Species and speciation in phylogenetic systematics, with examples from the North American fish fauna. Ann. Missouri Bot. Gard. 72: 596–635.

WILEY, J. R. 1968. Guide to the amphibians of Missouri. Missouri Speleol. 10: 132–172.

WILGERS, D. J., AND E. A. HORNE. 2006. Effects of different burn regimes on tallgrass prairie herpetofaunal species diversity and community composition in the Flint Hills, Kansas. J. Herpetol. 40: 73–84.

WILGERS, D. J., AND E. A. HORNE. 2007. Spatial variation in predation attempts on artificial snakes in a fire-disturbed tallgrass prairie. Southwest. Nat. 52: 263–270.

WILGERS, D. J., E. A. HORNE, B. K. SANDERCOCK, AND A. W. VOLKMANN. 2006. Effects of rangeland management on community dynamics of the herpetofauna of the tallgrass prairie. Herpetologica 62: 378–388.

WILLIAMS, E. C. 1947. The terrestrial form of the newt, *Triturus viridescens*, in the Chicago region. Nat. Hist. Misc. 5: 1–3.

WILLIAMS, K. L. 1970. Systematics of the Colubrid Snake *Lampropeltis triangulum* Lacepede. Unpublished Doctoral dissertation, Louisiana State University, Baton Rouge.

WILLIAMS, K. L. 1978. Systematics and Natural History of the American Milk Snake, *Lampropeltis triangulum*. Publications in Biology and Geology, Number 2. Milwaukee Public Museum Press, Milwaukee, Wisconsin.

WILLIAMS, K. L. 1988. Systematics and Natural History of the American Milk Snake, *Lampropeltis triangulum*. 2nd revised ed. Milwaukee Public Museum, Milwaukee, Wisconsin.

WILLIAMS, K. L. 1994. *Lampropeltis triangulum*. Cat. American Amphib. Reptl. 594.1–594.10.

WILLIMON, K., AND J. T. COLLINS. 1996. Geographic distribution: *Heterodon platirhinos*. Herpetol. Rev. 27: 88.

WILLIS, Y. L., D. L. MOYLE, AND T. S. BASKETT. 1956. Emergence, breeding, hibernation, movements and transformation of the Bullfrog, *Rana catesbeiana*, in Missouri. Copeia 1956: 30–41.

WILSON, L. D. 1970. The Coachwhip Snake, *Masticophis flagellum* (Shaw): taxonomy and distribution. Tulane Stud. Zool. Bot. 16: 31–99.

WILSON, L. D. 1973a. *Masticophis*. Cat. American Amph. Ret. 144.1–144.2.

WILSON, L. D. 1973b. *Masticophis flagellum*. Cat. American Amph. Reptl. 145.1–145.4.

WITTNER, D. 1978. A discussion of venomous snakes of North America. HERP (Bull. New York Herpetol. Soc.) 14(1): 12–17.

WOLFENBARGER, K. A. 1951. Systematic and Biological Studies on North American Chiggers of the genus *Eutrombicula* (Acarina, Trombiculidae). Unpublished Master's thesis, University of Kansas, Lawrence.

WOLFENBARGER, K. A. 1952. Systematic and biological studies on North American chiggers of the genus *Trombicula*, subgenus *Eutrombicula*. Ann. Entomolog. Soc. America 45: 645–677.

WOOD, R. D. 1985. Critical habitats for endangered and threatened herps of Kansas. Kansas Herpetol. Soc. News. 60: 13–15.

WOODBURNE, M. O. 1956. Notes on the snake, *Sistrurus catenatus tergeminus*, in southwestern Kansas and northwestern Oklahoma. Copeia 1956: 125–126.

WOODBURY, A. M. 1956. Uses of marking animals in ecological studies: marking amphibians and reptiles. Ecology 37: 670–674.

WOODBURY, A. M., AND D. M. WOODBURY. 1942. Studies on the rat snake, *Elaphe laeta*, with descriptions of a new subspecies. Proc. Biol. Soc. Washington 55: 133–142.

WOODMAN, N. C. 1959. The Systematic Status of *Natrix sipedon* in the Interior Highlands. Unpublished Master's thesis, University of Arkansas, Fayetteville.

WOOSTER, L. D. 1925. Nature studies (animals). Section IV. Reptiles and amphibians. Kansas St. Teachers Coll. Hays Bull. 15(4): 57–64.

WOOSTER, L. D. 1935. Notes on the effects of drought on animal populations in western Kansas. Trans. Kansas Acad. Sci. 38: 351–352.

WOOSTER, L. D. 1937. Nature Studies. John S. Swift Co., Inc. St. Louis, Missouri.

WRIGHT, A. H. 1931. Life Histories of the Frogs of the Okefinokee Swamp, Georgia. MacMillan, New York.

WRIGHT, A. H., AND A. A. WRIGHT. 1949. Handbook of Frogs and Toads. Comstock Publishing Company, Ithaca, New York.

WRIGHT, A. H., AND A. A. WRIGHT. 1952. List of the snakes of the United States and Canada by states and provinces. American Mid. Nat. 48: 574–603.

WRIGHT, A. H., AND A. A. WRIGHT. 1957. Handbook of Snakes of the United States and Canada. 2 Vols. Comstock Publishing Company, Ithaca, New York.

WRIGHT, B. A. 1941. Habit and habitat studies of the Massasauga Rattlesnake (*Sistrurus catenatus catenatus* Raf.) in northeastern Illinois. American Mid. Nat. 25: 659–672.

WRIGHT, J. 1996. Variations within populations of the Bullsnake. Reptile & Amphibian Magazine. September–October: 16–22.

WRIGHT, J. W., AND L. J. VITT (EDS.). 1993. Biology of Whiptail Lizards (Genus *Cnemidophorus*). Oklahoma Museum of Natural History, Norman.

YARROW, H. C. 1875. Report upon the collections of batrachians and reptiles made in portions of Nevada, Utah, California, Colorado, New Mexico, and Arizona during the years 1871, 1872, 1873 and 1874. Pp. 509–584 *In* Report upon Geographical and Geological Explorations and Surveys West of the One Hundredth Meridian. Vol. 5. Washington, D. C.

YARROW, H. C. 1883. Check list of North American reptilia and batrachia, with catalogue of specimens in U. S. National Museum. Bull. U. S. Natl. Mus. 24: 1–249.

YARROW, H. C., AND H. W. HENSHAW. 1878. Report upon the reptiles and batrachians collected during the years of 1875, 1876 and 1877 in California, Arizona and Nevada. Pp. 206–226 *In* Annual Report upon the Geographical Surveys of the Territory of the United States West of the One Hundredth Meridian. Appendix NN. Washington, D. C.

YEDLIN, I. N., AND G. W. FERGUSON. 1973. Variations of aggressiveness of free-living male and female Collared Lizards, *Crotaphytus collaris*. Herpetologica 29: 268–275.

YOUNG, E. A., AND M. C. THOMPSON. 1995. New county records for two reptiles in Kansas. Trans. Kansas Acad. Sci. 98: 80–81.

YOUNG, J. 1986. The survey of Schermerhorn Park Cave, Parts 1, 2, and 3, Cherokee County, Kansas. Kansas Caves 2: 24–33.

YOUNG, J., AND J. BEARD. 1993. Caves in Kansas. Kansas Geol. Surv. Educ. Ser. 9: *iv* + 1–47.

YOUNG, J. E., AND B. I. CROTHER. 2001. Allozyme evidence for the separation of *Rana areolata* and *Rana capito* and for the resurrection of *Rana sevosa*. Copeia 2001: 382–388.

YOUNG, R. A. 1973. Notes on the Graham's Water Snake, *Regina grahamii* Baird and Girard, in DuPage County, Illinois. Bull. Chicago Herpetol. Soc. 8: 4–5.

ZAPPALORTI, R. T. 1976. The Amateur Zoologist's Guide to Turtles and Crocodilians. Stackpole Books, Harrisburg, Pennsylvania.

ZAPPALORTI, R. T. 1990. Musk Turtle. Reptile & Amphibian Magazine. March–April: 44–48.

ZERWEKH, M. 2003. Large Northern Water Snake (*Nerodia sipedon*) from Kansas. J. Kansas Herpetol. 7: 12.

ZIMMERMAN, J. L. 1990. Cheyenne Bottoms: Wetland in Jeopardy. University Press of Kansas, Lawrence.

ZUCKERMAN, L. 1988. Fred the frog. Kansas Wildlife and Parks 46(1): 29.

ZUG, G. R., A. H. WYNN, AND C. RUCKDESCHEL 1986. Age determination of Loggerhead Sea Turtles, *Caretta caretta*, by incremental growth marks in the skeleton. Smithsonian Contrib. Zool. 427: 1–34.

ZWEIFEL, R. G. 1956. Two pelobatid frogs from the Tertiary of North America and their relationships to fossil and recent forms. American Mus. Novitates 1,762: 1–45.

ZWEIG, G., AND J. W. CRENSHAW 1957. Differentiation of species by paper electrophoresis of serum proteins of *Pseudemys* turtles. Science 126: 1,065–1,067.

INDEX
TO STANDARD COMMON NAMES AND SCIENTIFIC NAMES

A

Acris blanchardi, 47, 49, 231
aestivus, *Opheodrys*,125, 126, 224, 245
Agkistrodon, 237
   contortrix, 177, 179, 237
   piscivorus, 177, 180, 182, 237
albagula, *Plethodon*, 223
Alligator Snapping Turtle, 10, 190, 191, 194, 195, 247
Ambystoma,
   maculatum, 222
   mavortium, 16, 17, 227, 251
   texanum, 16, 18, 19, 227
   tigrinum, 16, 20, 21, 227
American Toad, 33, 36, 37, 46
americanus, *Anaxyrus*, 36, 37, 228
Anaxyrus, 14, 35, 227, 228
   americanus, 36, 37, 228
   cognatus, 38, 39, 228
   debilis, 9, 40, 41, 227, 228
   fowleri, 42, 43, 228
   punctatus, 10, 43, 44, 227, 228
   woodhousii, 45, 46, 228
Anole, Green, 10
Anolis carolinensis, 10
anthracinus, *Plestiodon*, 88, 89, 235
Apalone, 217, 246, 252
   mutica, 217, 219, 246
   spinifera, 217, 220, 221, 246
areolatus, *Lithobates*, 10, 60, 61, 231
Arizona, 112
   elegans, 10, 112, 113, 240
Aspidoscelis sexlineata, 101, 102, 233
attenuatus, *Ophisaurus*, 106, 107, 108, 232
atrox, *Crotalus*, 12

B

Barred Tiger Salamander, 13, 15, 16, 17, 251
bilineata, *Lacerta*, 10, 103, 104, 233
Blackhead Snake, Plains, 109, 139, 140, 242
Blackneck Garter Snake, 224
blairi, *Lithobates*, 62, 63, 231
blanchardi, *Acris*, 47, 49, 231
Blanchard's Cricket Frog, 33, 47, 49
blandingii, *Emys*, 11

Blanding's Turtle, 11
Blind Snake, New Mexico, 9, 109, 110, 111, 236
bombifrons, *Spea*, 33, 35, 229
Boreal Chorus Frog, 33, 56, 57, 229
Box Turtle, 190, 201
   Eastern, 190, 210, 211, 248
   Ornate, 190, 212, 213, 248
Broadhead Skink, 9, 14, 77, 92, 93, 236
Bronze Frog, 9, 33, 67, 68, 231
Brown Snake, 109, 160, 161, 163
Bullfrog, 33, 64, 65, 66, 153, 203
Bullsnake, 129

C

calligaster, *Lampropeltis*, 116, 117, 240
carolina, *Terrapene*, 210, 211, 248, 250, 251
carolinensis, *Anolis*, 10
carolinensis, *Gastrophryne*, 9, 72, 73, 229
Carphophis, 140
   vermis, 141, 142, 241
catenatus, *Sistrurus*, 177, 187, 189, 238
catenifer, *Pituophis*, 129, 130, 243
catesbeianus, *Lithobates*, 64, 65, 66, 231
Cave Salamander, 9, 15, 24, 26, 27
Checkered Garter Snake, 9, 109, 164, 165
Chelydra serpentina, 191, 192, 246, 247
Chihuahuan Night Snake, 10, 109, 150, 151
Chorus Frog, 47, 229
   Boreal, 33, 56, 57, 229
   Spotted, 33, 52, 53, 229
   Strecker's, 9, 33, 58, 59
Chrysemys, 201
   picta, 201, 202, 248
chrysoscelis, *Hyla*, 50, 51, 52
cinerea, *Hyla*, 11
clamitans, *Lithobates*, 9, 67, 68, 231
clarkii, *Pseudacris*, 52, 53, 229
Coachwhip, 109, 123, 124, 241
Coal Skink, 77, 88, 89, 235
cognatus, *Anaxyrus*, 38, 39, 228
Collared Lizard, Eastern, 14, 77, 79, 80
collaris, *Crotaphytus*, 79, 80, 233
Coluber, 7, 112
   constrictor, 114, 115, 224, 241

Common,
    Garter Snake, 109, 170, 171, 224
    Map Turtle, 9, 190, 203, 205, 249
    Mudpuppy, 15, 29, 31, 32, 226
    Musk Turtle, 190, 196, 199, 200, 248
    Snapping Turtle, 153, 190, 191, 192, 246
concinna, Pseudemys, 208, 209, 249
consobrinus, Sceloporus, 81, 86, 87, 233
constrictor, Coluber, 114, 115, 224, 241
contortrix, Agkistrodon, 177, 179, 237
Cooter, Eastern River, 190, 208, 209, 249
Cope's Gray Treefrog, 33, 50, 51, 52
Copperhead, 109, 177, 179, 181, 184, 187, 237
cornutum, Phrynosoma, 81, 83, 85, 233
Cottonmouth, 109, 177, 180, 182, 237
Crawfish Frog, 10, 33, 60, 61
Crayfish Snake, Graham's, 109, 158, 159
Cricket Frog, Blanchard's, 33, 47, 49
Crotalus, 238
    atrox, 12
    horridus, 10, 177, 182, 184, 238
    scutulatus, 11
    viridis, 177, 185, 186, 238
Crotaphytus collaris, 79, 80, 233
crucifer, Pseudacris, 3, 9, 54, 55, 230
cyrtopsis, Thamnophis, 224

D

debilis, Anaxyrus, 9, 40, 41, 227, 228
dekayi, Storeria, 160, 161, 245
Desmognathus, 11
Diadophis, 140
    punctatus, 142, 144, 239
Diamondback Rattlesnake, Western, 12
Diamondback Water Snake, 109, 153, 154, 155
dissecta, Rena, 9, 110, 111, 236
Dusky Salamander, 11

E

Earless Lizard, Lesser, 77, 81, 83
Earth Snake,
    Rough, 10, 109, 174, 175, 245
    Smooth, 9, 109, 175, 176, 245
Eastern,
    Box Turtle, 190, 210, 211, 248
    Collared Lizard, 14, 77, 79, 80
    Glossy Snake, 10, 109, 112, 113
    Gray Treefrog, 33, 50, 51, 52
    Hognose Snake, 10, 109, 130, 146, 147, 149, 239
    Mud Turtle, 223
    Narrowmouth Toad, 9, 33, 72, 73
    Newt, 9, 15, 21, 22, 23, 226

Racer, 109, 114, 115, 224, 241
River Cooter, 190, 208, 209, 249
Tiger Salamander, 15, 16, 20, 21
elegans,
    Arizona, 10, 112, 113, 240
    Thamnophis, 224
emoryi, Pantherophis, 127, 128, 242, 243
Emys blandingii, 11
erythrogaster, Nerodia, 152, 153
Eurycea,
    longicauda, 9, 24, 25, 227
    lucifuga, 9, 24, 26, 27, 227
    multiplicata, 223
    spelaea, 9, 24, 27, 28, 226
    tynerensis, 223

F

False Map Turtle, 190, 205, 207, 249
fasciatus, Plestiodon, 89, 91, 236
Five-lined Skink, 77, 89, 91, 92, 93, 236
flagellum, Masticophis, 123, 124, 241
Flathead Snake, 109, 137, 138, 242
flavescens, Kinosternon, 196, 198, 246, 247, 248
fowleri, Anaxyrus, 42, 43, 228
Fowler's Toad, 33, 42, 43
Fox Snake, Western, 224
Frog, 32, 225, 227, 250
    Blanchard's Cricket Frog, 33, 47, 49
    Boreal Chorus, 33, 56, 57, 229
    Bronze, 9, 33, 67, 68, 231
    Chorus, 47
    Crawfish, 10, 33, 60, 61
    Leopard, 153
    Northern Leopard, 11
    Pickerel, 33, 68, 69, 231
    Plains Leopard, 33, 62, 63, 71, 170, 231
    Southern Leopard, 33, 61, 63, 64, 70, 71, 231
    Spotted Chorus, 33, 52, 53, 229
    Strecker's Chorus, 9, 33, 58, 59
    Wood, 10, 11

G

Garter Snake,
    Blackneck, 224
    Checkered, 9, 109, 164, 165
    Common, 109, 170, 171, 224
    Plains, 109, 115, 168, 169
    Western Terrestrial, 224
Gastrophryne, 229
    carolinensis, 9, 72, 73, 229
    olivacea, 72, 74, 75, 229
Gecko, Mediterranean, 10, 77, 78, 232

geographica, Graptemys, 9, 203, 205, 249
Glass Lizard, Western Slender, 75, 77, 106, 107, 108
Glossy Snake, Eastern, 10, 109, 112, 113
Gopher Snake, 109, 115, 129, 130, 242
gracilis, Tantilla, 137, 138, 242
Graham's Crayfish Snake, 109, 158, 159
grahamii, Regina, 158, 159, 244
Graptemys, 201, 249
    geographica, 9, 203, 205, 249
    kohnii, 207
    ouachitensis, 205, 207, 249
    pseudogeographica, 205, 207, 249
Gray Treefrog, 14
    Cope's, 33, 50, 51, 52
    Eastern, 33, 50, 51, 52
Great Plains,
    Narrowmouth Toad, 33, 72, 73, 74, 75
    Rat Snake, 109, 115, 127, 128, 134, 242
    Skink, 14, 77, 94, 95, 234
    Toad, 33, 37, 38, 39
Green,
    Anole, 10
    Lacerta, Western, 10, 77, 103, 104
    Snake, Smooth, 223, 224
    Snake, Rough, 109, 125, 126, 224
    Toad, 9, 33, 40, 41
    Treefrog, 11
Grotto Salamander, 9, 15, 24, 27, 28
Ground,
    Skink, 77, 88, 99, 100, 234
    Snake, 109, 135, 136

H

Hemidactylus turcicus, 10, 77, 78, 232
Heterodon, 140, 238, 239
    nasicus, 10, 145, 147, 239
    platirhinos, 10, 147, 149, 239
Hognose Snake,
    Eastern, 10, 109, 130, 146, 147, 149, 239
    Western, 10, 109, 130, 145, 146, 147, 148, 149, 239
holbrooki, Lampropeltis, 118, 119, 240
Holbrookia maculata, 81, 83, 232
Horned Lizard, Texas, 77, 81, 83, 85
horridus, Crotalus, 10, 177, 182, 184, 238
hurterii, Scaphiopus, 223
Hurter's Spadefoot, 223
Hyla, 47
    chrysoscelis, 50, 51, 52, 230
    cinerea, 11
    versicolor, 50, 51, 52, 230
Hypsiglena, 140
    jani, 10, 150, 151, 241

I

Italian Wall Lizard, 10, 77, 103, 105, 106

J

jani, Hypsiglena, 10, 150, 151, 241

K

Kingsnake,
    Prairie, 109, 115, 116, 117
    Speckled, 109, 115, 118, 119
Kinosternon,
    flavescens, 196, 198, 246, 247, 248
    subrubrum, 223
kohnii, Graptemys, 207

L

Lacerta bilineata, 10, 103, 104, 233
Lacerta, Western Green, 10, 77, 103, 104
Lampropeltis, 112
    calligaster, 116, 117, 240
    holbrooki, 118, 119, 240
    triangulum, 120, 122, 240
lateralis, Scincella, 88, 99, 100, 234
laticeps, Plestiodon, 9, 92, 93, 236
lecontei, Rhinocheilus, 9, 131, 132, 239
Leopard Frog, 153
    Northern, 11
    Plains, 33, 62, 63, 71, 170, 231
    Southern, 33, 61, 63, 64, 70, 71, 231
Lesser Earless Lizard, 77, 81, 83
lineatum, Tropidoclonion, 172, 173, 243
Lined Snake, 109, 172, 173
Liochlorophis vernalis, 223
Lithobates, 60, 230, 231
    areolatus, 10, 60, 61, 231
    blairi, 62, 63, 231
    catesbeianus, 64, 65, 66, 231
    clamitans, 9, 67, 68, 231
    palustris, 68, 69, 231
    pipiens, 11
    sphenocephalus, 63, 64, 70, 71, 231
    sylvaticus, 10
Lizard, 76, 232, 251
    Eastern Collared, 14, 77, 79, 80
    Italian Wall, 10, 77, 103, 105, 106
    Lesser Earless, 77, 81, 83
    Prairie, 77, 81, 86, 87
    Texas Horned, 77, 81, 83, 85
    Western Slender Glass, 75, 77, 106, 107, 108
longicauda, Eurycea, 24, 25, 227
Longnose Snake, 9, 109, 131, 132, 239
Longtail Salamander, 9, 15, 24, 25

louisianensis, Necturus, 29, 30, 226
lucifuga, Eurycea, 24, 26, 27, 227

M

Macrochelys temminckii, 10, 191, 194, 195, 247
maculata,
    Holbrookia, 81, 83, 232
    Pseudacris, 56, 57, 229
maculatum, Ambystoma, 222
maculosus, Necturus, 29, 31, 32, 225, 226
Map Turtle, 201
    Common, 9, 190, 203, 205, 249
    False, 190, 205, 207, 249
    Mississippi, 207
    Ouachita, 190, 205, 207
marcianus, Thamnophis, 9, 164, 165, 243
Massasauga, 109, 177, 184, 187, 189, 238
Masticophis, 112
    flagellum, 123, 124, 241
mavortium, Ambystoma, 16, 17, 227, 251
Mediterranean Gecko, 10, 77, 78, 232
Milk Snake, 109, 120, 121, 122
Mintonius vulpinus, 224
Mississippi Map Turtle, 207
Mohave Rattlesnake, 11
Mud Turtle, 196
    Eastern, 223
    Yellow, 190, 191, 196, 198, 246, 248
Mudpuppy, 15, 29
    Common, 15, 29, 31, 32, 226
    Red River, 15, 29, 30, 226
multiplicata,
    Eurycea, 223
    Spea, 223
Musk Turtle, Common, 190, 196, 199, 200, 248
mutica, Apalone, 217, 219, 246

N

Narrowmouth Toad,
    Eastern, 9, 33, 72, 73
    Great Plains, 33, 72, 73, 74, 75
nasicus, Heterodon, 10, 145, 147, 239
Necturus, 226, 251
    louisianensis, 29, 30, 226
    maculosus, 29, 31, 32, 225, 226
Nerodia, 7, 73, 151, 244
    erythrogaster, 152, 153, 244
    rhombifer, 154, 155, 244
    sipedon, 7, 156, 157, 244
New Mexico,
    Blind Snake, 9, 109, 110, 111, 236
    Spadefoot, 223
Newt, Eastern, 9, 15, 21, 22, 23, 226
Night Snake, Chihuahuan, 10, 109, 150, 151

nigriceps, Tantilla, 139, 140, 242
Northern,
    Leopard Frog, 11
    Painted Turtle, 190, 201, 202
    Prairie Skink, 77, 97, 98
    Water Snake, 7, 109, 152, 153, 156, 157, 244
Notophthalmus viridescens, 9, 21, 22, 23, 226

O

obsoletus,
    Plestiodon, 94, 95, 234
    Scotophis, 3, 133, 134, 244
obtusirostris, Plestiodon, 96, 97, 235
occipitomaculata, Storeria, 9, 162, 163, 245
odoratus, Sternotherus, 196, 199, 200, 247, 248
Oklahoma Salamander, 223
olivacea, Gastrophryne, 72, 74, 75, 229
Opheodrys, 112
    aestivus, 125, 126, 224, 245
Ophisaurus attenuatus, 106, 107, 108, 232
ornata, Terrapene, 212, 213, 248
Ornate Box Turtle, 190, 212, 213, 248
Ouachita Map Turtle, 190, 205, 207
ouachitensis, Graptemys, 205, 207, 249

P

Painted Turtle, Northern, 190, 201, 202
palustris, Lithobates, 68, 69, 231
Pantherophis emoryi, 112, 127, 128, 242, 243
Peeper, Spring, 9, 33, 54, 55, 230
Phrynosoma cornutum, 81, 83, 85, 233
Pickerel Frog, 33, 68, 69, 231
picta, Chrysemys, 201, 202, 248
pipiens, Lithobates, 11
piscivorus, Agkistrodon, 177, 180, 182, 237
Pituophis, 112
    catenifer, 129, 130, 243
Plainbelly Water Snake, 109, 152, 153, 155,
        157, 244
Plains,
    Blackhead Snake, 109, 139, 140, 242
    Garter Snake, 109, 115, 168, 169
    Leopard Frog, 33, 62, 63, 71, 170, 231
    Spadefoot, 33, 35, 229
platirhinos, Heterodon, 10, 147, 149, 239
Plestiodon, 88, 234, 235, 236
    anthracinus, 88, 89, 235
    fasciatus, 89, 91, 236
    laticeps, 9, 92, 93, 236
    obsoletus, 94, 95, 234
    obtusirostris, 96, 97, 235
    septentrionalis, 97, 98, 235
Plethodon albagula, 223

Podarcis siculus, 10, 103, 105, 106, 233
Prairie,
    Kingsnake, 109, 115, 116, 117
    Lizard, 77, 81, 86, 87
    Rattlesnake, 109, 177, 184, 185, 186
Prairie Skink,
    Northern, 77, 97, 98
    Southern, 77, 96, 97
proximus, Thamnophis, 166, 167, 243
Pseudacris, 47, 229
    clarkii, 52, 53, 229
    crucifer, 3, 9, 54, 55, 230
    maculata, 56, 57, 229
    streckeri, 9, 58, 59, 229
Pseudemys, 201
    concinna, 208, 209, 249
pseudogeographica, Graptemys, 205, 207, 249
punctatus,
    Anaxyrus, 10, 43, 44, 227, 228
    Diadophis, 142, 144, 239

R
Racer, Eastern, 109, 114, 115, 224, 241
Racerunner, Six-lined, 77, 101, 102, 233
radix, Thamnophis, 168, 169, 243
Rat Snake,
    Great Plains, 109, 115, 127, 128, 134, 242
    Western, 109, 133, 134, 135
Rattlesnake, 188
    Mohave, 11
    Prairie, 109, 177, 184, 185, 186
    Timber, 10, 109, 177, 182, 184, 186, 187, 238
    Western Diamondback, 12
Red River Mudpuppy, 15, 29, 30, 226
Red-spotted Toad, 10, 33, 43, 44
Redbelly Snake, 9, 109, 160, 162, 163
Regina, 151
    grahamii, 158, 159, 244
Rena dissecta, 9, 110, 111, 236
Rhinocheilus, 112
    lecontei, 9, 131, 132, 239
rhombifer, Nerodia, 154, 155
Ribbon Snake, Western, 73, 109, 166, 167
Ringneck Snake, 109, 142, 144
River Cooter, Eastern, 190, 208, 209, 249
Rough,
    Earth Snake, 10, 109, 174, 175, 245
    Green Snake, 109, 125, 126, 224

S
Salamander, 15, 225, 252
    Barred Tiger, 13, 15, 16, 17, 251

    Cave, 9, 15, 24, 26, 27
    Dusky, 11
    Eastern Tiger, 15, 16, 20, 21
    Grotto, 9, 15, 24, 27, 28
    Longtail, 9, 15, 24, 25
    Oklahoma, 223
    Smallmouth, 15, 16, 18, 19
    Spotted, 222
    Western Slimy, 223
Scaphiopus hurterii, 223
Sceloporus, 81
    consobrinus, 81, 86, 87, 233
Scincella lateralis, 88, 99, 100, 234
Scotophis, 112
    obsoletus, 3, 133, 134, 244
scripta, Trachemys, 215, 216, 249
scutulatus, Crotalus, 11
semiannulata, Sonora, 135, 136, 241
septentrionalis, Plestiodon, 97, 98, 235
serpentina, Chelydra, 191, 192, 246, 247
sexlineata, Aspidoscelis, 101, 102, 233
siculus, Podarcis, 10, 103, 105, 106, 233
sipedon, Nerodia, 7, 156, 157
sirtalis, Thamnophis, 170, 171, 224, 243
Sistrurus catenatus, 177, 187, 189, 238
Six-lined Racerunner, 77, 101, 102, 233
Skink, 88
    Broadhead, 9, 14, 77, 92, 93, 236
    Coal, 77, 88, 89, 235
    Five-lined, 77, 89, 91, 92, 93, 236
    Great Plains, 14, 77, 94, 95, 234
    Ground, 77, 88, 99, 100, 234
    Northern Prairie, 77, 97, 98
    Southern Prairie, 77, 96, 97
Slender Glass Lizard, Western, 75, 77, 106, 107, 108
Slider, 190, 215, 216, 249
Slimy Salamander, Western, 223
Smallmouth Salamander, 15, 16, 18, 19
Smooth,
    Earth Snake, 9, 109, 175, 176, 245
    Green Snake, 223, 224
    Softshell, 190, 217, 219, 222, 246
Snake, 109, 232, 236, 251
    Blackneck Garter, 224
    Brown, 109, 160, 161, 163
    Checkered Garter, 9, 109, 164, 165
    Chihuahuan Night, 10, 109, 150, 151
    Common Garter, 109, 170, 171, 224
    Diamondback Water, 109, 153, 154, 155
    Eastern Glossy, 10, 109, 112, 113
    Eastern Hognose, 10, 109, 130, 146, 147, 149, 239
    Flathead, 109, 137, 138, 242
    Gopher, 109, 115, 129, 130, 242

Graham's Crayfish, 109, 158, 159
Great Plains Rat, 109, 115, 127, 128, 134, 242
Ground, 109, 135, 136
Lined, 109, 172, 173
Longnose, 9, 109, 131, 132, 239
Milk, 109, 120, 121, 122
New Mexico Blind, 9, 109, 110, 111, 236
Northern Water, 7, 109, 152, 153, 156, 157, 244
Plainbelly Water, 109, 152, 153, 155, 157, 244
Plains Blackhead, 109, 139, 140, 242
Plains Garter, 109, 115, 168, 169
Redbelly, 9, 109, 160, 162, 163
Ringneck, 109, 142, 144
Rough Earth, 10, 109, 174, 175, 245
Rough Green, 109, 125, 126, 224
Smooth Earth, 9, 109, 175, 176, 245
Smooth Green, 223, 224
Water, 75
Western Fox, 224
Western Hognose, 10, 109, 130, 145, 146, 147, 148, 149, 239
Western Rat, 109, 133, 134, 135
Western Ribbon, 73, 109, 166, 167
Western Terrestrial Garter, 224
Western Worm, 109, 141, 142
Snapping Turtle, 191
Common, 153, 190, 191, 192, 246
Alligator, 10, 190, 191, 194, 195, 247
Softshell, 217, 246
Smooth, 190, 217, 219, 222, 246
Spiny, 190, 217, 220, 221, 246
Sonora, 112
semiannulata, 135, 136, 241
Southern,
Leopard Frog, 33, 61, 63, 64, 70, 71, 231
Prairie Skink, 77, 96, 97
Spadefoot, 33
Hurter's, 223
New Mexico, 223
Plains, 33, 35, 229
Spea,
bombifrons, 33, 35, 229
multiplicata, 223
Speckled Kingsnake, 109, 115, 118, 119
spelaea, *Eurycea*, 9, 24, 27, 28, 226
sphenocephalus, *Lithobates*, 63, 64, 70, 71, 231
spinifera, *Apalone*, 217, 220, 221, 246
Spiny Softshell, 190, 217, 220, 221, 246
Spotted,
Chorus Frog, 33, 52, 53, 229
Salamander, 222
Spring Peeper, 9, 33, 54, 55, 230

Sternotherus odoratus, 196, 199, 200, 247, 248
Storeria, 151
dekayi, 160, 161, 245
occipitomaculata, 9, 162, 163, 245
Strecker's Chorus Frog, 9, 33, 58, 59
streckeri, *Pseudacris*, 9, 58, 59, 229
striatula, *Virginia*, 10, 174, 175, 245
subrubrum, *Kinosternon*, 223
sylvaticus, *Lithobates*, 10

T

Tantilla, 112, 242
gracilis, 137, 138, 242
nigriceps, 139, 140, 242
temminckii, *Macrochelys*, 10, 191, 194, 195, 247
Terrapene, 201, 248
carolina, 210, 211, 248, 250, 251
ornata, 212, 213, 248
Terrestrial Garter Snake, Western, 224
texanum, *Ambystoma*, 16, 18, 19, 227
Texas Horned Lizard, 77, 81, 83, 85
Thamnophis, 151
cyrtopsis, 224
elegans, 224
marcianus, 9, 164, 165, 243
proximus, 166, 167, 243
radix, 168, 169, 243
sirtalis, 170, 171, 224, 243
Tiger Salamander,
Barred, 13, 15, 16, 17, 251
Eastern, 15, 16, 20, 21
tigrinum, *Ambystoma*, 16, 20, 21, 227
Timber Rattlesnake, 10, 109, 177, 182, 184, 186, 187, 238
Toad, 32, 35, 225, 227, 228, 250
American, 33, 36, 37, 46
Eastern Narrowmouth, 9, 33, 72, 73
Fowler's, 33, 42, 43
Great Plains, 33, 37, 38, 39
Great Plains Narrowmouth, 33, 72, 73, 74, 75
Green, 9, 33, 40, 41
Red-spotted, 10, 33, 43, 44
Woodhouse's, 33, 37, 45, 46
Trachemys, 201
scripta, 215, 216, 249
Treefrog, 47
Cope's Gray, 33, 50, 51, 52
Eastern Gray, 33, 50, 51, 52
Gray, 14, 230
Green, 11
triangulum, *Lampropeltis*, 120, 122, 240
Tropidoclonion, 151
lineatum, 172, 173, 243
turcicus, *Hemidactylus*, 10, 77, 78, 232

Turtle, 190, 246, 250, 251
    Alligator Snapping, 10, 190, 191, 194, 195, 247
    Blanding's, 11
    Box, 190, 201
    Common Map, 190, 203, 205, 249
    Common Musk, 190, 196, 199, 200, 248
    Common Snapping, 153, 190, 191, 192, 246
    Eastern Box, 190, 210, 211, 248
    Eastern Mud, 223
    False Map, 190, 205, 207, 249
    Map, 9
    Mississippi Map, 207
    Northern Painted, 190, 201, 202
    Ornate Box, 190, 212, 213, 248
    Ouachita Map, 190, 205, 207
    Yellow Mud, 190, 191, 196, 198, 246, 248
tynerensis, *Eurycea*, 223

V

valeriae, *Virginia*, 9, 176, 245
vermis, *Carphophis*, 141, 142, 241
vernalis, *Liochlorophis*, 223
versicolor, *Hyla*, 50, 51, 52
Virginia, 151, 245
    striatula, 10, 174, 175, 245
    valeriae, 9, 176, 245
viridescens, *Notophthalmus*, 9, 21, 22, 23, 226
viridis, *Crotalus*, 177, 185, 186, 238
vulpinus, *Mintonius*, 224

W

Wall Lizard, Italian, 10, 77, 103, 105, 106
Water Snake, 75, 153
    Diamondback, 109, 153, 154, 155
    Northern, 7, 109, 152, 153, 156, 157, 244
    Plainbelly, 109, 152, 153, 155, 157, 244
Western,
    Diamondback Rattlesnake, 12
    Fox Snake, 224
    Green Lacerta, 10, 77, 103, 104
    Hognose Snake, 10, 109, 130, 145, 146, 147, 148, 149, 239
    Rat Snake, 109, 133, 134, 135
    Ribbon Snake, 73, 109, 166, 167
    Slender Glass Lizard, 75, 77, 106, 107, 108
    Slimy Salamander, 223
    Terrestrial Garter Snake, 224
    Worm Snake, 109, 141, 142
Wood Frog, 10, 11
Woodhouse's Toad, 33, 37, 45, 46
woodhousii, *Anaxyrus*, 45, 46, 228
Worm Snake, Western, 109, 141, 142

Y

Yellow Mud Turtle, 190, 191, 196, 198, 246, 248